AF601880

Fundamentals of Soil and Water Conservation Engineering

The Author

Dr. S.K. Gupta, is an Agricultural Engineer with specialization in Soil and Water Engineering. He obtained B. Tech (Agril. Engg.) and Master of Agricultural Engineering from the Punjab Agricultural University (PAU), Ludhiana and Ph.D. in Civil Engineering from Jawahar Lal Nehru Technological University, Hyderabad, Telangana, India. Dr. Gupta is Fellow of the National Academy of Agricultural Sciences, Indian National Academy of Engineering and Indian Society of Agricultural Engineering. Since his joining CSSRI in 1971, he has been Head, Division of Drainage and Water Management, Head, Indo-Dutch Network Project, Head, Division of Irrigation and Drainage Engineering and Project Coordinator (AICRP on Management of Salt Affected Soils and Use of Saline Water in Agriculture). After superannuation, Dr. Gupta worked as Emeritus Scientist (ICAR) and INAE Distinguished Professor at the same institute. Dr. Gupta conducted researches in various subjects in soil and water engineering notably irrigation water management, surface and subsurface drainage, hydrology of salt affected soils, leaching and use of saline water in agriculture. He has published more than 150 research papers in high impact factor journals and published 15 books. Currently, he is an independent Researcher pursuing his own interests in research and publishing. Dr. Gupta has been consultant to WAPCOS-Louis Berger, MoWR, MoRD, UP-Dutch Tube Well Project, Action Aid International, Synergics Hydro India-Oromia Water Works Enterprise, Ethiopia and others. He has been bestowed with Rafi Ahmad Kidwai Award by ICAR. Besides he is recipient of several awards from NAAS, ISAE, IE (India), CBIP and MoWR among other organizations. He is the Chief Editor of Journal of Water Management of the Indian Society of Water Management.

Fundamentals of Soil and Water Conservation Engineering

– *Author* –

S.K. Gupta

2020

Daya Publishing House®

A Division of

Astral International Pvt. Ltd.

New Delhi – 110 002

ISBN 9789390371167 (Int. Edition)

Published by : **Daya Publishing House®**
A Division of
Astral International Pvt. Ltd.
– ISO 9001:2015 Certified Company –
4736/23, Ansari Road, Darya Ganj,
New Delhi-110 002
Ph. 011-43549197, 23278134
E-mail: info@astralint.com
Website: www.astralint.com

Foreword

Due to human activities, global natural resources are degrading at an alarming rate and affecting the sustainable growth of local communities around the world. Over the years, many of the forestlands and their natural habitats have been disturbed to bring more land under agriculture to meet the food security needs of the growing population in the world. In addition, majority of the available water supplies worldwide are either overused or polluted adding additional environmental pressure on other natural resources such as plants and animals. For example, about 30% of fish stocks are depleted and about 900 cultivated plant species and 22% of the more than 8000 animal species are close to extinction. In developing countries like India, the average cultivated land per capita is 0.1 ha, requiring effective management of small farming systems to feed the growing population. According to a report prepared by the Indian National Academy of Agricultural Sciences, 126 M ha and 11 M ha of agricultural lands are severely affected by water and wind erosion in India, respectively. In addition, chemical degradation (22.76 M ha), physical degradation (6.41 M ha) and other forms of degradation (12.17 M ha) are adversely affecting the sustainability of agricultural production systems in India. Climate change is another major challenge affecting the productivity and farm incomes in India.

To preserve the health of natural resources (soil, water and air) and to sustain a productive and environmentally sound agriculture requires multidisciplinary approach. Therefore, agricultural graduates must be trained not only on future challenges facing the agriculture but also with skill sets to develop solutions by using multidisciplinary approach. This will require revamping of agriculture curricula for undergraduate and graduate programs all over the world. All agricultural students must be given applied and hand-on experiential learning knowledge on diverse subjects including agriculture, environment, natural resources, climate change, ecology, biodiversity, and innovative best practices on sustainable production systems. To preserve natural resources of ecologically sensitive zones of the world, innovative soil and water management practices are needed to conserve water and prevent soil erosion.

Water resources worldwide are becoming scarce because of mismanagement and/or inefficient use. Agriculture consumes about 70% of available freshwater worldwide, making it the largest user of freshwater. New water resources can be generated only from developing and using innovative water conservation and water harvesting techniques, wastewater treatment, and improving water-use efficiencies rather pumping more water from already depleted groundwater aquifers and developing large water development projects. To help agricultural graduates to develop innovative methods to properly manage soil and water resources, a sub-discipline of Soil and Water Engineering became part of the curriculum and is currently taught to all B.Sc. Agriculture (Honours) students. A better name for this discipline of Soil and Water would be Land and Water Management bringing inputs on environment, ecology, biodiversity, and climate change. The new book titled 'Fundamentals of Soil and Water Conservation Engineering' by Dr. S.K. Gupta addresses these issues and would be a good textbook to prepare students of agriculture to better manage soil and water resources.

I am happy to note that Dr. S.K. Gupta has taken the leadership to venture on this project, utilize his lifelong research innovations in writing this textbook for the benefit of agriculture graduates. I had the opportunity to read the contents of this book, which is written in well-defined 32 chapters under four thematic areas namely Engineering Survey, On-farm Development, Irrigation Water Management, and the Soil and Water Conservation Engineering. Land surveying discusses the various surveying techniques to undertake and plan soil and water conservation activities for productive and sustainable management of lands within agricultural watersheds. The other sections cover wide range of subjects namely land levelling, groundwater and pumps, open and underground water conveyance techniques, farm drainage, measurement of irrigation water, surface irrigation and irrigation efficiencies, sprinkler irrigation and drip irrigation. The soil and water conservation section includes causes and mechanics of water erosion, gully erosion and control, agronomic measures for erosion control, contour *bunding*, terraces for water erosion control, wind erosion and control, runoff estimation and grassed waterways, universal soil loss equation, rainwater harvesting, and farm ponds to innovatively manage soil and water resources. Contents of all chapters are written in response to the latest recommendations made by the Dean´s committee of the Indian Council of Agricultural Research for such a course.

I want to congratulate and commend the author for making valuable contribution by writing this comprehensive textbook for students and faculty of the colleges of agriculture in India and abroad. I have a special word of appreciation for Dr. Gupta, as I know him from college days where we spent four years together as students and later kept very close professional interactions during our entire life's professional careers. I am convinced that this textbook will help the teachers and students to have a better teaching/learning experience and will recommend highly. I am sure that the textbook will be a cherished possession of the faculty and students of agriculture and agricultural engineering and valuable academic asset for libraries of colleges and universities worldwide. In addition, it would a valuable resource for field practitioners, researchers, and policy makers because many topics/chapters of this textbook deal with several Government sponsored programs on watershed management implemented by several ministries, the *Rashtriya Krishi Vikas Yojana* (RKVY), the *Pardhan Mantri Sinchai Yojana*, and Mission on Micro-irrigation in India.

Ramesh Kanwar, PhD

Charles F. Curtiss Distinguished Professor of Water Resources Engineering
Iowa State University, Ames, Iowa, USA and
Vice Chancellor Emeritus, Lovely Professional University, Jalandhar, India.

Preface

Soil and water engineering in its broader sense is the application of engineering principles to the solution of soil and water management problems encountered in agriculture. Some of the engineering problems of consequence are: farm development, water resource development, irrigation and water management, drainage, and soil and water conservation. The most effective tool to scientifically design many of these practices is through the engineering surveys. Considering the importance of these issues, all agricultural universities have designed a course titled 'Fundamentals of Soil and Water Conservation Engineering' for the students of agriculture. All these subjects are taught in details to the students of agricultural engineering. The purpose of this book is to provide a professional text to the students of agriculture who need to know these issues in simple language without going into their basics. Another purpose of this book is to serve as a handbook sort of reference to agricultural engineers who often need to refresh their knowledge while working in the field.

The textbook includes subject matter on four engineering activities for increasing farm productivity and accordingly divided into 4 sections.

1. Engineering Survey
2. On-farm Development
3. Irrigation Water Management
4. Soil and Water Conservation Engineering

The first section divided into 10 chapter covers a wide range of issues on engineering survey. The issues respectively covered in these chapters are: Introduction to engineering survey, Chain types and measurement errors, Ranging, Areas and volumes. Chain survey: triangulation, Plane table survey, Compass survey, Dumpy level, Levelling and contouring and Introduction to theodolite.

The section on on-farm development has been grouped into 5 chapters namely Land levelling, Groundwater and pumps, Open channel conveyance, Underground pipe conveyance and Farm drainage. Clearly all the issues that help to develop a most productive farm with sustainable production are included in this section.

The irrigation water management section has been divided into 6 chapters. The first four chapters in this section deal with Classification of irrigation systems, Measurement of irrigation water, Surface irrigation and Irrigation efficiencies. The next two chapters relate to micro-irrigation namely the Sprinkler irrigation and the Drip irrigation and fertigation, the most efficient water and nutrient saving technologies.

The last section deals with the Soil and Water Conservation Engineering. It is divided into eleven chapters covering a wide spectrum of issues such as: Soil erosion: an introduction, Causes and mechanics of water erosion, Gully erosion and control, Agronomic measures for erosion control, Contour *bunding*, Terraces for water erosion control, Wind erosion and control, Runoff estimation and grassed waterways, Universal soil loss equation, Introduction to rainwater harvesting, and Farm ponds.

The textbook is written in a very clear and easy-to-understand language. Numerical problems and examples have been included to emphasize design principles and to facilitate an understanding of the subject matter. A glossary has been added for quick overview of the terms used in the book. The textbook designed and developed for the undergraduate/post graduate students of agricultural/ agricultural engineering has been profusely illustrated so that students are able to visualize the processes and phenomena being dealt with. Objective Type Questions for each section have been added. Subject index is included.

The author places on record the contributions of a galaxy of engineers and scientists that have contributed to the development of the science and art of soil and water engineering. Besides my own experiences, I have relied on the vast literature available on the subject matter included in this book. Although each issue has been prepared following a thorough review, only some important references are included at the end of textbook for brevity. In spite of the best efforts some references might have been left out through oversight. I express my heartfelt thanks to all the authors whose research paper, review articles, books including textbooks and lecture notes have been glanced through while framing the various chapters. The author expresses his unreserved thanks to his family members especially his spouse Sunita, who in spite of my failing to keep the promise of sparing quality time post superannuation, gave me wonderfully generous support and encouragement. Without her/their full support, it might not have been possible to bring this textbook to its logical end.

I believe that the current publication is being brought out at the right time since the soil and water management is receiving the attention of the Government. A number of programmes have been implemented and more are in pipeline. Notable among the programmes are the *Rashtriya Krishi Vikas Yojana* (RKVY), *Pardhan Mantri*

Sinchai Yojana, and Mission on Micro-irrigation. I believe that this textbook will meet the widely felt need of a comprehensive textbook and guide on soil and water engineering. I hope that it will find a place in the shelves of libraries of all the state agricultural universities and deemed universities and will also find a place in the personal collection of teachers, researchers, students and field functionaries. I solicit your learned feedback and contributions in the form of case studies or research results that might help to enrich the future editions of this textbook.

S.K. Gupta
Ex-Head Irrigation and Drainage Engineering
Independent Researcher

Contents

Section I

Engineering Survey

Chapter 1

Introduction to Engineering Survey

Introduction

Surveying is the art and science of determining the relative positions of different objects on the surface and below the surface of the earth by measuring the horizontal and vertical distances between them and by preparing maps on a suitable scale. On the other hand, the art of determining the relative vertical distances of different points on the surface of the earth is known as levelling. In levelling, the measurements are taken only in the vertical plane.

In practice, the survey comprises of:

- ☆ Measurement of horizontal and vertical distances between objects
- ☆ Measuring angles between lines or determining the direction of lines
- ☆ Establishing points by pre-determined angular and linear measurements

Besides, the actual survey measurements, mathematical calculations are also used to:

- ☆ Determine distances, angles, directions, locations, elevations, areas, and volumes using data of the survey
- ☆ Portraying the survey data graphically by constructing maps, profiles, cross sections, and diagrams

The collection of data by linear and angular measurements and elevation differences is called the field work. The processing of data plotting and computation of area and volume are called office work.

Objectives of Surveying

The aim of surveying is to prepare a plan or map to show the relative positions of the objects on the surface of the earth. The map is drawn to some suitable scale.It shows the natural features of a state/country such as towns, villages, roads, railways, and rivers, *etc.* Maps may also include details of different engineering works, such as roads, railways, irrigation, canals, *etc.* Following activities are undertaken to meet the objectives.

- To determine the relative position of any objects or points of the earth
- To determine the distance and angle between different objects
- To prepare a map or plan to represent an area on a horizontal plan
- To develop methods through the knowledge of modern science and the technology and use them in the field
- To solve measurement problems in an optimal way

Uses of Surveying

Surveying data are used for the following applications.

- To prepare a topographical map which shows the hills, valleys, rivers, villages, towns, forests, *etc.* of a place/region
- To prepare a cadastral map showing the boundaries of fields, houses, and other properties.
- To prepare an engineering map showing details of engineering works such as roads, railways, reservoirs, irrigation canals, *etc.*
- To prepare a military map showing the road and railway communications with different parts of the country. Such a map also shows strategic points that are important for the defense of the country.
- To prepare a contour map to determine the capacity of reservoir and to find the best possible route for roads, railways, *etc.*
- To prepare a geological map showing areas including existing underground resources.
- To prepare an archeological map including places where ancient relics exist.

Principles of Surveying

There are two fundamental principles (Venkatramaiah, 1996).

1. To work from the whole to the part
   - Control points: - triangulation of traversing
   - Triangulation divided into large triangle
   - Triangles- subdivided in to small triangles
   - To control and localize minor errors

   It may be noted that if we work from the part to the whole; small errors are magnified and may be uncontrollable at the end
2. To fix the position of new stations by at least two independent processes. The stations are fixed from points already surveyed using

- ☆ Linear measurement or
- ☆ Angular measurements or
- ☆ Both the linear and angular measurements.

Kinds of Surveying

Surveying is primarily divided into two types namely:

1. Plane surveying
2. Geodetic surveying

Plane surveying involves small areas and the curvature of the earth is not taken into account. In other words, it assumes that the earth's surface is flat. For small areas less than 250 km^2 plane surveying can safely be used. For constructions of most engineering projects such as canals, railways, highways, buildings, pipelines, *etc.*, this type of survey is used. It is worth noting that the difference between an arc distance of 18.5 km and the subtended chord lying in the earth's surface is 7 mm. Also the sum of the angles of a plane triangle and the sum of the angles in a spherical triangle differ by 1 second for a triangle on the earth's surface having an area of 196 km^2. To make computations in plane surveying, formulas of plane trigonometry, algebra, and analytical geometry are used.

Geodetic surveying involves large areas (in excess of 250 km^2) where the curvature of the earth or the true shape of the earth (spheroid) is important and needs to be taken into account. It is used to establish network of precisely located control points. Geodetic survey are used to map out entire continent, measure the size and shape of the earth or in carrying out scientific studies such as determination of the Earth's magnetic field and direction of continental drifts. The following clarification establishes the need for geodetic surveys.

The shape of the earth is thought of as a spheroid, although technically, it is not really a spheroid. In 1924, the convention of the International Geodetic and Geophysical Union adopted 41,852,960 ft as the diameter of the earth at the equator and 41,711,940 ft as the diameter at its polar axis. The equatorial diameter was computed on the assumption that the flattening of the earth caused by gravitational attraction is exactly 1/297. Therefore, distances measured on or near the surface of the earth are not along straight lines or planes, but on a curved surface. Hence, in the computation of distances in geodetic surveys, allowances are made for the earth's minor and major diameters from which a spheroid of reference is developed. The position of each geodetic station is related to this spheroid. The positions are expressed as latitudes (angles north or south of the Equator) and longitudes (angles east or west of a prime meridian) or as northing and easting on a rectangular grid.

Classification of Surveys

Surveys can be classified into various categories depending on the purposes, instruments and methods used and the nature of the field.

Classification Based on Instruments

Chain Survey

It is the simplest type of surveying in which only linear measurements are made with a chain or a tape (https:/civilseek.com/chain-surveying/). Angular measurements are not taken. Following instruments are used.

1. Chain and tape
2. Cross staff
3. Ranging rods
4. Offset rods
5. Arrows

Compass Survey

In compass survey, the angles are measured with the help of a magnetic compass.

1. Prismatic compass
2. Ranging rods
3. Offset rods

Chain and Compass Survey

In this survey linear measurements are made with a chain or a tape and angular measurements with a compass.

1. Prismatic compass
2. Chain and tape
3. Ranging rods
4. Offset rods

Plane Table Survey

It is a graphical method of surveying in which field works and plotting both are done simultaneously.

1. Plane Table with tripod stand
2. Alidade
3. Trough compass
4. 'U' Frame with plumb bob
5. Spirit level
6. Chain
7. Ranging rods

Theodolite Survey

In this survey, the horizontal/vertical angles are measured with a theodolite more precisely than compass and the linear measurements are made with chain or a tape.

Tacheometry Survey

A special type of theodolite, known as tacheometer, is used to determine horizontal and vertical distances indirectly.

Levelling Survey

This type of survey is used to determine the vertical distances and relative heights of points with the help of an instrument known as level.

1. Dumpy level
2. Tripod stand
3. Telescopic metric staff

Photogrammetric Survey

It utilizes the principles of aerial photogrammetry, in which measurements made on photographs are used to determine the positions of photographed objects.

EDM Survey

Electronic distance measuring instrument is a surveying instrument for measuring distance electronically between two points through electromagnetic waves. Although EDM stands for electronic distance measurement, in this type of survey, all measurements (length, angles co-ordinates) are made with the help of EDM instruments (*i.e.* total station).

Classification Based on the Method Used

Triangulation Survey

In order to make the survey manageable, the area to be surveyed is first covered with series of triangles. Lines are first run round the perimeter of the plot, then the details fixed in relation to the established lines. This process is called triangulation. The triangle is preferred as it is the only shape that can completely cover an irregularly shaped area with minimum space left.

Traverse Survey

If the bearing and distance of a place from a known point are known then it is possible to establish the position of that point on the ground. From this point, the bearing and distances of other surrounding points may be established. In the process, positions of points linked with lines linking them emerge. The process of establishing these lines is called traversing, while the connecting lines joining two points on the ground with known bearing and distance is known as traverse. A traverse station is each of the points of the traverse, while the traverse leg is the straight line between consecutive stations. Traverses may either be close or open. When a series of connected lines forms a closed circuit, *i.e.* when the finishing point coincides with the starting point of a survey, it is called a 'closed traverse'. When a sequence of connected lines extends along a general direction and does not return to the starting point, it is known as 'open traverse' or unclosed traverse.

Classification Based on Purposes

Topographic

It is used to produce maps and plans. Maps are usually on small scales and heights are presented in the form of points on the map or as a contour. Maps are used for navigational, geological, and geographical purposes. Plans usually drawn to scale are used for engineering design.

Engineering

These surveys are undertaken to provide special information for construction of civil engineering and building projects. This involves surveying work before, during and after the engineering works.

- *Before Engineering Works*: Also known as setting out is the survey carried out before the start of any construction. The proposed position of what is to be constructed (building, utility network, new road *etc.*) is marked on the ground (in plan and in height)
- *After Engineering Works*: As built surveying is carried out after the completion of any construction. As built drawings provide the actual location of buildings, utility networks, roads as constructed on-site. This information is very useful for the owners.

Cadastral

These are undertaken to define and record the boundary of properties, legislative area and even countries.

Geological Survey

In this type of survey both surface and subsurface surveying are conducted to locate different minerals and rocks. In addition, geological feature of the terrain such as folds are located.

Construction Survey

Surveys required to establish points, lines, grades, and for staking out engineering works (after the plans have been prepared and the structural design has been completed) is known as construction survey. It can be treated as a part of engineering survey as well.

Mine Survey

Mine surveys include both surface and underground surveys. It is conducted for the exploration of mineral deposits and to guide tunnelling and other operations associated with mining.

Archaeological Survey

It is conducted to locate relics of antiquity, civilization, Kingdom, forts, temples, *etc.*

Military Survey

It has a very important and critical application in the military. Aerial surveys are conducted for this purpose. It is conducted to locate strategic positions for the purpose of army operations

Classification Based on Nature of Field

Land Survey

Land Survey is done on land to prepare plans and maps of a given area. Topographical, city and cadastral surveys are some of the examples of land surveying.

Hydrographical Survey

The survey of water bodies to produce coastline, seabed maps for engineering navigation, water supply, or sub-aqueous construction is known as hydrographic surveys.

Route Survey

Refers to those controls, topographic, and construction surveys necessary for the location and construction of highways, rail, roads, canals, transmission lines, and pipelines.

Astronomic Survey

The surveys are conducted to determine latitudes, longitudes, azimuths, local time, *etc.* for various places on the earth by observing heavenly bodies (The sun or star).

Aerial Survey

An aerial survey is conducted to aircraft. Aerial cameras take photographs of the surface of the earth in overlapping strips of land. This is also known as photographic survey.

Agricultural Surveying

A kind of land survey, agricultural surveying is the simplest form of plane surveying. It is used to correctly locate the field boundaries and accurately compute the areas. Land levelling and grading can be perfectly accomplished if the differences in elevations are known. Alignments of irrigation and drainage channels can be effectively done. Surveying plays a vital role in planning soil conservation measures like contour *bunding*, graded *bunding*, bench terracing construction of farm ponds and percolation ponds *etc.* In addition to this, surveying plays a key role in laying underground pipeline system, alignment of irrigation channels, drainage systems, farm roads and farm stead construction *etc.*

Types of Instruments Used

1. Distance measurement: Tape (Steel, PVC), Chain, EDM (Electronic Distance Measurement)

2. Height Measurement: Level (optical, digital, laser)
3. Angle Measurement: Theodolite

Modern instruments include: Total Station (a complete surveying instrument), GPS (Global Positioning System) and other electronic instruments

Units of Measurement

The system of measurement used for surveying in India is Système International (SI). Throughout this text SI units will be used; the most commonly used in surveying being shown in Table 1.1. The S1 unit for an angle is the radian (rad), but most surveying instruments measure in degrees, minutes and seconds, which is known as the sexagesimal system (Basak, 2014). This is the only unit of measure that is not SI. This system uses angular notation in increments of 60 by dividing the circle into 360 degrees; degrees into 60 minutes; and minutes into 60 seconds. Therefore;

1 circle = 360° = 21,600´ = 1,296,000"

1° = 60´ = 3600"

1´ = 60"

Table 1.1: Commonly Used SI Units in Surveys

Quantity	*Unit*	*Symbol*
Length	kilometer, meter, millimeter	km, m, mm
Area	Square meter, hectare	m^2, ha
Volume	Cubic meter	m^3
Angle	Degrees Minutes Seconds	° ' "
Mass (Weight)	Kilogram	kg
Temperature	Celsius	°C

The Field Book

The notebook in which field measurements are noted is known as the 'field book'. It is a oblong book of size 200 mm × 120 mm opens lengthwise, and can be carried in the pocket. Field books are of two types:

1. Single-line
2. Double-line.

Single-line

In this type of field book, a single red line is drawn through the middle of each page. This line represents the chain line, and the chainages are written on it. The offsets are recorded, with sketches, to the left or right of the chain line. The recording of the field book is started from the last page and continued towards the first page.

The main stations are marked by △ and subsidiary stations by ⊖ and tie stations by (Figure 1.1).

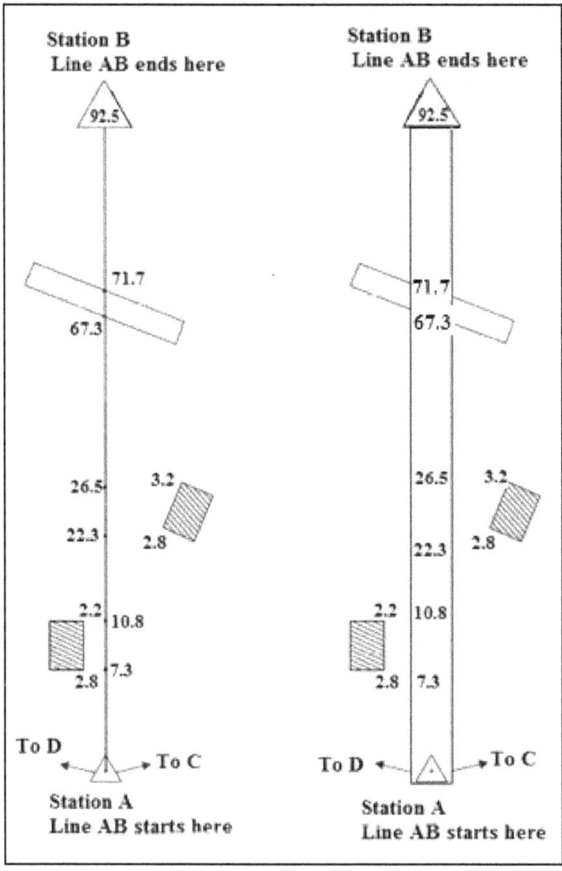

Figure 1.1: Typical Pages of a Single and Double Line Field Books.

Double-line

In this type of field book, two blue lines with a space of 15 to 20 mm are drawn through the middle of each page (Figure 1.1). The space between the two lines is utilized for noting the chainages. The offsets are recorded, with sketches, to the left or right of this column. The recording is begun from the last page and continued towards the first. The main stations, subsidiary stations and tie stations are marked in the same fashion as in the case of single-line field book.

Precautions during Entering the Field book

1. All measurements should be noted as soon as they are taken.
2. Each chain line should be recorded on a separate page. Normally it should start from the bottom of one page and end on the top of another. No line should be started from any intermediate position.

3. Clearly indicate the decimal portion of a measured value. For a number less than one, a zero should be placed before the decimal point, (*e.g.* 0.15 instead of.15). Always record significant zeroes to show precision of measurement, (*e.g.* 6.40 instead of 6.4 if measuring to the hundredths).
4. Over –writing should be avoided.
5. Figures and hand-writing should be neat and legible.
6. Index-sketch, object-sketch and notes should be clear. Always use standard abbreviations and symbols.
7. Reference sketches should be given in the field book, so that the station can be located whenever required.
8. The field book should be entered in pencil and not in ink.
9. If an entry is incorrect or a page damaged, cancel the page and start the entry from a new one.
10. Erasing a sketch, measurement or note should be avoided.
11. The surveyor should face the direction of chaining so that the left-hand and right-hand objects can be recorded without any confusion.
12. The field-book should be carefully preserved.
13. The field-book should contain the following:
    - ☆ Name
    - ☆ Location
    - ☆ Date of survey
    - ☆ Name of party members
    - ☆ Page index or chain line

Example 1.1

1. Enter the following information in the double-line field book

   Total length of line AB 96 m

   The offsets of a depression on the left as per the following chainage

   Chainage: 5, 10, 15, 20, 25; Offsets respectively: 16, 12, 10, 14, 20

   Chainage and offsets to an embankment towards the right

   Chainage: 5, 25, 45, 280; Offsets respectively: 13, 17, 19, 19.9

   A temple at chainage 6.5 m 6 m to the right

   A tree at offset of 10 m towards left of chainage 60 m

   Using a double line field book, the survey record is entered and shown in Figure 1.2.

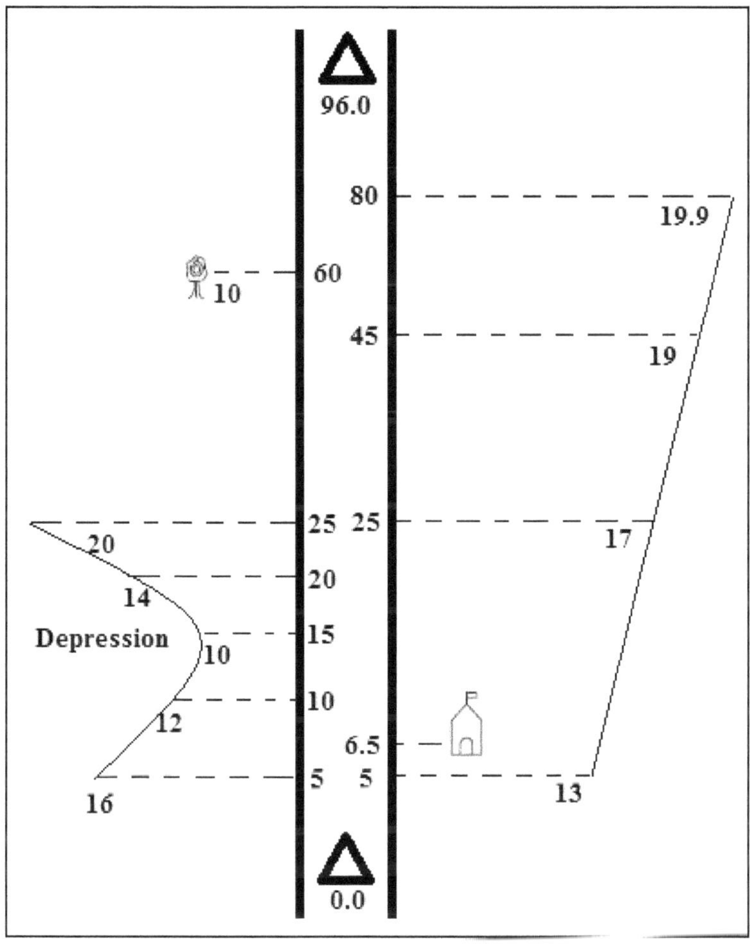

Figure 1.2: Entry of Chain Survey in the Field Book.

Chapter 2

Chain Types and Measurement Errors

Chain surveying is the simplest and the oldest form of surveying an area using only the linear measurements. It can be defined as the process of taking direct measurement, although not necessarily with a chain. It should be noted that plotting triangles requires no angular measurements, if the three sides are known. In this method the lengths of lines marked on the field are measured, while the details are measured by offsets and ties from these lines. In this survey after the field work, the rest work, such as plotting calculation *etc.* are done in the office.

Chain surveying is recommended if:

- ☆ The ground surface is more or less level
- ☆ The area to be surveyed is small
- ☆ A small-scale map is to be prepared and
- ☆ The formation of well-conditioned triangles is easy

Chain surveying is unsuitable when:

- ☆ The area is crowded with many details
- ☆ The area consists of too many undulations
- ☆ The area is very large
- ☆ The area is too wet
- ☆ The formation of well-conditioned triangles becomes difficult due to obstacles

The benefits of chain surveying are:

- ☆ The equipment used to conduct chain survey is simple to use
- ☆ The equipment used in chain survey can easily be replaced

- ☆ The method does not involve complicated mathematical calculation
- ☆ Only few people are needed to conduct the survey.

The major disadvantages of chain surveying are:

- ☆ Simple chain survey is subject to several errors that accumulate
- ☆ It is time consuming.

Equipment Used in Chain Surveying

The equipment used in chain survey are divided into three groups, namely

1. Those used for linear measurement. (Chain, steel band, linear tape)
2. Those used for slope angle measurement and for measuring right angle (*e.g.* Abney level, clinomater, cross staff, optical squares)
3. Other items (Ranging rods or poles, arrows, pegs *etc.*).

Some of the equipment is briefly described in the following sections.

Chain

The chain is usually made of steel wire, and consists of long links joined by shorter links. It is designed for hard usage, and is sufficiently accurate for measuring the chain lines and offsets of small surveys. Chains are made up of links. Surveying chains are fitted with brass handles at each end. Tally markers made of plastic or brass are attached at different positions or at each tenth link. To avoid confusion in reading, chains are marked in a similar manner from both end (*e.g.* Tally for 2 m and 18 m is the same) so that measurements may be commenced with either end of the chain.

There are different types of chains used in surveying. Depending upon the length of the chain, these are divided into following types.

- ☆ Metric chains
- ☆ Gunter's chain or surveyor's chain
- ☆ Engineer's chain
- ☆ Revenue chain
- ☆ Steel band or Band chain

Metric Chains

Metric chains are the most commonly used chain in India. These chains come in many lengths such as 5, 10, 20 and 30 m. Most commonly used is 20 m chain.

Gunter's Chain or Surveyor's Chain

Gunter chain comes in standard 66 ft. This chain consists of 100 links, each link being 0.66 ft or 7.92 in. The length 66 ft is selected because it is convenient in land measurements as follows.

10 square Gunter's chains = 1 acre

10 Gunter chains = 1 furlong (1/8 miles or 220 yards)

80 Gunter chains = 1 mile

Engineer's Chain

This chain comes in 100 ft length. It consist of 100 links each link being 1ft long. At every 10 links a brass ring or tag is provided for indication of 10 links. Readings are taken in feet and decimal.

Revenue Chain

The standard size of this type of chain is 33 ft. A total of 16 links are provided, each link being $2\frac{1}{16}$ ft (2.0625 ft). This chain is commonly used in cadastral survey.

Steel Band or Band Chain

These types of chain consist of a long narrow strip of steel of uniform width of 12 to 16 mm and thickness of 0.3 to 0.6 mm (Figure 2.1). This chain is divided by brass studs at every 20 cm. Instead of brass studs, band chain may be provided which have graduated engraving in centimeter. For easy use and workability, band chains are wound on steel crosses or metal reels from which they can be easily unrolled. These steel bands are available in 20 m and 30 m length. The operating tension and temperature for which it has been graduated are (or should be) indicated on the band.

Figure 2.1: A Pictorial View of Steel Band.

Construction Details of a Metric Chain

The chain consists of several small parts for easy handling or convenience of measurements. A 20 m long metric chain has 100 links (30 m having 150 links). Thus, every link of this type of chain is 0.2 m (200 mm). For reading fractions and also for accurate reading, tallies are provided at every meter for chains of length 5 and 10 meter and at 5 meter for chains of length 20 and 30 m. In case of chains of lengths 20 and 30 m, small brass rings are provided at every one meter except the place where tallies are provided (Figure 2.2). At the ends of the chain, brass handles with swivel joint are provided so that it can be easy to roll or unroll the chain without twisting and knots. The total length of the chain is marked on the brass handle at the ends.

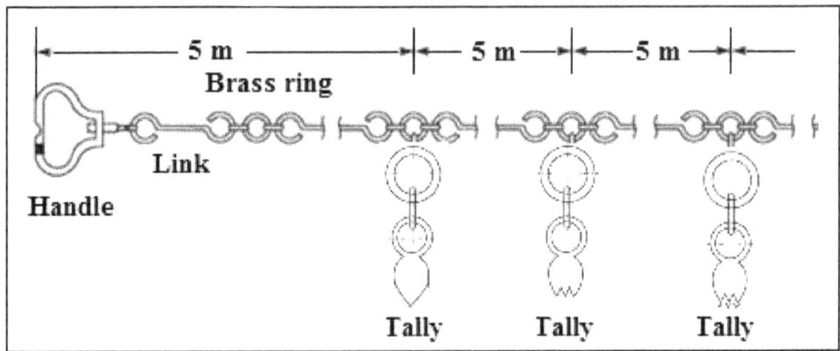

Figure 2.2: Constructional Details of a 30 m Metric Chain.

Other Equipment

Tapes

Tapes are used where greater accuracy of measurement is required, such as the setting out of buildings and roads. They are 15 m or 30 m long marked in meters, centimeter and millimeters. Tapes are classified into four types:

Linen or Linen with Steel Wire Woven into the Fabric

It is very light and handy. It is commonly used to take subsidiary measurements, such as offset. These tapes are liable to stretch during use and should be frequently tested for length. If wet it may shrink. They should never be used for works that require high accuracy.

Fiber Glass Tapes

Also known as synthetic tapes, these are manufactured from glass fiber with a PVC coating. They are graduated every 10 mm and figured every 100 mm. Meter figures are shown in red. They are convenient for measuring short lengths. These are much stronger than linen and are not likely to stretch during use.

Steel Tapes

The tape made of steel or stainless steel, is available in various lengths up to 100 m (20 m and 30 m being the most common). The outer end is provided with a ring. The length of the tape includes the metal ring. The tape is marked with a line at every five millimeters, centimeters, decimeters, and meter. These are much more accurate, and are usually used for setting out buildings and structural steel works. These are encased in steel or plastic boxes with a recessed winding lever or mounted on open frames with a folding winding lever. When the button release device is pressed, the tape automatically rewind in to the case.

Arrows

They are also called as marking or chaining pins and are used to mark the end of chain during the process of chaining. Arrow consists of a piece of steel wire about 0.5 m long, and is used for marking temporary stations. A piece of coloured cloth, white or red ribbon is usually attached or tied to the end of the arrow to be clearly seen on the field. They are pointed at one end that is inserted in to the ground. The other end is in the shape of a ring.

Pegs

Pegs are made of wood having dimensions of 50 mm × 50 mm and some convenient length. These are tapered at one end. They are used for points which are required to be permanently marked, such as intersection points of survey lines. They are driven in to the ground with about 4 cm lengths, remaining being above the ground. Pegs are driven with a mallet and nails are set at the tops.

Ranging Rod

These are poles of circular or octagonal in section 2 m, 2.5 m or 3 m long, painted with characteristic red and white or black and white bands which are usually 0.2 m or 0.5 m long. These are tipped with a pointed steel shoe to enable them to be driven into the ground. They are made of seasoned timber such as teak, blue pine, *sisam* or deodar. Now a days instead of timber, mild steel hollow pipes are used. They are used in the measurement of lines with the tape, and for marking any points which need to be seen.

Ranging Poles

These are similar to the ranging rods but are heavier. They vary in length from 4 m to 6 m or more. Used in the case of very long lines.

Offset Rods

Similar to the ranging rods, they are usually 3 m long and are divided into parts each 0.2 m length. Top is provided with an open ring for pulling or pushing the chain through a hedge. It has two short narrow vertical slots. It is used for aligning short offsets.

Laths and Whites

Laths are used for ranging long lines, or used over uneven ground where the ranging rod is not visible due to obstructions. These are unusually 1.0 m long, light in weight, cheap, coloured white so that these are easily visible at a great distance.

Whites on the other hand are thin strip of bamboo, 40 cm to 1 m in length. When the ranging rods are not available or are insufficient for the site, whites are used. One end is sharp and the other end is split to insert pieces of white papers. They are also useful for temporary marking of counter points.

Plumb Bob

A plumb consists of a specially designed weight made of brass, steel, or other materials, including plastic. It is fixed at one end of a coarse string made of twisted cotton or nylon threads. Precisely machined and balanced bobs have pointed tips (Figure 2.3). The plumb bob is used during measurements of distance along slopes where points are needed to be transferred to the ground. It is also used to test the verticality of ranging poles.

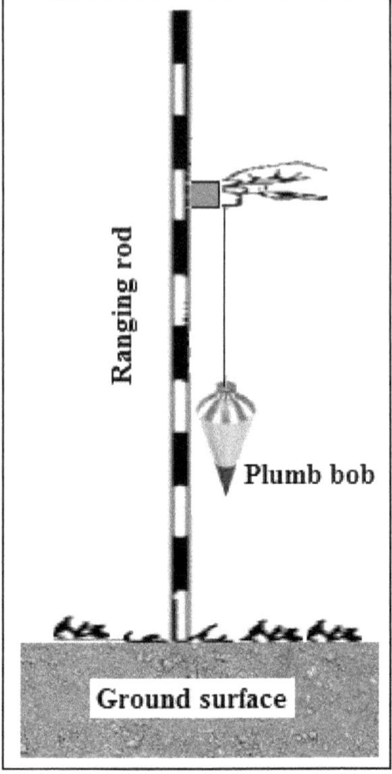

Figure 2.3: A Plumb Bob Used to Ensure Verticality of a Ranging Rod.

Optical Square

Optical Square is a simple sighting instrument used to set out right angles. It is used where greater accuracy is required. There are two kinds of optical square, one using two mirrors and the other a prism. The mirror method is based on the fact that a ray of light is reflected from a mirror at the same angle as that at which it strikes the mirror. Because of practical difficulties in using squares with mirrors, they have been replaced by squares with prisms called prismatic squares. Two major kinds of prismatic squares are: single prismatic squares and double prismatic squares. The prism of the single prismatic square is fitted in a metal frame with a handle. Attached to the handle is a hook to which a plumb bob can be connected (Figure 2.4). The prism square is more accurate than mirror types.

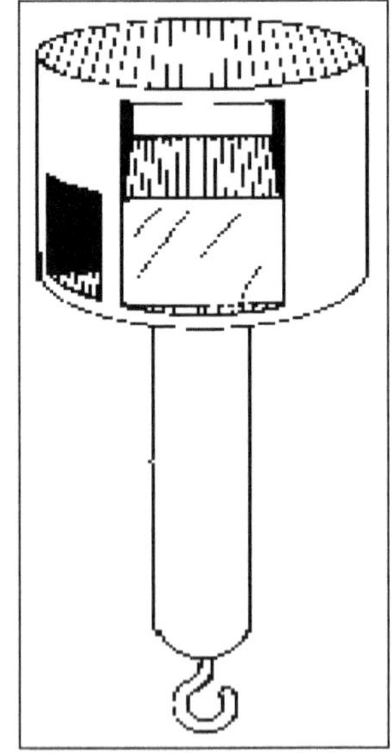

Figure 2.4: A Line Diagram of an Optical Square.

Cross Staff

It is an instrument consisting of a staff carrying a brass circle divided into four equal parts or quadrants by two lines intersecting each other at the center. The two pairs of vanes are set at right angle to each other with a wide and narrow slit in each vane. The instrument is mounted upon a pole, so that when it is set up, it is at normal eye level. It is also used for setting out lines at right angle to the main chain line. There are several kinds of cross staff such as: open cross staff, French cross staff, an octagonal brass box with slits cut in each face so that opposite pairs form a straight line and adjustable cross staff.

Clinometer

This instrument is used to measure angles of ground slopes (slope angle). They are of several forms, the common form is the Watking's Clinometer, which consist of a small disc of about 60 mm diameter. A weighted ring inside the disc can be made to hang free. By sighting across this graduated ring, one can read the angle of slopes. It is less accurate than an Abney level.

Abney Level

Named after its inventor, Sir William Abney, an Abney level is a small surveying instrument that consists of a sighting tube and a spirit level linked to a protractor that allows angles and slopes to be measured. It consists of a rectangular, telescopic tube (without lenses) about 125 mm long with a graduated arc attached. A small bubble is fixed to the vernier arm. Once the image of the bubble is seen reflected in the eyepiece, the angle of the line of sight can be read off with the aid of the reading glass. The most common current use of an Abney level is to measure the height of trees in forestry.

Precautions in using Chain Surveying Instruments

For accuracy and long lasting use of the surveying equipment, following points should be kept in view.

- After use in wet weather, chains should be cleaned, and steel tapes should be dried and wiped with an oily rag.
- A piece of colored cloth should be tied to arrows (or ribbon – attached) to enable them to be seen clearly on the field.
- Ranging rods should be erected as vertical as possible at the exact station point.
- The operating tension and temperature for which steel bands/tapes are graduated should be kept in view and necessary corrections applied.
- Linen tapes should be frequently tested for length (standardized) and always after repairs.
- Always keep tapes reeled up when not in use.

General Procedure for Chain Survey

Reconnaissance

Walk over the area to be surveyed and note the general layout, the position of features and the shape of the area.

Choice of Stations

Decide upon the framework to be used and drive in the station pegs to mark the selected stations.

Station Marking

Station marks, where possible should be tied - in to some permanent objects so that they may be easily replaced if moved or easily found during the survey. In soft ground, wooden pegs may be used while rails may be used on roads or hard surfaces.

Witnessing

It means sketching the immediate area around the station that show existing permanent features, the position of the stations and its description and designation. Measurements are then made from at least three surrounding features to the station point and recorded on the sketch. The aim of witnessing is to re-locate a station at a later date even by other surveyors.

Offsetting

Offsets are usually taken perpendicular to chain lines in order to dodge obstacles on the chain line.

Sketching and Other Details

A sketch of the layout should be made on the last page of the chain book besides entering the date and the name of the surveyor.

Criteria for Selecting Survey Lines/Offsets

During reconnaissance, the following points must be borne in mind as the criteria to provide the best arrangement of survey lines.

Few Survey Lines: The number of survey lines should be kept to a minimum but must be sufficient for the survey to be plotted and checked.

Long Base Line: A long line should be positioned right across the site to form a base on which to build the triangles.

Well-Conditioned Triangles: The triangles with angles greater than 30° and not exceeding 150° are preferable so that the arcs used for plotting should intersect as close as 90°. It provides sharp definition of the stations point.

Check Lines: Every part of the survey should be provided with check lines that are positioned in such a way that they can be used for off setting too. It helps to save any unnecessary duplication of lines.

Short Offsets: To survey lines, short offsets should be selected. It will help to reduce the labour requirement as short distances can be measured by one person instead of two people, if tape is used.

Besides try to avoid obstacles such as steep slopes and rough ground. Stations should be positioned on the extension of a check line or triangle. Such points can be plotted without the need for intersecting arcs.

Errors in Surveying

Despite the best chain survey equipment and methods used, it is still impossible to take observations that are completely free of small variations caused by errors which must be guided against or their effects corrected. Errors in chaining may be classified as (Venugopala Rao, 2015):

- ☆ Personal errors
- ☆ Compensating errors, and
- ☆ Cumulating errors.

Personal Errors

Also known as gross errors are caused as a result of taking a wrong reading, wrong recording, reading from wrong end of chain *etc.* by either the surveyor or his assistants. These types of errors also include miscounting the numbers of tape length, wrong booking, sighting wrong target, measuring anticlockwise reading, turning instruments incorrectly, displacement of arrows or station marks *etc.* Gross errors can occur at any stage of the survey while observing, booking, computing or plotting. These errors can be serious errors as these cannot be detected easily but would have a damaging effect on the results if left uncorrected. Gross errors can be eliminated only by careful methods of observing, booking and constantly checking both operations.

Random or Compensating Errors

These errors may be sometimes positive and sometimes negative. Hence these are likely to get compensated when large numbers of readings are taken. The magnitude of such errors can be estimated by theory of probability. The following are the examples of such errors:

- ☆ Incorrect marking of the end of a chain.
- ☆ Fractional part of chain may not be correct although the total length may be correct.
- ☆ Graduations in tape may not be exactly same throughout.
- ☆ In the method of stepping while measuring sloping ground, plumbing may be crude.

Random errors cannot be removed from observation but methods can be adopted to ensure that they are kept within acceptable limits. In order to analyze random errors or variable, statistical principles must be used and in surveying their effects may be reduced by increasing the number of observations and finding their mean. It is commonly assumed that random variables are normally distributed.

Systematic or Cumulative Errors

The errors that occur always in the same direction are called cumulative errors. In each reading the error may be small, but when large numbers of measurements are made they may be considerable, since the error is always on one side. Examples of such errors are:

1. Bad ranging
2. Bad straightening
3. Erroneous length of chain
4. Temperature variation

5. Variation in applied pull
6. Non-horizontality
7. Sag in the chain, if suspended for measuring horizontal distance on a sloping ground.

Systematic errors have the same magnitude and sign in a series of measurements that are repeated under the same condition, thus contributing negatively or positively to the reading. It makes the readings shorter or longer. For example Errors 1, 2, 6 and 7 are always +ve since they make measured length more than actual. These errors are caused by badly adjusted instrument and the physical condition at the time of measurement. Expansion of steel and frequent changes in electromagnetic distance measuring (EDM) instrument, are just some of these errors.

This type of error can be eliminated from a measurement using corrections (*e.g.* effect of tension and temperature on steel tape). Another method of removing systematic errors is to calibrate the observing equipment and quantify the error allowing corrections to be made to the observations. Observational procedures by re-measuring the quantity with an entirely different method using different instrument can also be used to eliminate the effect of systematic errors.

For accuracy, it is necessary that chains are tested from time to time. It can be done by measuring the chain with respect to

- Steel tape
- By standard chain
- Permanent test gauge
- Pegs driven in the field at required distances
- Permanent test gauge made with dressed stones

Testing and Adjustment of Chain

As the chain is made of metal, it may undergo many changes due to temperature effect or human error *etc.* A tolerance level is given for all the chain lengths as follows:

5 m chain = + or – 3 mm

10 m chain = + or – 3 mm

20 m chain = + or – 5 mm

30 m chain = + or – 8 mm

Within these tolerances, chains should be accurate within 2 mm.

Chain Lengths

Chain lengths may increase or decrease over a prolonged period of use. It can happen due to the following reasons.

Chain length can be shortened due to:

- Bending of links
- Sticking of mud in the rings

If chain is found short, then

- Straighten the links
- Replace the small rings with big one
- Insert additional rings
- Flatten the circular rings

Chain length increases due to

- Opening of small rings
- Wearing of surfaces

If chain is found long, then

- Close the joints of the rings
- Reshape the elongated rings
- Remove one or two rings
- Replace worn out rings

Corrections in Chain Surveying

The expression that "a tape is standard at a certain temperature and under a certain pull" means that under these conditions the actual length of the tape is exactly equal to its nominal length. Since the tape is not used in the field under standard conditions, it is necessary to apply following five corrections for the measured lengths of the chain/tape:

- Corrections for absolute length of the chain
- Corrections for temperature
- Corrections for tension or pull
- Corrections for slope or verticality of the line
- Corrections for sag.

A correction is plus or positive when the uncorrected length is to be increased, and minus or negative when it is to be decreased in order to obtain true length.

Correction for Absolute Length

It is the usual practice to express the absolute length of a tape as its nominal or designated length plus or minus a correction. The correction for the measured length can be calculated.

Let l is the designated length of the tape and la the actual length of the tape, then

Correction in the chain length, $C = la - l$ (2.1)

Corrected length of the tape, $la = l + C$ (2.2)

If the total length measured by the chain is L, m then

Total correction, $C_a = L\,C/l$ (2.3)

Corrected length = L + C_a = L + LC/l = L((l+C)/l) (2.4)

Thus, corrected length = L (Corrected length of chain/ Nominal length of the chain) (2.5)

The sign of the correction (C_a) will be the same as that of C. It may be noted that L and l must be expressed in the same units and the unit of C_a is the same as that of C.

Correction for Area

If A is the area measured by the incorrect chain, then

Corrected area, Ac = $(L+Ca)^2 = L^2 (L+C) (l)^2$ (2.6)

Since L^2 can be written as area, we get

Ac = A $((l+C) (l)^2$

Ac = A $((l+C) (l)^2$

Thus, corrected area = L (Corrected length of chain/ Nominal length of the chain)^2 (2.7)

Similarly, corrected volumes can be calculated by the formula

Vc = V $((1+C/l)^3$ (2.8)

Thus, corrected Volume, Vc = V (Corrected length of chain/Nominal length of the chain)^3

Example 2.1

A distance of 1200 m was measured with a 20 m chain. After the measurements chain was found to be 60 mm longer. If the length of chain was perfectly correct at the beginning of measurement, what is the true length of the distance measured?

Since the chain was perfectly correct in the beginning, one can assume that it went wrong somewhere in between the measurements. So the average correction is given as:

C = (0+60)/2 = 30 mm = 0.03 m.

Correction for measured length = 1200 × (0.03/20) = 1.8 m

True length = 1200 +1.8 = 1201.8 m

Example 2.2

A survey was conducted with a 20 m chain and a map of the field was prepared to a scale of 1cm = 5 m. The area of the plan was found to be 91.1 cm^2. It was observed that the chain length was 20.05 m instead of 20 m. What is the true area of the field?

Initial length of the chain = 20 m

Correction in the chain length = 20.05- 20.0 = 0.05 m

Measured area = 65 m^2

Actual area = 91.1 × 5^2 = 2277.5 m^2

Corrected area = 2277.5 $(20.05/20)^2$ = 2283.2 m^2

Correction for Temperature

Since the length of a tape increases with increasing temperature or reduces with decreasing temperature, the measured distance may be too small or too large at a temperature which is not the same at which the length of the chain was standardized. The correction due to change in temperature is given as:

$$C_t = \alpha (T_m - T_o) L \quad (2.9)$$

Here C_t is the correction for temperature, m, α is the coefficient of thermal expansion, T_m is the mean temperature during measurement and T_o is the temperature at which the tape is standardized and L is the total measured length in m. The sign of the correction is plus or minus according to T_m is greater or less than T_o. The coefficient of thermal expansion is often defined as the fractional increase in length per unit rise in temperature (m/m per unit rise in temperature). The coefficient of expansion for steel varies from 10.6×10^{-6} to 12.2×10^{-6} per degree centigrade and that for invar from 5.4×10^{-7} to 7.2×10^{-7}. If the coefficient of expansion of a tape is not known, an average value of 11.4×10^{-6} for steel and 6.3×10^{-7} for invar may be assumed.

Example 2.3

A line was measured with a steel tape which was exactly 30 m long at 20°C and found to be 652.340 m. If the temperature during measurement was 30°C, find the true length of the line. Take coefficient of expansion of the tape per °C = 0.0000118.

Temperature correction per tape length = $C_t = \alpha (T_m - T_o) L$

Here L = 30 m: T_o=20°C; T_m = 30°C; α = 0.0000118

C_t = 0.0000118 (30-20) 30 = 0.00354 m (+ ve)

Hence the length of the tape at 30°C = 30 + C_t = 30 + 0.00354 = 30.00354 m.

Now true length of a line = (L′/L) × Measured length.

True length = (30.00354/30) × 652.340 = 652.417 m.

Correction for Pull (or Tension)

Similar to temperature, length of the chain also depends upon the pull, increasing with increasing pull and decreasing with decreasing pull than the standard pull at which chain was standardize. The correction, when the pull used during measurement is different from that at which the tape is standardized, is given as:

$$C_p = (P - P_o)L/AE \quad (2.10)$$

Here C_p is the correction for pull, m, P is the pull applied during measurement, Newtons (N), P_o is the pull under which the tape is standardized, N, L is the measured length in m, A is the cross-sectional area of the tape, m^2, and E is the modulus of elasticity of steel.

The value of E for steel may be taken as 19.3 to 20.7×10^{10} N/m^2 and that for invar 13.8 to 15.2×10^{10} N/m^2. The sign of the correction is plus, as the effect of the pull is to increase the length of the tape and consequently, to decrease the measured length of the line.

Example 2.4

In the above example 2.3, steel tape standardized at 45 N was pulled during the measurement at 95 N. The measured length was 652.40 m. The tape was uniformly supported during the measurement. Find the true length of the line if the cross-sectional area of the tape was 0.02 cm^2 and the modulus of elasticity = 20×10^6 N per cm^2.

Temperature correction per tape length from example 2.3 = 0.00354 m (+ ve)

Pull correction per tape length = $(P_m - P_o)L/AE = (95 - 45) \times 30/(0.02 \times 20 \times 10^6)$ = 0.00375 m (+ve)

Sag correction per tape length = 0

Combined correction = 0.00354 + 0.00375 m = 0.00729 m

True length of the tape = 30.00729 m

True length of the line = (30.00729/30) × 652.40 = 652.559 m.

Correction for Sag

The sag correction is required only when the tape is suspended during measurement. When a tape is stretched over two points of support, it takes the form of a catenary, although in practice, is assumed to be a parabola. The sag correction is the difference in length between the arc and the subtending chord (Figure 2.5). Since the effect of the sag on the tapes is to make the measured length too great this correction is always subtractive.

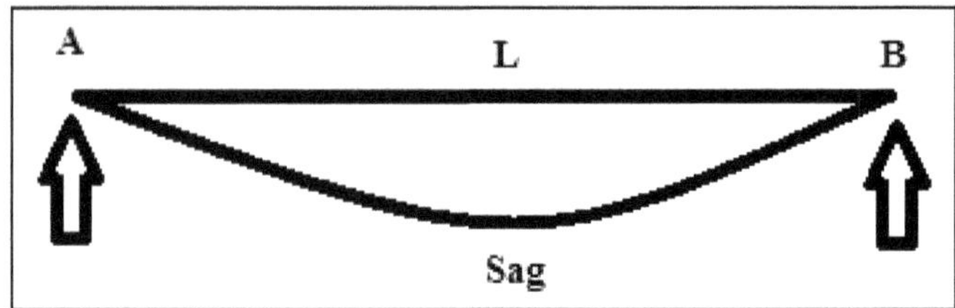

Figure 2.5: Schematic of Sag Correction in Chain Survey.

It is given by the formula

$$C_s = L\,(m \times g \times L)^2/24P^2 = L(M \times g)^2/24P^2 \quad (2.11)$$

Such that C_s is the sag correction for a single span, m, L is the distance between supports, m, m is the mass of the tape, kg/m, M is the total mass of the tape, kg, P is the applied pull, Newtons (N). The correction for sag C_s is always subtracted from observed distance.

Example 2.5

A 30 m tape is suspended between the ends under a pull of 120 N. The mass of the tape is 0.9 kg. Find the corrected length of the tape.

Correction for sag = $L (M \times g)^2/24 P^2$

$L = 30$ m; $M = 0.9$ kg; $P = 120$ N.

$C_s = 30 \times (0.9 \times 9.81)^2/24 \times 120^2 = 2338.54/345600 = 0.00677$ m.

Corrected length of the tape = $L - C_s = 30 - 0.00677 = 29.993$ m.

Normal Tension

The normal tension is the tension at which the effects of pull and sag are neutralized, *i.e.* the elongation due to increase in tension is balanced by the shortening due to sag. It may be obtained by equating the corrections for pull and sag.

Correction for Slope

This correction is required when the points of support are not exactly at the same level. Let S represents the slope distance between two points A and B, h be the difference in elevation and H the horizontal distance between the points, all in the same units. Let θ be the slope of the terrain (Figure 2.6). Then, the horizontal distance is given as:

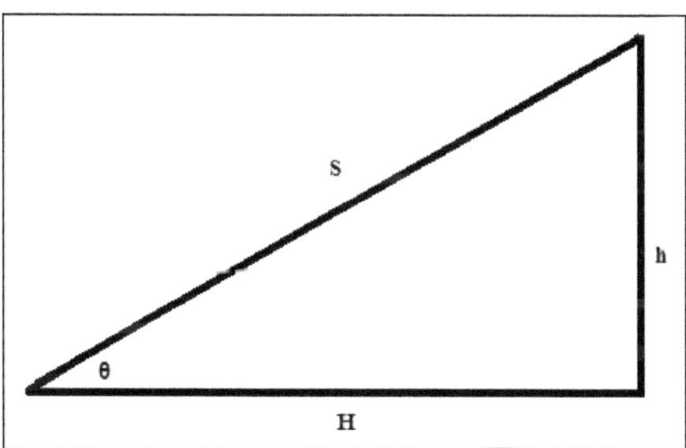

Figure 2.6: Schematic Diagram Showing the Measurements Up or Down the Slope.

$$H = S \cos \theta \quad (2.12)$$

$$H = \sqrt{(S^2 - h^2)} \quad (2.13)$$

Here, S is the length along the slope, H is the horizontal distance and θ is the subtended angle.

Slope correction, C_s for measured distance S is given as

$$C_s = H - S = \sqrt{(S^2 - h^2)} - S \text{ (always subtractive)} \quad (2.14)$$

If h is too small as compared to S, then

$$C_s = S (1 - h^2/S^2)^{1/2} - S \quad (2.16)$$

Expanding and retaining only the first term, we get

$$C_s = S (1 - h^2/2S^2) - S = h^2/2S \quad (2.17)$$

The exact equation from eq. (2.14) can be written as:

$$C_s = H - S = S\,(1 - \cos\theta) \text{ (always subtractive)} \quad (2.18)$$

Correction for Vertical Allignment

If the intermediate points are not in correct alignment with the ends of the line or they are not on the line to be measured, a correction, C_v known as correction for alignment has to be subtracted from measured distance. It is given by:

$$Cv = d^2/2S \quad (2.19)$$

Here d is the perpendicular distance by which the other end of the tape of length L is out of alignment. Note that d is assumed to be always small as compared to measured distance and hence we can use the simplified formula.

Example 2.6

The downhill end of the 50 m tape is held 90 cm too low. What is the horizontal length?

Correction for slope = $h^2/2S$

Here h = 0.9 m; S = 50 m

The required correction = $0.9^2/(2 \times 50) = 0.0081$ m.

Hence the horizontal length = 50 – 0.0081 = 49.992 m

Example 2.7

A survey line PQ is run on a terrain having distance 240 m along a downward slope 1 in 12 from P to A. Find the horizontal distance between P and A if the distances were measured with a 20 m tape which was 5 cm too long.

Since slope is 1 m down in each 12 m, total drop = 20 m

$H = \sqrt{(240^2 - 20^2)} = \sqrt{(260)(220)} = 239.17$

Since the chain was 5 cm too long, the actual distance is

Ha = 239.17 (20.05/20) = 239.77 m

Example 2.8

What is the true length, if a 100 m tape is held 1.2 m out of line?

Correction for incorrect alignment= $d^2/2S$ (-ve)

Here d = 1.2 m; S = 100 m.

Correction = $1.2^2/(2 \times 100) = 0.0072$ m.

True length = 100 – 0.0072 = 99.993 m.

Reduction to Mean Sea Level

In plane surveying, horizontal distances are reduced to sea level when it is required to count them into equivalent distance at another elevation, such as the plane coordinate system or the average elevation of a survey for which the variation in elevation over the area is large (Figure 2.7). Reduced distance at mean sea level, d is given as:

$$d = Dh/R \tag{2.20}$$

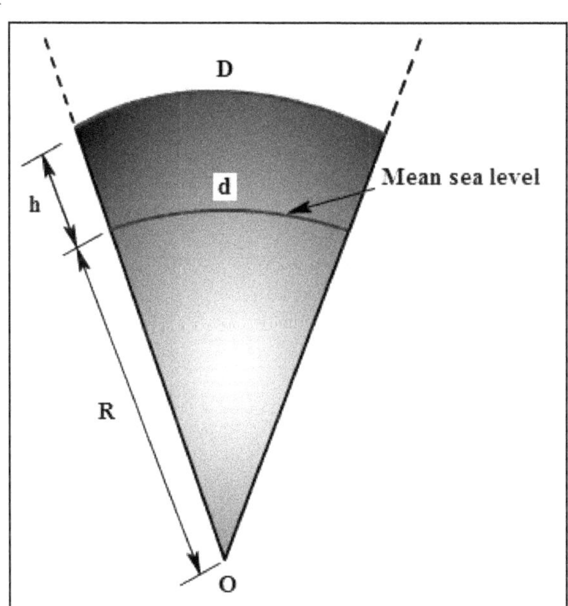

Figure 2.7: Reduction of Distances to Mean Sea Level.

Here, D is the measured distance on the surface of the earth, R is the mean radius of the earth and h is the mean height of the terrain above the mean sea level (msl). The correction for altitude may be positive for tunneling work or mining work going below the mean sea level. For surface works it is positive.

Chapter 3

Ranging

If a survey line is longer than a chain length, it is absolutely necessary to align intermediate points on the chain line to ensure that the measurements are made along the straight line. Method of locating or establishing intermediate points on a straight line between two fixed point or two survey stations is known as ranging. Two methods of ranging are:

- Direct method
- Indirect method also known as reciprocal ranging

Direct Methods

This method is used when two ends of survey stations or survey lines are inter-visible. It can be accomplished by using any of the following two methods.

Ranging by Eye

Consider two points A and D which are inter-visible to each other. In this method, if we want to locate a point B on ground which is in line with AD, ranging can be done by eye-judgment. Ranging rods are erected vertically at each end of the survey line A and D (Figure 3.1).The surveyor stands 1-2 m behind the ranging rod while the assistant holds the ranging rod vertically at the intermediate station, B and/or C. Assistant then moves the ranging rod under the guidance of surveyor in such a way that ranging rod held by assistant is in the line AD at point B between A and D (Figure 3.1). Any other point say C can be located in a similar manner. Surveyor has to guide assistant by using some hand signals so that ranging rod comes in the line. To avoid errors due to the ranging rods not being vertical, the lower end of the rod are cited for alignment. Flags at the top can be used if distances involved are large.

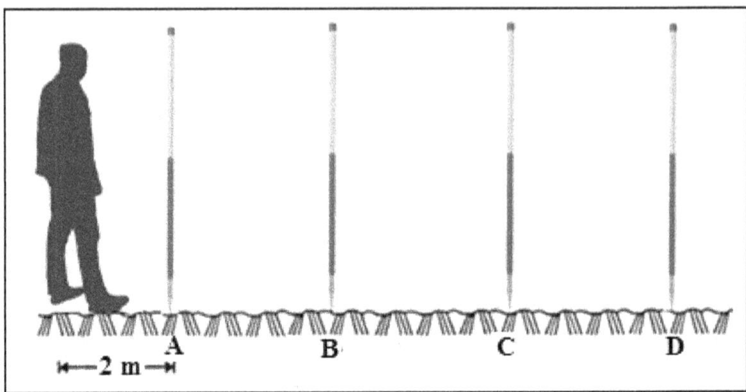

Figure 3.1: Methodology for Ranging by Eye.

The hand signals usually used by a surveyor to direct the assistant to move to the desired position are as follows.

- Slow sweeps with right hand: Move slowly a small distance to the right.
- Rapid sweeps with right hand: Move rapidly a long distance to the right.
- Slow sweeps with left hand: Move slowly a small distance to the left.
- Rapid sweeps with left hand: Move rapidly a long distance to the left.
- Right arm extended: Go on moving to the right.
- Left arm extended: Go on moving to the left.
- Right arm up and moved to the right: Make the rod vertical by moving its top to the right.
- Left arm up and moved to the left: Make the rod vertical by moving its top to the left.
- Both hands above head and then brought down: The position of the rod is correct.
- Both arms extended forwarded horizontally and then brought down: Fix the rod.

The surveyor may use a handkerchief to signal when the assistant has to read the signals from a far off distance.

Ranging by a Line Ranger

Line ranger is a small reflecting instruments used to range long lines (Figure 3.2 left). It is a light and easy to use instrument for ranging. It consists of 2 plane mirrors or 2 right-angled isosceles prisms placed one above the other. Diagonals of two prisms are silvered so as to reflect light. Lower prism is fixed while the upper prism is moveable. Instrument is provided with handle at bottom which eases the use of the instrument. The surveyor takes approximate position at an intermediate station say P with his face at right angles to the line and with the instrument held near the level of the eye. The rays of light from A enter the upper prism and after being reflected from the hypotenuse enters the eye at right angles to the line AB. Similarly rays from B enter the lower prism and after being reflected from the hypotenuse is also seen by the same eye. Thus the images of the rods at A and B are seen directly in the upper and lower prisms respectively. Initially the two images may not coincide (Figure 3.2 middle). The observer will continue to move forward and backward at right angles to the line AB until both the images coincide exactly. If the two images coincide, the required point P is then vertically below the center of the instrument (Figure 3.2 right). A pebble is dropped from the handle of line ranger to mark the point on the ground or a plumb bob is used (Figure 3.2 left).

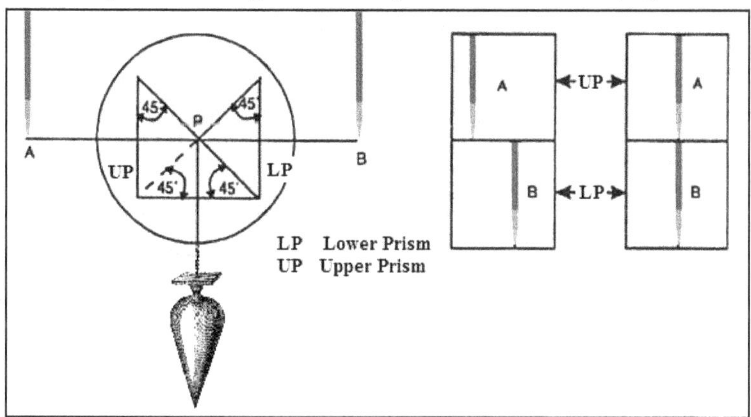

Figure 3.2: Ranging by a Line Ranger.

Indirect Ranging

This method is used when two ends of survey stations or survey line are not inter-visible (Anderson and Mikhail, 2012). Let A and B be the ends of a survey line to be measured as a rising ground between them (Figure 3.3). Two chain men with ranging rods take the positions M1 and N1 such that they are as nearly in line with A and B as they could judge. The chain men at M1 can see N1 and B. The chain men at N1 could see M1 and A. First chain men at N1 direct M1 to M2 so that he comes in the line with A and N1. Then the chain man at M2 directs N1 to N2 such that he comes in line with B and M2. The process is repeated so that they align each other successively directing each other until they are both finally in the line AB at M and N respectively.

Reciprocal Ranging Across a Valley or Depression

X and Y are the two stations at the ends of a valley. Let the intermediate points A, B, C ....F, G...etc. are to be fixed in line with X and Y across the valley. To do this, the surveyor at X will direct the assistant at A to be in line with Y so that the top of ranging rod at A is brought in line with the bottom of the ranging rod at Y (Figure 3.4). The ranging rod is then fixed at A. The surveyor at X again directs his assistant to move downwards to B so that the top of the ranging rod at B is in line of sight pointing to the bottom of rod at A. This process is continued till a point is reached near the lowest portion which is invisible from X. The surveyor then goes

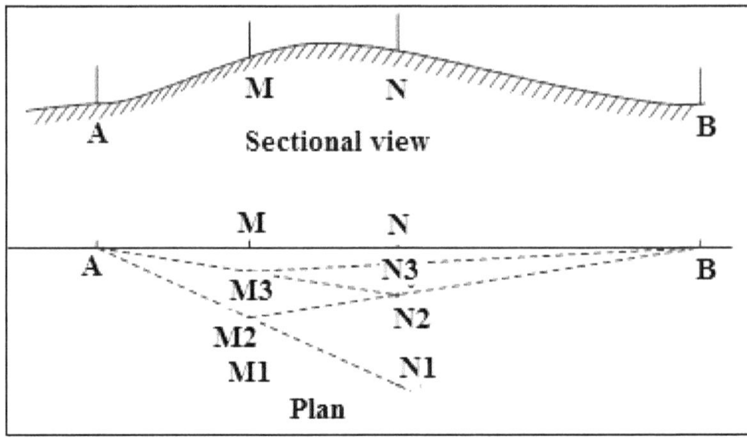

Figure 3.3: Reciprocal or Indirect Ranging.

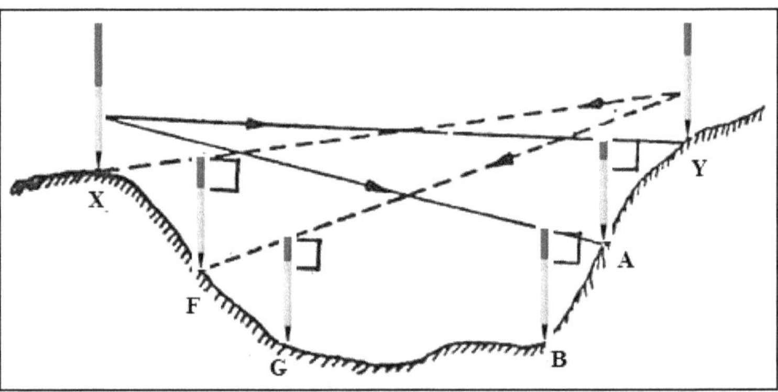

Figure 3.4: Ranging Across a Valley.

to the other end-station Y. He then direct the assistant to move to the position F such that top of the ranging rod at F is in line of sight pointing F such that top of the ranging rod at F is in line of sight pointing to the bottom of the rod at X. The process is repeated as before. X, F, G ...B, A and Y are then in the same straight line.

Chapter 4

Areas and Volumes

The calculation of areas is a fundamental calculation to be performed by a surveyor, whether it be the calculation of the areas of a parcel of land, or of a lake or a forest, or an area to be paved or a construction area. In the metric system, area is measured in square meters (m^2) or in hectares (ha). One hectare (ha) = 10,000 m^2. Normally, two types of areas are determined. One is the areas of regular figures, such as rectangles and triangles. The other is the areas of irregular figures such as lakes or properties bounded by features such as rivers. Most of the times, surveyor is also expected to calculate the volumes such as for lakes and earthwork *etc.* This chapter describes various methods of calculating areas and volumes of both regular and irregular plane figures.

Areas of the Regular Figures

Some of the common regular figures encountered in surveys are given in Figure 4.1. The figures from 1 to 8 respectively are square, rectangle, triangle (3, 4, 5), parallelogram, trapezium and polygon. There are two ways to determine the area. One is the direct use of the formulae given in Table 4.1. The other is to convert the regular figure into several other regular figures and calculate the area by summing the area of each regular figure. In segment number 8 of Figure 4.1, a regular polygon has been divided into 3 triangles. The area of the polygon is the sum of the areas of the three triangles. This method can be used for any regular closed area which can be divided into figures of regular shapes.

Area by Coordinates

The method of area calculation by coordinates is very handy when one is able to determine the coordinates of each point defining the shape of the region of the cross-section. The method of area calculation by coordinate's method is explained through Figure 4.2 and Table 4.2. The coordinates of the regular figure ABCD are given and reproduced in Table 4.2.This method is applicable if the first control point is also the last control point as shown in Table 4.2 in last row 6.

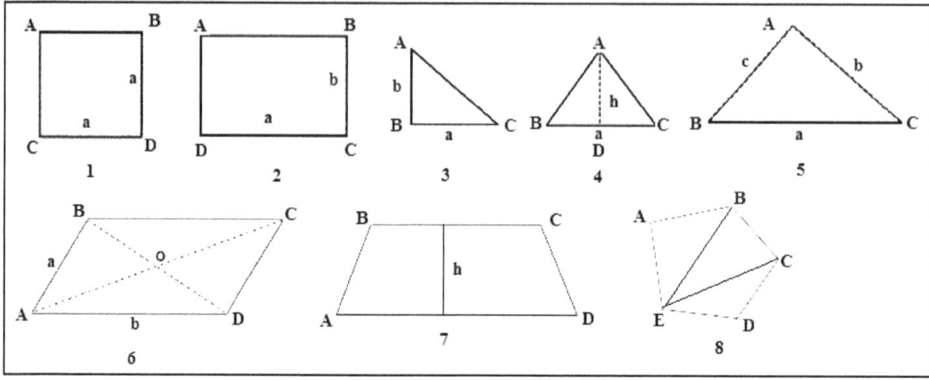

Figure 4.1: The Description of some of the Regular Figures for Calculating the Areas.

Now use the following formula (See Table 4.2):

$$\text{Area} = \{[(NA \times EB) + (NB \times EC) + (NC \times ED) + (ND \times EA)] - [(EA \times NB) + (EB \times NC) + (EC \times ND) + (ED \times NA)]\}/2 \quad (4.1)$$

Table 4.1: Areas of Regular Figures (Refer Figure 4.1)

Number in Figure 4.1	*Name*	*Area*	*Remarks*
1	Square	a^2	Square of the length of any side
2	Rectangle	a × b	Multiplication of length of base to height
3	Right angle triangle	½ (a×b)	Half the base times the length of the side making the right angle with the base
4	Triangle	½ (a×h)	Half the base times the perpendicular height
5	Triangle	S(S-a)(S-b)(S-c)	S = (a+b+c)/2
6	Parallelogram	a×b× sin A **or** ½×AC×BD×sin BOC	Two formulae are used. Sine of angle A and angle BOC respectively in the two formula
7	Trapezium	{(AD+BC)/2} × h	½ the sum of parallel sides times the perpendicular height
8	Polygon or other regular figure	–	Divide in regular figures or use coordinate method

Using this equation, the result will be positive, if one travels in clockwise direction. If one travels in the anticlockwise direction, the result will be negative. However, the absolute value will be same in both cases. Thus, one can use any direction but be sure to use absolute value of the result.

Using the formula for the data given in Table 4.2, we have:

$$\text{Area}=\{[(25 \times 35+15 \times 20+5 \times (-20)+5 \times (-25)] - [(-25) \times 15+35 \times 5+20 \times 5+(-20) \times 25]\}/2$$

Table 4.2: Area Calculations Using Coordinate Method

Point	*X-coordinate (Easting, E)*	*Y-coordinate (Northing, N)*	*N × E*	*E × N*
A	-25 (EA)	25 (NA)		
B	35(EB)	15(NB)	875	-375
C	20(EC)	5(NC)	300	175
D	-20(ED)	5(ND)	-100	100
A	-25(EA)	25(NA)	-125	-500
			Σ950	Σ-600

Area=[950-(-600)]/2

Area=1550/2

Area=755.00 m^2.

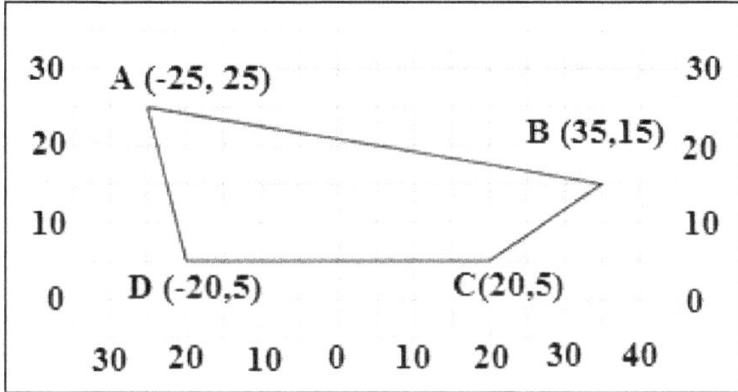

Figure 4.2: Coordinates of a Regular Figure for Area Evaluation.

Area of Figures with Irregular Boundaries

When survey boundaries include irregular figures, area calculations are made using any of the following methods.

1. Graphical method
2. Midpoint ordinate rule
3. Average ordinate rule
4. Trapezoidal rule
5. Simpson's rule

Graphical Method

The area of any complex figure, whether it has straight lines or curved boundaries, can be calculated by estimating how many squares are covered by the figure. The simplest method is to place over the area to be measured, a grid of known dimensions (say 5 mm grid squares) on tracing paper. Estimate by eye, the number of squares and part squares covered by the area (Figure 4.3). Knowing the area of each square and the scale of the map, one can calculate the area. The accuracy of the results of this method obviously depends upon the care with which the grid squares are estimated. It also depends upon the grid size. Finer the grid more accurate will be the results. However, if a very fine grid is used, the counting can be time-consuming and tedious and this aspect must be balanced against the desired accuracy.

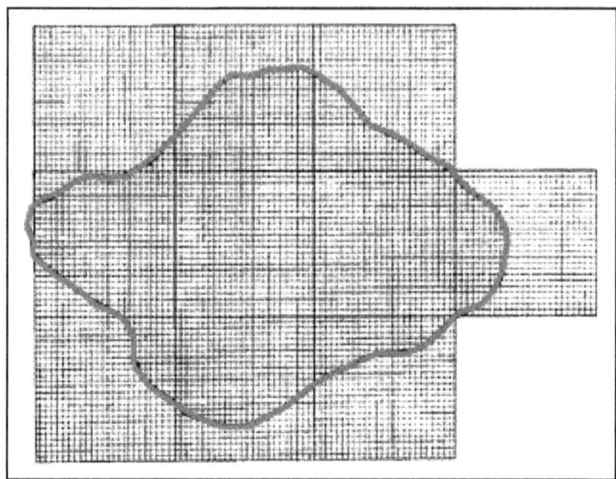

Figure 4.3: Graphical Method to Calculate Areas of Irregular Shape.

Let us assume that an irregular lake on a map scale of 1:100,000 is placed on a 5×5 mm grid. The total squares counted are 240 mm^2. The area of the lake is calculated as follows:

The area of each grid is 25 mm^2

Therefore, the area of the lake on the grid is $240 \times 25 = 6000$ mm^2

Multiplying by the scale $= 6000 \times 100000^2 = 6 \times 10^{13}$ mm^2

Since 1 $m^2 = 1000^2$ $mm^2 = 10^6$ mm^2

Therefore area in $m^2 = 6 \times 10^{13}/10^6 = 6 \times 10^7$ m^2

Since 1 ha $= 10^4$ m^2

Area of the lake in ha $= 6 \times 10^7/10^4$ ha $= 6000$ ha

Mid-Point Ordinate Rule

This rule states that area of the irregular figure is given by the summing of all the ordinates taken at midpoints of each division multiplied by the length of the base line having the ordinates. Mathematically,

$$\text{Area} = ([O1 + O2 + O3 + \ldots + On] \times L)/n \quad (4.2)$$

L is the length of baseline ($n \times d$) such that n is the number of equal parts, the baseline is divided and d is the common distance between the ordinates (Figure 4.4).

Example 4.1

The perpendicular offsets were taken at 10 m interval from a survey line to an irregular boundary line. The ordinates measured at midpoint of the division are 10, 13, 17, 16, 19, 21, 20 and 18 m. Calculate the area enclosed by the midpoint ordinate rule.

Given: Ordinates O1 = 10, O2 = 13, O3 = 17, O4 = 16, O5 = 19, O6 = 21, O7 = 20, O8 = 18.

Common distance, d =10 m

Number of equal parts of the baseline, n = 8

Length of baseline, L = n × d = 8 × 10 = 80 m

Area = [(10+13+17+16+19+21+20+18) × 80]/8 = 1340 m^2

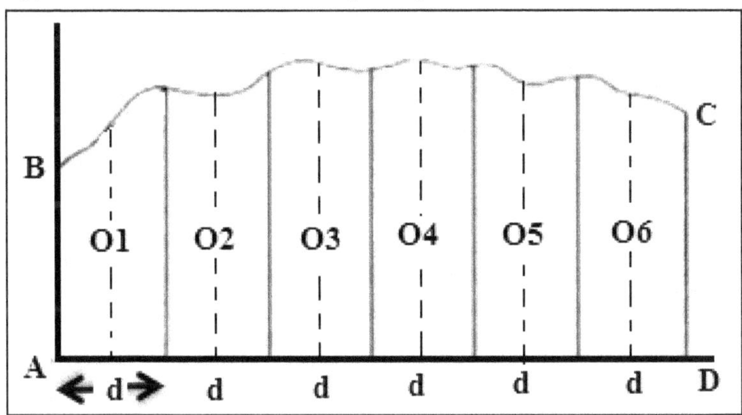

Figure 4.4: Midpoint-Ordinate Rule for Area Measurement.

Average Ordinate Rule

The rule states that the area is given by the average of all the ordinates taken at each of the division of equal length multiplied by baseline length divided by number of ordinates. Mathematically, it is given as:

$$\text{Area} = [(O1 + O2 + O3 + \dots + On) \times L]/(n+1) \quad (4.3)$$

Here, O1, O2, O3, O4....On are the ordinate taken at each division.

Note: Number of ordinates will be equal to number of divisions plus 1.

Example 4.2

The following perpendicular offsets were taken at 10 m interval from a survey line to an irregular boundary line.

9, 12, 17, 15, 19, 21, 24, 22, 18

Calculate area enclosed between the survey line and irregular boundary line.

Area = [(O1+ O2+ O3+ .... + O9) × L]/(n+1)

= [(9+12+17+15+19+21+24+22+18) × 8 × 10]/(8+1)

= 139538 m^2

Trapezoidal Rule

When confronted with an irregular boundary, general survey practice is to locate the boundary by a series of offsets (Figure 4.5). The selected interval (d) between offsets is chosen such that the length of the boundary between offsets is assumed straight, and the area is divided into a series of trapezoids. Since the theory states that the irregular boundary approximates a straight line, the interval between offsets is kept short to get higher accuracy. Moreover, even if first or last ordinate happens to be zero, they need be included in the calculations in this method as well as in the Simpson's rule.

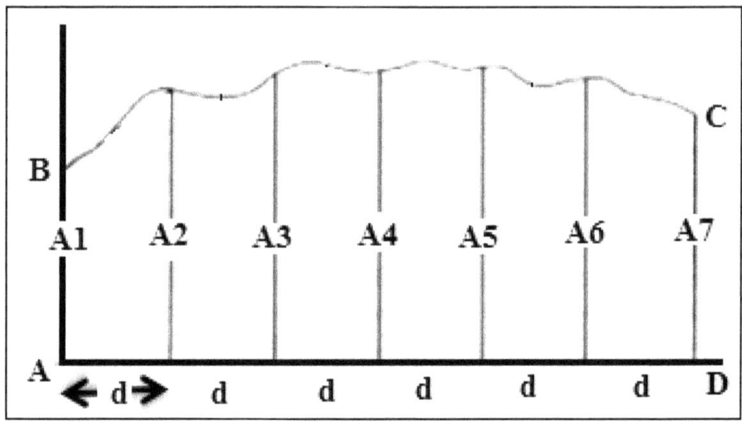

Figure 4.5: Illustration of Trapezoidal Rule to Calculate Area.

Area of Trapezoid (1) = [(A1 + A2)/2] d

Area of Trapezoid (2) = [(A2 + A3)/2] d

Area of Trapezoid (3) = [(A3 + A4)/2] d

Area of Trapezoid (4) = [(A4 + A5)/2] d

Area of Trapezoid (5) = [(A5 + A6)/2] d

Area of Trapezoid (6) = [(A6 + A7)/2] d

We can find the total area by summing up the areas of all the trapezoids.

Area = ½ × d (A1 + 2A2 + 2A3 + 2A4 + 2A5 + 2A6 + A7)

This can be simplified to

$$\text{Area} = d\,(½A1 + A2 + A3 +.+ An\text{-}1 + ½An) \tag{4.4}$$

$$\text{Area} = \text{Interval}\,(½[\text{first ord.} + \text{last ord.}] + \Sigma\ \text{intermediate ords.}) \tag{4.5}$$

This formula is also known as the End Area Rule.

Example 4.3

The following offsets are taken from a chain line to an irregular boundary. Use the trapezoidal rule to calculate the area. All chainages and offsets are in meters.

Chainage	0	25	50	75	100	125	150
Offset	4.6	6.0	7.5	5.5	6.3	5.0	3.0

Common distance, d = 25 m

Area = $d[0.5(O_1+O_7) + O_2+O_3+O_4+O_5+O_6]$

= 25[0.5(4.6+3.0)+ 6.0+7.5+5.5+6.3 +5.0]

Area = $25 \times 37.1 = 852.5\ m^2$

Simpson Rule

Another formula, known as Simpson's Rule, is also used to compute areas of irregular figures. It states that, sum of the first and last ordinates plus twice the sum of remaining odd ordinates plus four times the sum of remaining even ordinates, multiplied by 1/3rd of the common distance between the ordinates gives the area.

$$\text{Area} = (D/3)\ (\text{Offset1} + \text{Offsetn} + 4\Sigma\text{even offsets} + 2\Sigma\text{odd offsets}) \quad (4.6)$$

Simpson's Rule, which usually gives greater accuracy, assumes that the irregular boundary is composed of a series of parabolic arcs. If the irregular boundary closely approximates a straight line, then the trapezoidal rule should be used. It is essential that the figure under consideration be divided into an even number of equal strips; *i.e.* there must be an uneven number of offsets. If the numbers of ordinates are even, the area of last division may be calculated separately and added to the result obtained by applying Simpson's rule to the remaining ordinates.

The relative merits of trapezoidal and Simpson rules may be compared in the following manner:

Trapezoidal Rule	*Simpson's Rule*
The boundary between the ordinates is considered to be straight	The boundary between the ordinates is considered to be an arc of a parabola
There is no limitation. It can be applied for any number of ordinates	To apply this rule, the number of ordinates must be odd. However, one can circumvent this limit by separately calculating the area of the last division and use Simpson method for the remaining offsets. The area for the last division is added to find out the total area
It gives an approximate result	It gives a more accurate result

Example 4.4

The following offsets are taken from a chain line to an irregular boundary. Use the Simpson rule to calculate the area. All chainages and offsets are in meters.

Chainage	0	25	50	75	100	125	150
Offset	4.6	6.0	7.5	5.5	6.3	5.0	3.0

Common distance, d = 25m

Area = $d/3[(O_1+O_7) + 2(O_3+O_5)+4(O_2+O4+O_6)]$

= 25/3[(4.6+3.0)+2(7.5+6.3)+4(6.0+5.5+5.0)]

Area = 843.3 m^2

Example 4.5

The following offsets were taken at 20 m intervals from a survey line to an irregular boundary line

3.50, 4.30, 6.75, 5.25, 7.50, 8.80, 7.90, 6.40, 4.40, 3.25 m

Calculate the area enclosed between the survey line and the irregular boundary line by the trapezoidal and Simpson's rule

The trapezoidal rule

Area=20/2{3.50+3.25+2(4.30+6.75+5.25+7.50+8.80+7.90+6.40+4.40)}

= 20/2{6.75+102.60} = 1090.35 m^2

Simpson's Rule

Since the number of ordinates are even, we apply Simpson's rule for offsets O1 to O9 and the area between O9 and O10 is calculated by the trapezoidal rule.

A1=20/3{3.50+4.40+4(4.30+5.25+8.80+6.40)}+2(6.75+7.50+7.90)

= 20/3(7.90+99.00+44.30)= 1008.00 m^2

A2= 20/2(4.40+3.25)= 76.50 m^2

Total area= A1+ A2 =1008.00+76.50 = 1084.5 m^2

Example 4.6

The following offsets were taken from a survey line to a curved boundary line:

Distance, m	0	5	10	15	20	30	40	60	80
Offset, m	2.5	3.8	4.6	5.2	6.1	4.7	5.8	3.9	2.2

Find the area between the survey line and the curved boundary line by (a) Trapezoidal Rule and (b) Simpson's Rule.

Here, the intervals between the offsets are not regular throughout the length. So the section is divided into three compartments. Let,

$\Delta 1$ = Area of the 1st section

$\Delta 2$ = Area of the 2nd section

$\Delta 3$ = Area of the 3rd section

Here,

d1 = 5 m; d2 = 10 m; and d3 = 20 m

Trapezoidal Rule

$\Delta 1 = 5/2\ \{2.50 + 6.10 + 2(3.80 + 4.60 + 5.20)\} = 89.50\ m^2$

$\Delta 2 = 10/2\{6.10 + 5.80 + 2(4.70)\} = 106.50\ m^2$

$\Delta 3 = 20/2\ \{5.80 + 2.20 + 2(3.90)\}\ 158.00\ m^2$

Total Area = 89.50 + 106.50 + 158.00 = 354.00 m^2

By Simpson's Rule

$\Delta 1 = 5/3\{2.50 + 6.10 + 4(3.80 + 5.20) + 2\ (4.60)\} = 89.66\ m$

Similarly $\Delta 2 = 102.33\ m^2$ and

$\Delta 3 = 157.33\ m^2$

Total area = 89.66 + 102.33 + 157.33 = 349.32 m^2

Instruments for Area Measurements

Planimeter

This instrument is used to determine the area of a given map/figure. It can measure the areas of all regular and irregular shapes. It is the best and most expeditious method and gives accurate results compared to other methods (Bhavikatti, 2010). There are two types of planimeters: the Amsler polar planimeter and the rolling planimeter. All one has to do is to pivot the instrument at any point and then run the end pin all over the perimeter of the area to be measured. It has got a flexible linkage which allows it to move without any effort in all directions. One problem is that of scale. The planimeter gives the area measurement in its own scale or square meter and one needs to convert it to the desired scale. If the planimeter scale is not known, then trace a known area say a small square within the map. By measuring the area of the square, one can calculate the scale ratio of the map. For example if the square is 300 square feet and the planimeter measures it as 1.5 square meter then 1 m^2= 200 feet square. Thus, multiplying the final reading of the planimeter in m^2 by 200 will give the reading in square feet.

Digitizing

In this method, a device called a digitizer is used to convert point locations on a graphic image (irregular figure) to plane (x, y) coordinates. This process enables irregular area to be rapidly and precisely computed. The digitizer is used in conjunction with a digitizing table, command menu, visual display screen and computer. The figure to be digitized is placed on the digitizing table, and any preliminaries carried out, for example creation of design file, setting window area and any other software requirement. A start point or datum is selected on the irregular figure and, using the digitizer, the figure is traced. The (x, y) E, N coordinates are recorded with the digitizer by manually positioning the intersection of the crosshairs on the cursor over the intended point and pressing the record button. With an irregular figure, the digitizer is operated in continuous mode and points are recorded at constant intervals in distance or time and fed into the computer. On return to the start point, the computer calculates and provides the area. The method is based on calculation of areas by coordinates method discussed previously.

Example 4.7

Calculate the area of the Figure 4.6 using average ordinate method, trapezoidal rule and Simpson rule.

The figure comprises of a regular rectangle and irregular boundary. The area of the rectangle is common and so it is calculated separately and will be added later in the area calculated by each method.

Area of the rectangle = 50 × 40 = 2000 m^2

First offset = 53-40 = 13

Last offset = 49-40 =9

L = n × d = 12 × 6=72

Average ordinate method: Using eq. (4.3)

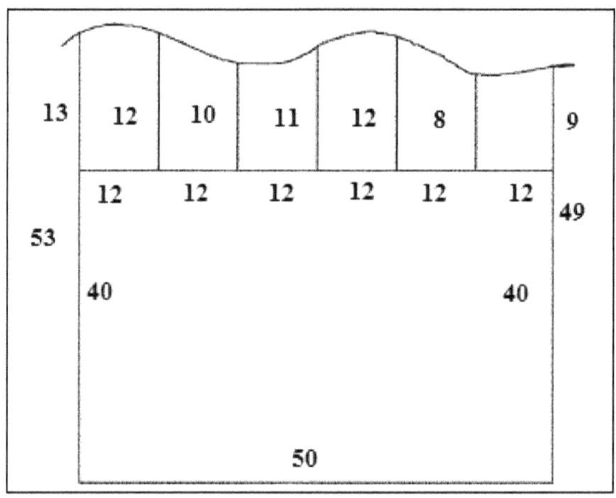

Figure 4.6: Example Figure to Calculate Area.

Area = [(13+ 12+ 10+11+12+8+9) × 72]/7 = 761.1 m^2

Total area = 2000+761.1 = 2761.1 m^2

For Trapezoidal rule, we use eq. (4.5)

Area = Interval (½[first + last] + Σ intermediates)

Area = 12(1/2(13+9) +12+10+11+12+8)) = 768 m^2

Total area = 2000+768 = 2768 m^2

For Simpson rule, we use eq. (4.6)

Area = d/3. (Offset1 + Offsetn + 4Σeven offsets + 2Σodd offsets)

Area = 4. (13+9 + 4 × 31 + 2 × 22) = 760 m^2

Total area = 2000+760 = 2760 m^2

Volume Measurement

The excavation, removal and dumping of earth is a frequent operation in agricultural engineering or civil engineering projects. For example, trenches are to be dug in contour trenching, the earth being stored at some convenient place, usually alongside the trench. Since the payments are made on the calculated volume of material handled, it is essential that surveyors have a grasp of the principles of estimating volumes for earthworks (Basak, 2014). In volume calculations, we in fact compute end areas and then multiply by a third dimension to find the volume. Direct formulae to calculate the volume of some of the regular shapes shown in Figure 4.7 are available. The formulae to calculate the volumes of these shapes are given in Table 4.3. Few general methods of calculating the volume or earthworks are discussed in the following sections.

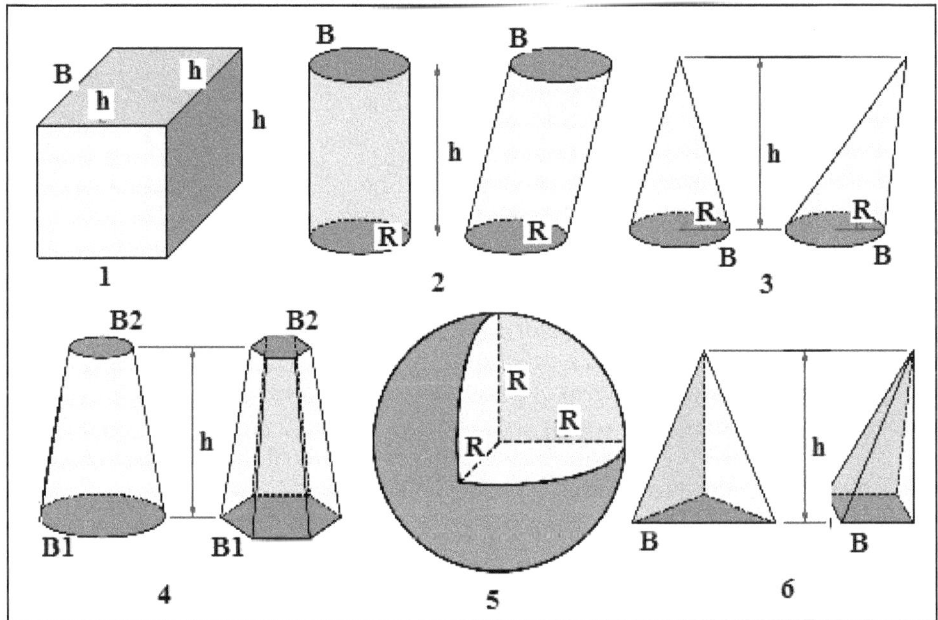

Figure 4.7: Few Common Shapes Encountered in Field Engineering Works.

Average End-Area Method

The area between the two surfaces is referred to as the end-area. The volume (V) of earthwork (cut or fill) between adjacent cross-sections is obtained by multiplying the average of the end-areas (A1 and A2) by the distance (L) between the sections.

$$V = [(A1+A2)/2]\ L \quad (4.7)$$

Table 4.3: Areas of Regular Figures (Refer Figure 4.7)

Number in Figure 4.7	*Name*	*Volume*
1	Cube	$B \times h = h^3$
2	A right and oblique cylinder	$B \times h = \ R^2 \times h$
3	A right and oblique cone	$(1/3)\ (B \times h) = (1/3)\ (R^2 \times h)$
4	Frustum	(h/3) (B1-B2+ B1+B2)
5	Sphere	$(4/3)\ (\ R^3)$
6	Pyramid	$(1/3)\ (B \times h)$

This formula will be exact only if the end areas (A1 and A2) are equal. However, the area will be approximate, if the end areas are not equal. If one end-area has a value of zero, as in the case of a cone or pyramid, the earthwork volumes will be different if calculated with the help of eq. (4.7) compared to the actual volumes calculated with the help of formulae given in Table 4.3. The prismoidal method is used to accurately determine the earthwork volumes in such cases.

Prismoidal Method

A prismoid is a solid having ends that are parallel and sides that are trapezoidal. The ends and sides of the prismoid are not congruent, (*i.e.* they do not have the same size or shape). Most earthwork solids obtained from cross-sections fit this description. The volume (V) of a prismoidal shape is calculated from the two end-areas (A1 and A2), the area (Am) of a section midway between A1 and A2, and the distance (L) between the two outer sections. The midpoint area is determined from averaging the corresponding linear heights and widths of the two end-areas and not by averaging the end areas.

$$V = 1/6 \times (A1+A2+4Am) \times L \qquad (4.8)$$

This method is commonly used to measure the volume of farm ponds, when bottom dimensions and side slopes are known. It may be noted that, eq. (4.8) is simplification of Simpson rule, if number of observations are only 3.

Example 4.8

Compute the volume of earth work involved in the construction of a farm pond of the following size: bottom 6 × 5 m. Side slope 2: 1, depth of pond 4 m. Work out the cost of earth work also if it costs Rs. 60 per m^3.

Size of pond at bottom = 6 × 5 m

Area at bottom = 30 m^2

Size of pond at ground level:

Length of pond = 6 + 8 + 8 = 22 m

Breadth of pond = 5 + 8 + 8 = 21 m

Cross sectional area of pond at ground level = 21 × 22 = 462 m^2

Area of pond at mid height = 14 × 13 = 182 m^2

Using prismoidal rule, we have V = 2/2[30+462+2 × 182] = 856 m^3

Cost of earth work = 60 × 856 = Rs. 51360

Contour Method

As an alternative to the determination of volumes from cross-sections, it is possible to calculate volumes using the horizontal areas contained by contour lines. The areas contained within the contours are determined using a planimeter, or a digitizer. The volume is then calculated by either trapezoidal or end area or Simpson's Rule as discussed previously for the areas. It is illustrated by the following examples.

Example 4.9

The areas within contour lines for a reservoir are as follows.

Contour, m	200	198	196	194	192	190	188	186	184
Area, m^2	738	708	693	506	241	98	50	31	4

Calculate the volume of water by trapezoidal and Simpson rules.

Volume by trapezoidal rule

V = d {(738 + 4)/2 +Σ Intermediates}

V = 2 (371+ 2331)

V = 5404 m^3

Volume by Simpson's Rule

V = d/3 {(A1 + AL + 2 Σodds Areas + 4 Σeven Areas)}

V = 2/3 (738 + 4 + 2 × 984 + 4 × 1343)

V = 5381m^3

Example 4.10

The areas enclosed by the contours in the lake are as follows:

Contour, m	270	275	280	285	290
Area, m^2	2050	8400	16300	24600	31500

Calculate the volume of water between the contours 270 m and 290 m by trapezoidal and Prismoidal formula.

Volume according to trapezoidal formula:

=5/2{2050+31500+2(8400+16300+24600)}

=330,250 m^3

Volume as per prismoidal formula:

=5/3{2050+31500+4(8400+24600)+ 2 × 16300}= 330916.7 m^3

Example 4.11

An embankment 10 m wide and side slopes 1:1 is required to be made on a ground which is level in a direction transverse to the center line. The central heights at 40 m intervals are: 0.90, 1.25, 2.15, 2.50, 1.85, 1.35, and 0.85. Calculate the volume of earth work according to trapezoidal and prismoidal methods.

The areas of cross-sections are calculated by the following formula:

$\Delta = (b+sh) \times h$

$\Delta 1 = (10+1 \times 0.90) \times 0.90 = 9.81 \text{ m}^2$

$\Delta 2 = (10+1 \times 1.25) \times 1.25 = 14.06 \text{ m}^2$

$\Delta 3 = (10+1 \times 2.15) \times 2.15 = 26.12 \text{ m}^2$

$\Delta 4 = (10+1 \times 2.50) \times 2.50 = 31.25 \text{ m}^2$

$\Delta 5 = (10+1 \times 1.85) \times 1.85 = 21.92 \text{ m}^2$

$\Delta 6 = (10+1 \times 1.35) \times 1.35 = 15.32 \text{ m}^2$

$\Delta 7 = (10+1 \times 0.85) \times 0.85 = 9.22 \text{ m}^2$

Volume according to trapezoidal formula:

$V = 40/2\{9.81+ 9.22+2(14.06+26.12+31.25+21.92+15.32)\}$

$= 20\{19.03+217.34\} = 4727.4 \text{ m}^2$

Volume calculated by prismoidal formula:

$V = 40/3\{10.22+9.58+4(14.06+31.25+15.32)+ 2(26.12+21.92)\}$

$= 40/3\ (19.80+ 242.52+96.08) = 4778.7 \text{ m}^2$

Volumes from Spot Heights

With this method earthwork involved in the construction of large tanks, borrow pits or even large recreation areas can be determined by calculating the amount of cut or fill and the mean height (where cut equals fill). Having located the outline of the area on the ground, the surveyor divides the area into squares; that is, a grid is placed over the area. The smaller the grid interval, the more precise are the results. It is because the grade between grid intersections is considered constant. A 10m × 10m grid generally suffices to have a good estimate of volume on most projects.

The volume of each individual grid square is computed by averaging the height above (or below) the proposed cut-fill RL of the four corners of each grid and multiplying it by the grid area.

For the Figure 4.8 (left), the volumes are:

Volume of grid 1= $(A + B + F + E)/4 \times$ area of grid

Volume of grid 2= $(B + C + D + E)/4 \times$ area of grid

Volume of grid 3= $(F + G + H + E)/4 \times$ area of grid

Volume of grid 4= $(H + I + D + E)/4 \times$ area of grid

Total Volume = Volume1 + Volume2 + Volume3 + Volume4

$$= \text{Area of grid} \times \tfrac{1}{4}\,(A+B+F+E+B+C+D+E+F+G+H+E+H+I+D+E)$$

$$= \text{Area of grid} \times \tfrac{1}{4}\,(A+2B+C+2D+4E+2F+G+2H+I) \qquad (4.9)$$

This gives us the general formula of:

$$V = A1\,(\Sigma h1 + 2\Sigma h2 + 3\Sigma h3 + 4\Sigma h4)/\Sigma m \qquad (4.10)$$

Where, A1 is the area of full grid square, h1is the heights that are used once (See values in circles with number 1), h2 is heights that are used twice (See values in circles with number 2), h3 is the heights that are used three times and h4 is heights that are used four times (See values in circles with number 4). Σm is the summation of number of times each value is used.

The conventional method for indicating the RL of the spot height is to write the value diagonally across the intersection of the two lines, with the whole meter values on the left hand side and the decimals of a meter on the right hand side (Figure 4.8 right). Here the height of the corner on the top left is 14.95 m.

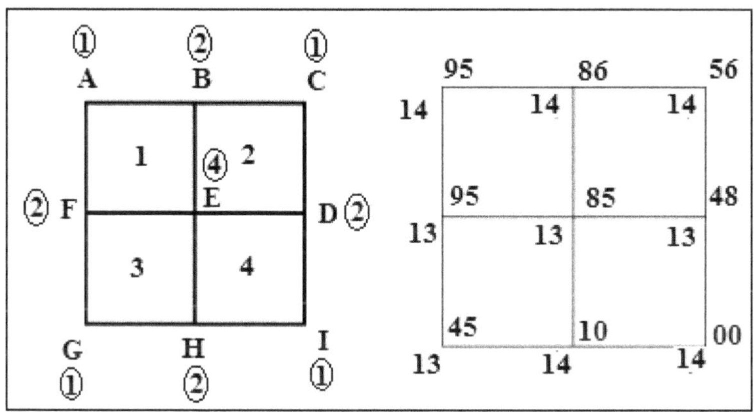

Figure 4.8: Spot Height of a Grid to Calculate Volume: Illustration (Left) and Data Example (Right).

Similarly, it can be shown that the mean height will be given as:

$$\text{Mean Height} = (\Sigma h1 + 2\Sigma h2 + 3\Sigma h3 + 4\Sigma h4)/4n \qquad (4.11)$$

Such that n is number of grids. Other notations being the same as discussed in the previous equation. Volume is then given as:

$$V = A1\,(\text{Mean height} - \text{proposed RL}) \qquad (4.12)$$

Example 4.12

A 10 meter grid was set out over an excavation site and levels taken (Figure 4.8 right). Formation RL is proposed to be 11 m. Compute the volume of earth to be removed.

$$\text{Volume} = A1\,(\Sigma h1 + 2\Sigma h2 + 3\Sigma h3 + 4\Sigma h4)/\Sigma m$$

The values of various parameters are calculated using Figure (4.8 right) as follows.

h1	h2
14.95 – 11.00 = 3.95 14.56 – 11.00 = 3.56 14.00 – 11.00 = 3.00 13.45 – 11.00 = 2.45 S = 12.96× 1= 12.96	14.86 – 11.00 = 3.86 13.48 – 11.00 = 2.48 13.95 – 11.00 = 2.95 14.10 – 11.00 = 3.10 S = 12.39× 2 =24.78
h3	h4
S = 0 × 3 = 0	13.85 – 11.00 = 2.85 S = 2.85 × 4 = 11.40

Sm = 1+ 1+1+1+2+2+2+2+4 = 16

The above calculations can also be performed in the following tabular form:

Spot Height	*Formation Level*	*Reduced Height*	*m*	*Reduced Height × m*
14.95	11.0	3.95	1	3.95
14.56	11.0	3.56	1	3.56
14.00	11.0	3.00	1	3.00
13.45	11.0	2.45	1	2.45
14.86	11.0	3.86	2	7.72
13.48	11.0	2.48	2	4.96
13.95	11.0	2.95	2	5.90
14.10	11.0	3.10	2	6.20
13.85	11.0	2.85	4	11.40
			Sm =16	Sh1 +2Sh2 + 3Sh3 + 4Sh4= 49.14

V = 400 (12.96 + 24.78 + 0 + 11.40)/16 = 1229 m^3.

We can also calculate the mean height using spot heights and eq. (4.11) as:

(56.96 + 112.78 + 0 + 55.4)/4n = 14.07 m

The volume then can be given by eq. (4.12) as:

Volume = 400 (14.07 -11.0) = 1228 m^3

Chapter 5

Chain Survey: Triangulation

What is Triangulation Survey?

Triangulation surveying is the tracing and measurement of a series or network of triangles to determine distances and relative positions of points spread over an area. It is achieved by measuring the length of one side of each triangle and deducing its angles and length of other two sides by observation from this bascline. In triangulation, entire area to be surveyed is covered with a framework of geometric figures composed of triangles. Horizontal angles and a limited number of sides (base lines) are measured. By using angles and base line lengths, triangles are solved trigonometrically and the positions of stations (vertices) are calculated. For the triangle, the length of the first line, which is measured precisely, is known as the base line. The other two computed sides are used as new baselines for two other triangles interconnected with the first triangle. By extending this process, a chain or network of triangles can be spread over the entire area.

Triangulation is preferred for hills and undulating areas, since it is easy to establish stations at reasonable distances apart, with inter-visibility. In plain and crowded areas the method may not be suitable as the inter-visibility of stations is affected. The difficulty can only be overcome by building towers, which is quite expensive. The main disadvantage of triangulation is the accumulation of error in the lengths and direction of lines, since both of them, for successive lines depend upon the computations of the preceding lines. It necessitates the use of check bases.

Objectives of Triangulation

- ☆ To establish accurate control for plane and geodetic surveys of large areas by terrestrial methods
- ☆ To establish accurate control for photogrammetric surveys of large areas

- To assist in the determination of the size and shape of the earth by making observations for latitude, longitude and gravity, and
- To determine accurate locations of points in engineering works such as :
  - Fixing center line and abutments of long bridges over large rivers
  - Fixing center line, terminal points, and shafts for long tunnels
  - Transferring the control points across wide sea channels, large water bodies, *etc.*
  - Detection of crustal movements, *etc.*
  - Finding the direction of the movement of clouds.

Classification

The system of triangulation is classified as:

- First order (primary): to determine the shape and size of the earth, to cover a vast area like a country
- Second order (secondary): network within first order triangulation, for a region/province
- Third order (tertiary): within second order triangulation, for detailed engineering and location surveys

A system consisting of triangulation stations connected by a chain of triangles is called triangulation system or triangulation figure (Figure 5.1). The most common type of figures used in a triangulation system are:

- Single chain of triangles
- Double chain of triangles
- Braced quadrilaterals
- Centered triangles and polygons
- A combination of above systems.

Geometric conditions that must be fulfilled by these figures in triangulation system are:

- The sum of interior angles should be $(2n-4)x90^{\circ}$, where n = no. of sides of the figure
- If all the angles are measured at a station, their sum should be 360°.
- The length of sides calculated through more than one routes should agree.

It is impossible to fulfill all the geometric conditions, owing to the errors, until the field measurements have been adjusted.

Chain Triangulation

The principle of chain surveying is triangulation. It is the simplest kind of surveying and is adopted when level of high accuracy is not required. In this system sides of various triangles are measured directly in the field and no angular

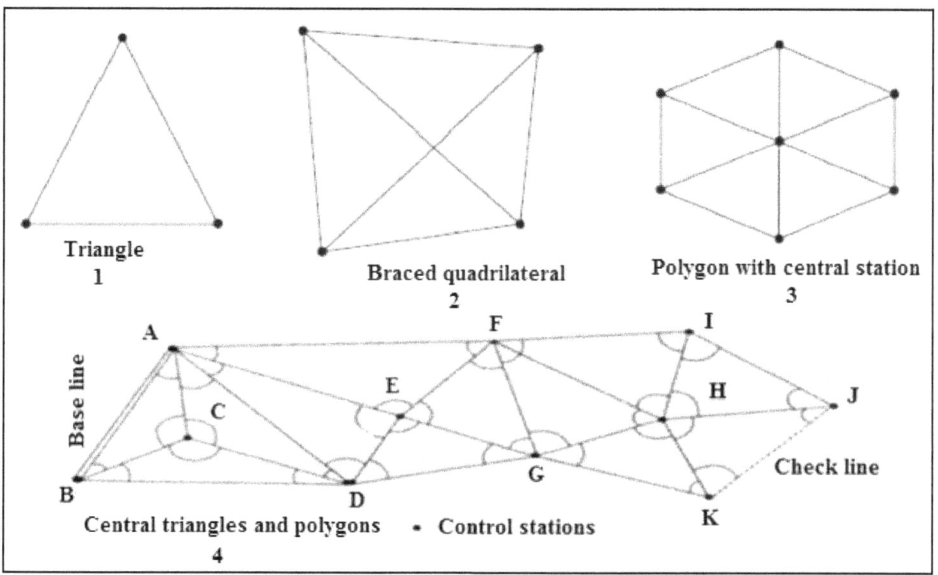

Figure 5.1: A System of Polygons with Control Stations.

measurements are taken. Since the triangle is a simple plane geometrical figure, it can be plotted from the measured length of its sides alone. Therefore, the principle of chain survey is based on triangulation.

In chain surveying, a network of triangles is preferred. Preferably all the sides of a triangle should be nearly equal, thus each angle being nearly 60°. It will ensure minimum distortion due to errors in measurement of sides and also during plotting. Such an ideal condition is not always met. Therefore, attempt are made to have well-conditioned triangles in which no angle is smaller than 30° and no angle is greater than 120°.

The area of a field in chain triangulation is measured as follows:

- ☆ Let ABCDE is the area to be measured (Figure 5.2). Fix the pegs at A, B, C, D and E
- ☆ Divide the area into three triangles ADE, ABD and BCD by joining AD and BD
- ☆ Measure the lengths AB, BC, CD, DE, EA, AD and BD.

Calculate the area of the triangles. The sum of the areas of the three triangles is the area of the given field.

Area of Δ = 0.5 [S (S-a) (S-b) (S-c)] (5.1)

Here S = (a+b+c)/2 and a, b and c are the lengths of the three sides of the triangle.

The area of the given field = Δ AED +Δ ABD+ Δ BDC

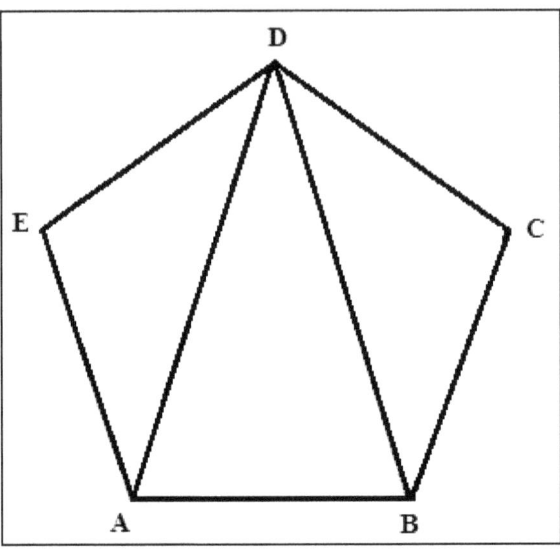

Figure 5.2: Calculation of Area of a Polygon.

Well-Conditioned and Ill-Conditioned Triangles

A triangle is said to be well-conditioned when no angle in it is less than 30° or greater than 120°. An equilateral triangle is considered to be the best-condition or ideal triangle. Well-conditioned triangles are preferred because their apex points are very sharp and can be located by a single 'dot'. In such a case, there is no possibility of relative displacement of the plotted point.

A triangle in which an angle is less than 30° or more than 120° is said to be ill-conditioned. These triangles are not used in chain surveying. It is because their apex points are not sharp and well defined which is why a slight displacement of these points may cause considerable error in plotting.

Operations in Triangulation Survey

The field work of triangulation is carried out in the following well defined four operations:

1. Reconnaissance
2. Station preparation: Marking the stations on the ground.
3. Baseline measurement
4. Measurement of angles.

Besides field work, triangulation survey involves specifications, design of stations and signals, and reduction and adjustment of the observations.

Reconnaissance

Before beginning the survey work, the surveyor should walk over the whole area to be surveyed in order to examine the ground and determine the possible arrangement of survey framework. This preliminary inspection is called reconnaissance. During this investigation, the surveyor examines the terrain and also assesses the inter-visibility of the main survey stations. He should ensure that the whole area is enclosed by main survey lines, and also that it is possible to form well-conditioned triangles. He should observe various objects and boundary lines carefully and select the survey lines in such a manner that the objects can be located by short offsets. The base line should preferably be taken through the center of the area and on fairly level ground. The surveyor prepares index sketch or key plan. Besides, he should collect miscellaneous information regarding access to various triangulation stations, transport facilities and availability of labour.

Marking the Stations on the Ground

Survey stations are the points at the beginning and the end of a chain line. They may also occur at any convenient points on the chain line. Stations are categorized as:

1. Main stations
2. Subsidiary stations and
3. Tie stations

Main stations are taken along the boundary of an area constituting the 'main survey lines'. The main survey lines should cover the whole area to be surveyed. These stations serve as the controlling points. The main stations are denoted by letters A, B, C, D, *etc.*

Subsidiary stations on the main survey lines or any other survey lines are known as "Subsidiary stations". These stations are taken to run subsidiary lines for dividing the area into triangles, for checking the accuracy of triangles and for locating interior details. These stations are denoted by letters S_1,S_2,S_3, *etc.*

Tie stations are also subsidiary stations taken on the main survey lines. Lines joining the tie stations are known as tie lines. Tie lines are mainly taken to fix the directions of adjacent sides of the chain survey map. These are also taken to form 'chain angles' in chain traversing, when triangulation is not possible. Sometimes tie lines are taken to locate interior details. Tie stations are denoted by letters T_1, T_2, T_3. *etc.*

After reconnaissance, the stations are marked on the ground by wooden pegs. These pegs are generally 2.5 cm square and 15 cm long, and have pointed ends. They are driven into the ground leaving about 2.5 cm above the ground. The station point is marked with a cross so that it can be traced if the wooden peg is removed by somebody.

The following criteria are adopted while selecting triangulation stations.

- ☆ The stations should be so selected that the general principle of surveying may be strictly followed.
- ☆ The stations should be inter-visible. For this purpose the station points should be on the highest ground such as hill tops, house tops, *etc.*
- ☆ Easily accessible with instruments.

- Form well-conditioned triangles.
- The base line should be the longest of the main survey lines.
- Lengths of sights are neither too small nor too long.
- These should be located at commanding positions so as to serve as control for subsidiary triangulation, and for possible extension of the main triangulation scheme.
- Useful for providing intersected points and also for detail survey.
- In wooded country, the stations should be selected such that the cost of clearing and cutting, and building towers, is minimal.
- Grazing line of sights should be avoided, and no line of sight should pass over the industrial areas to avoid irregular atmospheric refraction.
- The subsidiary stations should be suitably selected for taking check lines.
- The tie stations should be suitably selected to fix the directions of adjacent sides.
- Stations should not pose many obstacles to chaining. If rivers, canals, ponds, ditches, buildings, *etc.* lie on the chain line, these obstacles should be avoided in chaining operation by applying some fundamental geometric rules.
- The survey lines should not be very close to main roads, as survey work may then be interrupted by traffic.

Reference or Index Sketch

After preliminary inspection of the area, the surveyor should prepare a neat hand sketch showing the arrangement of the framework and approximate position of the objects. He should note the names of the stations on the sketch maintaining some order (clockwise or anticlockwise). The field work should be executed according to this index sketch. The names and sequence of chain lines should be followed as directed in the index sketch. The 'base line' should be clearly indicated in the index sketch. As a precaution against station pegs being removed or missed, a reference sketch should be made for all main stations. It is nothing but a hand sketch of the station showing at least two measurements from some permanent objects. A third measurement may also be taken.

Measurement of Baseline and Offsets

Boundary Line

The line joining the main stations is called boundary lines or chain line.

Base Lines

The line on which the framework of the survey is built is known as the 'base line'. It is main and longest line, which passes approximately through the center of the field along a fairly level ground. It should be measured very carefully and accurately as all the other measurements, to show the details of the work, are taken with respect to this line. The magnetic bearings of the base line are taken to fix the north line of the map.

- Most important part of triangulation
- Aligned and measured with great accuracy
- Forms the basis of computations of triangulation system
- Equipment used are: standardized tapes, Hunter's short base, tacheometric measurements, EDM

Tie or Subsidiary Lines

A tie line joins two fixed points on the main survey lines of the triangle. It helps in taking offsets of the objects falling within the triangle and which are too far away from the main line. It helps to locate the interior details. The position of each tie line should be close to some features, such as paths, building *etc.* A check line or tie line is never extended beyond the main lines.

Check Line

A check line is also a tie line which helps in checking the accuracy of the work after plotting on a drawing sheet. A check line also termed as a proof line joins the apex point of a triangle to some fixed point on its base or any other side of the triangle. A check line is measured to check the accuracy of the framework. The length of a check line, as measured on the ground should agree with its length on the plan. Sometimes this line is also used to locate interior details.

Offsets

These are the lateral measurements from the base line to fix the positions of the different objects of the work with respect to base line. There are two kinds of offsets:

1. Perpendicular offsets
2. Oblique offsets

If the measurements are taken at right angle to the survey line, the offsets are called perpendicular or right angled offsets (Bhavikatti, 2010). Perpendicular offsets may be taken in the following ways:

- By setting a right angle in the ratio 3:4:5 (Figure 5.3). The included angle between the two sides AB and BC is a right angle if $AC^2=AB^2+BC^2$.
- By setting a perpendicular by swinging a tape from the object to the chain line. The point of minimum reading on the tape will be the base of the perpendicular (Figure 5.4). Herein $PB < PA < PC$ or any other point on the line XY.
- By setting a right angle with the help of builder's square or tri-square
- By setting a right angle by cross-staff or optical square.

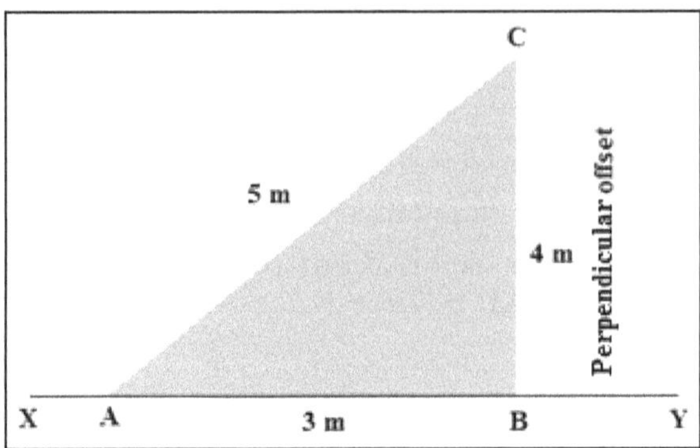

Figure 5.3: Drawing a Perpendicular Offset Using a Chain/Tape.

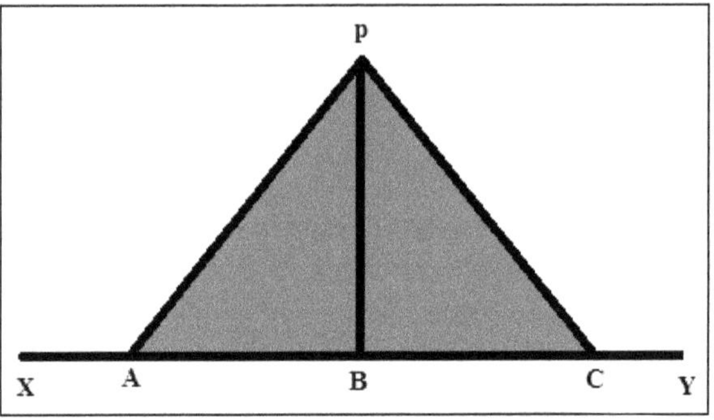

Figure 5.4: Drawing a Perpendicular Offset from a Station Using a Chain/Tape.

The above described methods can also be used to draw a line parallel to the chain line. If XY is the chain line and Q is a point through which a line parallel to the chain line is to be drawn. From Q, perpendicular QP is drawn on XY at P. Point R is now selected on XY and RS is drawn perpendicular to XY at R in such a way that RS=PQ; QS is joined which is now parallel to XY (Figure 5.5 left).

In Figure 5.5 (right), point R is selected on XY. QR is joined and bisected at O. Another point P on XY is selected and PO is joined. Now PO is extended to S so that PO=OS. QS is joined. QS is parallel to XY.

Any offset not perpendicular to the chain line is said to be oblique. Oblique distance is always greater than the perpendicular distance (Figure 5.6 middle).

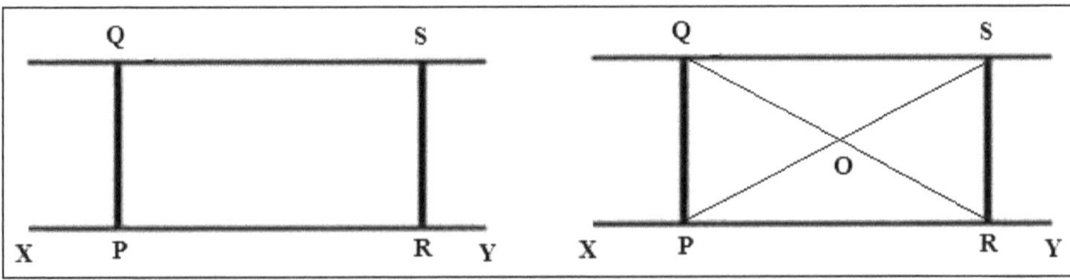

Figure 5.5: Procedures to Draw a Line Parallel to the Chain Line.

Oblique offsets are taken when the objects are at a long distance from the chain line or when it is not possible to set up a right angle due to some difficulties. Suppose AB is a chain line and P is the corner of a building. Two points 'a' and 'b' are taken on the chain line. The chainages of 'a' and 'b' are noted. The distances 'aP' and 'bP' are measured and noted in the field book. Then 'aP' and 'bP' are the oblique offsets. When the triangle abP is plotted, the apex point P will represent the position of the corner of the building (Figure 5.6 right).

Perpendicular offsets are preferred for the following reasons:

- They can be taken very quickly
- The progress of survey is not hampered
- The entry in the field book is easy
- The plotting of the offsets is easy

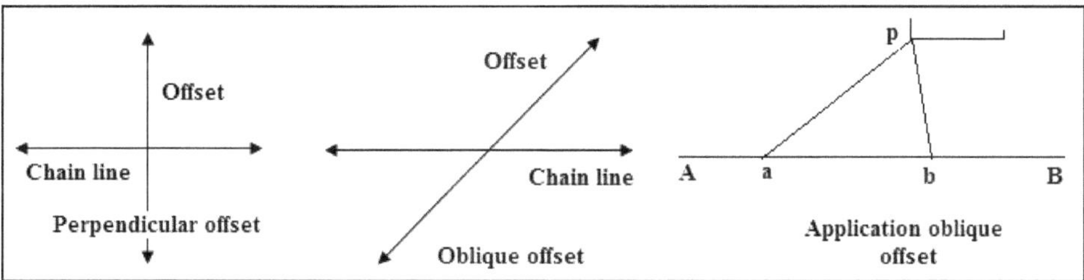

Figure 5.6: A Perpendicular Offset (Left) and Oblique Offsets (Middle and left).

The distance on the chain line at which the foot of the offset is taken is known as a 'chainage'. According to length, the offsets having length greater than or equal to 15.0 m are termed as 'long offset' and the offsets of length less than 15.0 m are known as 'short' offsets. Short offsets are more accurate than long offsets.

Number of Offsets

There is no hard and fast rule regarding the number of offsets taken. The offsets should be taken according to the nature of the object, the objective being to correctly represent the object. Some guidelines are:

1. When the boundary of the object is approximately parallel to the chain line, perpendicular offsets are taken at regular intervals.
2. When the boundary is straight, perpendicular offsets are taken at both ends of it.
3. When the boundary line is zigzag, perpendicular offsets are taken at every point of bend to represent the shape of the boundary accurately. In such a case, the interval of the offsets may be irregular.
4. When a road crosses the chain line perpendicularly, the chainage of the intersection point is to be noted.
5. When a road crosses a chain line obliquely, the chainages of intersection points are noted. At least one offset is taken on both sides of the inter-

section points. More offsets may be taken depending on the nature of the road.

6. When the building is small, its corners are fixed by perpendicular or oblique offsets and the other dimensions are taken directly on the field and noted in the field book.
7. When the building is large, zigzag in shape and oblique to the chain line, then the corners are fixed by perpendicular or oblique offsets. The full plan of the building is drawn on a separate page along with all the dimensions. This page should be attached with the field book at the proper place.
8. When the object is circular, perpendicular offsets are taken at short and regular intervals.

Limiting Length of Offsets

The maximum length of the offset should not be more than the length of the tape used in the survey. Generally, the maximum length of offset is limited to 15 m. However, this length also depends upon the following factors:

- ✰ The desired accuracy of the map
- ✰ The scale of the map
- ✰ The maximum allowable deflection of the offset from its true direction and
- ✰ The nature of the ground

Running Survey Line

Ranging and chaining is started from the base line, which is measured carefully. These measurements are noted in the field book showing the offsets to the left or right according to their position. Then the other survey lines are ranged and chained maintaining the sequence of the traverse. The offsets and other field records are noted simultaneously (Figure 5.7). The check lines and tie lines are also measured and noted at the proper place. The station marks are preserved carefully until field work is completed. The magnetic bearings of the base line are measured by prismatic compass, if necessary.

Measurement of Horizontal Angles

Following instruments/methods are used to measure horizontal angles:

- ✰ Optical/digital theodolites for primary and secondary triangulation
- ✰ Transit theodolites for tertiary triangulation
- ✰ Method of repetition in tertiary and secondary triangulation
- ✰ Method of reiteration in primary triangulation

Degree of Accuracy

Degree of accuracy is determined at the beginning of any survey work. It is worked out according the following factors:

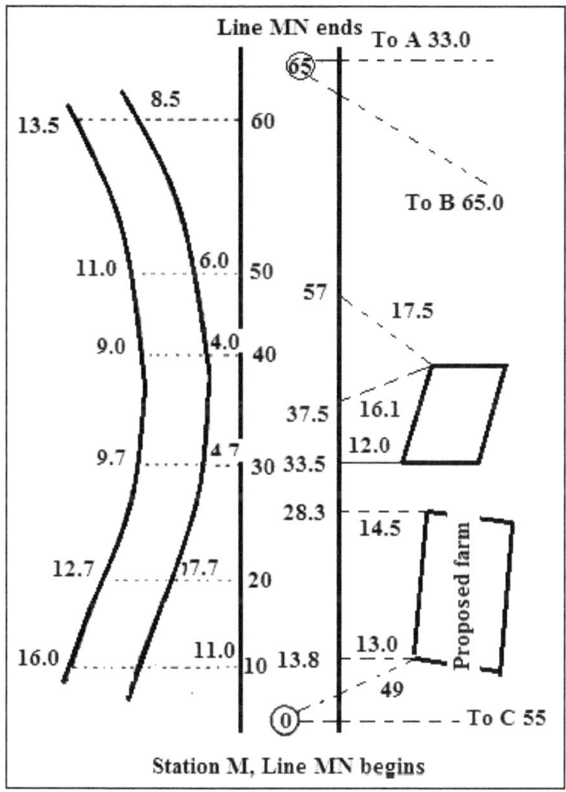

Figure 5.7: Field Book Noting of a Chain Survey.

- ☆ Scale of plotting
- ☆ Permissible error in plotting

During reconnaissance survey, the lengths of the main survey lines are approximately determined by the pacing method. One pace or walking step of a man is considered to equal 80 cm. The permissible error in plotting may be obtained from the concerned department. Then the degree of accuracy in measurement is ascertained. Following example illustrates the issue.

Suppose the scale of plotting is 5 m to 1 cm. Then, 1 cm on the map is equal to 500 cm on the ground. If the allowable error is 0.02 cm then 0.02 cm on the map = 500 × 0.02 = 10 cm on the ground. In that case, the measurement should be taken nearest to 10 cm.

Conventional Symbols

In any map the objects are shown by symbols rather than by names. As such, a surveyor should know the standard conventional symbols for some common objects (Mishra, 2018). Some of these symbols are shown in Figures 5.8 and 5.9.

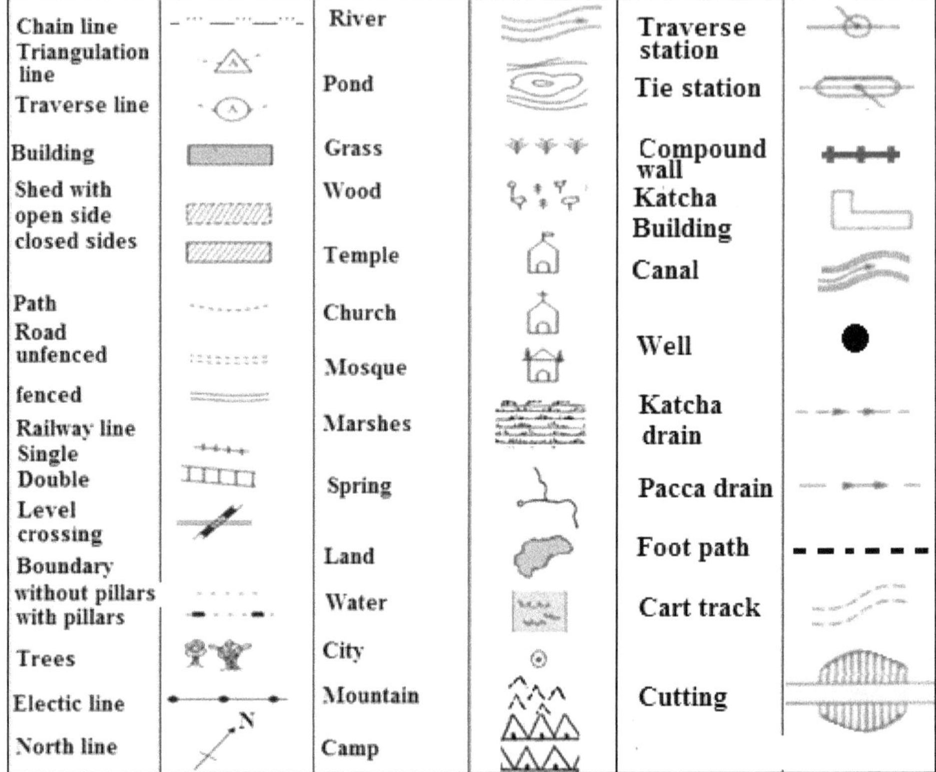

Figure 5.8: Some Conventional Symbols Used in Surveying.

Equipment for Chain Survey

The following equipment is required for conducting chain survey.

1. Metric chain (20 m) = 1 no.
2. Arrows = 10 nos.
3. Metallic tape (15 m) = 1 no.
4. Ranging rods = 3 nos.
5. Offset rod = 1 no
6. Clinometer = 1 no
7. Plumb bob with thread = 1 no
8. Cross staff or optical square = 1 no
9. Prismatic compass with stand = 1 no.
10. Wooden pegs = 10 nos.
11. Mallet = 1 no
12. Field book = 10 nos.
13. Good pencil = 1 no

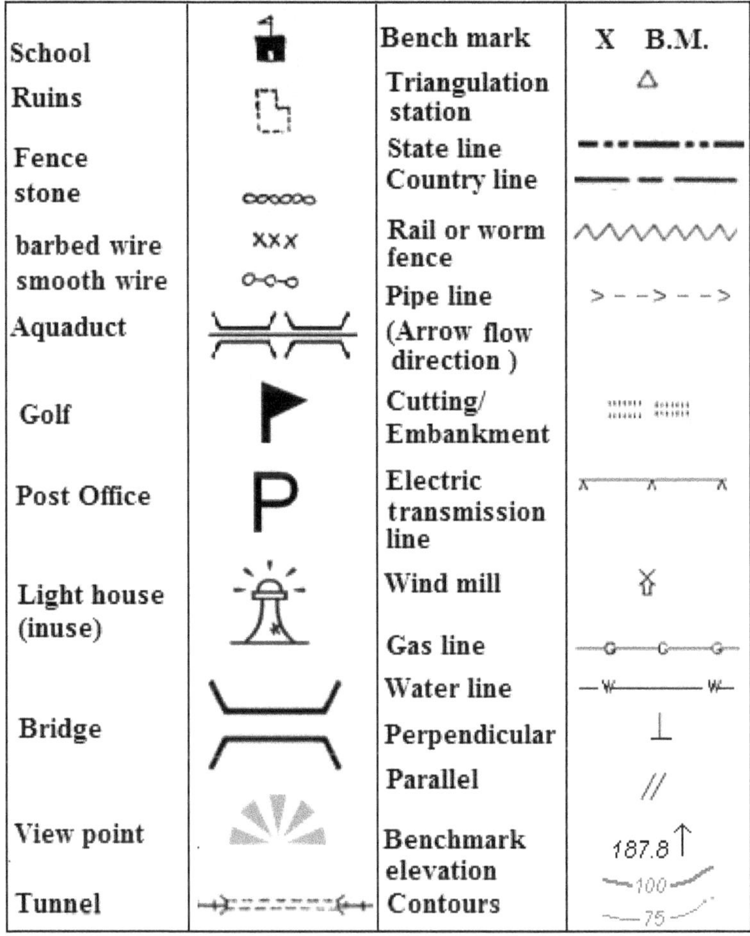

Figure 5.9: Additional Conventional Symbols Used in Surveying.

14. Pen knife = 1 no.
15. Eraser (rubber) = 1 no.

Equipment for Plotting

1. Drawing board (normal size – 1000 mm × 700 mm)
2. Tee-square
3. Set-square (450 and 600)
4. Protractor
5. Cardboard scale – set of eight
6. Instrument box

7. French curve
8. Offset scale
9. Drawing paper of good quality (normal size – 880 mm × 625 mm)
10. Pencils of good quality – 2 H, 3 H or 4 H
11. Eraser (rubber) of good quality
12. Board clips or pins
13. Ink of required shade
14. Colour of required shade
15. Inking pen (or Hi-tech pen) and brushes
16. Handkerchief, knife, paperweight, *etc.*
17. Mini drafter

Preparation of Map

After completing the field work, the next step is to plot the data to get the plan of the area. Steps involved in plotting are:

1. Depending upon the area of the field and the area of drawing sheet, a scale is decided. A suitable scale is chosen so that the area can be accommodated in the space available on the map.
2. A margin of about 2 cm from the edge of the sheet is drawn around the sheet.
3. The title block is prepared on the right hand bottom corner.
4. The north line is marked on the right-hand top corner, and should preferably be vertical. When it is not convenient to have a vertical north line, it may be inclined to accommodate the whole area within the map.
5. A suitable position for the base line is selected on the sheet so that the whole area along with all the objects it contains can be drawn within the space available on the map.
6. The framework is completed with all survey lines, check lines and tie lines. If there is some plotting error which exceeds the permissible limit, the incorrect lines should be resurveyed.
7. Until the framework is completed in proper form, the offsets should not be plotted.
8. The plotting of offsets should be continued according to the sequence maintained in the field book.
9. The main stations, substations, chain line, objects, *etc.* should be shown as per standard symbols.
10. The conventional symbols used in the map should be shown on the right-hand side.
11. The scale of the map is drawn below the heading or in some suitable space. The heading should be written on the top of the map.

12. Unnecessary lines, objects *etc.* should be erased.
13. The map should not contain any dimensions.

Inking the Map

The inking should begin from the left-hand-side and proceed towards the right-hand-side, and from the top towards the bottom.

Colouring the Map

In general, colour washing of engineering survey maps is not recommended. However, if it is necessary, the colour shades should be very light, and according to the colour conventions. The colouring should also be started from the left-hand-side towards the right and from the top towards the bottom.

Large-Scale and Small-Scale Maps

When 1 cm of a map represents a small distance, it is said to be a large-scale map. For example,

If 1 cm = 1 m, then representative fraction (RF) = 1/100

When 1 cm of the map represents a large distance, it is called a small-scale map. For example,

1 cm = 100 m, then RF = 1/10000

A map having a RF of less than 1/500 is considered to be large-scale. A map of RF more than 1/500 is said to be small-scale.

Chapter 6

Plane Table Survey

Plane table surveying is a graphical method of surveying in which both the field works and the plotting are done simultaneously. It is simple and cheaper than theodolite survey. It is most suited to map/survey small areas. It is generally used for surveys where high precision is not required.

Merits

- It is most suitable for preparing small scale map or surveying small areas
- It is more rapid and less costly than other types of survey
- Field book is not necessary, eliminating the possibility of errors in booking
- As surveying and plotting are done simultaneously in the field, chances of omissions of any detail are minimal
- As the instruments used are simple, not much skill for operation of instruments is required
- It is particularly suitable for magnetic area where prismatic compass is not reliable
- Contour and irregular objects may be represented accurately
- The plotting details can immediately be compared with the actual objects present in the field Thus, errors as well as accuracy of the plot can be ascertained along with the progress of survey
- Irrelevant details get omitted in the field itself.

Demerits

- Plane table is essentially a tropical instrument. It is not suitable to work in wet climate or in places where high winds predominate

- There are several accessories to be carried out and are likely to be lost
- It is not suitable for accurate work
- If the sun is bright, plotting may be difficult due to the strain on the eyes.

General Equipment

The following equipment is used for conducting the plane table survey.

- Plane table
- Alidade for sighting (telescopic or simple)
- Plumb bob and U – frame or plumbing fork
- Compass
- Spirit level
- Chain
- Ranging rods
- Tripod
- Drawing sheet and drawing tools
- Paper clips or screws
- Water proof cover.

Plane Table

It is made of well-seasoned wood. It varies in size, the common sizes vary from 40 cm × 30 cm to 75 cm × 60 cm or 45 cm square, 60 cm square, *etc.* The board is mounted on a tripod (Figure 6.1) using a levelling head or a ball-and-socket arrangement so that it can be leveled and revolved about a vertical axis and may be clamped in any position. A sheet of drawing paper, called plane table sheet, is fastened to the board.

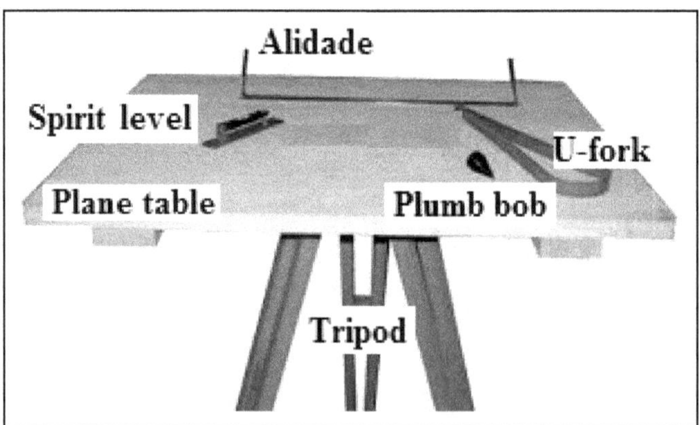

Figure 6.1: Plane Table Fixed on a Tripod Stand (Note some of the other equipment placed on the table).

Alidade

There are two kinds of alidades, a simple and a telescopic alidade. A simple alidade has two edges. While one edge is straight, the other one is beveled. The straight edge is made of brass or gunmetal and works as a ruler. The beveled edge is called the fiducial edge. It has two vanes at both its ends. The vanes are hinged and can be folded when the instrument is not in use. One of the vanes is the sight vane or the eye vane (Figure 6.2). It has three equidistant narrow slits. The surveyor looks through these slits towards the object or station. The other vane which is known as object vane is open and a fine thread is stretched between the top and bottom slit. This thread is used to establish a horizontal line of sight parallel to the ruler. The problem with simple alidade compass is that it can be used only when the elevations of the objects are low.

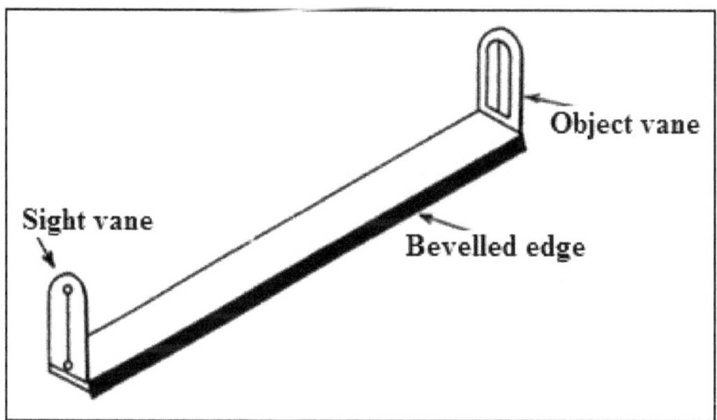

Figure 6.2: A View of the Simple Alidade

While the plane alidade can only take horizontal sights, the telescopic alidade is used to take inclined sights. The telescopic alidade also increases the accuracy of the sight taken. A small spirit level tube is provided with the telescope so that the instrument can be leveled with the working station. A scale is mounted on the horizontal axis. Lines are drawn along the straight ruler. It is also mounted with a vertical circle that gives us the angle of the object with the horizontal axis.

Plumbing Fork and Plumb Bob

Plumbing fork is a U-shaped metal frame that has two different types of arms. One is horizontal (upper hand) and the other is inclined at a certain angle (lower hand). The upper arm has a pointer at the end while the lower arm has a hook from which plumb bob is suspended. When the plumbing fork is kept on the table it has to pass through a particular point. The plumb bob helps in centering.

Compass

Compasses are of two types namely prismatic compass and surveyor's compass. Both the compasses are used to navigate the surveying terrain. These are used to determine the bearings of the land that has been traversed. Both prismatic and surveyor's compass have the following parts:

- ☆ Magnetic needle
- ☆ Graduated circle
- ☆ Vanes for line of sight
- ☆ A box to keep them in place

Spirit Level

When plane table surveying is carried out it is really essential to make the working station (surface) horizontal. Spirit level is used for this purpose. A bubble (mercury is usually used) is put inside the tube. The tube has an upward curvature which makes the bubble remain in center unless the surface is inclined.

Water Proof Cover

Water Proof cover protects the sheet from rain.

Operations Involved in Plane Table Survey

Fixing/Levelling the Plane Table

Fix the plane table to the tripod stand. Arrange the drawing sheet on the plane table using paper clips or thumb screws. The setting up the plane table includes the following three operations.

- ☆ Centering the plane table
- ☆ Levelling the plane table
- ☆ Orientation of plane table

Centering of Plane Table

The table should be set up at a convenient height for working say about 1m. The legs of tripod should be spread well apart and firmly fixed in to the ground. The table should be approximately leveled by tripod legs and judging by the eye. Then the operation of centering is carried out by means of U-frame and plumb bob. The plane table is exactly placed over the ground station by U-frame and plumb bob. In this way, it is possible to arrange the plotted point exactly over the ground point.

Levelling the Plane Table

The process of levelling is carried out with the help of level tube. The bubble of level tube is brought to center in two directions, which are right angles to each other. This is achieved by moving legs.

Orientation of Plane Table

Whenever we are using more than one instrument station, orientation is essential. The process of keeping the plane table always parallel to the position, which is occupied at the first station, is known as orientation. It can be done by using compass or back sighting. In this case, the plane table is rotated such that plotted lines in the drawing sheet are parallel to corresponding lines on the ground.

Methods of Plane Table Survey

Four approaches are commonly used in surveying namely radiation method, intersection method, traversing method and resection method.

Radiation Method

It is the simplest method but is only effective if the whole survey is completed from one single station (Venkatramaiah, 1996). It means the table is set in such a position from where all the other points of the field are easily visible. The procedure is as follows:

- ☆ A point P is selected in such a fashion that all the other points (A B C D E) are easily seen from P (Figure 6.3)
- ☆ Centering, levelling, and orientation is completed prior to surveying
- ☆ Select a point "P" on the sheet so that it is exactly over station "P"
- ☆ Mark the direction of the magnetic meridian
- ☆ At first, by putting the alidade on point P a line of sight for station A is drawn
- ☆ After measuring the distance of PA on the field, the measurement is put on paper to a suitable scale
- ☆ Similarly, points B, C, D, and E are sighted and the distances PB, PC, PD and PE on ground respectively are measured
- ☆ Points a, b, c, d, and e are joined on paper (Figure 6.3).

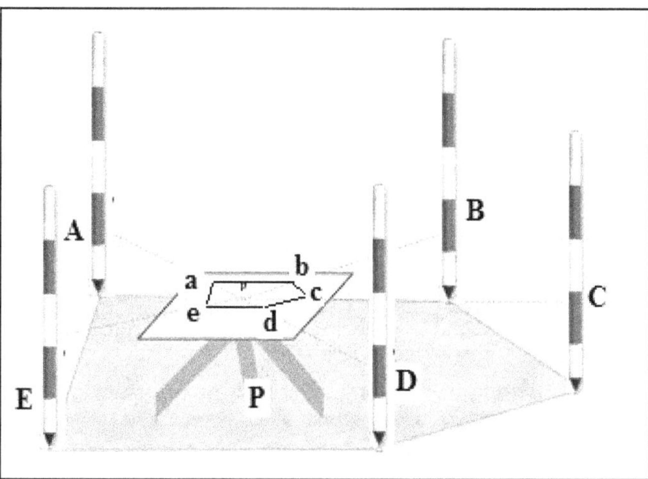

Figure 6.3: Radiation Method of Plane Table Survey.

Intersection Method

In this method we can locate the point by plotting two rays from two known stations. As such, the plane table is shifted to a known distance in a particular direction marked on the ground and the line of sights are drawn to make intersection of the radial lines already drawn from the first set up of the instrument. The following procedure is adopted.

- ✰ Select two points O1 and O2 such that the points to be plotted are visible from the two stations.
- ✰ Set the table on O1 and locate on the sheet (Figure 6.4).
- ✰ Pivot on O1 to bisect O2.
- ✰ Measure the distance and locate O2 on the sheet to a convenient scale.
- ✰ Now O1 O2 is known as the base line.
- ✰ From O1, bisects the objects A and B and draw rays.
- ✰ Shift the table to 'O2' and place the alidade along O2 and O1 and rotate the table till O1 is bisected. Clamp the table.
- ✰ From O2 bisect the objects A and B and draw rays.
- ✰ The intersections of rays drawn from O1 and O2 will give the points a and b representing the two corners A and B of the object (Figure 6.4).
- ✰ To check the accuracy measure AB and compare with plotted distance AB.
- ✰ The same procedure is used for other features such that each point is bisected from two stations.

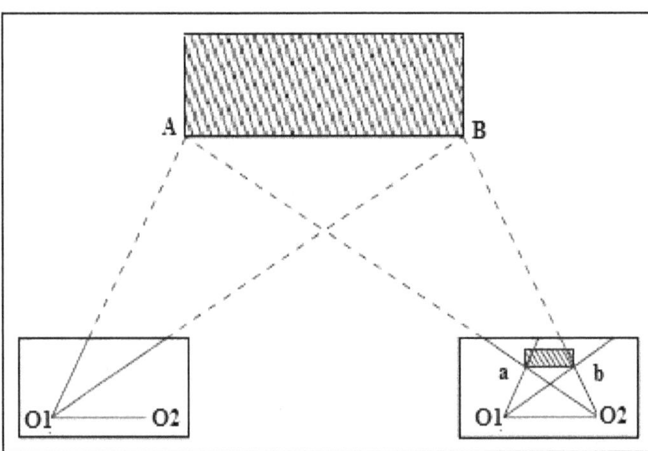

Figure 6.4: Schematic Sketch to Explain the Intersection Method.

Traverse Method

Traversing is the connection of series of straight lines. In case of traversing, plane table is located at one point and is moved from one to the next station until whole area is traversed and plotted on the drawing sheet. The following step wise procedure is used.

- ✰ Select the traverse stations A,B,C,D,E *etc.* on the ground (Figure 6.5).
- ✰ Set the table on starting station 'A' and perform temporary adjustments.

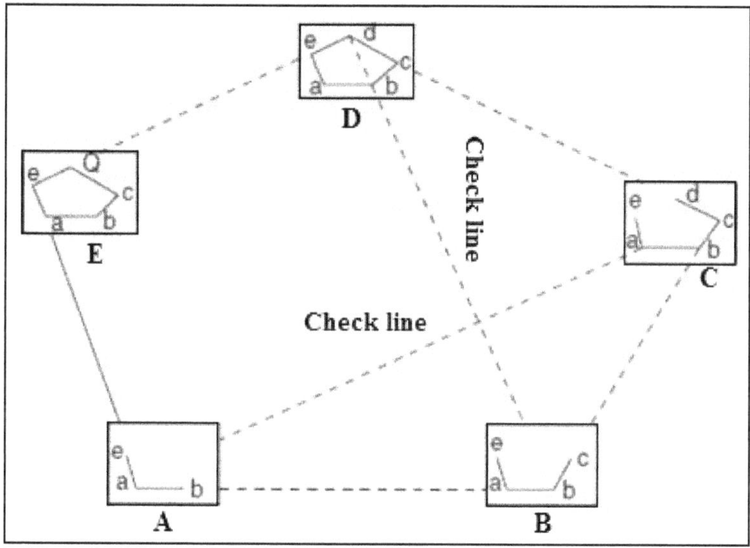

Figure 6.5: Schematic Illustration of the Traversing Method.

- Mark the magnetic meridian.
- Locate A on the sheet as 'a'.
- Pivot on 'a' bisect the next station B and draw a ray
- Measure the distance AB and locate 'b' on the sheet with a suitable scale.
- Shift the table to next station B, set the table over B, and do temporary adjustments.
- Place the alidade along 'ba' and bisect A for doing orientation of plane table.
- Pivot on b, bisect c, draw a ray
- Measure the distance BC and locate 'c' on the sheet with the suitable scale.
- Repeat the same procedure at every successive station until the traverse is completed (Figure 6.5).
- Make checks in between to ensure the accuracy of the survey.

Resection Method

In the resection method of plane table surveying, location of plane table is unknown. If a, b and c are the plotted positions of objects A, B and C respectively, to locate instrument station P on the paper, the orientation of table is achieved with the help of a, b, c. The resectors Aa, Bb, Cc are drawn to get the 'p', the plotted position of P. Hence in the resection method major effort is to ensure suitable orientation. It is why it is also called method of orientation. The method can be conducted by two field conditions as follows.

- The three-point problem
- The two-point problem

In the first method, three points and their positions in the field are known. Plane table is placed at a position from where all the three points are visible. By sighting those three points it is possible to locate the point where equipment is located. This can be achieved by any of the four methods namely tracing method, Lehmann method, analytical method and graphical method. In the two-point problem, two points are sighted from other point corresponding to the points given in plane table sheet. Here two cases may be applied *i.e.* when the points can be occupied by the plane table and when the plane table cannot occupy the controlling stations. Further discussion on these methods and issue is not included for the sake of brevity being outside the scope of the course.

Avoiding Errors in Plane Table Survey

Some of the steps to avoid or minimize errors in plane table survey are as follows.

1. The table should be accurately centered with plumb bob and fork.
2. The table must be accurately oriented and then clamped after it is shifted from one station to the other. Otherwise the lines drawn from the new stations will not correspond with the previous ones.
3. The alidade should be properly centered on the station point on paper.
4. Papers often get contracted or expanded so it is best to use paper clips.
5. The rays should be drawn using the straight ruler only.
6. It is better to check the distances manually from time to time for better accuracy and efficiency.
7. To distinguish clearly, small letters like p, q, r, *etc.* should be used on the paper to represent the corresponding points P, Q, R, *etc.* on the ground.

Chapter 7

Compass Traversing

Introduction

In chain surveying, the area to be surveyed is divided into a number of triangles (Chain triangulation). But triangulation may pose insurmountable problems especially under the following situations:

- ☆ If the surveying area is large
- ☆ If the plot for surveying has numerous obstacles and undulations which prevents chaining
- ☆ If the surveying has to be completed in a time bound manner

Under all these conditions, compass surveying is usually adopted. It is called as traversing method of survey. In traversing, the framework consists of a number of connected lines. The lengths are measured by chain or tape and the directions identified by angle measuring instruments. In one of such methods, the compass is used to measure the angles. Hence, the process is known as compass traversing. Thus, compass surveying is the branch of surveying in which the position of an object is located using angular measurements determined by a compass and linear measurements using a chain or tape. Traversing in an anticlockwise direction is always convenient while running the survey lines. Some of the commonly used units and terms to which a surveyor should be familiar are described as under.

Units of Angle

Degree Unit

A circle is divided in to 360 equal degrees written as 360°

1 degree = 60 minutes, 1 minute = 60 seconds

For example, 2 degrees 5 minutes 7 seconds is written as $2^\circ 5' 7''$

Radian Unit

A circle is equal to 2π

1° degree = $\pi/180$ radians

Example 7.1

Convert the angle 6.23456° to degrees minutes and seconds

Subdivide the angle from the whole degrees: 6.23456 - 6 = 0.23456

Multiply the decimal value by 60: $0.23456 \times 60 = 14.0736$ (14 is the whole minutes)

Subdivide the whole minutes from the number: $14.0736 - 14 = 0.0736$

Multiply the decimal value by 60: $0.0736 \times 60 = 4.41$ (4 is the whole seconds)

Thus, $6.23456^{\circ} = 6^{\circ}\ 14'\ 4''$

Example 7.2

Convert the angle $39^{\circ}\ 42'\ 17''$ to the degree decimal unit

Divide the seconds fraction (17″) by 60: $17/60 = 0.2833$

Add the value to the minute's fraction (42′): 42.2833

Divide the minute's fraction (42.2833) by 60: $42.2833/60 = 0.7047$

Add the value to the degrees such that: $39^{\circ}\ 42'\ 17'' = 39.7047$

Important Terms

True Meridian: The line or plane passing through the geographical north pole (NP) and geographical south pole (SP) and any point on the surface of the earth, is known as the 'true meridian' or 'geographical meridian' (Figure 7.1). The true meridian at a station is constant. The true meridians passing through different points on the earth's surface are not parallel, but converge towards the poles. In surveys where areas are small, the true meridians passing through different points are assumed parallel.

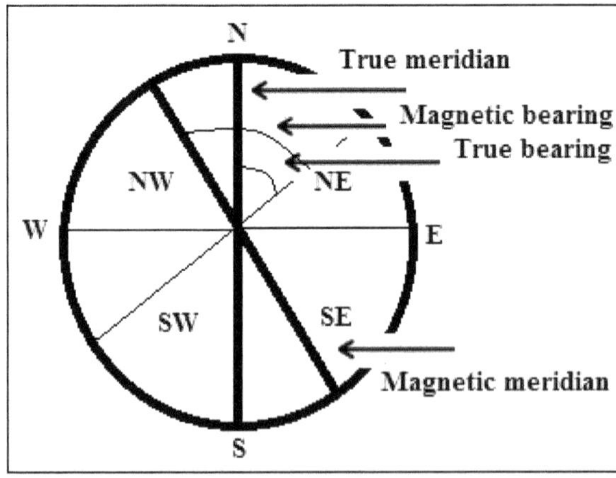

Figure 7.1: Definition Sketch of Magnetic and True Meridians and Bearings.

Azimuth: The angle between the true meridian and a line is known as 'true bearing' of the line (Figure 7.1). It is also known as the 'azimuth'.

Magnetic Meridian: When the magnetic needle is suspended freely and balanced properly, unaffected by magnetic substances, it indicates a direction. This direction is known as magnetic meridian (Figure 7.1).

Magnetic Bearing or Bearing: The angle between the magnetic meridian and a line is known as the 'magnetic bearing' or simply the 'bearing' of the line (Figure 7.1). Magnetic bearings are designated by two systems namely Whole circle bearing (WCB), and Quadrant bearing (QB).

Whole Circle Bearing (WCB): The magnetic bearing of a line measured clockwise from the North Pole towards the line, is known as the 'whole circle bearing', of that line. Such a bearing may have any value between 0° and 360°. The whole circle bearing of a line is obtained by prismatic compass.

Quadrantal Bearing (QB): The magnetic bearing of a line measured clockwise or anticlockwise from NP or SP (whichever is nearer to the line) towards the east or west is known as QB. This system consists of 4-quadrants NE, SE, NW, SW. The values lie between 0-90° but the quadrants should always be mentioned. Quadrant bearings are obtained by the surveyor's compass. The WCB and QB bearings can be determined as per the following conversion tables.

Conversion of WCB to QB

Case	*WCB between (Degrees)*	*RB*	*Quadrant*
1	0-90	WCB	N-E
2	90-180	180-WCB	S-E
3	180-270	WCB-180	S-W
4	270-360	360-WCB	N-W

Conversion of RB to WCB

Case	*RB in Quadrant*	*Rule of WCB*	*WCB between (Degrees)*
1	N-E	WCB = RB	0-90
2	S-E	WCB =180-RB	90-180
3	S-W	WCB=180+RB	180-270
4	N-W	WCB =360-RB	270-360

Arbitrary Meridian: Sometimes, for the survey of small area, a convenient direction is assumed as a meridian, known as the 'arbitrary meridian'. For example; the starting line of a survey can be taken as the arbitrary meridian.

Arbitrary Bearing: The angle between the arbitrary meridian and a line is known as the 'arbitrary bearing' of the line.

Grid Meridian: Sometimes, several lines parallel to the true meridian are assumed for a particular zone. These lines are termed as 'grid lines' and the central line the 'grid meridian'.

Grid Bearing: The bearing of a line with respect to the grid meridian is known as the 'grid bearing' of the line.

Reduced Bearing (RB): The whole circle bearing of a line converted to quadrantal bearing is termed as reduced bearing. Thus, the reduced bearing is similar to the quadrantal bearing. Its value lies between 0° and 90°, but the quadrants should be mentioned for proper designation.

Fore and Back Bearing: The bearing of a line measured in the direction of the progress of survey is called the 'fore bearing' (FB) of the line (Figure 7.2). The bearing of a line measured in the direction opposite to the survey is called the 'back bearing' (BB) of the line (Figure 7.2). In the WCB system, the difference between the FB and BB should be exactly 180°, and the negative sign when it is more than 180°.

$$BB = FB \pm 180^\circ \tag{7.1}$$

Use the +ve sign when FB<180°

Use the –ve sign when FB> 180°

In the quandrantal bearing (*i.e.* reduced bearing) system, the FB and BB are numerically equal but the quadrants are just opposite. For example, if the FB of AB is N 30° E, then its BB is S 30° W.

Magnetic Declination: The horizontal angle between the magnetic meridian and true meridian is known as 'magnetic declination'. When the north end of the magnetic needle is pointed towards the west side of the true meridian, the position is termed 'Declination West'. When the north end of the magnetic needle is pointed towards the east side of the true meridian, the position is termed 'Declination East'. The true and magnetic bearings are related as per the following relations:

True bearing = Magnetic bearing ± Declination (+ sign when declination east and - sign when declination is towards west)

Magnetic bearing = True bearing ± Declination (+ sign when declination west and - sign when declination is towards east)

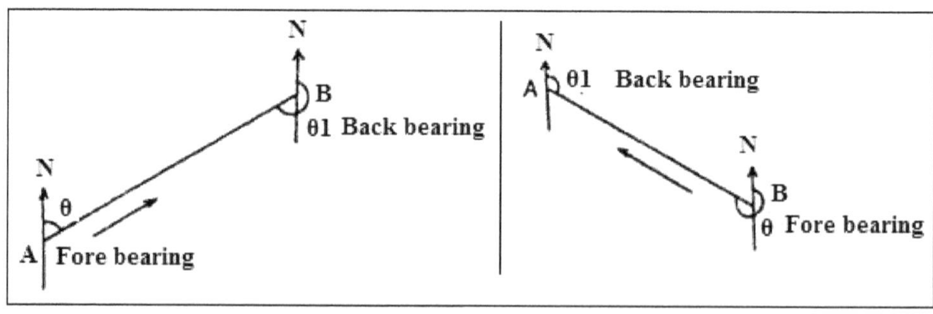

Figure 7.2: Definition Sketch of Fore and Back Bearings.

Isogonic and Agonic Lines: Lines passing through points of equal declination are known as 'isogonic' lines. The line passing through points of zero declination is said to be the 'agonic' line. The Survey of India Department has prepared a map of India in which the isogonic and agonic lines are shown properly as a guideline to conduct the compass survey in different parts of the country.

Variation of Magnetic Declination: The magnetic declination at a place is not constant. It varies due to the following reasons:

Secular Variation: The magnetic meridian behaves like a pendulum with respect to the true meridian. After every 100 years or so, it swings from one direction to the opposite direction, and hence the declination varies. This variation is known as 'secular variation'.

Annual Variation: The magnetic declination varies due to the rotation of the earth with its axis inclined in an elliptical path around the sun during a year. This variation is known as 'annual variation. The amount of variation is about 1 to 2 minutes.

Diurnal Variation: The magnetic declination varies due to the rotation of the earth on its own axis in 24 hours. This variation is known as 'diurnal variation'. The amount of variation is found to be about 3 to 12 minutes.

Irregular Variation: The magnetic declination is found to vary suddenly due to some natural causes, such as earthquakes, volcanic eruptions and so on. This variation is known as 'irregular variation'.

Dip of the Magnetic Needle: If a needle is perfectly balanced before magnetization, it does not remain in the balanced position after it is magnetized. This is due to the magnetic influence of the earth. The needle is found to be inclined towards the pole. This inclination of the needle with the horizontal is known as the 'dip of the magnetic needle'. It is found that the north end of the needle is deflected downwards in the northern hemisphere and the south end is deflected downwards in the southern hemisphere. The needle is just horizontal at the equator. To balance the dip of the needle, a rider (brass or silver coil) is provided along with it. The rider is placed over the needle at a suitable position to make it horizontal.

Local Attraction

A magnetic needle indicates the north direction when freely suspended or pivoted. But if the needle comes near some magnetic substances, such as iron ore, steel structures, electric cables conveying current; *etc.* it is found to be deflected from its true direction, and does not show the actual north (Basak, 2014). This disturbing influence of magnetic substances is known as 'local attraction'. To detect the presence of local attraction, the fore and back bearings of a line are taken. If the difference of the fore and back bearings of the line is exactly 180°, then there is no local attraction. If the FB and BB of a line differ from 180°, then the needle is said to be affected by local attraction, provided there is no instrumental error. To compensate for the effect of local attraction, the amount of error is found out and is equally distributed between the fore and back bearings of the line. As an example, let us consider a case when

Observed FB of AB = 60°30′

Observed BB of AB = 240°0′

Calculated BB of AB = 60°30° + 180°0′ = 240°30′

Corrected BB of AB = 1/2 (240°0′ + 240°30′) = 240°15′

Hence, corrected FB of AB = 240°15′ – 180°0′ = 60°15′

Correction for Local Attraction

Method I

The interior angles of a traverse are calculated from the observed bearings. Then an angular check is applied. The sum of the interior angles should be equal to (2n – 4) × 90° (n being the number of sides of the traverse). If it is not so, the total error is equally distributed among all the angles of the traverse. Then, starting from the unaffected line, the bearings of all the lines are corrected by using the corrected interior angles. This method is not employed normally, being quite laborious.

Method II

In this method, the interior angles are not calculated. Herein, the unaffected line is first detected. Then, commencing from the unaffected line, the bearings of the other affected lines are corrected by finding the amount of correction at each station. This is an easy method, and one which is generally employed. If all the lines of a traverse are found to be affected by local attraction, the line with minimum error is identified. The FB and BB of this line are adjusted by distributing the error equally. Then, starting from this adjusted line, the fore and back bearing of other lines are corrected.

Principle of Compass Survey

The principle of compass surveying is traversing, which involves a series of connected lines. The magnetic bearings of the lines are measured by prismatic compass and the distances of the lines are measured by chain. This survey does not require the formation of a network of triangles. Interior details are located by taking offsets from the main survey lines. Sometimes subsidiary lines may be taken for locating these details. Compass surveying is not recommended for areas where local attraction is suspected due to the presence of magnetic substances like steel structures, iron ore deposits, electric cables conveying current, and so on.

Traversing

As already stated in the last section, surveying which involves a series of connected lines is known as 'traversing.' The sides of the traverse are known as 'traverse legs'. In traversing, the lengths of the lines are measured by chain and the directions are fixed by compass or theodolite or by forming angles with chain and tape.

A traverse may be of two types – closed and open.

Closed Traverse

When a series of connected lines forms a closed circuit, *i.e.* when the finishing point coincides with the starting point of a survey, it is called a 'closed traverse' (Figure 7.3 left). Here ABCDEFA represents a closed traverse. Closed traverse is suitable for the survey of boundaries of ponds, forests estates, *etc.*

Open Traverse

When a sequence of connected lines extends along a general direction and does not return to the starting point, it is known as 'open traverse' or 'unclosed traverse' (Figure 7.3 right). Here ABCD represents an open traverse. Open traverse is suitable for the survey of roads, rivers, coast lines, *etc.*

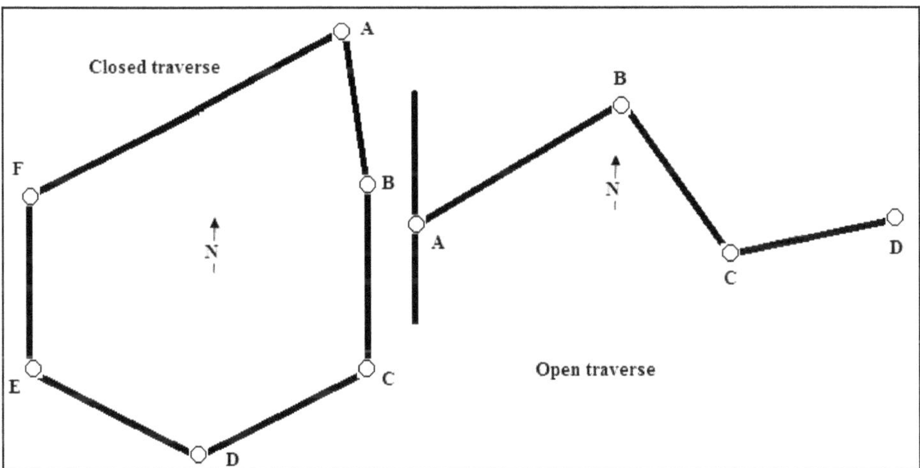

Figure 7.3: Closed and Open Traverses.

Methods of Traversing

Traverse survey may be conducted by the following methods:

- Chain traversing (by chain angle)
- Compass traversing (by free needle)
- Theodolite traversing (by fast needle) and
- Plane table traversing (by plane table)

Chain Traversing

Chain traversing is mainly conducted when it is not possible to adopt triangulation. In this method, the whole work is done with chain and tape. No angle measurement is used and the directions of the lines are fixed entirely by linear measurements. Angles fixed by linear or tie measurements are known as chain angles. The method is not suited where high accuracy is required. Also this method is not used where an angle measuring instrument such as a compass, sextant or theodolite is available.

Compass Traversing

In chain and compass traversing, the magnetic bearings of the survey lines are measured by a compass and the lengths of the lines are measured either with a chain or with a tape. The direction of magnetic meridian is established at each traverse station independently. The method is also known as tree or loose needle method. The observed bearings are verified and necessary corrections for local attraction are applied. In this method, closing error may occur when the traverse is plotted. This error is adjusted graphically.

Theodolite Traversing

In such traversing, the horizontal angles between the traverse legs are measured by theodolite. The lengths of the legs are measured by chain or by employing the stadia method. The magnetic bearing of the starting leg is measured by theodolite. Then the magnetic bearings of the other sides are calculated. The independent coordinates of all the traverse stations are then found out. This method is very accurate.

Plane Table Traversing

In this method, a plane table is set at every traverse station in the clockwise or anticlockwise direction, and the circuit is finally closed. During traversing, the sides of the traverse are plotted according to some suitable scale. At the end of the work, any closing error is adjusted graphically.

Check on Closed and Open Traverses

Check on Angular Measurements

- ☆ The sum of the measured interior angles should be equal to $(2N - 4) \times 90^\circ$ or $(N - 2) \times 180^\circ$ where N is the number of sides of the traverse.
- ☆ The sum of the measured exterior angles should be equal to $(2N + 4) \times 90^\circ$ or $(N + 2) \times 180^\circ$.
- ☆ The algebraic sum of the deflection angles should be equal to 360°.

Right-hand deflection is considered positive and left-hand deflection negative.

Check on Linear Measurement

- ☆ The lines should be measured once each on two different days (along opposite directions). Both measurements should tally.
- ☆ Linear measurements should also be taken by the stadia method. The measurements by chaining and by the stadia method should tally.

In open traverse, the measurements cannot be checked directly. But some field measurements can be taken to check the accuracy of the work.

Taking Cut-off Lines

Cut-off lines are taken between some intermediate stations of the open traverse. After plotting the traverse, the distances and bearings are noted from the map. These distances and bearings should tally with the actual records from the field. As illustrated in Figure (7.3), in addition to the taking of bearing of AB at station A, bearing of AD can also be measured, if possible. Similarly, at D, bearing of DA can be measured and check applied. If the two bearings differ by 180°, the work may be accepted as correct. Similar check is made at DG as the work progresses.

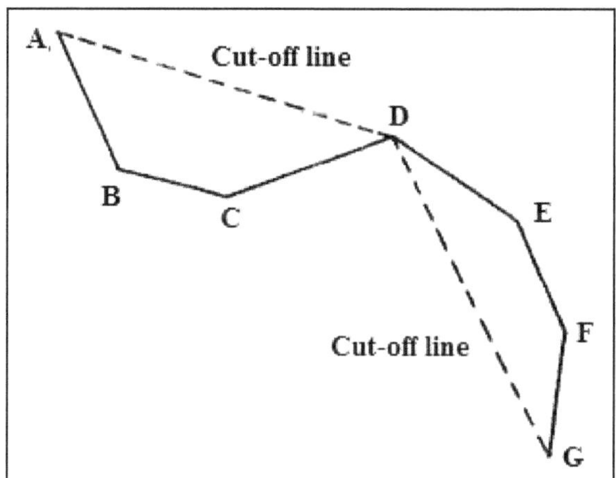

Figure 7.3: Cut-off Lines to Check the Accuracy of Open Traverse.

Taking an Auxiliary Point

Suppose ABCDEFG is an open traverse. A permanent point M1 is selected on one side of it (Figure 7.4). The magnetic bearings of this point are taken from the traverse stations A,B,C,D, *etc.* If the survey is accurate and so is the plotting,

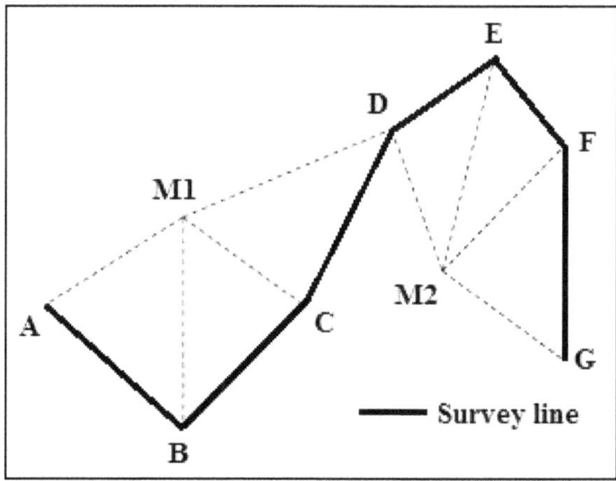

Figure 7.4: Auxiliary Point Method to Check the Accuracy of Open Traverse.

all the measured bearings of M1 when plotted should meet at the point M1. The permanent point M1 is known as the 'auxiliary point'. In a similar way, checking is done at auxiliary point M2.

Example 7.3

What is the sum of the interior angles of a closed polygon traverse that has a) 6 sides b) 8 sides and c) 12 sides?

Sum of interior angles is given as $(n - 2) \times 180^{\circ}$

For question a we have n= 6

Therefore interior angles are: $(6 - 2) \times 180 = 720^{\circ}$

Similarly

b) $(8 - 2) \times 180 = 1080^{\circ}$

c) $(12 - 2) \times 180 = 1800^{\circ}$

Example 7.4

The interior angles in a five sided closed polygon traverse were measured and found to be: A = 139°10″11″, B = 126°17′43″, C = 94°28′30″, D = 71°04′59″ and E = 108°58′31″. Compute the angular mis-closure.

Sum of interior angles is given as $(5 - 2) \times 180 = 540^{\circ}$

Sum angles = 139°17′43″+ 126°10″11″ + 94°04′59″+ 71°28′30″ + 108°57′31″ = 539°58′54″

Mis-closure = 540°- 539°58′54″ = 00°01′06″

Types of Compass

A compass is a navigational instrument used extensively since ancient times for direction setting and for navigating across the oceans. It helps to determine direction relative to Earth's magnetic poles. It consists of magnetized pointer that is free to be aligned to magnetic north or true north or sometimes to an arbitrary direction based on the location of the celestial bodies. Magnetic North refers to the pole of Earth's magnetic field. It is the direction of the north tip of the Earth's magnetic field and true north refers to the geographic North Pole.

There are two types of compass:

1. The prismatic compass
2. The surveyor's compass

The Prismatic Compass

It is a sophisticated device designed for highly accurate surveys. It is a portable magnetic compass which can either be used as a hand instrument or can be fitted on a tripod. In this compass, the readings are taken with the help of a prism, which allows the user to read the compass bearings while seeing distant objects. A prismatic compass essentially consists of a circular non-magnetic metallic compass box of diameter 8 to 10 cm. The other components include: a magnetic needle, graduated ring, adjustable mirror, object vane, eye vane, glass cover and a horse hair (Figure 7.5). The magnetic needle is made of a broad, magnetized iron bar and is pointed at both the ends. The aluminum ring is graduated from 0° to 360°clockwise with 0° is marked at the south, 90° at the west, 180°at north and 270° at the east. A rider of brass or silver coil is provided with the needle to counterbalance its dip. The sight vane and the reflecting prism are fixed diametrically opposite to the box. The sight vane is hinged with the metal box and consists of horsehair at the center. The prism consists of a sighting slit at the top and two small circular holes, one at bottom of the prism and the other at the side of the observer's eye. It is also provided with two dark glasses, red glass for sighting luminous objects at night and the blue glass for reducing the strain on the observer's eye in bright daylight. A mirror is provided with the sight vane. The mirror can be lowered or raised, and can also be inclined. A brake pin is provided just at the base of the sight vane. If pressed gently, it stops the oscillations of the ring. A lifting pin is provided just below the sight vane. When the sight vane is folded, it presses the lifting pin. The lifting pin then lifts the magnetic needle out of the pivot point to prevent damage to the pivot head. A glass cover is provided on top of the box to protect the aluminum ring from dust.

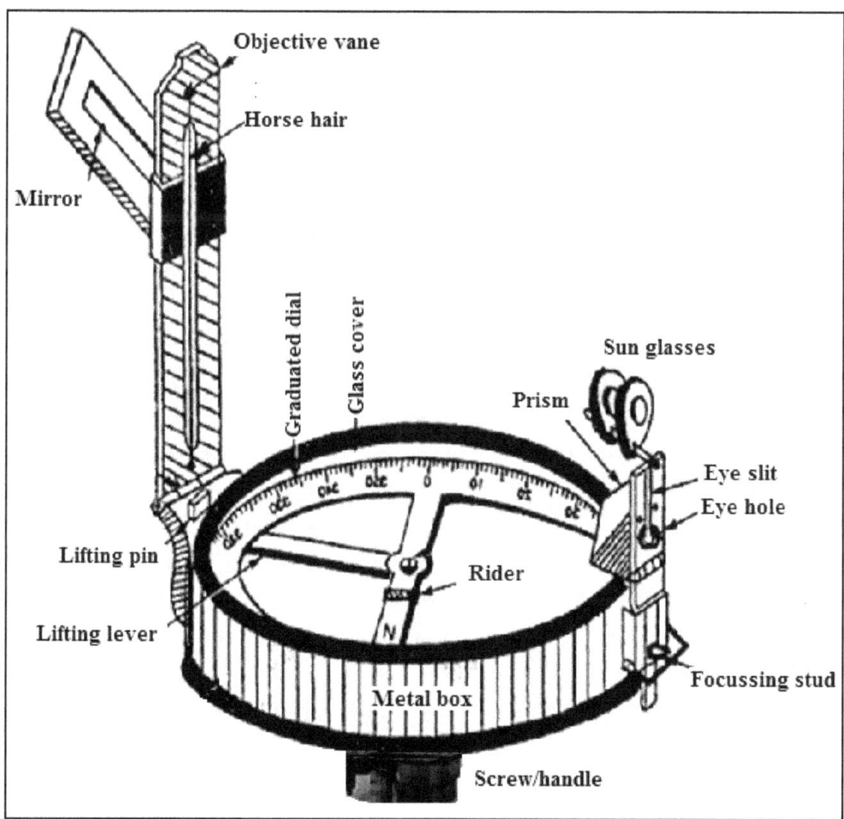

Figure 7.5: Components of a Prismatic Compass.

Before using the instrument some permanent and temporary adjustments are required.Permanent adjustments are done only if the internal parts of the prism are disturbed or damaged. They are:

- ☆ Adjustments in levels
- ☆ Adjustment of pivot point
- ☆ Adjustment of sight vane
- ☆ Adjustment of needle

The temporary adjustments are done at each station. The prismatic compass is temporarily set on the tripod over the station at which the bearing of survey line is to be measured. The following steps are than taken.

- ☆ **Centering:** It is the process of fixing the compass exactly over the station. Centering is usually done by adjusting the tripod legs. Also a plumb-bob is used to judge the accurate centering of instruments over the station.
- ☆ **Levelling:** The instrument needs to be leveled in both cases *i.e.* used in hand or mounted over a tripod. If it is used in hand, the graduated disc should swing freely and appears to be completely level in reference to the top edge of the case. If the tripod is used, they usually have a ball and socket arrangement for levelling purpose.
- ☆ **Focusing the prism:** Slide the prism up or down to focus so as to make the readings clear and readable.

Let AB be the survey line as shown in Figure 7.6, the bearing of which is required to be measured. The instrument is set at A and a ranging rod is fixed at

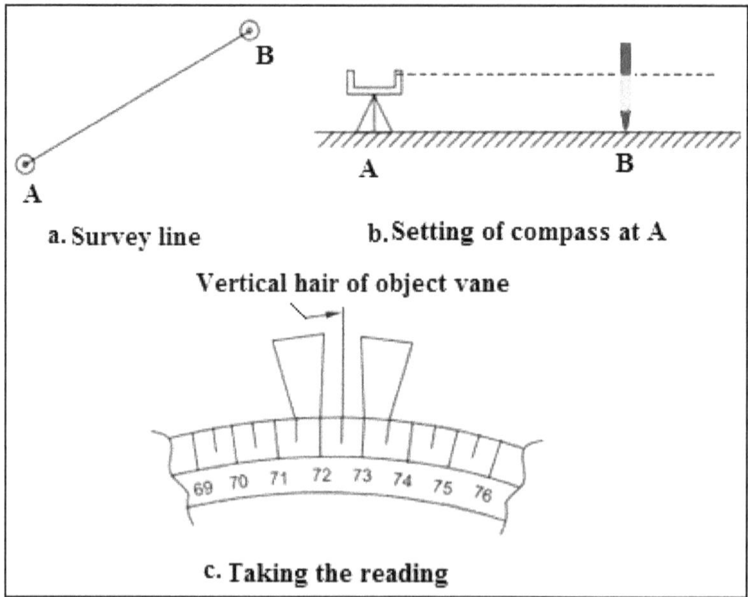

Figure 7.6: Taking the Compass Reading of Survey Line AB (Top left), Setting the Compass at A (Top right) and taking the Reading (Bottom).

B. The compass is turned so that line of sight is aligned in the direction of AB by making eye slit of observation vane, vertical hair of object vane and ranging rod at B in same horizontal line. Wait for oscillation of graduation ring to dampen, with the use of brake pin if necessary. The viewing prism is focused by moving it vertically with the help of focusing stud. The reading of the image of hair line as observed through prism is noted indicating the whole circle bearing of survey line. The process is repeated to check the repeatability of measurements. This bearing is called fore-bearing of line AB. The procedure of taking the reading of a survey line is illustrated in Figure 7.6.

Traversing

The instrument is successively set at each station of the traverse. The bearings of survey lines in a traverse in the direction of progress of survey is called the fore bearing while the bearing of the same survey line from the end station (station B on line AB) is termed back bearing. Fore bearing and back bearing of each line is taken and recorded in the field notebook. The observational errors in this survey tend to compensate as each bearing is observed independently. Distances between each survey stations are measured using a chain/tape. The offset points are located either by procedure followed in chain surveying or by angular measurement with compass.

Local external field at a place can influence the readings seriously. Besides the local attraction as discussed previously, the actual measurements of bearings can also be disturbed if the surveyors, keeps bunch of iron keys, iron knives or buttons, steel framed spectacles near the instrument. Even the chains and arrows, used in surveying, near the compass can also affect bearings. Proximity of such objects should be avoided.

The Surveyor's Compass

Surveyor's compass consists of a circular brass box containing a magnetic needle which swings freely over a brass circle which is divided into 360°. They are usually mounted over a tripod and leveled using a ball and socket mechanism. The horizontal angle is measured using a pair of sights located on north – south axis of the compass. Although surveyor's compass is similar to the prismatic compass, The difference between prismatic and surveyor's compass are listed in Table 7.1.

In the surveyor's compass also two types of adjustments namely temporary and permanent as described in the prismatic compass are made. The surveyor's compass is mainly used in mine surveying. For other types of the survey, one normally does not prefer to use this kind of compass.

Calculation of Included Angles

Having conducted the compass survey as described, next step in plotting the survey results on maps is to calculate the included angle between two consecutive survey lines of the traverse.

If the whole circle bearings of two lines at a station where these lines intersect are recorded, then the included angle between these lines will be equal to the difference

Table 7.1: Salient Differences between a Prismatic and Surveyor Compass

Prismatic Compass	*Surveyor's Compass*
There is a prism at the viewing end. As the reflecting prism carries the eye vane also, sighting of the object and taking the reading are done simultaneously	There is no prism on it. Slit consists of an eye-vane (in place of prism) with a fine sight slit.
The viewing and sighting are accomplished simultaneously	The object is sighted first and then the surveyor goes round the instrument to read the graduation on the graduated ring corresponding to the north end of the needle
The graduated circle is fixed to the broad type of needle. Hence it will not rotate with the line of sight	The graduated aluminium ring is attached to the circular box. It is not fixed to the magnetic needle. Hence it rotates with line of sight
A mirror is attached to the object vane	No mirror is attached to the object vane
The magnetic needle does not act as an index	The magnetic needle acts as an index while reading
Graduations are in whole circle bearing	Graduations are in quadrantal system
Graduations are marked inverted since its reflection is read through the prism	Graduations are marked directly without inversion
It can be used both hand held as well as on a tripod	It can be used only with a tripod
Least count of compass is 30 minutes	Least count of surveyor compass is 15 minute
The ring is graduated 0 to 360° with the zero at S	The ring is graduated from 0° to 90° in four quadrants. 0° is marked at the north and south, and 90° at the east and west. The letters E (east) and W (west) are interchanged from their true positions. The figures are written the right way up. In certain surveyor compass, the graduation may be made from 0 to 360° in anti-clock-wise direction with the zero at N

between the whole circle bearings of two lines. If the difference is less than 180° the included angle is interior angle and if it is more that 180° it is the exterior angle between the two lines forming the traverse. Let us consider the example given schematically in Figure (7.7 left).

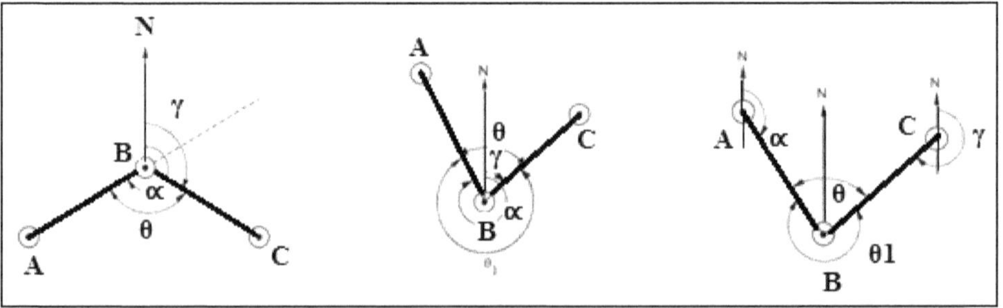

Figure 7.7: Assessing the Interior Angle Case 1 (Left), Case 2 (Middle) and Practical Explanation of Case 2 (Right).

If the back bearing (BA in terms of WCB) of line AB, *i.e.* (α) = 240° and fore bearing (FB) of line BC, (γ) = 120°. Then the included angle ABC, $\theta = \alpha - \gamma = 240° - 120° = 120°$. Thus, one can define the rule as:

If both the bearings are measured from a common point (B) then included angle can be obtained by subtracting FB of next line (BC) from the BB of previous line (AB) or FB of BA.

Now If the WCB at point of intersection of survey lines AB and BC (*i.e.* at station B) are not known. Let us consider that fore bearing of line AB (*i.e.* WCB of line AB at A) and back bearing of line BC (*i.e.* WCB of line BC at C) are known, then the included angle at station B between survey lines AB and BC (Figure 7.7 middle) can be obtained as follows.

WCB of AB at B = Back bearing of line AB at B = 150° + 180 ° = 330°.

Back bearing of line BC at C = 220°.

WCB of BC at B = Fore bearing of line BC at B = 220° – 180° = 40°.

Therefore, applying the rule described previously, Included angle $\theta 1$ = 330° – 40° = 290°

Since it is more than 180°, it is exterior angle as described in Figure 7.7 (right).

Hence, Interior angle $\theta = 360° - \theta 1 = 360° - 290° = 70°$.

Example 7.5

Find out the angles of the triangle (Figure 7.8), if following information from compass survey is available

Line	*Fore Bearing*	*Back Bearings (Calculated)*
AB	40° 0'	BB of AB =220°
BC	110°0'	BB of BC =290°
CA	280°0'	BB of CA =100°

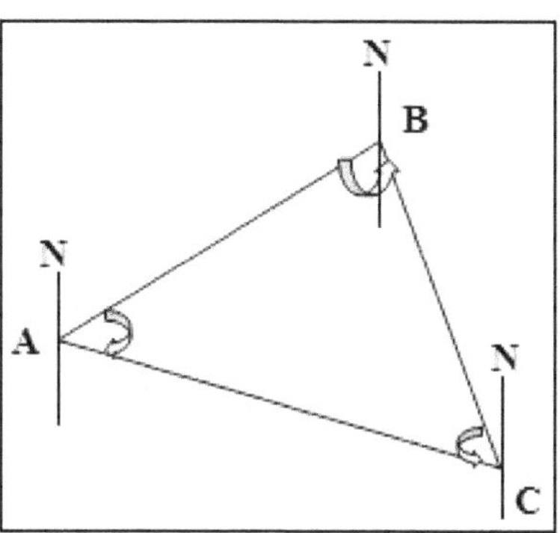

Figure 7.8: A Schematic Diagram of the Triangle for Example 7.5.

Formula to calculate included angle is given as:

Angle = BB of previous line - FB of next line

Angle A = BB of CA – FB of AB = 280° - 180° - 40° = 60°

Angle at B = BB of AB - FB of BC = 40° +180° -110° = 110°

Angle at C = BB of BC - FB of CA =110° + 180° - 280° = 10°

Sum= 60° +110° +10° = 180°

Check: (2n-4) 90° = (6-4) 90°=180° (Here n is the number of sides of the traverse)

Example 7.6

Determine the included angles in Figure 7.9 for the fore bearings observed during the survey in a clockwise direction.

Using the formula: BB of previous line - FB of next line

We calculate the angles A, B, C and D as follows:

Angle A = BB of – FB of AB = 280° - 180° – 40° = 60°

Angle B = BB of AB – FB of BC = 40° + 180° – 70° = 150°

Angle C = BB of BC – FB of CD = 70° + 180° – 210° = 40°

Angle D = BB of AD – FB of DC = FB of DA = 280° – (210° -180°) = 250°

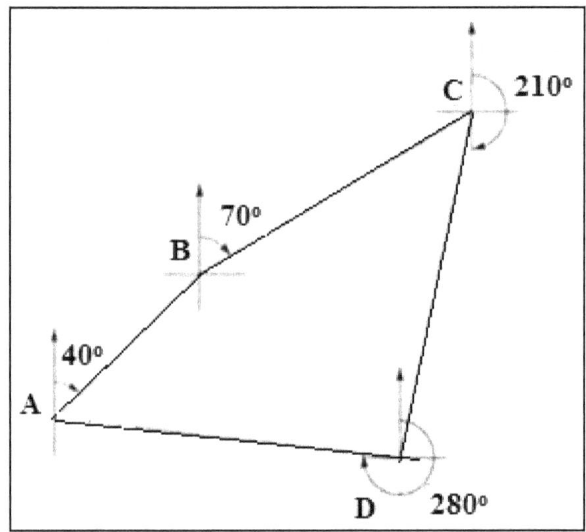

Figure 7.9: Compass Bearing of Various Points of the Polygon.

Since angle D is more than 180°, it is external angle and hence

Angle D = 360° -250° = 110°

Total = 60° + 150° + 40° +110° = 360°

Check sum of Angle = (2n-4) × 90° = (2 × 4 -4) × 90° = 360°

Chapter 8

Dumpy Level

Levelling: Definition

For executing any engineering project, it is essential to determine the elevations of different points along the alignment of proposed project. The art of determining relative altitudes of points on the surface or beneath the surface of earth is called levelling. The fundamental principle of levelling lies in finding out the separation of level lines passing through a point of known elevation (Benchmark, BM) and that through an unknown point (whose elevation is required to be determined). The purpose of this chapter is to explain the levelling instruments "Level" and "Staff" besides the other equipment.

Commonly Used Terms in Levelling

Datum: Datum plane is an arbitrarily assumed level surface or line with reference to which levels of other lines or surface are calculated.

Reduced Level (RL): Height or depth of a point above or below the assumed datum is called reduced level.

Benchmark: To begin with the survey, a spot in the area is first selected whose elevation from the Mean Sea Level is already known. Such a spot can be found by using a Survey of India topological map (also known as topo-sheet) for the area. Normally, it is possible to find at least one spot in the vicinity whose elevation will be marked on the topo-sheet. This point is referred as the benchmark or in short BM. The importance of the BM is that the elevations of other points are calculated from it. Three kinds of benchmarks are commonly used in surveying.

Great Trigonometrically Survey (GTS) Benchmark: It is established carefully at safe location with correction for curvature and refraction. It is established by Great Trigonometrical Survey, a Government of India organization. GTS benchmarks are established all over the country with highest precision survey, the datum being mean sea level.

A bronze plate is provided on the top of a concrete pedestal with elevation engraved on it. It also serves as a benchmark. It is well protected with masonry structure built around it so that its position is not disturbed by animals or by any unauthorized person. The positions of GTS benchmarks are shown in the published topo-sheets.

Permanent Benchmark: It is established at a safe location by Government Departments like Public Works Department, Irrigation *etc.* These are carefully transferred from the GTS benchmark.

Temporary Benchmark (TBM): In many engineering projects the difference in elevations of neighbouring points is more important than their reduced level with respect to mean sea level. In such cases, a relatively permanent point, like plinth of a building or corner of a culvert, are taken as benchmarks, their level assumed arbitrarily such as 100.0 m, 300.0 m, *etc.* Example: a benchmark established by an NGO for the watershed survey work is called temporary benchmark. This type of benchmark is also established at the end of the day's work, so that the next day work may be continued from that point. Such point should be on a permanent object so that it can be easily identified on the following day.

Arbitrary Benchmark: It is an imaginary benchmark in which elevation of any point is assumed without being transferred from any benchmark. Example: For a survey of any dam, if an elevation of a nearby stone is assumed to be as 100 m, that stone is called arbitrary benchmark.

Mean Sea Level (MSL): It is the average level of the ocean's surface, calculated as the arithmetical mean of hourly tide levels taken over an extended period. It is simply the mean average of all the high and low tides in the ocean. MSL adopted by Survey of India is now Bombay

Level Surface: The surface which is parallel to the mean spheroidal surface of the earth is known as level surface (Figure 8.1). For example; water surface with no motion.

Level Line: A line lying throughout on a level surface is a level line. This is normal to the plumb line at all points.

Horizontal Surface: At the instruments axis, the horizontal surface is tangent to the level surface. Over short distances (<100 m) the horizontal surface and the level surface will coincide. For long levelling lines the effects of the gravity field must be considered

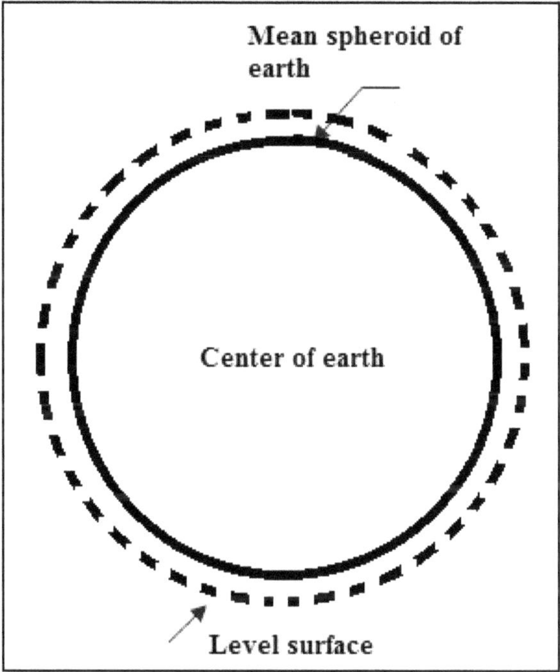

Figure 8.1: Definition of a Level Surface.

Vertical Plane: A vertical plane is any plane containing a vertical line. A vertical line at any point is a line normal to the level surface at that point. A plumb line is an example of a vertical line.

Line of Collimation or Line of Sight: The line of collimation or the line of sight is the line joining the intersection of cross-hairs to the optical center of the object glass and its continuation.

Axis of Telescope: The axis of a telescope is the line joining the optical center of the object glass to the center of the eye piece.

Dumpy Level

The dumpy level is the simplest and most inexpensive of the machines (such as alidades, theodolites *etc.*) used for surveying and levelling. Dumpy level, an optical surveying and levelling instrument, consists of a telescope tube firmly secured in two collars fixed by adjusting screws to the stage by the vertical spindle. It is used to determine relative elevations of different points of a surveying area. The telescope can rotate only in a horizontal plane. Moreover, it is easy to use and learn its operation. As such, it is a natural choice as a surveying method. The major parts of a dumpy level are (Figure 8.2):

Foot Screw: Three in number are used for temporary setting (levelling) of dumpy level.

Focus Screw: It is used for focusing object for clear vision.

Eye Piece with Cross Hair Focus Screw: It is used for focusing cross hairs for clear vision.

Level Tube: It is used for temporary setting of dumpy level, to see whether the dumpy has been properly leveled.

Objective Lens: It ensures clear vision of farther objects.

Telescope: The whole tube housing eye piece, objective lens *etc.*

Compass: In new instruments, the base of the dumpy levels also incorporates a 360° compass.

Besides, we need a tripod and a staff as other major accessories for the dumpy level survey.

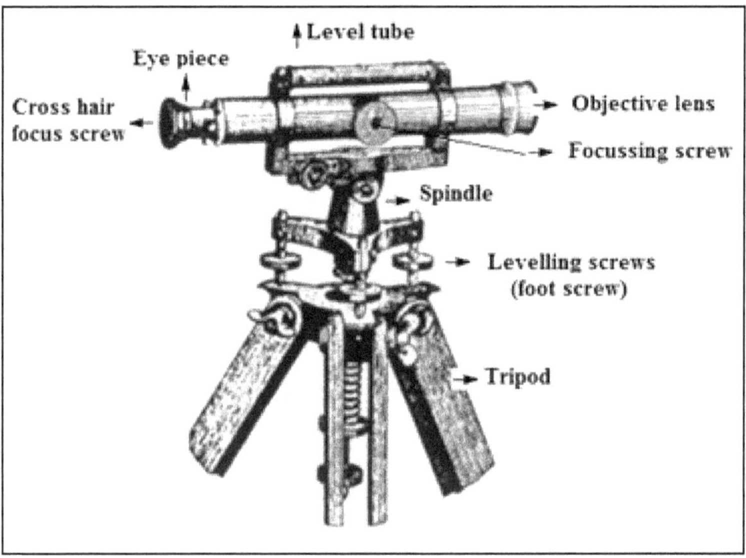

Figure 8.2: A View of a Dumpy Level Fixed on a Tripod Stand.

Tripod

It is a stand of dumpy level. The tripod is composed of aluminum or plastic, with three extendable/lockable legs, a base plate with a screw fitting with which the level head is attached. A canvas carrying strap and a belt to secure the legs together are provided for easy handling and transport.

Staff

It is a kind of scale, which is used to read the level at a point to calculate the reduced level (Figure 8.3). The staff, a wooden scale, is available in 4 m, 5 m and 6 m lengths having rectangular cross section. The lower end of the staff is shod with metal to protect it from wear and tear. It is usually the point of zero measurement from which all other graduations are numbered. Staff are either solid (having single piece of 3 m height) or folding (of 4 m or height into two or three pieces). The values are notched on the staff in black and red. The numbers in the black are cm units and those in red are meter units. This staff is capable of measuring a fairly wide range of elevations – between 1 cm and 4 m with a least count of 0.5 cm. Five mm black and five mm white strip is marked on the staff. In the left of staff black color number 1,2,3... shows completion of 10 cm, 20 cm, 30 cm... In the right of staff red color number 1,2,3... shows completion of 1m, 2 m, 3 m.

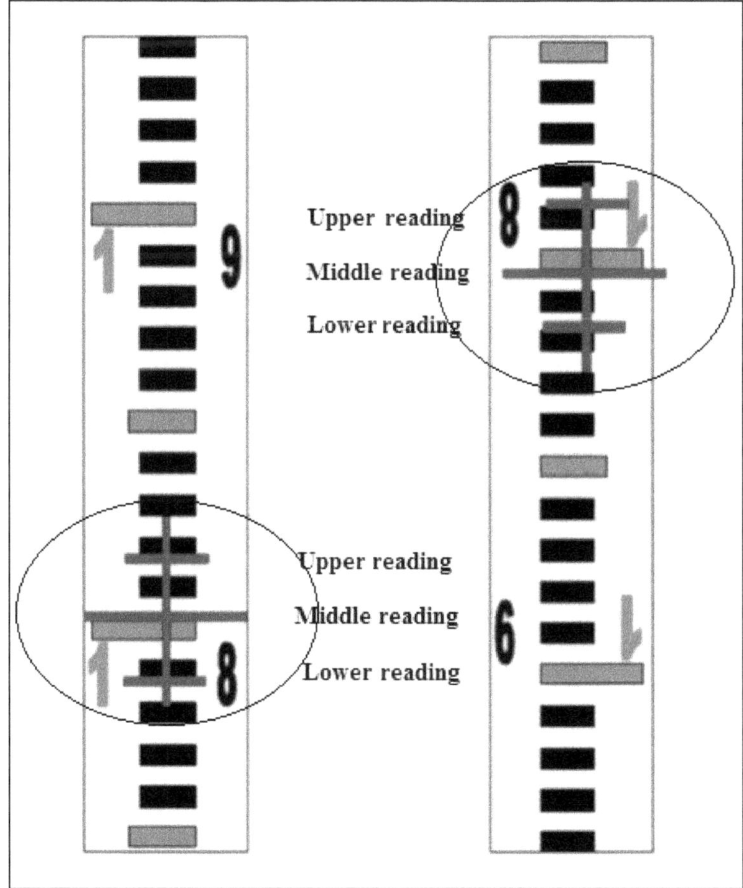

Figure 8.3: Dumpy Staff in Normal Position (Left) and as seen through the Dumpy (Right).

Setting up a Temporary Benchmark (TBM)

Mark an easily identifiable permanent feature nearby the survey area. Make a careful note with a precise description describing the location and nature of the TBM, preferably with a note annotated on a map and a digital photograph. Use geographical positioning system, if possible. Then note down its latitude and longitude as well. Transfer its elevation from a nearby permanent benchmark.

Permanent Adjustment

An instrument from manufacturer is generally available in perfect adjustments. However, the instrument may get out of adjustments due to worn out and loose fittings or due to mishandling. So it is always better to check the instruments occasionally especially before any precise survey work is undertaken. Three fundamental lines in a level instrument are:

- ☆ Vertical axis
- ☆ Axis of the level tube
- ☆ Line of sight

These fundamental lines are related to each other as follows:

- ☆ Axis of the level tube is perpendicular to the vertical axis
- ☆ Horizontal cross hair should lie in a plane perpendicular to the vertical axis, so that it will lie in a horizontal plane when the instrument is properly leveled.
- ☆ The line of sight is parallel to the axis of the level tube. Also, the optical axis, the axis of the objective lens and the line of sight should coincide.

In a properly adjusted dumpy level, desired relations among fundamental lines should be met with.

Temporary Adjustment of Dumpy Level

Once the instrument is permanently in adjustment then temporary adjustment of the level is performed at each set up of level before the actual work of taking observations begins. It comprises of a sequence of events.

- ☆ Setting up level
- ☆ Levelling of level
- ☆ Focusing and elimination of parallax

Fixing the Dumpy Level on the Tripod

Spread out the tripod stand of the dumpy level on the ground such that the steel plate on top of the tripod is level. Gently but firmly press the tripod down into the ground so that it does not move. The steel plate on top of the tripod has a hollow screw on it (Figure 8.4). On the dumpy, a hole is provided at the base. Place the dumpy so that the screw on the tripod ts into this hole and the dumpy is securely fastened on to the tripod.

Levelling the Dumpy Level

Before using the dumpy, it is necessary to ensure that the instrument is completely level. Three round screws are provided at the bottom end of the dumpy (Figure 8.2). Half open these 3 screws. Set the telescope parallel to any two-foot screw. Then move both the screws, one in clockwise and the other one in an anticlockwise direction. It is continued until the bubble of level tube comes at center. Try to move both the screws in equal measure (Figure 8.5 left).

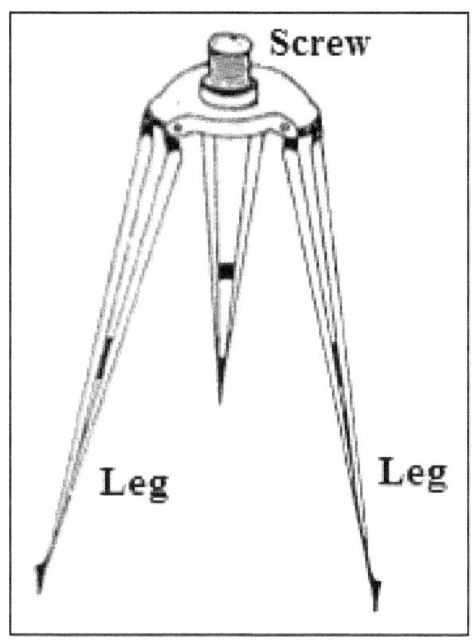

Figure 8.4: The Tripod with Fitting Screw.

Set the telescope perpendicular to the selected screws (Figure 8.5 right). Since the bubble might have lost its center position, now move the third screw in clockwise or anticlockwise until bubble of level tube comes at center.

Now again set the dumpy to first step and set the bubble at the center of the bubble tube as described earlier. Now set the dumpy to second step and set the bubble at the center of the bubble tube as described earlier. These steps are repeated until bubble is not disturbed at the center of the level tube. When this state is achieved, the bubble will always remain centered irrespective of direction in which it is moved. The dumpy level in this position is ready for further use.

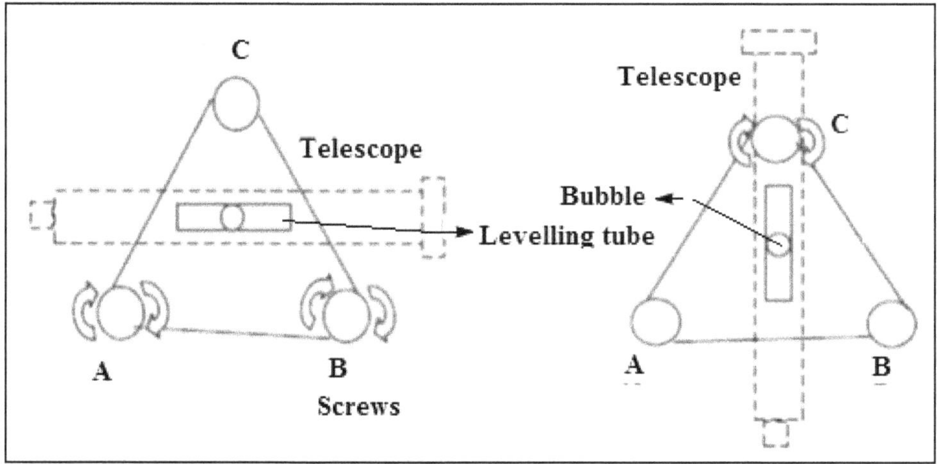

Figure 8.5: Line Diagram Illustrating the Setting Procedure of a Dumpy Level.

Elimination of Parallax

The dumpy has a built-in telescope. At one end of the telescope is an eye piece or the viewnder.It is at this place, we place our eye to see the place to survey. At the other end is the lens on which image of what we are surveying is captured. On looking through the view finder, we see three lines on the lens: two lines running upright and one running across these two lines. Parallax is a condition arising when the image formed by the objective is not in the plane of the cross hairs. It has to be ensured by rotating the eye-piece for clear cross hairs and by rotating focusing screw provided for objective focus. Once leveled and telescope adjusted, the dumpy is moved around very gently, lest the settings are disturbed (Venugopala and Akella, 2015).

Taking Readings

Align the telescope in the direction of the object. Keep the staff at the point where reading is to be taken. Focus at the staff placed on the spot through focusing screw for clear vision. Now read the staff reading. Depending upon the kind of survey, we can read the middle or all the three reading- upper reading, lower reading and middle reading.

In most old dumpy levels, the scale on the staff appears upside-down when viewed from the view finder (Figure 8.3). Thus, readings in such a case should be taken carefully. In reality, when we place the staff on the ground and view it with our bare eyes, we see the numbers increase as we move our eye up the staff from the bottom. However, when we see the same staff through the view finder we will observe that the numbers decrease as we move up the staff, as though we have turned the staff upside down. The numbers also appear upside down. The number 9 will appear to be 6 and vice versa. So this distinction must be kept in view while taking the readings. Moreover, whenever measurements are taken on the staff from the dumpy telescope, always take the numbers as ascending from the top to the bottom.

Surveying: Dumpy Level

Nomenclature of various readings taken at different stations is illustrated in Figure (8.6). Normally, the staff is kept at a distance of 10 to 200 m from the dumpy so that the reading can be taken easily. Procedure of filling survey readings in field book is illustrated through the example 8.1.

Back Sight Reading: It is first reading taken at a point of known elevation from a survey station. Whenever the dumpy is shifted to new position first reading should always be noted in back sight.

Intermediate Sight Reading: Excluding the first and the last readings, other readings taken from survey stations, are noted as intermediate readings.

Foresight Reading: It is last reading taken from a survey station. Whenever the dumpy is likely to be shifted to a new position the last reading from the station is noted as foresight.

Station

Any point whose elevation is to be determined is known as station or a point which is to be established at a given elevation. It is the point where the staff is held and not the point where the level is set up.

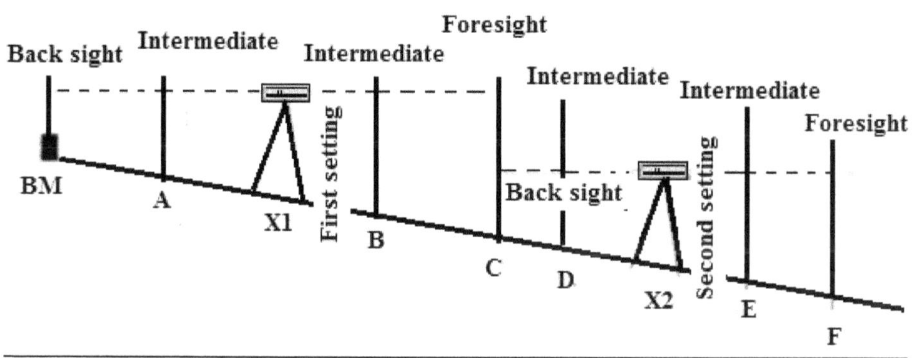

Figure 8.6: Nomenclature of Various Readings at different Stations.

Change Point

A change point (CP) is a point which shows the shifting of the level. It is a point on which fore and back sights are taken. It is why it is also known as turning point (TP). Any stable and well defined object such as boundary stone, curb stone, rail, rock, *etc.*, is used as a change point. A benchmark may also be taken as a change point.

Height of Instrument (HI)

A surveying term used in levelling for the height of the line of sight of a levelling instrument above the adopted datum. In trigonometric levelling for the height of the center of the theodolite above the ground or station mark, in stadia surveying for the height of the center of the telescope of the transit or telescopic alidade above the ground or station mark, and in differential levelling for the height of the line of sight of the telescope at the levelling instrument when the instrument is level.

In a particular set up of the instrument height of instrument, which is the elevation of the line of sight, is constant.

Level Field Book

A level field book, also called level book is used for taking down each staff reading during levelling and subsequently, used for finding out the elevation of points/stations. Although two types of level books are available depending upon the method to be used by the surveyor. But, in general a level book contains columns of both the types together (Table 8.1) and it is for a surveyor to use only the relevant columns.

Height of Instrument Method

As already mentioned, in any particular set up of an instrument the height of instrument (HI) is constant. It is given as: HI=known elevation +BS. The elevation of unknown points then can be obtained by subtracting the staff readings at the desired points from the height of instrument. RL= HI - IS or FS.

This is the basic principle behind the height of instrument method for finding the reduced levels.

Table 8.1: A Page of a Level Field Book

Name of the Surveyor									Page No.	
Levelling from ______ to ______										
Instrument No. ______							Conducted by:			
Station	Distance (m)	Bearings		Staff Reading			Height of Instrument		Reduced Level	Remarks
		Fore	Back	Back (BS)	Inter (IS)	Fore (FS)	Rise	Fall		

Filling the Field Book

The following examples illustrate the various kinds of reading and the entry of these readings in the field book.

Example 8.1

The dumpy is at station X1 (Figure 8.7). First reading on benchmark is taken as 2.01 m. RL of benchmark is 100 m. The readings at locations P, Q, R and S are 2.75, 1.82, 0.86, and 0.43 m, which have been taken from the same dumpy station. Fill the reading in the field book and calculate the RL of point P, Q, R and S. If the dumpy is now shifted to another station, where the new reading at stations S and T will be entered in the field book?

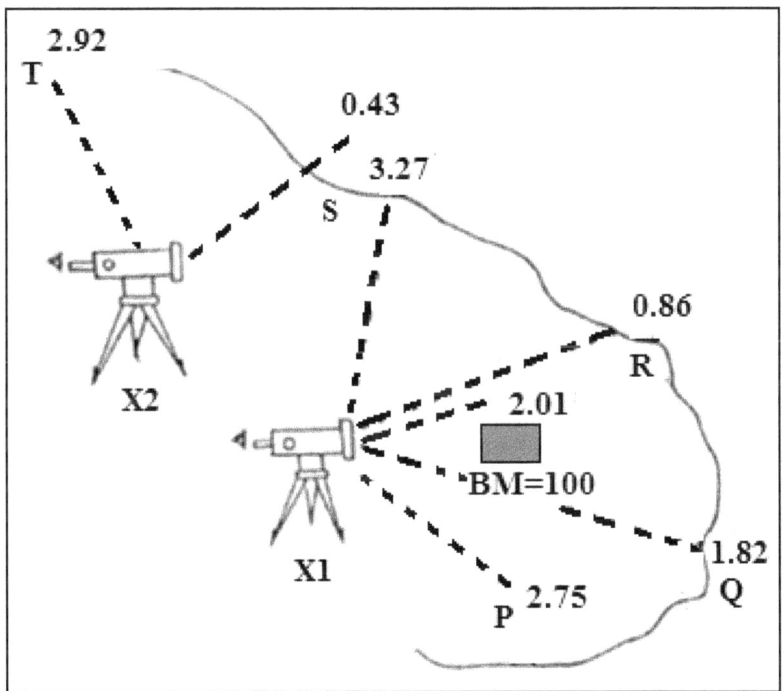

Figure 8.7: Survey Layout and Observed Level Readings for Example 8.1.

The reading on the staff at the BM is referred as back sight or simply BS because the staff is placed at a spot whose elevation is already known. Since the dumpy has been completely leveled, the height of its telescope is equal to the elevation of the reading on the staff visible through the viewnder. Since the elevation of the BM is 100 m (filled in the column of RL) and the reading on the staff is 2.01 m. Then, the elevation of the dumpy = BM+BS = 100 +2.01 = 102.01 (Table 8.2). It is also called the Height of the Instrument (HI).

Table 8.2: Filling of the Field Book for Example 8.1

Dumpy Location	*Station*	*Back Sight*	*Intermediate Sight*	*Fore Sight*	*HI*	*RL*
X1	BM	2.01			102.01	100.0
	P		2.75		102.01	99.26
	Q		1.82		102.01	100.19
	R		0.86		102.01	101.15
	S			0.43	102.01	101.58
X2		3.27			104.85	
	T			2.92		101.93
Check for station X1		5.28		3.35		1.67

Since the instrument remains level so long as it remains at the same spot, the HI also will remain the same.

Now shift the staff at P and take the reading. Suppose the reading is 2.75 m (Intermediate sight). Then elevation at P = Elevation of dumpy (HI) – reading on staff = 102.01 – 2.75 = 99.26 m.

Now keep the staff at Q. The reading here is 1.82 (Intermediate reading). Then elevation at Q = Elevation of dumpy (HI) – reading on staff = 102.01 – 1.82 = 100.19 m.

Now keep the staff at R. The reading here is 0.86 (Intermediate reading). Then elevation at R = Elevation of dumpy (HI) – reading on staff = 102.01 – 0.86 = 101.15 m.

Now keep the staff at S. The reading here is 0.43. Since now it is proposed to shift the instrument, it is the last reading from this set. So this reading is entered as foresight. Then elevation at S = Elevation of dumpy (HI) – reading on staff = 102.01 – 0.43 = 101.58 m.

Now the instrument is set at point X2. Level the dumpy all over again. Point S will now serve as the new BM, since we know its elevation (101.58 m). After levelling the dumpy, take a reading by keeping the staff at S. This is now the new Back Sight (BS). Now find out the height of the dumpy (new HI) at this spot as follows:

$$HI = BS + R.L. = 101.58 + 3.27 = 104.85$$

Now keep the staff at T and take a reading. Suppose the reading is 2.92 m. Now if we finish our survey at this point, the last reading is entered as foresight.

Then elevation at T = Elevation of dumpy (HI) – reading on staff = 104.85 – 2.92 = 101.93m.

Formulae Used

$$HI = BS + RL \tag{8.1}$$

$$RL = HI - IS \text{ or } FS \tag{8.2}$$

Arithmetic Check

The following equality is used to check the entries and the calculations.

$$\Sigma BS - \Sigma FS = \text{Last RL} - \text{First RL} \quad (8.3)$$

ΣBS is summation of back sight column

ΣFS is summation of fore sight column

For Table 8.2, the test reveals that

5.28 – 3.35 = 101.93 – 100

1.93 = 1.93 (Since both are equal, entries and calculations are correct)

Example 8.2

Calculate RL of the points A to E using the information on BS, IS and FS given in the Table below (in italics). RL of the BM may be taken as 300 m above MSL. Also check your calculations.

Dumpy Location	*Station*	*Back-Sight*	*Inter-mediate Sight*	*Fore-Sight*	*HI**	*RL*	*Calculations (Values indicate column numbers)*
1	*2*	*3*	*4*	*5*	*6*	*7*	*8*
X1	BM	*1.65*			**301.65**	**300.00**	**6 = 3+7**
	A		2.45		**301.65**	**299.20**	**7 = 6-4**
X2	B	*1.25*		*3.70*	**301.65**	**297.95**	**7 = 6-5**
	C		2.25		**299.20**	**297.00**	**6= 7+3 (of previous row) 7= 6-4**
	D		2.75		**299.20**	**296.45**	**7 = 6-4**
	E			*1.90*	**299.20**	**297.30**	**7=6-5**
Check		2.90		5.60		-2.7	

* Note: Height of instrument remains the same until the instrument is changed to a new station

The values are filled in last two columns as HI and RL and the calculations given in column 8 indicating the column used in the calculations.

Check is given as: $\Sigma BS - \Sigma FS$ = Last RL - First RL

2.90 - 5.60 = 297.3 - 300 = -2.70

Rise and Fall Method

For the same set up of an instrument, staff reading is higher at a lower point and less for a higher point. Thus, staff readings provide information regarding relative rise and fall of terrain points. It is the basic premise behind rise and fall method for finding out elevation of unknown points.

The method essentially comprises of determining the difference of elevation between consecutive points by comparing each point to the point immediately preceding it. The difference between the staff reading indicates a rise/fall according to the staff reading at the point. The RL is then found adding the rise to, or subtracting the fall from the reduced level of preceding point.

Arithmetic check

$$\Sigma BS - \Sigma FS = \Sigma rise - \Sigma fall = last\ RL - first\ RL \qquad (8.4)$$

This method is relatively complex and is not easy to use. It is also time consuming. Visualization is necessary regarding the nature of the ground. On the other hand, it is good since complete check is there for all readings. This method is preferable for check levelling where numbers of change points are more.

Allowable error = $\pm 5 \sqrt{n}$ mm; where n = no. of instrument positions

Example 8.3

The survey values of few points are given the following table. Complete the last 3 columns under the head reduction to find out RL based on rise and fall method. Check calculations using the guidelines.

		Booking			*Reduction*		
Station	*Remark*	*BS*	*IS*	*FS*	*Rise*	*Fall*	**RL**
1	BM	2.554					**52.554**
	A		1.786		**0.768**		**53.322**
	B		0.927		**0.859**		**54.181**
	C		1.983			**1.056**	**53.125**
	D			3.589		**1.606**	**51.519**
	D	1.305					
	E		1.422			**0.117**	**51.402**
	F			0.571	**0.851**		**52.253**
		3.859		4.160	**2.478**	**2.779**	
		3.859 -4.160 = -0.301			**2.478-2.779 = -0.301**		**52.253-52.554 = -0.301**

Check shows that

$$\Sigma BS - \Sigma FS = \Sigma rise - \Sigma fall = last\ RL - first\ RL = -0.301$$

Calculating Distances using Dumpy Level

Dumpy level can also be used to measure approximate distances. As such, it can also be used to quickly calculate the slopes. As previously explained, a dumpy level is provided with two stadia hairs above and below the cross hair, called the upper and lower stadia. These are equidistant from the cross hair. To measure the distances following procedure is used.

- Level the instrument as before
- Take a reading on the crosshair. Let us call the reading as Middle Reading (MR)
- Take the readings on the upper and lower stadia hairs. These can be called Upper Reading (UR) and Lower Reading (LR) respectively. If the

instrument is level, the difference between UR and MR will be the same as that between MR and LR

- ✰ Take the difference between the UR and the LR

The horizontal distance, HD is then given as:

$$HD = N\ (UR\text{-}LR) \qquad (8.5)$$

For most instruments, the value of N is 100. If it is different for a particular instrument then take that value of N in the calculations.

The procedure is illustrated in Figure 8.8. For example, if the UR and LR at point A in Figure 8.8 are 3.1 m and 1.5 m respectively, then the distance between the instrument and A is:

$(3.1 - 1.5) \times 100 = 160$ m

Note that the middle reading is 2.3 and the difference between UR and MR and MR and LR is same.

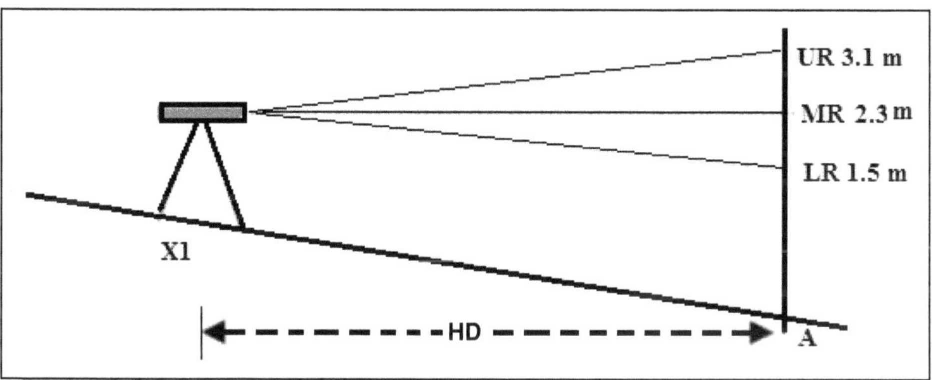

Figure 8.8: Calculation of Horizontal Distance

Calculation of Slope using Dumpy Level

Slope between any two points (say A and B) is given as the ratio of the vertical interval and the horizontal distance between the two points. Mathematically

$$\text{Slope} = VI/HD \qquad (8.6)$$

Here VI is difference in the RL of station A and RL of station B.

Slope may be expressed as a ratio (1:2 *etc.*), as percent ($100 \times VI/HD$) and as angle θ since $\tan\theta = VI/HD$.

Example 8.4

The Dumpy Level readings (m) at 2 points A and B are as follows:

	A	B
Upper Reading (UR)	1.72	1.78
Middle Reading (Cross hair)	1.70	1.75
Lower Reading (LR)	1.68	1.72

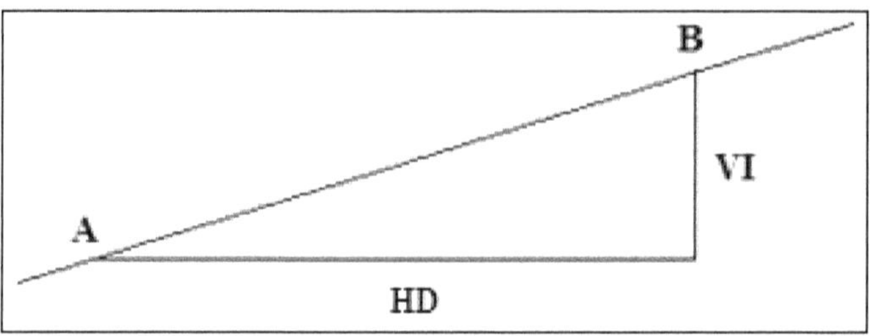

Figure 8.9: Schematic Diagram Illustrating the Determination of Slope.

Calculate the approximate per cent slope between A and B. Assume that both the stations are in the same direction of the dumpy level in a straight line. What is the slope if station A is on one and station B on the other side of the dumpy level?

Distance of A from the dumpy = (1.72 – 1.68) 100 = 0.04 × 100 = 4 m

Distance of B from the dumpy = (1.78 – 1.72) 100 = 0.06 × 100 = 6 m

HD = 6 – 4 = 2 m

The level difference (from middle readings) = 1.75-1.70 = 0.05 m

Per cent slope = 0.05 × 100/2 = 2.5 per cent

In the second event HD = 6+4 = 10 m

Since the middle readings are the same, VI = 0.05.

Per cent slope = 0.05 × 100/10 = 0.5 per cent

Example 8.5

A dumpy level is placed in between stations A and B forming a straight line. The readings shown in italics were taken. Calculate the HD, RL, distance of the stations from the dumpy level and slope of the land (per cent) between the two stations.

Dumpy Location	*Stations*	*BS*	*IS*	*FS*	*UR*	*LR*	*HI*	*RL*	*Distance (m)*
X1	BM	1.7					101.7	100	
	A		2.1		2.5	*1.7*	101.7	99.6	**80**
	B			2.9	3.7	*2.1*	101.7	98.8	**160**

The column filled- in are given in bold in the same table.

Now the total distance or HD = 240 m

VI = RL(A) – RL(B) = 99.6 – 98.8 = –0.8 m (Minus sign shows that slope is from station B to station A).

Slope = 100 × 0.8/240 = 0.33 per cent.

Basic Rules of Levelling

- ☆ Always start and finish a levelling run on a Benchmark (BM or TGBM) and close the loops

- ☆ Keep fore sight and back sight distances as equal as possible
- ☆ Keep lines of sight short (normally < 50 m)
- ☆ Never read below 0.5 m on a staff to avoid refraction
- ☆ Use stable, well defined change points
- ☆ Beware of shadowing effects while crossing water

Marking Contours with Dumpy Level

A contour, as already defined, is an imaginary line joining different points which have the same elevation. Suppose, a contour of 102 m is proposed to be marked with the help of a dumpy level. If the BM is assumed to be 100 m and the height of instrument is 103.5 m. move the staff away from the BM in such a way that the RL of the points are higher than the BM. The reading at these points should be read as 103.5 – 102.0 = 1.5 m. Now move the staff to another point where the reading is 1.5 m. If the reading is more than 1.5 m, then move the staff to a higher point. If it is less than 1.5 m., move the staff towards a lower point or a dip. Repeat this until you find a point whose reading is exactly 1.5 m. This is the second point at the same elevation (102 m) and hence is on the contour line of 102 m. Now find a third point and then a fourth and so on until you get the whole contour. The practical application of this method is to identify the submerged areas of water storage structures. Herein, the level at the exit weir is first determined. Staff is then moved to all points having the same reading as the exit weir. Marking or plotting these points will give the submergence area of the water storage structure.

Transferring the Benchmark

A permanent BM can normally be located on a topo-sheet. However, it is likely that this BM is located far away from the watershed. In such cases, any permanent feature of the land, such as a bridge, building or a pillar can be taken as a temporary benchmark and its elevation can be assumed to be 100. All elevations during survey are calculated with respect to this benchmark. Soon after the survey, the RL of the temporary benchmark is identified through dumpy level survey starting from the permanent benchmark. Assume, for instance, that the RL of the permanent benchmark is 180 m. Through the dumpy level survey we find that the RL of the temporary benchmark (assumed to be 100 m) is actually 160 m above MSL. An addition of (160-100) =60 m is made to all elevations worked out in the earlier survey. This process is called transferring the benchmark.

Testing the Dumpy Level

Keep the dumpy on one spot. Now select any two points which are 30 m away from this one. Find out the difference in elevation between these two points. Now shift the dumpy to a new position. Again measure the difference in elevation between the same two points from this new position of the dumpy. If some difference is observed between the elevations of these two points, then there is something wrong with the dumpy level.

Precautions

Dumpy Level

- ☆ It is a precision instrument, and needs to be handled carefully. When not in use, the instrument should always be kept in its box.
- ☆ During rains, make sure that instrument is properly covered. Best is to unscrew the head and place it in its box.
- ☆ If the level head has become wet, ensure that it is dried before being stored in its box.
- ☆ After the survey is complete, carefully unscrew the instrument and place it in its correct position within the box. Close the lid and make sure the catch is secure. Do not throw or drop the box.

Staff

- ☆ Always pull out the sections one at a time.
- ☆ Put the staff back in its sleeve after use.
- ☆ Keep mud and grit off it as much as possible, as this will scratch the painted markings.
- ☆ After the survey, wipe off each section of the staff gently with the supplied cloth.

Tripod

- ☆ Undo the catches on the tripod legs and carefully move to retract the legs.
- ☆ Clamp the catches back and fasten the belt.
- ☆ Ensure that the tripod does not get dented or damaged, as this may make it unusable.

Chapter 9

Levelling and Contouring

Levelling is defined as 'a branch of surveying which deals with the measurement of relative heights of different points on, above or below the surface of the earth'. Thus, in levelling, the measurements (elevations) are taken in the vertical plane. It can also be put simply as the process to determine the difference in elevation between two points.

Types of Levelling

- Simple levelling
- Differential levelling
- Fly levelling
- Profile levelling
- Reciprocal levelling
- Cross sectional levelling
- Barometric levelling
- Trigonometric levelling
- Hypsometric levelling

Simple Levelling

When the difference in the elevation of two nearby points visible from a single position of the level is required then simple levelling is performed (Figure 9.1). It is the simplest operation in levelling. Suppose A and B are two such point and level is set approximately mid-way between A and B may not necessary on the line joining them. Let the respective reading on A and B be 2.340 and 3.315. The difference between them is 3.315-2.340=0.795 m.

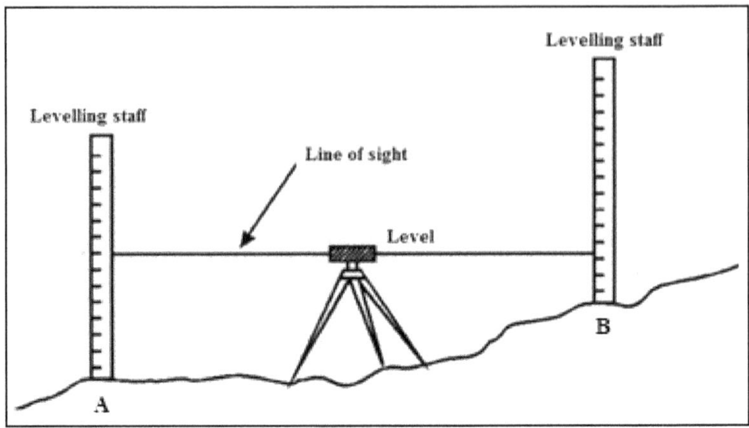

Figure 9.1: A View of the Simple Levelling.

Differential Levelling

This method is used in order to find out the difference in elevation between two points.

- If they are far apart
- If the difference in elevation between them is too great.

In such cases it is necessary to set up the level in several positions and to work in a series of stages. The method of simple levelling is employed at each of the successive stages (Figure 9.2). The process is also known as compound continuous levelling.

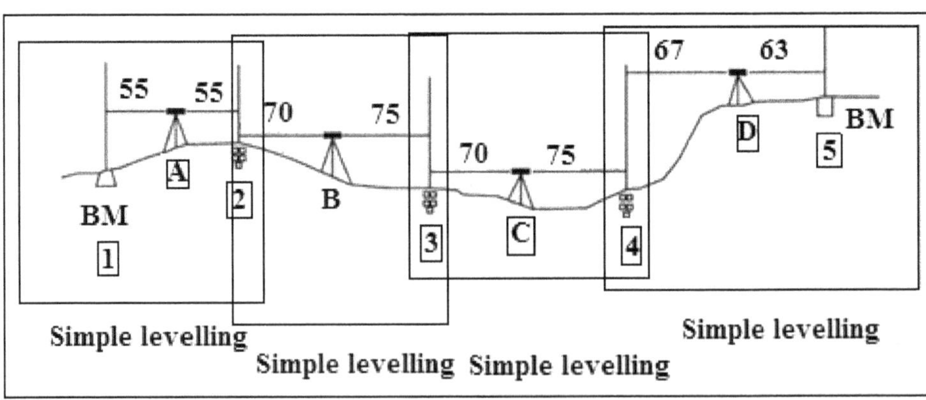

Figure 9.2: Pictorial Depiction of Differential Levelling.

Fly Levelling

It is performed when the work site is very far away from the bench mark. The surveyor starts by taking BS at BM and proceed towards worksite till a suitable place is reached for temporary BM. All works are done with respect to temporary BM. At the end of the day, the surveyor comes back to original BM. This is called fly levelling. It is low precision method and is generally used to find/check approximate levels and used during reconnaissance survey.

Profile Levelling

The operations involved in determining the elevation of ground surface at small spatial interval along a line is called profile levelling. Profile levelling yields elevations at definite points along a reference line, provides the needed data for designing the facilities such as highways, railroads and transmission lines *etc.* The route along which a profile is run may be single straight line or a broken line or a series of straight lines connected by curves, as in case of a rail road, highway or canal. After getting the RL of various points the profile is drawn (Figure 9.3). The main purpose of profile levelling is to provide data from which the depth of fill or cut required to bring the existing surface up to, or down to, the grade elevation required for the highway can be determined.

Normally vertical scale is much larger than horizontal scale for the clear view of the profile. It is also called the longitudinal levelling or sectioning.

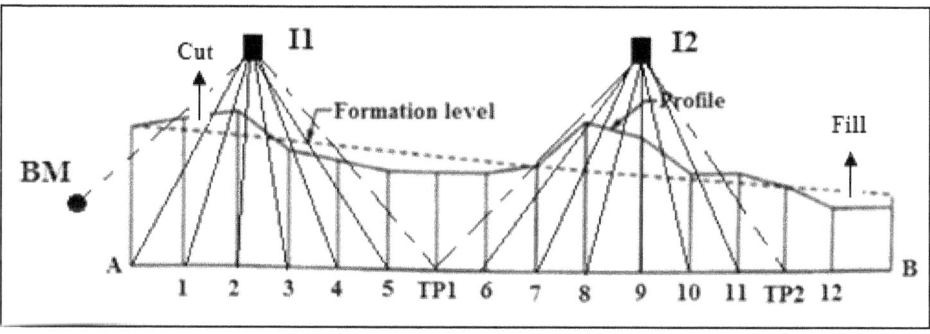

Figure 9.3: Profile Levelling.

To begin with the profile levelling, a level is placed at a convenient location not necessarily along the line of observations (say I1). The instrument is positioned in such a way that first back sight is taken clearly on a BM. Then, observations are taken at regular intervals (say at 1, 2, 3, 4) along the central line and foresight to a properly selected turning point (say TP1) (Figure 9.3). The instrument is then re-positioned to some other convenient location (say I2). After proper adjustment of the instrument, observations are started from TP1 and then at regular intervals (say at 5, 6 *etc.*) terminating at another turning point, say TP2. Staff readings are also taken at salient points where marked changes in slope occur. The distances as well as directions of lines are also measured.

Reciprocal Levelling

When levelling is required across river/pond where it may not be possible to set-up levels in between the points, then this method is applied to get rid of various errors. Two set-up of the level are used, one at either end (Figure 9.4). Set the level near the point on one side of the bank and take readings on both A and B points. Shift the level to point B on the other side of the bank and again take the readings at both the points A and B. Then the true level difference, D between the points is given as:

$$D = \{(b1-a1) + (b2-a2)\}/2 \qquad (9.1)$$

Here a1 and b1 are the readings from the first set up near point A for two point A and B respectively while a2 and b2 are the readings from set up near the point B (Figure 9.4). The errors due to curvature, refraction and imperfect adjustments (shown as e) are eliminated while taking the average.

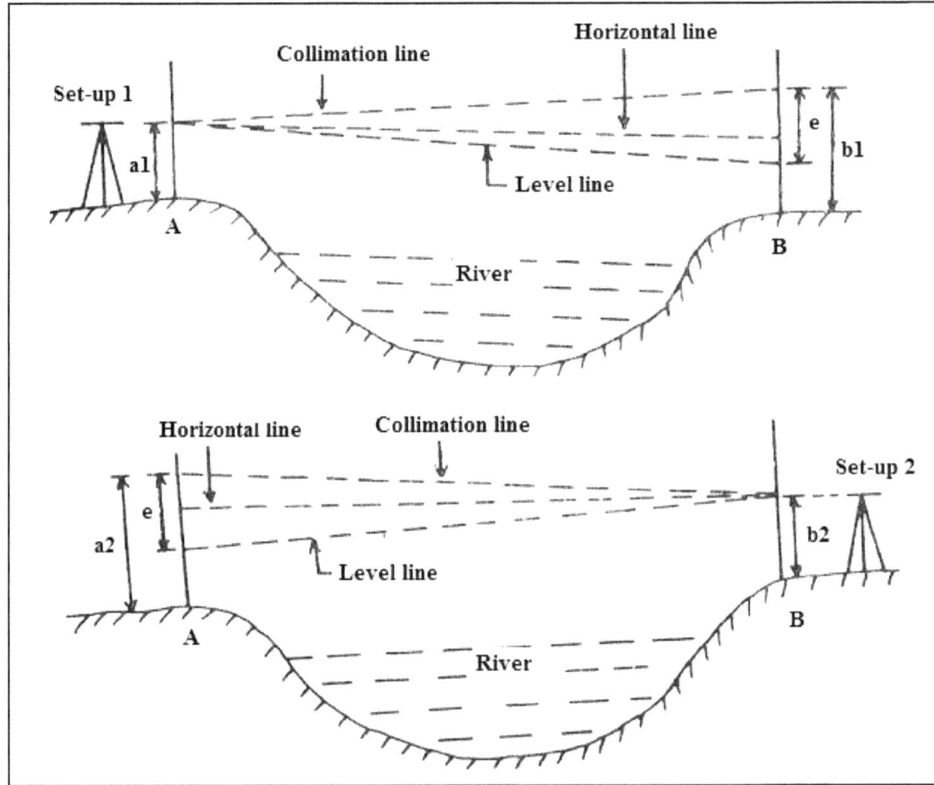

Figure 9.4: Set-up for Reciprocal Levelling First (Top) Second (Bottom).

Cross-sectional Levelling

In many projects, terrain information transverse to the longitudinal section (through profile levelling) is also required such as for highways, railways, and canals *etc.* While in profile levelling, the elevations of a series of points length wise along the line are determined, in cross-section levelling, elevations of points on a succession of lines running at right angles to the length wise line of the highway are determined. The cross sections must extend a sufficient distance on each side of the center line to provide a view of the surrounding terrain. If, for any reason, a cross-section is run in any other direction, the angle with the center line is also noted (Figure 9.5). Readings are taken at equal intervals on both sides of the center line and at significant changes in the terrain. The observations are then recorded as being to the left or right of the center line. Reduction of levels, plotting *etc.* is done as in case of profile levelling.

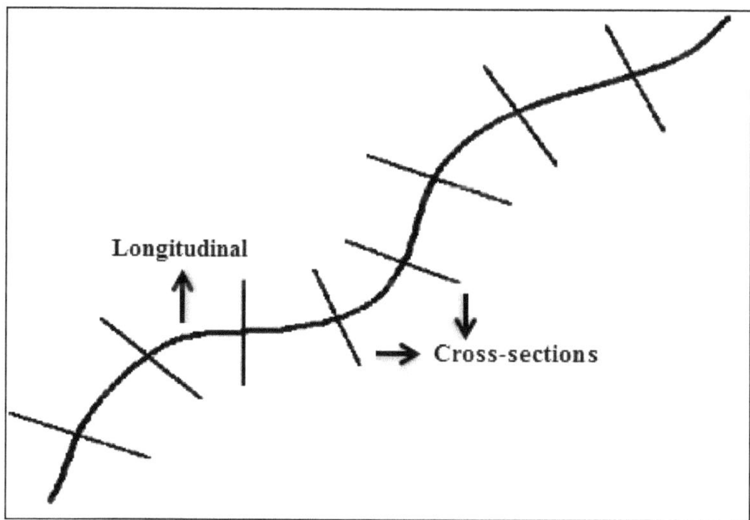

Figure 9.5: A Schematic View of Cross-sectional Levelling.

Barometric Levelling

The levels are determined on the basis of atmospheric pressure difference. Altimeter is commonly used in this survey. It provides very rough estimates.

Trigonometric Levelling

The process of levelling in which elevations of point or the difference between points is measured from the observed horizontal distances and vertical angles in the field is called trigonometric levelling (Figure 9.6). In this method, trigonometric relations are used to find the elevation of a point from angle and horizontal distance. It is also termed as indirect levelling.

Hypsometric Levelling

In this kind of levelling, principle that boiling point of water reduces with increasing altitude is used to calculate the elevations. Hypso-meter also called thermo-barometer is used in hypsometric levelling.

Main Sources of Errors in Levelling

The main sources of error in levelling can be ascribed to:

- Instrumental errors

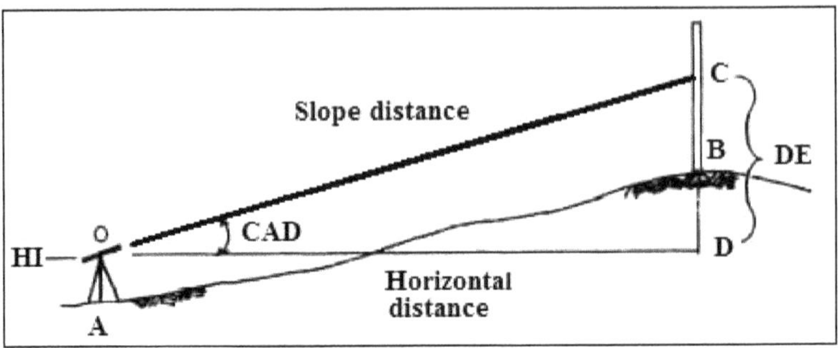

Figure 9.6: Trigonometric Land Levelling.

- ✰ Errors of manipulation
- ✰ Errors due to settlement of level and staff
- ✰ Errors due to natural causes
- ✰ Common mistakes in levelling

Instrumental Errors

It can occur due to imperfect adjustment due to which line of collimation may not be truly horizontal when the instrument is leveled. If the level is in perfect adjustment, the line of collimation should be parallel to the bubble axis so that it becomes horizontal when the bubble is centered. The error due to the inclination of the line of collimation is proportional to the distance of the staff from the instrument. Other causes are: defective level tube being too sluggish or over sensitive, shaky tripod that might render the instrument unstable resulting in erroneous readings and incorrect graduations of the staff.

Errors of Manipulation

The errors may be caused due to careless levelling of the instrument, the bubble not being central at the time of taking reading, improper removal of the parallax and the staff not being held vertical.

Errors due to Settlement of Level and Staff

If the level is set up on soft ground it may gradually settle down in the interval between taking a back sight and the fore sight reading. It will make the fore sight reading smaller than it should be. Settlement of staff especially at the change points during the time the instrument is shifted and set up will result in back sight becoming greater than it should be.

Error due to Natural Causes

The effect of curvature of the earth and refraction are small and opposite. As such, these are not taken into account. Nonetheless taking readings during wind storms and on hot sunny days may result in significant errors.

Common Mistakes in Levelling

- ✰ The levelling staff not being fully extended.
- ✰ The staff held upside down.
- ✰ Reading the staff upwards instead of downwards.
- ✰ Taking reading coincident with one of the stadia-hairs (top or bottom) instead of central-hair.
- ✰ Omitting an entry.
- ✰ Entering a reading in the wrong column.
- ✰ Reading the wrong number of meters and decimeters.

Permissible Limits of Error in Levelling

The permissible closing error may be expressed as $E = C \sqrt{km}$

Here, E is the error in meters; C is the constant, the value of which depends upon the type of instrument, nature of the terrain, atmospheric conditions and care and judgment of the level man, and km is the distance in kilometers. For rough levelling value of E is $\pm 0.100 \sqrt{Km}$, ordinary levelling: $E = \pm 0.025 \sqrt{Km}$, accurate levelling: $E = \pm 0.012 \sqrt{Km}$ and precise levelling: $E = \pm 0.006 \sqrt{Km}$

Contouring

A contour line is defined as "an imaginary line passing through points of equal reduced levels". A contour line may also be defined "as the intersection of a level surface with the surface of the earth". Thus, contour lines on a plan illustrate the topography of the area. The process of tracing contour lines on the surface of the earth is called contouring.

A map showing contour lines is known as contour map. A contour map gives an idea of the altitudes of the surface features as well as their relative positions in the plan. It serves the purpose of both, a plan and a section. Contour lines are drawn as fine and smooth free hand curved lines. Sometimes they are represented by broken lines as well. They are inked either in black or in brown colour. It is better to use a drawing pen than a writing pen while drawing contour lines. As far as possible, draw contour lines using the French curves. Conventionally, every fifth contour is made thicker than the rest. The elevation of contours is written in a uniform manner either on the higher side or in a gap left in the line. When the contour lines are too long the elevation are written at two or three places along the contour. On small scale maps, it will suffice to figure every fifth contour. A map with only a few contour lines will be flat (and often low lying). If a map has lots of contours it is a mountainous or hilly area. The actual pattern of the lines can be used to interpret more details about the area.

Contour Interval and Horizontal Equivalent

The constant vertical distance between two consecutive contours is called the contour interval. The horizontal distance between any two adjacent contours is termed as the horizontal equivalent. The contour interval is constant between the consecutive contours while the horizontal equivalent is variable and depends upon the slope of the ground.

The contour interval is selected on the basis of the following factors.

The nature of the ground: In flat and uniformly sloping country, the contour interval is small, but in broken and mountainous region, the contour interval should be large otherwise the contours will come too close to each other.

The purpose and extent of the survey: Contours interval is small if the area surveyed is small and the maps are required to be used for the design work or for determining the quantities of earth work *etc.* Wider interval shall have to be kept for large areas and comparatively less important works.

The scale of the map: The contour interval should be in the inverse ratio to the scale *i.e.* the smaller the scale, the greater the contour interval.

Time and expense of field and office work: The smaller the interval, the greater is the amount of field -work and plotting-work (office work).

The guidelines to decide the values of the contour interval for various purposes are as follows.

- For large scale maps of flat country, for building sites for detailed design work and for calculation of quantities of earth work: 0.2 to 0.5 m.
- For reservoirs and town planning schemes: 0.5 to 2 m.
- For location surveys: 2 to 3 m.
- For small scale maps of broken country and general topographical work: 3 m, 5 m, 10 m or 25 m.

Characteristic of Contours

- All points in a contour line have the same elevation.
- Flat ground is indicated where the contour are widely separated and the steep ground where they run close together.
- A uniform slope is indicated when the contour lines are uniformly spaced and a plane surface when they are straight, parallel and equally spaced.
- A series of closed contour lines on the map represent a hill, if the higher values are inside (Figure 9.7 left).
- A series of closed contours on the map indicate a depression, if the higher values are outside (Figure 9.7 right).
- Contour lines cross ridge or valley lines at right angles. If the higher values are inside the bend or loop in the contour, it indicates a "Ridge". (Figure 9.8 left).
- And if the higher values are outside the bend, it represents a "Valley" (Figure 9.8 right).
- Contour lines cannot end anywhere but close on themselves either within or outside the limits of the map.

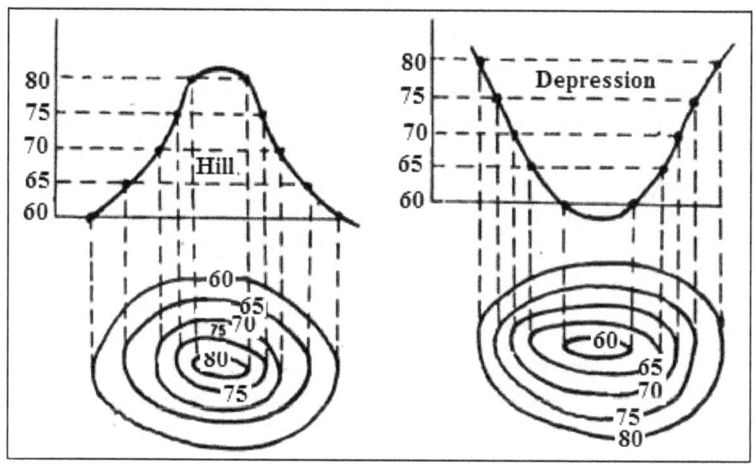

Figure 9.7: The Contours as Appear on the Map: Hill (Left) and Depression (Right).

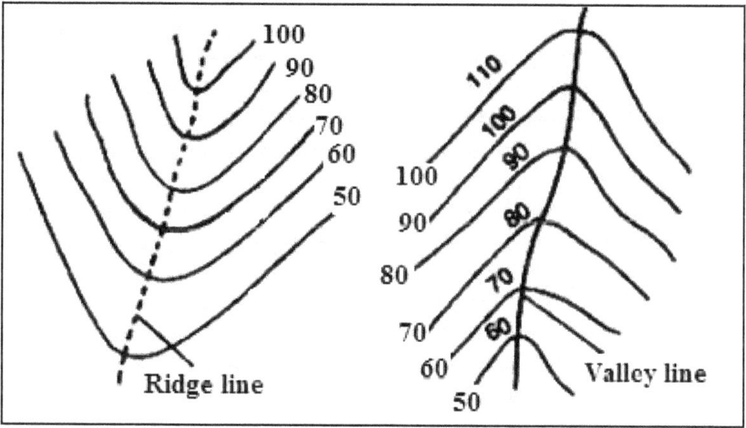

Figure 9.8: The Contours as Appear on the Map: Ridge Line (Left) and Valley Line (Right).

- ☆ Contour lines cannot merge or cross one another on map except in the case of an overhanging cliff. (Figure 9.9 left).
- ☆ Contours never run into one another except in the case of a vertical cliff (Figure 9.9 right). In this case, several contours coincide and the horizontal equivalent becomes zero.
- ☆ A depression between summits is called a saddle. It is represented by four sets of contours as shown in Figure 9.10. It represents a dip in a ridge or the junction of two ridges. In the case of a mountain range, it takes the form of a pass. Line passing through the saddles and summits gives watershed line.

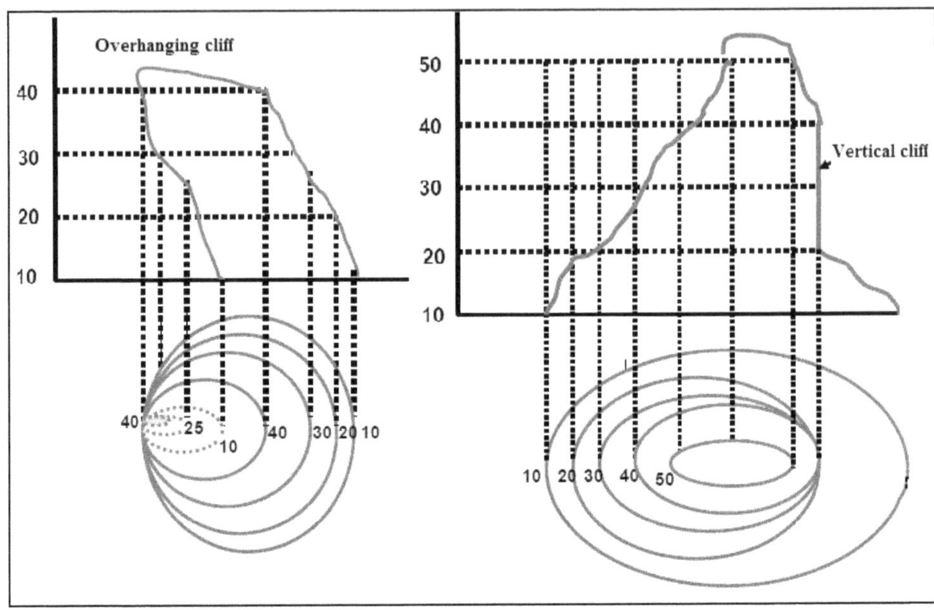

Figure 9.9: The Contours as Appear on the Map: Overhanging Cliff (Left) and Vertical Cliff (Right).

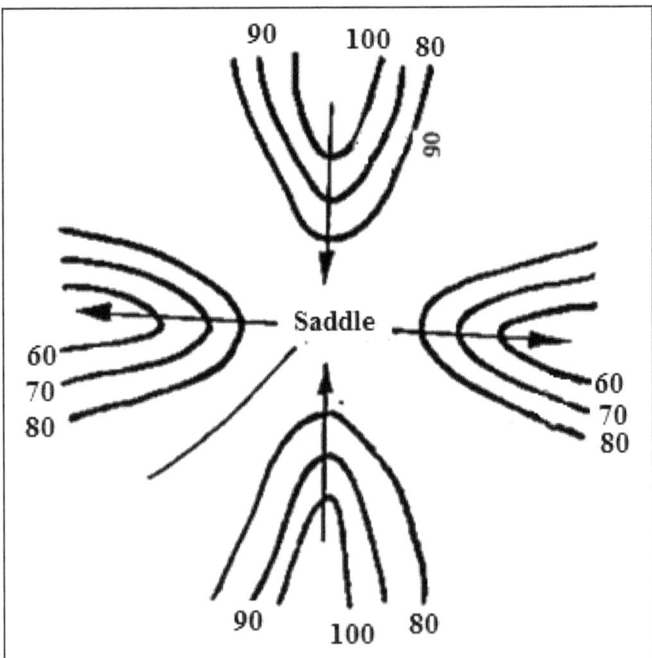

Figure 9.10: The Contours as Appear on the Map for a Saddle.

Purpose of Contouring

Contour survey is carried out at the beginning of any engineering project such as a road, a railway, a canal, a dam, a building *etc.* In agricultural engineering, it is required to design structures such as farm ponds and contour bunds or operations like land levelling and similar other activities. A contour map helps to:

- ☆ Get information about the ground whether it is flat, undulating or mountainous.
- ☆ To locate the physical features of the ground such as a pond depression, hill, steep or small slopes.
- ☆ Select the most economical or suitable sites for various structures.
- ☆ Decide alignment of a canal so that it follows a ridge line.
- ☆ To mark the alignment of roads and railways so that the quantity of earthwork both in cutting and filling should be minimum.
- ☆ To find the capacity of a reservoir and volume of earthwork especially in a mountainous region (See section on volume calculations).
- ☆ To trace out the given grade of a particular route.

Contour Gradient for Alignment of Roads, Railways and Canals

A contour gradient is defined as a line joining the points on different contours along the same gradient. Figure 9.11 show a contour map on which the contour lines are at 2 m intervals. The ground is sloping in an upward direction from A to B. Suppose one is interested in tracing the path of a road with a ruling gradient of 1 in 30. Let us begin at point A on the 40 m contour line. Since the contour intervals is 2 m and the gradient 1 in 30, the horizontal distance between successive points

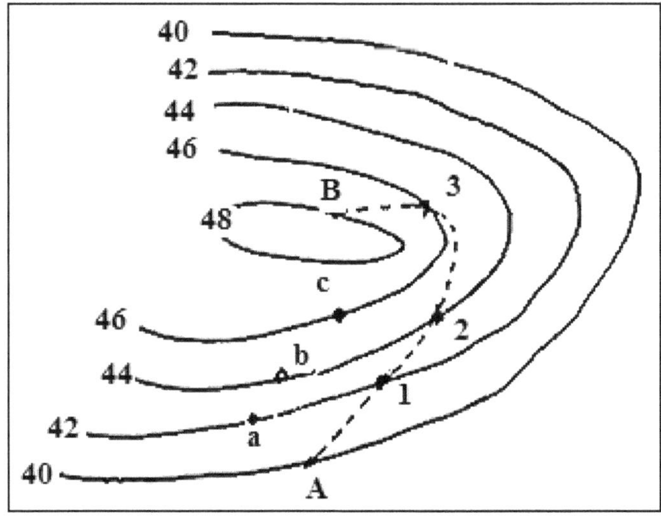

Figure 9.11: Tracing the Path with a known Gradient.

on consecutive contours is 60 m (2 × 30). With A as center and radius equal to 60 m, draw an arc cutting the 42 m contour at 1. With 1 as center and the same radius, draw an arc intersecting the 44 m contour at 2 and so on for successive contours. Join these points which lie on the desired gradient. It may be noted that each of the arcs described will intersect the next contour at two points *viz.* 1 and a, 2 and b, 3 and c *etc.* on 42, 44, 46 m contour *etc.* The points following the desired route such as 1, 2, 3, *etc.* should be joined.

Drawing the Contours

Two methods are used to locate the contours namely the direct method, and the indirect method.

Direct Method

In this method, the contours are directly traced out in the field by locating a number of points on each contour. These points are then surveyed and plotted on plan and the contours drawn through them. Although the method is most accurate but very slow, tedious and time consuming since lot of time is wasted in searching points of the same elevations. Nonetheless direct method have few positive features and limitations.

- ✰ The method is most accurate but is very slow and tedious.
- ✰ It is used for small areas where great accuracy is desired.
- ✰ It is not very useful when the area is hilly.
- ✰ The calculation work of reducing the levels is comparatively more since the number of points in command from one set -up of the level is very less.

A temporary BM is established near the area to be surveyed with reference to a permanent BM through fly levelling. The level is then set up in such a position so that the maximum number of points can be commanded from the station. The height of instrument is determined by taking a back sight on the BM and adding it to the RL of the bench mark. The staff readings required to fix points on the various contours are calculated from the height of instrument. For example, if the height of instrument is 62.58 m, then the staff readings required to locate the 62, 61 and 60 m contours are 0.58, 1.58 and 2.58 m respectively. The staff is held on an approximate position and then moved up or down the slope until the desired reading is obtained. The point is marked with a peg. Similarly various other points are marked on each contour. The line joining all these points gives the required contour. It is easier to locate one contour at a time. Having fixed the contours within the range of the instrument, the level is shifted and set up in a new position. The new height of instrument and the required staff readings are then calculated in a similar manner and the process repeated till all the contours are located. The position of the contour points are located suitably either simultaneously with levelling or afterwards. A theodolite or a compass or a plane table traversing is usually adopted for locating these points. The points are then plotted on the plan and the contours drawn by joining the corresponding points.

Direct Method by Radial Lines

This method is suitable for small areas where a single point at the center can command the whole area. Radial lines are laid out from the common center by theodolite or compass and their positions are fixed up by horizontal angles and bearings. Temporary bench marks are first established at the center and near the ends of the radial lines. The contour points are then located and marked on these lines as explained in the previous approach. Their positions are determined by measuring their distances along the radial lines. They are then plotted on the plan and the contours drawn by joining all the corresponding points (Figure 9.12).

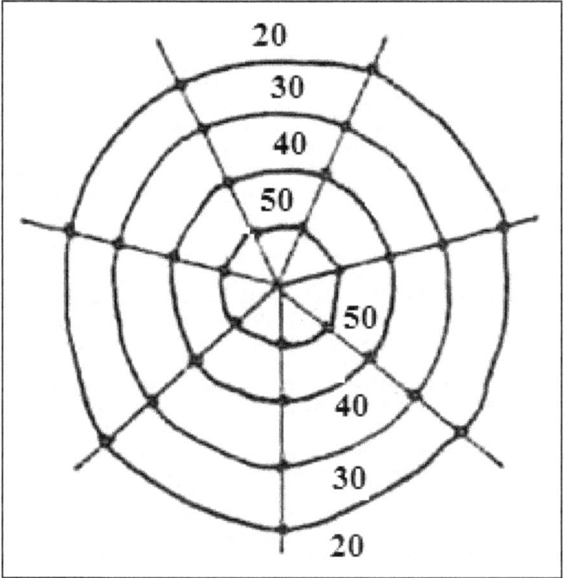

Figure 9.12: Method of Drawing Contours by Radial Lines.

Indirect Method: Contouring by Spot Levels

In this method, the points surveyed are not necessarily on the contour lines. The spot levels are taken along the series of lines laid out over the area. Besides, spot levels of the several representative points representing hills, depression, ridge and valley lines, and the changes in the slope all over the area to be contoured are also observed. Their positions are then plotted on the plan and the contours drawn by interpolation. This method is commonly employed being cheaper, quicker and less tedious as compared with the direct method. It can be done by any of the three methods.

Grid Method

Mark square grid on the land to be surveyed. The size of the grid depends on the extent of survey. Generally for small works a 1m × 1m grid is selected while for large works a larger grid size is used. It may vary from 5 m to 30 m depending upon the nature of the ground and the contour interval. Levels are taken at all the corners of the square and if necessary at the intersection of the diagonal or at important points within the squares which may be located by measurements from the corners. Levels on the intersection of diagonals can be used for verification of the interpolation. Contour map is plotted in the office by interpolating points of equal elevation based on the levels taken on the corners of the square as in Figure 9.13.

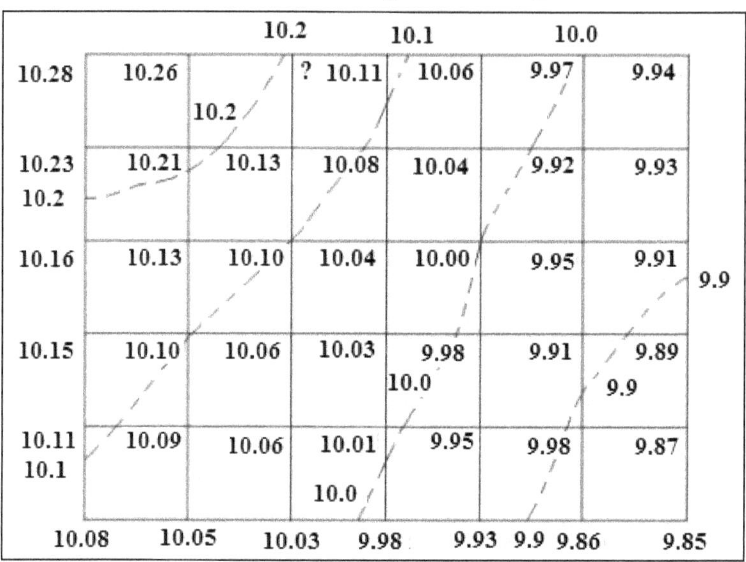

Figure 9.13: Grid Method of Contouring.

Cross-section Method

This method is most suitable for survey of long narrow strips such as a road, railway and canal *etc.* Cross -section are run transverse lo the center line of the work and representative points are marked along the lines of cross-section. The cross-section lines need not necessarily be at right angles to the center line of the work. The spacing of the cross -sections depends upon the topography of the area and the nature of the survey, the common value is 20 to 30 m in hilly and 100 m in flat terrain. The levels of the points along the section line are plotted on the plan and the contours are then interpolated as usual (Figure 9.14).

Tacheometric Method

Tacheometer is transit theodolite having a diaphragm fitted with two stadia wires, one above and other below the central wire. The tacheometer is used for both the vertical as well as for the horizontal measurements. The horizontal distance between the instrument and the staff -station is determined by multiplying the difference of the staff readings of the upper and lower stadia wires with the stadia constant of the instrument, which is usually 100. This method is most suited to hilly areas as the number of stations which can be commanded by a tacheometer is far more than those by a level. A number of radial lines are laid out at a known angular interval and representative points are marked by pegs along these radial lines. Their elevations and distances are then calculated and plotted on the plan.

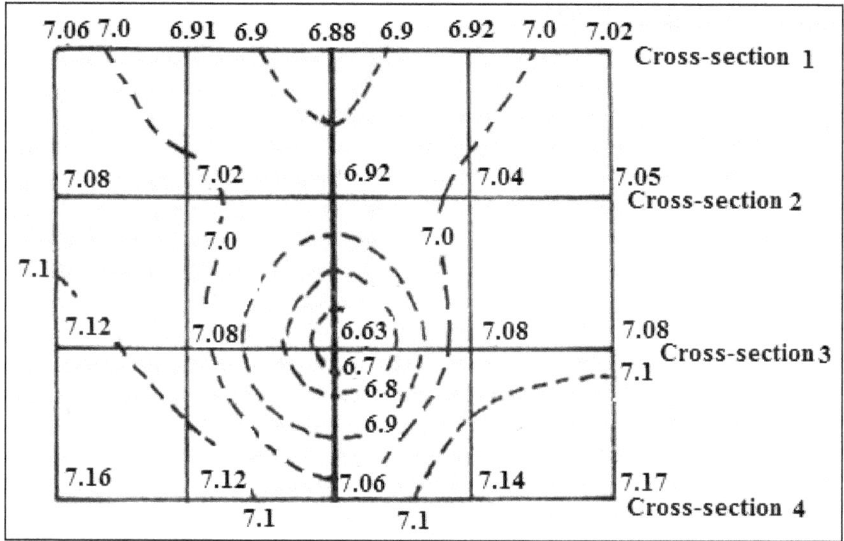

Figure 9.14: Cross-section Method of Contouring.

The contour lines are then drawn through interpolation. A comparative statement of direct and indirect methods of contouring is given as follows:

Direct Method	*Indirect Method*
Most accurate but slow and tedious	Not so accurate but rapid and less tedious
Expensive	Cheaper
Not suitable for hilly area	Suitable for hilly area
Calculations are done during the field work	Calculations are not required in the field
Calculations cannot be checked after contouring	Calculation can be checked as and when required

Interpolation of Contours

The process of spacing the contours proportionally between the plotted ground points is termed as interpolation of contours. It is an essential part of indirect contouring as only the spot levels are available in this method. The intermediate contours are also interpolated in direct contouring if the interval is large. During interpolation of contours, the ground between any two points is assumed to be uniformly sloping. There are three methods of interpolation.

- ☆ By estimation
- ☆ By arithmetical calculation
- ☆ Graphical method

By Estimation

The positions of the contour-points between ground points are estimated roughly, and the contours are then drawn through these points. This is a rough method and is suitable for small scale maps.

Arithmetical Interpolation

Although accurate, it is very time consuming method and is used for small areas where accurate results are required.

Let us take the case of two points A and B as shown in Figure 9.15. Let us also take that the two points are 30 m apart and the reduced levels A and B are 22.32 m and 24.90 m respectively. If the contour interval as 1 m, contours of 23 and 24 m will pass through the two points. Thus, these points need to be interpolated. The level difference between A and B is 2.58 m. The level difference between A and 23 and A and 24 m contours is 0.68 m and 1.68 m respectively. Therefore the horizontal distance between A and 23 m contour is given as (0.68/2.58) × 30 and that between A and 24 m contour as (1.68/2.58) × 30. These distances are then plotted to scale on the map (Figure 9.15).

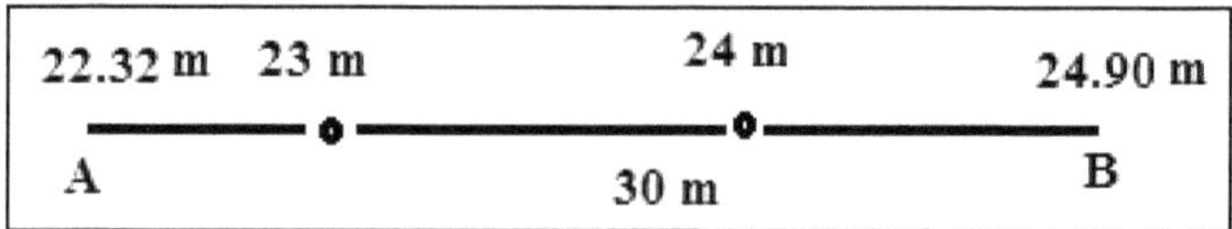

Figure 9.15: Arithmetic Interpolation of Contours between Two Spot Levels.

Graphical Method

Graphical method of interpolation is simpler as compared to arithmetical method. The results obtained are quite accurate. One of the methods is described as follows.

In this method, for the known contour interval, say 2 m, a number of parallel lines spaced at 0.2 m are drawn on a tracing cloth. Usually, the lines are drawn at one tenth distance of the contour interval. Every tenth line is made thick. Now to interpolate contours between two points A and B having elevations of 51.2 m and 53.6 m respectively, place the tracing cloth so that the point A is on the 6th line from the bottom. It is assumed that lines on the tracing cloth begin from 50 m (Bottom line, 50 × 0.2 × 6 = 51.2). Now, move the tracing cloth until B is on the 8th line above the 52 m thick line. The intersections of the 52 m thick line on line AB give the position of the point on the 52 m contour. Prick a pin on the plan. Note that only one contour of 52 m will pass through between points A and B, which has now been marked. Similar procedure is used to complete the whole exercise.

Chapter 10

Introduction to Theodolite

Introduction

Theodolite is used to measure the horizontal and vertical angles. It is more precise than a magnetic compass. While a prismatic compass measures the angle up to an accuracy of 30′, a vernier theodolite can measure angles up to an accuracy of 10″ to 20″ depending upon the kind of theodolite. In this chapter, discussion centers around the details of vernier type theodolite, being the most popular and widely used theodolite.

Classification

The theodolites are mainly classified into three broad categories (Bhavikatti, 2010).

- ☆ Vernier theodolite
- ☆ Digital theodolite
- ☆ Total Station

Vernier theodolite is also known as transit theodolite as it can be rotated in a vertical plane. In this type of instrument, observations are taken by using the principle of vernier caliper.

Digital theodolite provides the value of observation directly in viewing panel. The precision of this type of instrument varies in the order of 1″ to 10″.

Total station is an electronic instrument. In this instrument, all the parameters required to be observed during surveying can be obtained. The value of observation gets displayed in a viewing panel. The precision of this type of instrument varies in the order of 0.1″ to 10″.

Each type of theodolite is peculiar in its construction and mode of operation. However, inherent fundamentals of all the theodolites are similar

Uses of Theodolite

A theodolite is used for the following purposes.

- ☆ Measure horizontal angle
- ☆ Measure vertical angle
- ☆ Measure deflection angle
- ☆ Measure magnetic bearing
- ☆ Measure the horizontal distance between two points
- ☆ Find vertical height of an object
- ☆ Finding difference of elevation between various points
- ☆ Ranging a line

Vernier Theodolite

Salient parts of a typical vernier theodolite are: levelling head, shifting head, lower plate, upper plate, plate levels, standard (or a frame), vernier frame, telescope, vertical circle, altitude bubble and various screws. Some of these and some additional parts are shown in Figure 10.1.

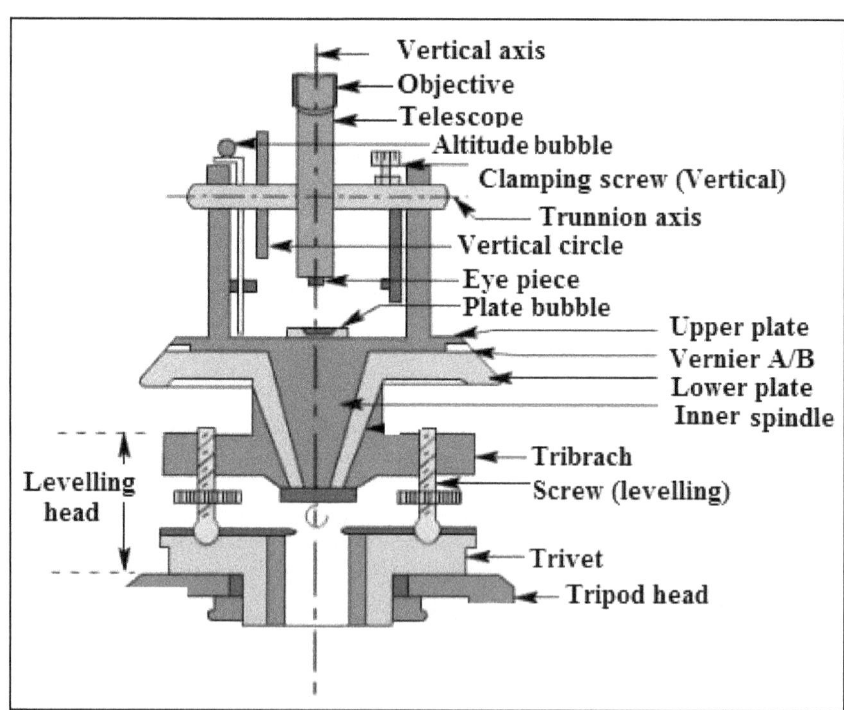

Figure 10.1: Schematic View of a Theodolite Showing its Various Parts.

Commonly Used Terms

Some of the commonly used terms are defined as follows.

Axis of Telescope: The axis is an imaginary line passing through the optical center of the object glass and optical center of eyepiece.

Axis of the Bubble Tube: It is an imaginary line tangential to longitudinal curve of bubble tube at its middle point.

Centering: The setting of theodolite exactly over a station marked by means of plumb bob is known as centering.

Changing Face: The operation of bringing the vertical circle from one side of the observer to the other is known as changing face. It is achieved by revolving the telescope by 180° in vertical plane about horizontal axis. Also revolving the telescope in horizontal plane about the vertical axis.

Double Sighting: It is the process of measurement of a horizontal angle or a vertical angle twice; once with the telescope in the normal condition and once with the telescope in the inverted condition. Double sighting is also called double centering.

Face Left: It means that the vertical circle of theodolite is on the left of the observer at the time of taking reading. The observation taken is known as face left observation.

Face Right: When the vertical scale of a theodolite is on the right of the observer, the position is called face right and observation taken as face right observation.

The two observation when averaged help to eliminate the collimation error.

Horizontal Axis: It is the axis of rotation of the telescope in the vertical plane.

Line of Collimation: It is an imaginary line passing through the optical center of the objective glass and its continuation.

Permanent Adjustment: When the desired relationship between fundamental lines is disturbed, then some procedures are adopted to establish this relationship. This adjustment is known as permanent adjustment.

Swinging the Telescope: It means to turn the telescope in a horizontal plane. It is 'right swing' when the telescope is turned clockwise and 'left swing' when the telescope is turned anticlockwise.

Telescope Inverted: The telescope is said to be inverted when its vertical circle is to the right of the observer and the bubble is downward.

Telescope Normal: The telescope is said to be normal or direct when its vertical circle is to the left of the observer and the bubble is upward.

Temporary Adjustment: The setting of the theodolite over a station at the time of taking any observation is called temporary adjustment.

Transiting: The method of turning the telescope about its horizontal axis in a vertical plane through 180° is termed as transiting. In other words, transiting results in a change of face.

Vertical Axis: It is the axis of rotation of the telescope in the horizontal plane.

Temporary Adjustments

At each station point, before taking any observation, it is necessary to carry out some operations. The set of these operations in sequence is done to make the instrument ready to take observation. These are known as temporary adjustment and comprises of the following operations.

1. Setting up the theodolite over a station
2. Levelling up
3. Elimination of parallax.

Setting Up

The setting operation consists of fixing the theodolite with the tripod stand along with approximate levelling and centering over the station. For setting up the instrument, the tripod is placed over the station. First, the approximate centering of the instrument is done by moving the tripod legs radially or circumferentially as per need so that the center of the tripod head lies above the station point and its head is nearly level as seen by the eye. The instrument is then fixed with the tripod by screwing through trivet. The height of the instrument should be such that observer can see through telescope conveniently. After this, a plumb bob is suspended from the bottom of the instrument to ensure that plumb bob lands on or very near to the station mark.

It may be noted that due to radial movement of the legs, plumb bob gets shifted in the direction of the movement of the leg without seriously affecting the level of the instrument. On the other hand, when the legs are moved sideways or circumferentially, the plumb does not shift much but the level gets affected. Sometimes, the instrument and the tripod have to be moved bodily for centering. It must be noted that the centering and levelling of instrument is done recursively. Finally, exact centering is done by using the shifting head of the instrument. During this, first the screw-clamping ring of the shifting head is loosened and the upper plate of the shifting head is slid over the lower one until the plumb bob is exactly over the station mark. After the exact centering, the screw clamping ring is tightened. The whole operation involved in placing the vertical axis of the instrument exactly over the station mark is known as centering.

Levelling Up

Having centered and approximately leveled the instrument, accurate levelling is done with the help of foot screws with reference to the plate levels, so that the vertical axis shall be truly vertical. To level the instrument the following operations are undertaken.

1. Turn the upper plate until the longitudinal axis of the plate level is roughly parallel to a line joining any two of the levelling screws.
2. Hold these two levelling screws between the thumb and first finger of each hand uniformly so that the thumb moves either towards each other or away from each other until the bubble comes to the center.
3. Turn the upper plate through 90° (until the axis of the level passes over the position of the third levelling screw).
4. Turn this levelling screw until the bubble comes to the center.
5. Rotate the upper plate through 90° to its original position and repeat step (2) till the bubble comes to the center.
6. Turn back again through 90° and repeat step 4.
7. Repeat the steps 2 and 4 till the bubble is central in both the positions.
8. Now rotate the instrument through 180°. The bubble should remain in the center of its run, provided it is in correct adjustment. The vertical axis will then be truly vertical.

Elimination of Parallax

Parallax is a condition arising when the image formed by the objective is not in the plane of the cross hairs. Unless parallax is eliminated, accurate sighting is not possible. Parallax can be eliminated in two steps.

Focusing the Eye-Piece

Point the telescope to the sky or hold a piece of white paper in front of the telescope. Move the eyepiece in and out until a distant and sharp black image of the cross-hairs is seen.

Focusing the Object

Telescope is now turned towards object to be sighted and the focusing screw is turned until image appears clear and sharp.

Measurement of Horizontal Angles

Set up: Level the instrument over the angle vertex

Back Sight: Aim the instrument at the point that marks the left-hand side of the angle and lock the motion

Zeroing: Set the reading at zero

Foresight: Free the motion and aim at the point that marks the right-hand side of the angle

Reading: Read the angle

Measurement of Vertical Angles

Set up: Level the instrument at the angle vertex

Zenith sight: Aim the instrument vertically toward the zenith and lock the motion

Zeroing: Set the reading at zero

Lower Sight: Aim the instrument toward the lower sight and take the reading

Calculation: Calculate the elevation or depression angle from the zenith angle

Common Errors

The errors in surveying may be attributed to instrumental errors, errors due to natural causes and due to personal mistakes (Figure 10.2).

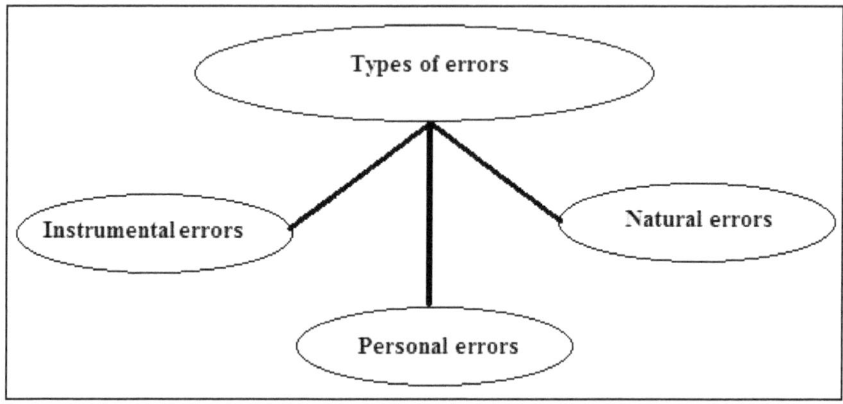

Figure 10.2: Common Types of Errors in Surveying.

Instrumental Errors

- ☆ Non adjustment of plate bubble
- ☆ Line of collimation not being perpendicular to horizontal axis
- ☆ Horizontal axis not being perpendicular to vertical axis
- ☆ Line of collimation not being parallel to axis of telescope
- ☆ Eccentricity of inner and outer axis
- ☆ Graduation not being uniform
- ☆ Vernier being eccentric

Personal Errors

- ☆ Inaccurate centering
- ☆ Inaccurate levelling
- ☆ Slip of the instrument if not firmly screwed
- ☆ Working wrong tangent screw
- ☆ Parallax not properly eliminated
- ☆ Inaccurate bisection of the point sighted
- ☆ Non-vertically of the ranging rod

- ☆ Other errors such as mistake in setting the verniers, mistake in reading the scales and verniers, mistake in reading wrong verniers and mistake while booking the readings

Natural Errors

- ☆ High temperature causes error due to irregular refraction
- ☆ High winds cause vibration in the instrument, and this may lead to wrong readings on verniers
- ☆ The sun shining on the instrument.

Section II

On-Farm Development

Chapter 11

On-Farm Land Levelling

Land Levelling Defined

Many fields appear smooth but may be too uneven to uniformly apply irrigation water. For a flat country, it may be almost impossible to say whether a land requires land levelling or not. Some of the observations made during irrigation of a field may reflect on the need of land levelling.

- ☆ Water does not run down the border strip or a furrow at a uniform speed
- ☆ Water is deeper on one side of a border strip than on the other
- ☆ Water ponds up at places
- ☆ Water breaks over furrows
- ☆ Some parts dry out before or stay wet longer than the rest of the field
- ☆ Water moves in small streams along with soil down slope

As a consequence of one or more of the above mentioned causes, one ends up in an uneven crop growth, reduced crop yields, and loss of water. A properly graded land surface ensures unobstructed smooth flow of water into the land, without eroding the soil and ensuring uniform distribution of water throughout the field. Land levelling has been defined as a process by which the surface relief of a field is modified to a desired grade and to certain specifications that provide a more suitable surface for efficient application of irrigation water and for improved drainage conditions. Land levelling is one of the most intensive practices that is applied to agricultural lands and often requires moving a lot of soil from one point to another.

Benefits of Precision Land Levelling

Effective land levelling whether through conventional or laser land levelling results in the following benefits over non-levelling:

- Graded/level and smooth soil surface resulting in more uniform distribution of water
- Reduced time of application to irrigate the field
- More uniform moisture environment for crops resulting in more uniform germination, growth and maturation of crops
- Reduced seeds, fertilizers, chemicals and fuel used in cultural operations
- Improved field trafficability
- Enables to use larger fields resulting in increase in the farming area and improved operational efficiency. Increasing field sizes from 0.1 ha to 0.5 ha increases the farming area by 5 per cent to 7 per cent. Besides it reduces operating time by 10 per cent to 15 per cent.
- Amount of water required for transplanting rice can be reduced
- Reduction of weeds by up to 40 per cent resulting in nearly 75 per cent decrease in the labour required for weeding
- Higher crop yields up to 10-20 per cent
- Helps to apply 'Precision Farming' techniques
- Improves nutrient use efficiencies
- Improved field drainage
- Uniform leaching of salt affected soils
- Reduced production of green house gases (GHG) as a result of reduced electric and fossil fuel consumption.

When Not to Level the Fields

Although land levelling is essentially required in an intensive agricultural system, yet there are few conditions under which it is advisable not to undertake land levelling.

Exposure of Unproductive Sub-soils

The productivity of land upon levelling may be considerably reduced if the fertile top soil layer is quite thin (soils are not deep enough) or if a significant proportion of the ground surface needs to be cut that may expose unproductive sub-soil. Therefore, a land levelling operation should never be planned without knowing the soil-profile conditions and the maximum cut that can be made without seriously affecting the productivity. In case land levelling is essentially required, intense fertilizer application should be undertaken to improve the productivity of sub-soil as a compensatory measure. The other option that can be adopted is to scrap the fertile soil and place it in one corner of the fields. After the completion of land levelling operation, it can be spread on the surface. An adequate depth of topsoil must cover the entire field after the land levelling and smoothening. In both cases, one needs to work out the economics of levelling.

Excessive Earth Moving

When the topography is highly undulating and irregular, then the volume of earth work may become too large turning the land levelling an economically non-viable proposition. As a thumb rule, earth moving exceeding 1500 m^3/ha may be uneconomical. Under these conditions, one may consider switchover from surface to some pressurized irrigation system such as sprinkle and trickle irrigation. As an alternative, it may be prudent to earmark such areas for other purposes such as pastures.

Steps in Land Grading Design

The total design of grading of fields in a farm land generally involves desk work as well as field work. It may be mentioned that desk work is involved both before and after undertaking field survey. The various steps in the overall land grading exercise are summarized as follows:

Pre-survey Work (Desk Work)

This included the following three activities.

1. Selection of irrigation method
2. Demarcation of field lay-out
3. Selecting slopes appropriate for farming

Prior to levelling design, the land development program must be planned so that the location of the filed boundaries, irrigation water supply system, drains and the farm roads are known. The levelling plans for individual fields must provide for places from where soil can be cut and used to meet the additional requirement (borrow pits). Similarly, places may be earmarked where excess earth if any, can be disposed.

An initial decision as to the method of surface irrigation to be employed will dictate field slope. Basins are designed to be level in both the longitudinal and transverse field directions. Borders should have zero cross-slope (may be 0.3 per cent if border width is less than 10 m) and may have advance slopes of up to 2 or 3 per cent, depending upon the crop and the soil conditions. As per general guidelines, the desirable forward grade is 1.0 per cent. Slopes exceeding 0.5 per cent should be used only where soil depths will not permit cuts to reduce the slope. Above this slope, the maximum non-erosive stream is very small, field efficiencies are low, labor requirement is more, and erosion hazards from natural rainfall are high. Furrow irrigation systems work well with advance slopes up to 1 to 3 per cent and cross-slopes of 0.5 to 1.5 per cent. When the cross slope exceeds 0.5 per cent then only the furrow irrigation method may be used. Safe limits of longitudinal and cross slopes for different soil types are given in Table 11.1.

Irrigation grades should generally be as flat as possible but still provide good surface drainage. Land levelling with slopes higher than 0.1 per cent is recommended to minimize problems of surface drainage in arable fields. It has been observed that the surface drainage efficiency increases with the slope up to 0.3 per cent. If the average natural slopes are greater than the prescribed ranges in this section, terraces or benches should be planned. To avoid excessive cuts to eliminate cross slopes, the fields should be leveled in strips at different elevations, separated by low ridges. This practice of grading is known as bench levelling. Earth work is done along the width of benches. The amount of earthwork is governed by the magnitude of the diagonal slope at right angles to the direction of irrigation.

Table 11.1: Recommended Longitudinal and Cross Slopes for Efficient Irrigation

Type of Soil	*Longitudinal Slope (per cent)*
Heavy (clay) soils	0.05 to 0.20
Medium (loamy) soils	0.20 to 0.40
Light (sandy) soils	0.25 to 0.65

Field Work (Survey Work)

1. Land clearing
2. Staking the plots for survey
3. Studying the soil profile

Besides, it also involves post-survey desk work, which includes:

1. Topographic map preparation
2. Design levelling method

Land Clearing

Prior to making the land grading survey, it is important to remove heavy vegetative growth from the land. Land clearing consists of removing of some or all the trees, bushes, vegetations, trash and boulders from the area specified for land grading.

Field Survey

With the field boundaries known or established, the most important step is to survey the area. The general practice is to establish a grid system over the field and set stakes at the grid points. The usual grid space is 25 m in each direction. Grid spacing such as 30 × 30 m, 20 × 20 m, and 15 × 15 m are also used, depending on the nature of the surface relief and the precision required in levelling. Each grid point is at the center of the grid square represents nearly equal area. For convenience in identification, the row lines are lettered and the column lines numbered (Figure 11.1). After all the points have been staked, level are run to determine the ground elevation at each stake using a dumpy level and a level rod or by adopting the usual survey procedures (See Engineering Survey Section of this Book). The first rod reading is made on a bench mark (BM) of known or assumed elevation. Any permanent point on or near the farm may be taken as the bench mark. In case of surveys involving large areas, it is desirable to establish the elevation of the bench mark with reference to mean sea level (MSL) by running a line of differential levels from a point of known elevation, like a nearby railway platform or other points.

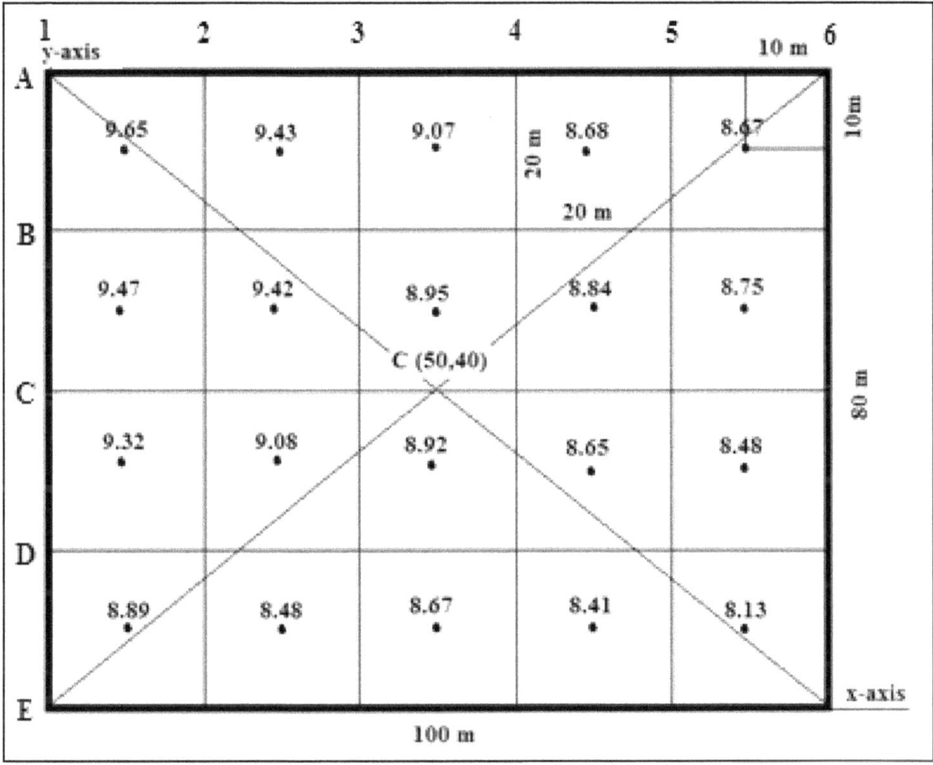

Figure 11.1: Elevation Map of a Rectangular Field.

To make the survey information more readily understood, contour lines are drawn at suitable intervals. The contour intervals are usually based on the average land slope. The recommended contour intervals for different ranges of land slopes are as follows:

Land Slope Range (Per cent)	*Contour Interval (cm)*
0 – 1	6 – 15
1 – 2	15 – 30
2 – 5	30 – 60
5 – 10	60 – 150

The contour lines should clearly indicate the direction and degree of slope, ridges, depressions and other topographical features. With the help of contour map the land is divided into fields that can be graded and irrigated individually to the best advantage. A topographical map also provides a basis for planning the irrigation system.

Soil Profile Conditions

Soil profile should be thoroughly studied before undertaking the levelling work. The soil profile map should depict soil texture, physical properties including infiltration rate, irrigation properties and the soil bulk density of the surface and the subsurface soils. Soil profile map should also provide the details of hard pans, rock or other material that might limit the depth of cut.

Phases of Land Levelling Operations

Land levelling operations may be grouped into three phases:

- ✰ Rough grading
- ✰ Land levelling
- ✰ Land smoothing

Rough grading is the removal of abrupt irregularities such as mounds, dunes and rings, and filling of pits, depressions and gullies. Land levelling also known as land grading or land forming or land shaping requires moving large quantities of earth over considerable distance. Levelling operations leaves irregular surfaces due to dumping of the soil. These irregularities are removed and a plane surface obtained by land smoothing which is the final operation in land levelling. Often these three phases are accomplished together but all of these may not be required in a particular setting. In some setting only one operation may suffice depending upon the requirement. If rough grading and/or land smoothing can accomplish the job then why to go for land levelling, being much more expensive.

Land Levelling

The factors that influence the design of a land levelling operations are:

- ✰ Depth and characteristics of the soil which may control the depth of cut
- ✰ Land slopes, which influence grades
- ✰ Direction of irrigation
- ✰ Method of irrigation
- ✰ Surface irregularities, which influence the division of a field into smaller fields or benches
- ✰ Water supply, which influences length or runs, grades, width of border strips and sometimes method of irrigation.
- ✰ Crops to be grown, which may determine the method of irrigation
- ✰ Economic considerations
- ✰ Farmer preferences
- ✰ Equipment available

Design Methods

Four basic methods of land levelling design are:

1. The plane method
2. The profile method

3. The plan inspection method
4. The contour adjustment method

While the first one is known as analytical method, the remaining three are designated as manual methods because these are hit and trial methods. A number of variations of each method are also in common use. Each has its advantages and disadvantages but when used intelligently, all provide satisfactory results. The merits of an individual method are often based on the experience of the engineer or designer. An experienced person might be able to decide effectively the formation parameters by using a topographic map by eye-judgment in a few trials.

Plane Method

Plane method is the most widely used and very useful method for developing a good quality levelling job. It is so designated because the resulting land surface has a uniform field slope and a uniform cross slope resulting in a true plane surface. Following are the design steps for the plane method.

- Sub-divide the entire field into sub- fields depending upon the topography so that each sub-division can be economically developed into a plane surface
- Determine the area of each sub-field
- Determine elevation at all the grid points in the field, either from available contour map of the field or by conducting fresh survey (Figure 11.1).
- Determine average elevation of the field by dividing the sum of the elevations of all the grid points by the number of points.
- Determine the centroid of the area. Since the field is a rectangle, the centroid is located at the points where the two diagonals of the rectangle meet. Assign the average elevation to this point. A plane passing through this point with this elevation will result in a cut-fill ratio of 1.0 or nearly this value.
- Determine the desired cut-fill ratio. Lower the plane by some amount to achieve the desired cut/fill ratio
- Fit the plane using least square approach or find out the elevations through manual calculations
- Compute the earth work
- Undertake the land levelling operation

Some of the relevant issues are discussed in the following sections utilizing the information given in Figure 11.1 or otherwise.

Determination of Centroid

Centroid of an object is a point where its center of gravity is located. The centroid of a rectangular field is located at the point of intersection of its diagonals. The centroid of a triangular field is located at the intersection of the lines drawn from its corners to the mid-points of the opposite sides. For irregular fields centroid is determined by geometric decomposition. Geometric decomposition comprises of decomposing the figure into simple geometric figures for which centroid can be easily identified (Table11.2). It is a widely used method because the computations are simple, and requires only basic mathematical principles. For any figure, to obtain the centroid C_i, the area of each figure A_i is determined. All holes that extend outside the compound shape are treated as negative values. Following equation is then use to compute the centroid.

$$C_x = \Sigma C_{ix} A_{ix} / \Sigma A_{ix} \tag{11.1}$$

$$C_y = \Sigma C_{iy} A_{iy} / \Sigma A_{iy} \tag{11.2}$$

Here C is the distance and A the area. The subscript i refers to segment of the figure 1, 2, 3 *etc.* and x and y relate to the x-axis and y-axis. For example C_{ix} refers to the distance of the centroid of ith segment from the x-axis. C_x and C_y are the distances of centroid of the whole area from x-axis and y-axis respectively.

Step-by-step Procedure

1. Divide the given compound shape into various primary figures *i.e.* rectangles, circles, semi-circles, triangles *etc.* Agricultural fields may mostly be covered by these shapes. In dividing the compound figure, include parts with holes. These holes are to be treated as solid components with negative values. Make sure to break down every part of the compound shape. However, in case some small portion is left out, one can continue with the process and finally apply the fitted plane to the whole area including the left out area.
2. Determine the area of each figure. Table 11.2 shows the area for different basic shapes and location of their centroid. After determining the area, give name to each such as area one, area two, area three, *etc.* Make the area negative for designated areas that are not to be included and act as holes.

Table 11.2: Area Formulae and Location of Centroid for Common Shapes

Shape	*Area*	*Distance from x-axis*	*Distance from y-axis*	*Pictorial Presentation*
Rectangle	bh	d/2	b/2	
Triangle	bh/2	h/3	Depends on the kind of triangle equi-lateral, isosceles, scalene and acute-angled, obtuse-angled, right-angled triangle	

Shape	*Area*	*Distance from x-axis*	*Distance from y-axis*	*Pictorial Presentation*
Right triangle	bh/2	h/3	b/3	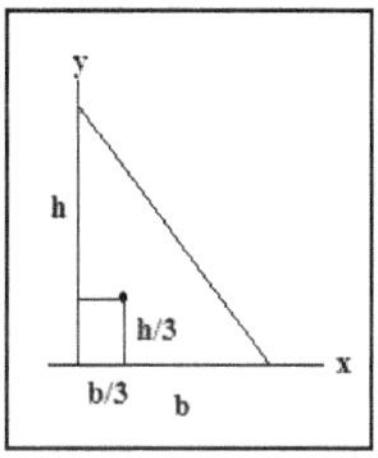 y h h/3 x b/3 b
Semi-circle	$\pi r^2/2$	$4r/(3\pi)$	0	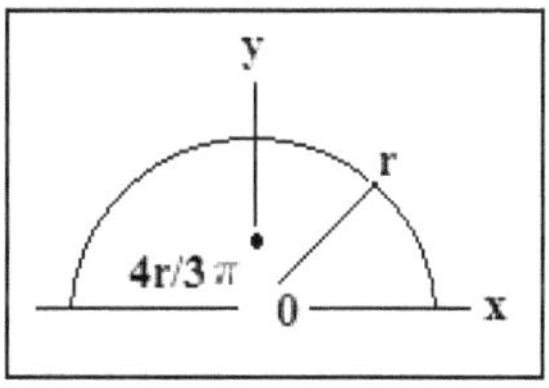 y r 4r/3π 0 x
Quarter circle	$\pi r^2/4$	$4r/(3\pi)$	$4r/(3\pi)$	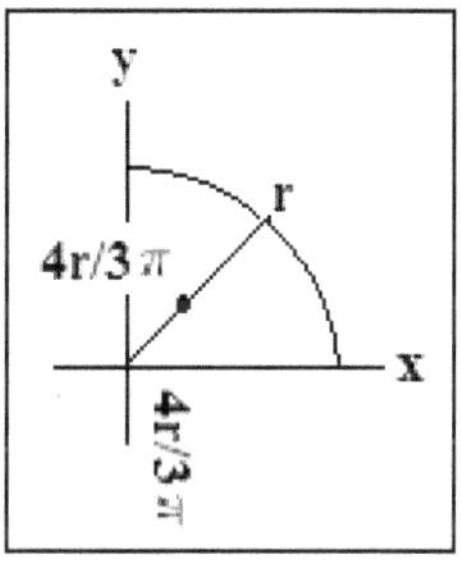 y r 4r/3π x 4r/3π
Semi-circular arc	πr	$2r/\pi$	0	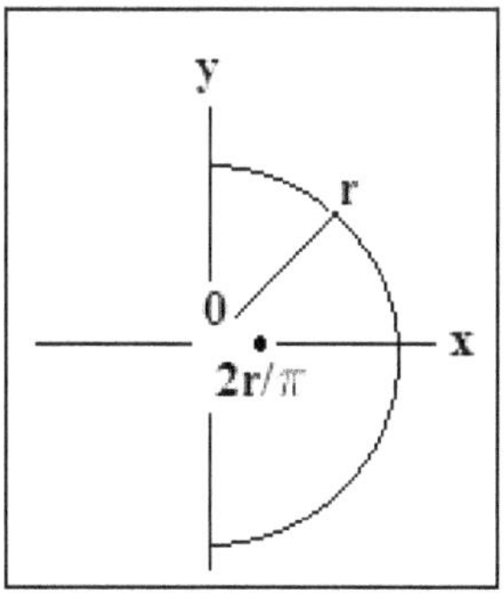 y r 0 x 2r/π

3. Identify x-axis and y-axis for the given figure. If x and y-axis are missing, draw the axis at the most convenient places. The axis can be positioned in the middle, left, or right.
4. Get the distance of the centroid of each divided primary figure from the x-axis and y-axis. Find out the moments of the area around x-axis and y-axis and use Eq. (11.1) and (11.2) to determine the centroid of the whole figure.

Example 11.1

Calculate the centroid of the area shown in Figure 11.2.

Method 1

Divide the compound shape into basic shapes. In this case, it is a rectangle and right triangle (Negative). Name the two divisions as Area 1 and Area 2.

Solve for the area of each division. The dimensions are 380 × 340 for the rectangle, 100 × 90 for the right triangle. Make sure to negate the values for the right triangle.

Area A1and A2 are calculated as follows.

A1 = 380 × 240 = 91200 m^2

A2 = 1/2 x100 x90 = 4500 m^2

ΣA = 91200 – 4500 = 86700 m^2 (Note A2 is negative)

Distance of the area A1 from the x-axis = 120 m

Distance of the area A1 from the y-axis = 190 m

Distance of the area A2 from the x-axis = 210 m

Distance of the area A2 from the y-axis = 346.7 m

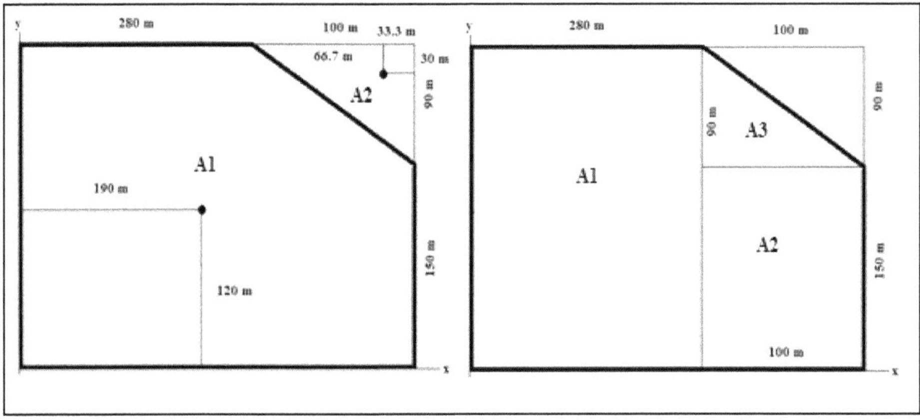

Figure 11.2: Determination of Centroid of a Irregular Figure.

Taking moment against y axis, we have

91200 x190 – 4500 x346.7 = 17328000 – 1560150 = 15767850

Distance of the centroid from y-axis = 15767850/86700 = 181.9 m

Taking moment against × axis, we have

91200 x120 – 4500 x210 = 10944000 – 945000 = 9999000

Distance of the centroid from x-axis = 9999000/86700 = 115.3 m

Method 2

Let us now divide the irregular area into regular figures each having the positive areas without hole.

Area 1 = 280 × 240 = 67200 m^2

Area 2 = 100 × 150 = 15000 m^2

Area 3 = 1/2 x100 x90 = 4500 m^2

ΣA = 67200 +15000+4500 = 86700 m^2

Taking moment against y axis, we have

67200 x140 + 15000 × 330 + 4500 x313.3 = 9408000 + 4950000 + 1409850 = 15767850

Distance of the centroid from y-axis = 15767850/86700 = 181.9 m

Taking moment against x axis, we have

67200 x120 + 15000 × 75 + 4500 x180 = 8064000 + 1125000 + 810000 = 9999000

Distance of the centroid from x-axis = 9999000/86700 = 115.3 m

Example 11.2

The grid elevations of a rectangular field are given in Table below. Find out the centroid of the rectangular field and compute its elevation for cut fill ratio equal to 1.0.

Stations	*Elevation of Stations at Lines*					*Total*
	1	*2*	*3*	*4*	*5*	
A	8.56	8.34	8.02	7.84	7.76	40.52
B	7.37	7.24	7.98	7.68	7.57	37.84
C	8.22	8.04	7.94	7.56	7.48	39.24
D	7.92	7.84	7.76	7.31	7.02	37.85
Total	32.07	31.46	31.70	30.39	29.83	155.45

The centroid can be located with sufficient accuracy by assuming that each stake in the field represents the same area. Since in this case grid is uniform, numbers of station are only used in the calculations. Following Table is prepared.

Line	*Distance from x-axis*	*No. of Stations*	*Col 2 × Col 3*	*Line*	*Distance from y-axis*	*No. of Stations*	*Col 6 × Col 7*
1	2	3	4	5	6	7	8
A	10	5	50	1	10	4	40
B	30	5	150	2	30	4	120
C	50	5	250	3	50	4	200
D	70	5	350	4	70	4	280
		20	800	5	90	4	360
			800/20 = 40			20	1000
							1000/20 = 50

The centroid will be 50 m away from the x-axis and 40 m away from the y-axis.

Sum of the elevations of 20 stations = 155.45

Average elevation = 155.45/20 = 7.77 m

This elevation if assigned to the centroid will result in a cut-fill ratio of 1.0.

Cut-fill Ratio

Experience has shown that the cut-fill ratio in any land levelling operation should be greater than one *i.e.* volume of cut should be more than fill. The principal reasons for a cut-fill ratio to be more than 1.0 are the compaction from equipment in the cut area which reduces the volume and also the compaction in the fill area which increases the fill volume. Another reason for the need to have additional fill material results from imperfections in the levelling operation. An argument usually put forth is that on level ground between stakes, the operator has an optical illusion of a dip in the middle, and therefore in filling, crowning often occurs, which requires additional fill material than what is estimated. It is rarely possible to estimate exactly the cut-fill ratio. It usually varies from 1.2 to 1.6. In extreme cases of heavy or light textured soils and deep or shallow excavation, the ratio may be as low as 1.1 and as high as 2.0 (USDA, 1983). For peat or muck soils, the ratio must be about 2.0 (Table 11.3).

Table 11.3: Cut-Fill Ratios for Various Soil Types

Soil Type	*Cut-Fill Ratio*
Fine textured (clayey)	1.3:1 to 1.4:1
Medium textured (clay-loam)	1.2:1 to 1.3:1
Coarse textured (sandy)	1.1:1 to 1.2:1
Organic soils	1.7:1 to 2.0:1

A settlement correction for the whole field is more convenient to apply with the plane method of computing cuts and fills. The settlement allowance or amount of lowering of elevation of the centroid may range from 0.3 to 1.0 cm for compact soils and 1.5 to 4.5 cm for loose soils. It may be pointed out that a small change in elevation will cause a considerable change in the cut-fill ratio.

Besides, what is stated in the above paragraphs, in instances where borrow is to be taken from the field from construction works such as roads, irrigation channels and filling of pits, the plane surface can be lowered by the amount of earthwork required. Similarly, if the earth excavated from drainage channels or other sources is to be utilized in making land levelling fills, the elevation of the plane surface should be raised by the quantity of earth available.

Fitting the Plane

To work out the design of the plane, two approaches are followed. The first one is to decide the slope to which the land is to be graded and make the subsequent calculations manually. In the other approach, the slope of the land is obtained based upon least square method such that for the given conditions the amount of excavation is minimum.

First Approach

The elevation of the centroid is determined as explained in the previous sections. The elevation of the nearest stake is determined from the known elevation of the centroid and values and directions of the slopes as explained in example 11.3.

Second Approach

The plane method is a simple least squares or linear regression fit of field elevations to a two-dimensional plane. Subsequent adjustments are made in the elevation of the centroid of the plane to compensate for cut-fill ratio. If the field has a basic X-Y orientation, the plane equation is written as:

$$H(X, Y) = AX + BY + C \tag{11.3}$$

Here E is the elevation of the X, Y coordinate; A, B are the regression coefficients; and C is the elevation of the origin or reference point for the calculations of field topography. The first step in evaluating the constants, A, B and C, is to determine the weighted average elevations of each grid point in the field. The purpose of the weighing is to adjust for any boundary stakes that represent larger or smaller areas than given by the standard grid dimension. The weighing factor is defined as the ratio of actual area represented by a grid point to the standard area. The grid point area is assumed to be the proportional area surrounding the stake. For a rectangular shape having the equal weighing factor of 1.0 for each stake, the slope of any line in the X or Y direction which best fits the natural ground surface, is determined by the least squares method. The slope of the best fit line through the average X-direction elevation (Hj) is A, which is found by:

$$A = \frac{\Sigma\, DjHj - \frac{!\Sigma Dj\ !\Sigma Hj\ !}{n}}{\Sigma\, Dj - \frac{!\Sigma Dj}{n}} \text{ (Summation over j = 1,m)} \tag{11.4}$$

For the best fit slope in the Y-direction, the slope B, is given as:

$$B = \frac{\Sigma(DiHi) - \frac{\{(\Sigma Di)(\Sigma Hi)\}}{n}}{\Sigma(Di^2) - \frac{\{(\Sigma Di)^2\}}{n}} \text{ (Summation over i = 1,n)} \tag{11.5}$$

Here A and B are slopes in the X and Y directions respectively, dimensionless, Dj and Di are the distances from the reference lines in the X and Y directions, m, Hj and Hi are the average elevation of the grid point at Dj and Di distances respectively, m and n are the total numbers of grid points. The slopes so determined must be checked to see if they are within satisfactory limits. For example, if the irrigation system is a border irrigation system, the cross-slope should be nearly zero. Finally, the average field elevation corresponds to the elevation of the field centroid (X, Y) is used to estimate the value of C as follows:

$$C = Hc \pm AX \pm BY \qquad (11.6)$$

Here Hc is the elevation of the centroid. The signs would depend upon the location of the origin with respect to centroid and direction of slopes. The plane so determined and passing through the centroid having the average elevation of all the grid points will give equal amounts of cut and fill. In order to have cut more than the fill, the centroid is lowered from the original level and the grid elevations are calculated to have the desired cut-fill ratio. It is a trial and error procedure.

Volumes of Cut and Fill

The elevations of other points are determined as per designed grades from any of the two approaches discussed in the previous section. The cut or fill at all grid points is computed by comparing the existing and proposed formation levels. The cuts and fills may be marked to the nearest cm since construction tolerances do not warrant a high degree of precision. When the proposed elevation is less than the existing, the difference indicates the depth of cut, otherwise the depth of fill in m. Usually, the cuts and fills are shown by the -ve and +ve signs. When cuts and fills at each stake are summed up, the cut-fill ratio is given by the sum of cuts divide by sum of fills. The volume of cut and fill for each square grid can be determined separately by the "Four Point" method.

$$Vc = \frac{L^2(\Sigma C)^2}{4(\Sigma C + \Sigma F)} \qquad (11.7)$$

$$Vf = \frac{L^2(\Sigma F)^2}{4(\Sigma C + \Sigma F)} \qquad (11.8)$$

Here Vc is the volume of cut, m^3, Vf is the volume of fill, m^3, L is the grid spacing and ΣC and ΣF are the cuts and fills on the four corners of the grid square, m.

The steps described in this section are repeated for each grid and volume summed to determine the total earth work in cutting in cubic meters. The value so arrived is multiplied by the prevailing rate of earth work per cum to determine the cost of earth work. By dividing the total cost by the total area of the field, one can arrive at the cost per hectare.

Example 11.3

The four-corner method is applied to calculate cut and fill volumes for a grid with an area 400 m^2. It has the following levelling requirements in meters: C 0.073, F 0.032, F 0.145 and C 0.083. Work out the volume of the cut and fill.

F = 0.032 +0.145 = 0.177 m

C = 0.073 +0.083 = 0.156

Using eq. (11.7), volume of cut, Vc is given as

$Vc = (400/4)\,(0.177)^2/(0.156+0.177) = (100 \times 0.0313)/0.333 = 3.13/0.333 = 9.40\ m^3$

Using eq. (11.8), volume of fill, Vf is given as

$Vf = (400/4)\,(0.156)^2/(0.156+0.177) = (100 \times 0.0243)/0.333 = 2.43/0.333 = 7.30\ m^3$

Example 11.4

For the data given in Figure 11.1 reproduced as Figure 11.3, calculate the centroid, the elevation of the centroid so that cut fill ratio equal to 1.0. Proceed to fit the plane of best fit to the data set. Calculate the cuts and fills and comment upon the calculations.

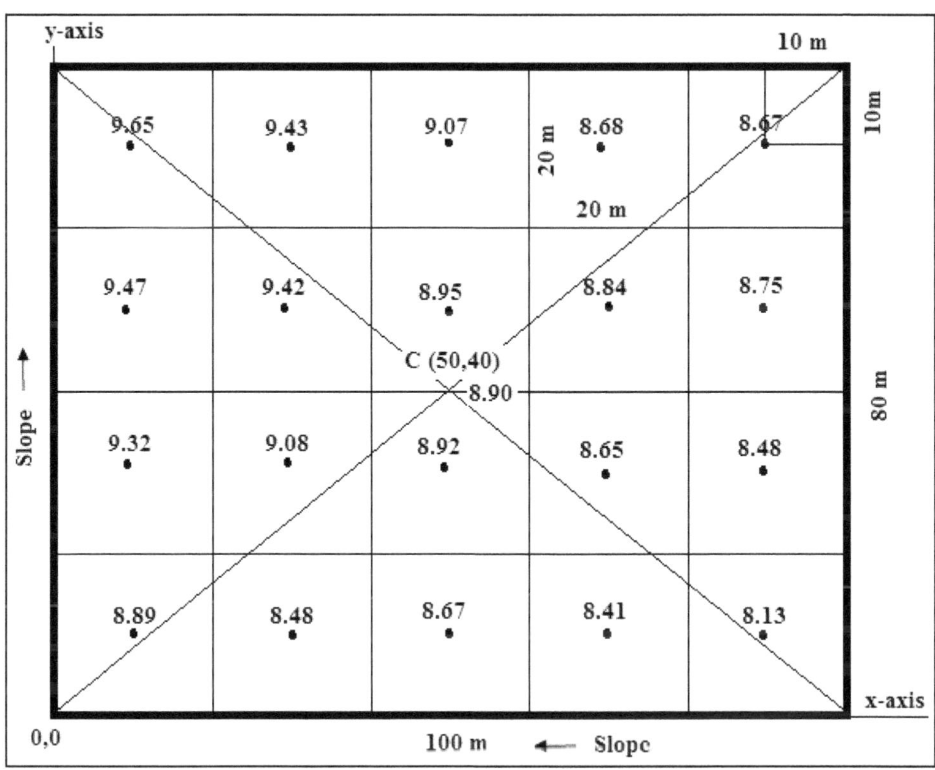

Figure 11.3: The Centroid and the Directions of Slope as Assessed through eq. (11.4) and (11.5)

For the data presented in Figure 11.1, we have

Total elevations of 20 stations = 177.96

Average elevation = 177.96/20 = 8.898 = 8.90 (slightly raised)

Since the area is a rectangle, its centroid can be determined by drawing the two diagonals (Figure 11.1). The point where these intersect is the centroid. The coordinates of the centroid are C (50,40) as shown in Figure (11.3). The elevation of this point if set to 8.90, will yield the cut-fill ratio as 1.0 or slightly less than 1.0 because the centroid level has been raised.

To use eq. (11.4) and (11.5), we prepare a table and make the following calculations.

Dj	*Hj (j=1,m)*	*Average Hj*	*Di*	*Hi (i=1,n)*	*Average Hi*
10	37.33	9.33	10	42.58	8.52
30	36.41	9.10	30	44.45	8.89
50	35.61	8.90	50	45.43	9.09
70	34.58	8.64	70	45.50	9.10
90	34.03	8.51			Hi =35.60
Dj =250		Hj = 44.48	Di =160		

$(Dj)^2 = 10^2 + 30^2 + 50^2 + 70^2 + 90^2 = 16500$

$(1/5)(Dj)^2 = 1/5\ (250 \text{x} 250) = 12500$

$Hj = 9.33 + 9.10 + 8.90 + 8.64 + 8.51 = 44.48$

$(1/5)\ Dj\ \ Hj = 1/5\ (250 \text{x} 44.48) = 2224$

$(Dj.Hj) = 10 \text{x} 9.33 + 30 \text{x} 9.10 + 50 \text{x} 8.90 + 70 \text{x} 8.64 + 90 \text{x} 8.51 = 93.3 + 273.0 + 445.0 + 604.8 + 765.9 = 2182$

$A = (2182 - 2224)/(16500 - 12500) = -0.011$

It means slope decrease as the x increases in the x-direction @ 1.1 per cent (Figure 11.3)

$(Di)^2 = 10^2 + 30^2 + 50^2 + 70^2 = 8400$

$(1/4)(Di)^2 = 1/4\ (160 \text{x} 160) = 6400$

$Hi = 35.60$

$(1/4)\ Di\ \ Hi = 1/4\ (160 \text{x} 35.60) = 1424.2$

$(Di.Hi) = 10 \text{x} 8.52 + 30 \text{x} 8.89 + 50 \text{x} 9.09 + 70 \text{x} 9.10 = 85.2 + 266.7 + 454.5 + 637.0 = 1443.4$

$B = (1443.4 - 1424.2)/(8400 - 6400) = 0.0096 = 0.96$ per cent

The value of C is given from eq. (11.6) as:

$8.90 + 1.1(50/100) - 0.96\ (40/100) = C$

$C = 8.90 + 0.55 - 0.384 = 9.07$

Thus, incorporating all the values in the equation of the plane, we have

$H(X, Y) = -0.011X + 0.0096Y + 9.07$

Check by calculating the value of centroid

$H(50,40) = -0.011 \text{x} 50 + 0.0096 \text{x} 40 + 9.07 = 8.90$

To determine the cut or the fill at each point, we use equation of the plane to find out the elevations at various stakes as given in the following Table (bold values). Cut and fills are calculated by subtraction. The cuts are marked with negative while fills are marked with positive signs. Summing the total cuts and fills reveal that

The total cuts = 0.98

The Total fills = 1.16

The fills are more than the cuts (cut-fill ratio is less than 1.0, which is undesirable). It has happened in this case because while assigning the value to the centroid, it was raised by 0.002 m or 2 mm *i.e.* from 8.898 to 8.90 m. Moreover, cuts and fills are also reported to nearest cm values. To have a higher cut-fill ratio, the centroid needs to be lowered than 8.90 m.

9.65 **9.63** -0.02	9.43 **9.41** -0.02	9.07 **9.19** +0.12	8.68 **8.97** +0.29	8.67 **8.75** +0.08
9.47 **9.44** -0.03	9.42 **9.22** -0.20	8.95 **9.00** +0.05	8.84 **8.78** -0.06	8.75 **8.56** -0.19
9.32 **9.25** -0.07	9.08 **9.03** -0.05	8.92 **8.81** -0.11	8.65 **8.59** -0.06	8.48 **8.37** -0.11
8.89 **9.06** +0.17	8.48 **8.84** +0.40	8.67 **8.62** -0.05	8.41 **8.40** -0.01	8.13 **8.18** +0.05

The Profile Method

The method is so named because the designer works with profiles on grid lines rather than with elevations on a map (Figure 11.4). Therefore, in this method, using the grid point elevation, ground profiles are plotted along the grid lines. The method is especially useful in levelling design for relatively flat lands or land with undulating topography. Essentially it consists of a trial and error method of adjusting grades on plotted profiles until the irrigation criteria are met and an earth work balance is attained. It is easier to select grades on a profile that will provide balanced cut and fill with a reasonably short haul. This method is widely used. The profiles are commonly plotted in one direction. The datum lines are kept at proposed profiles to calculate cuts and fills until the profiles indicate a satisfactory cut- fill ratio. When the slopes are to be given in both direction two way profiles are useful. The following stepwise procedure is used.

- ✰ Staking the field and running levels on a pre-decided grid. Any existing physical features between grid lines such as ridges, ditches, fence lines, *etc.*, are also taken.
- ✰ Plot the grid map and the map is designated as before levelling map. It is at this point that the engineer must study the map thoroughly and use his experience to find out areas that require different methods or direction of irrigation depending upon soils, topography, water supply, field drainage, crops grown, method of irrigation to be used and the farmer's preferences. Determine the best direction of irrigation.
- ✰ Establish maximum and minimum grade line slope limits. The cut and fills are balanced, cut being more than the fill. Besides showing final elevations,

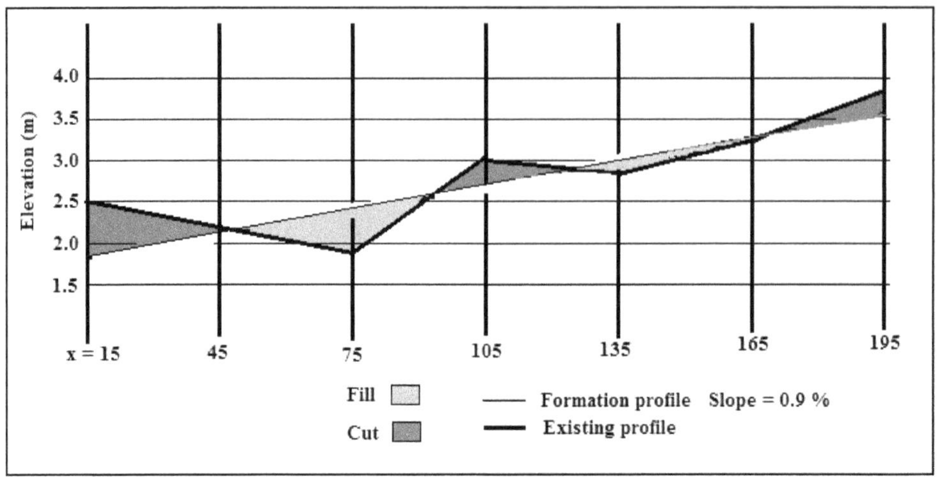

Figure 11.4: Levelling Design Using Profile Method.

the after levelling map should show lateral ditch layout, necessary structures, number and direction of borders and the field drainage.

☆ After the design is completed, estimates are made of the earthwork and its cost.

The Plan Inspection Method

The grid point elevations are noted on the plan, and the design grade elevations are determined by inspection after the careful study of the topography. It is largely a trial and error procedure. In selecting the formation level the designer must simultaneously consider the down field slope, cross slope, earth work balance and haul distance. The desired cut-fill ratio and volumes of earthwork are estimated from the summation of cuts and fills. The assumed slopes are altered until a satisfactory balance of the cut and fill volumes are obtained and the down field and cross slopes are within safe limits.

The Contour Adjustment Method

The contour adjustment method of land levelling consists of trial and error adjustments of the contour lines on a plan map. The method is especially adapted to the smoothening of steep lands that have to be irrigated. A contour map is drawn and the proposed ground surface is shown on the same map by drawing new contour lines (Figure 11.5). The uniformity of downfield slope is controlled by the uniformity of the horizontal spacing between contours, and the cross slopes can be examined by scaling the distance between contours at right angles to the direction of irrigation. The cuts and fills are estimated at the grid points by interpolation between contour lines and by taking the difference in elevation between the original and the new surface or the areas of cut and fill between the existing and formation contours.

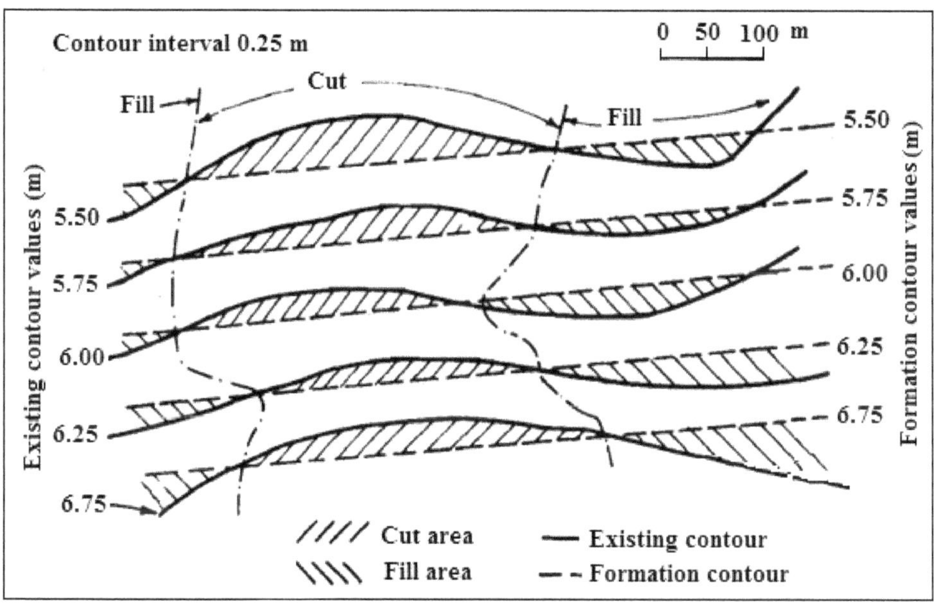

Figure 11.5: Levelling Design Using Contour Adjustment Method.

Shifting the Plan to the Field

To execute the designed map on to the field, the plan is marked on the field site.

1. Determine the required number of cut and fill stakes as per the number of cut and fill points on the plan.
2. Drive the cut and fill stakes into the ground at desired cut and fill points.
3. Mark two lines say one blue and the other red on each stake before driving in. The blue line indicates formation level and the red ground level at the stake point. Note that the blue line will be above the red line at "fill" points and below it at "cut" points. Use two different colour flags for differentiating cut and fill stakes for the convenience of the operator.
4. Care should be exercised not to disturb the grid stakes while performing operations.
5. While performing operations, see that the areas of cut are as close as possible to fill areas in order to keep hauling distance to a minimum as discussed in the next section.

Haulage of Earth

Besides the cut-fill consideration, another important issue is the magnitude of haulage of the excavated earth (expressed in say, ton-meters). The haulage depends on the location of the cut-area from which the earth is moved, and the location of the fill-area; and the route that is followed for this operation. The assignment of the cut and fill areas and the routing to connect the two areas should be optimized. The empty rehauling trips should be minimized. Linear programming based computer programmes are used in large scale land levelling works.

Equipments for Land Levelling

Rough Levelling

Before the actual land levelling work is initiated and if the land is too uneven then it may be worthwhile to include rough levelling as one of the pre-requisites. It is important to remove heavy vegetative growth from the land. Land clearing consists of removing of some or all the trees, bushes, vegetations, trash and boulders from the area specified for land grading. Rough grading is the removal of abrupt irregularities such as mounds, dunes and rings, and filling of pits, depressions and gullies. A bulldozer is commonly used for this purpose. A crawler tractor or pneumatic tyre based tractor equipped with dozer blades, ripper, brush rakes and U-blade is used. A crawler tractor gives superior traction on various soil types and ground conditions encountered during field operations. Brush rakes are used to clear vegetation and debris from the soil without removing the top soil. Dozer blades are used for cutting and pushing earth to short distances, normally when the haul distances do not exceed 25 m.

Land Levelling

To begin with the actual operation, it is useful to plough the land so that it may be easier to cut and transport the soil. Commonly used methods for land levelling operation are the following:

- ☆ Manual levelling
- ☆ Conventional levelling using animal or mechanical power
- ☆ Laser land levelling

Manual Levelling

An irrigator is able to observe the locations of high and low spots on the field during an irrigation event. Pre-sowing irrigation is commonly used to observe such locations. A farmer is then able to manually move the soil from high spots to low spots. It is not a onetime affair but is spread over a number of years as the fields are slowly leveled. The process is performed using spades or any handheld farm equipment. Clearly such operations are limited to small fields. Because of severe limitations and difficulty in carrying out the operations, such operations are now performed using animal/tractor power.

Animal Drawn or Power Driven Equipment

Conventional levelling is often performed with the help of animal or power driven equipments. The operation is carried out on year to year basis without any major engineering or technical inputs as discussed previously in this chapter. The cost of land levelling using these methods is quite less. The farmers have incorporated this operation within their seeding and planting schedules. As such, the cost of the operation remains hidden. There is wide diversity of equipment used for this purpose. Over the years many improved tools have emerged especially the tractor operated ones. Some of the commonly used animal drawn equipment are wooden float, wooden U leveler, bulk scraper and scoop *etc.* The tractor drawn equipment include scraper, terrace blade, carrier type scraper, automatic type leveler and land plane *etc.* These equipment are either individually owned and operated by the farmer or are hired from the neighbour. It may take about 12 days to level a hectare of land using draft animals, 7 to 8 days for a 2-wheeled tractor and 8 hours for a 4-wheeled tractor with a rear mounted tractor blade (IRRI, 2013).

Laser Land Levelling

Traditional methods of levelling are cumbersome, time consuming, and expensive, so more and more farmers are turning to modern methods of land levelling. Amongst these, the laser land levelling has emerged as one of the most significant advances in on-farm land levelling and surface irrigation technology (Rickman, 2002). All the benefits of land levelling mentioned in an earlier section are upscaled with a laser leveller resulting in additional benefits in terms of higher productivity and water saving than levelling with conventional techniques. Major limitations of this technology are:

- High cost of the equipment/laser instrument
- Need for skilled operator to set/adjust laser settings
- Less suited to irregularly sized and shaped fields
- Limited number of custom hiring operators
- Inappropriately selected land levelling equipment by few service providers

Laser Land Levelling Equipment

A laser land levelling system is shown in Figure 11.6. It has four essential elements.

- The laser emitter or generator
- The laser sensor or receiver

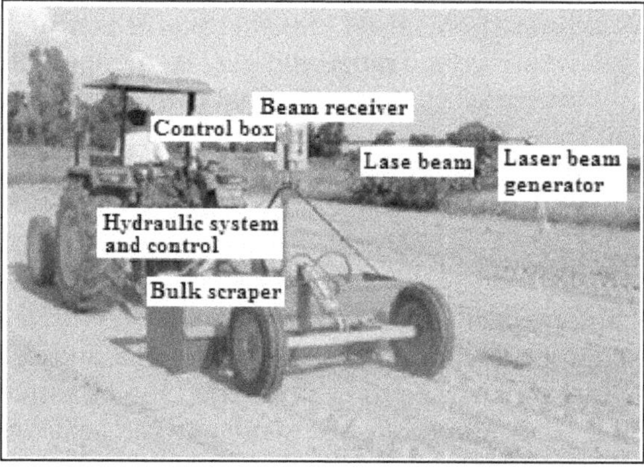

Figure 11.6: A Laser Controlled Land Levelling Equipment.

- ☆ The electronic and hydraulic control system
- ☆ The tractor and grading implement

The Laser Emitter

Figure 11.7: Laser Bean Generator.

The laser emission device (Figure 11.7) is a battery operated laser beam generator. The laser emitter is located on a tripod or a tower-like structure on or near the field and at an elevation such that the laser beam rotates above any obstructions on the field as well as the levelling equipment itself. It rotates at a relatively high speed on an axis normal to the field plane. This rotating beam creates a plane of laser light above the field which acts as the levelling reference in contrast to the elevation survey in conventional land levelling techniques. This reference plane is not affected by the earth movement, does not require a field survey to establish the high and low spots and does not require the operator to judge the magnitude of cuts and fills. Since the distance between the laser beam and the earth surface is defined, the deviations from this distance become the cuts and fills. There is little or no need for the exhaustive engineering calculations as in the conventional approach. Various beam generators are equipped with self-adjustment mechanisms that allow the alignment of the plane of the beam at desired longitudinal or latitudinal slope. Various kinds of generators are:

Manual Levelling Lasers

These lasers require the operator to manually level the unit in both the X-axis and Y-axis by using the units' screws and bubble vials. The operator relies on the bubbles for accuracy. These lasers are capable to achieve a maximum accuracy of 1 cm at 100 m.

Semi Self-Levelling Lasers

These lasers adjust themselves automatically within a range using a compensator. To get the prescribed range, the laser is equipped either with a circular bubble with a bull's eye, or electronic lights that turn green when one reaches the self levelling range. These lasers are quite accurate and have a shut-off feature if the laser is bumped or goes out of the self levelling range. They can achieve accuracy of at least 1 cm at 100 m.

Fully Self-Levelling Lasers

These lasers automatically find and maintain level within a specified range. These lasers are equipped with an electronic level vial and servomotors, which helps to level the instrument electronically. When the instrument is leveled, the laser starts spinning. They are the easiest to use and can achieve accuracy of up to 2.5 mm at 100 m.

Split-beam Lasers

These lasers emit simultaneous horizontal and vertical beams to establish both level and plumb reference lines. On the basis of slope adjusting facility, available beam generators are grouped under the following three categories:

- ✰ No slope adjustment, resulting in a table top field
- ✰ Longitudinal slope adjustment, resulting in desired slope in the direction of irrigation
- ✰ Both longitudinal and transverse slope adjustment

The cost of equipment accordingly varies with first kind being the cheapest while the third kind being the costliest.

The Laser Receiver or Sensor

The beam is received by a light sensor mounted on a mast attached to the land grading implement (Figure 11.8). In practice, the sensor is a series of detectors situated vertically and with up or down movement of the grading implement, the light is detected above or below the central detector. This information through the control system actuates the hydraulic system to raise or lower the implement until the light strikes the central detector. It is in this manner that the sensor on the mast is continually aligned with the plane on the laser beam and thereby references the moving equipment with the beam. The sensitivity of the laser sensor system is at least 10 to 50 times more precise than the visual judgment and manual hydraulic control of an operator on the tractor. Consequently, the land levelling operation is correspondingly more accurate. The skill of the operator is substantially less critical to the levelling.

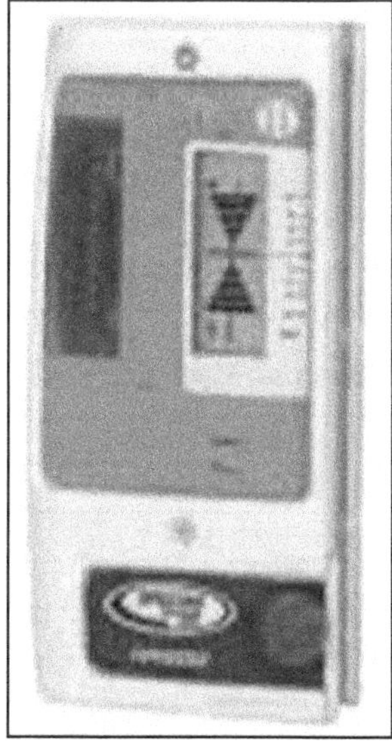

Figure 11.8: A Laser Receiver.

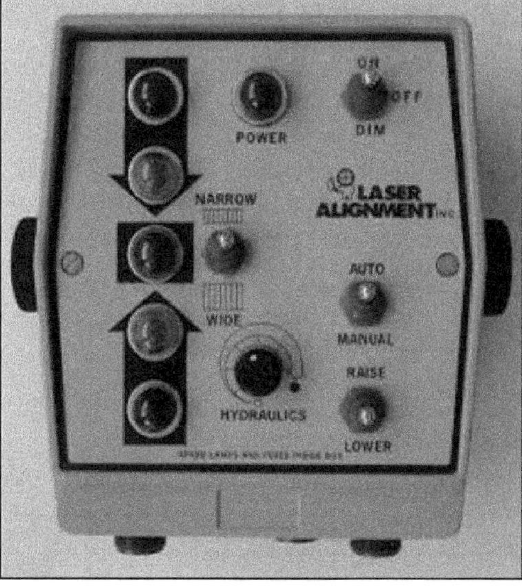

Figure 11.9: Laser Leveler Control Panel.

Control Panel

The control box, mounted on the tractor within easy reach of the operator, accepts and processes the signals transmitted by the receiver. It displays these signals to indicate the leveler position relative to the finished grade. The three control box switches are On/Off, Auto/Manual, and Manual Raise/Lower (It allows the operator to manually raise or lower the drag bucket, Figure 11.9). When the control box is set to automatic, it provides electrical output for driving the hydraulic valve. The operator can adjust the settings on the receiver and can override the receiver when he needs to pick up a bucketful of soil and transport it to another section of the field.

Hydraulic Control System

The hydraulic system of the tractor is used to supply oil to raise and lower the levelling bucket. As the hydraulic pump is a positive displacement pump, it always pumps more oil than required. Thus, a pressure relief valve is provided in the system to return the excess oil to the tractor reservoir. The electronic and hydraulic control systems generally have two operational modes. In the observation mode, the mast itself moves up or down as per undulations encountered in the field as the operator drives the equipment. The monitor in the tractor yields elevation data from which one can determine the average field elevations and slopes. In this mode, system operates as a self-contained surveying system. In the planing mode, the mast position is fixed relative to the implement blade which is then raised or lowered in response to the land topography. The beam plane is located at appropriate distance above the field and at the desired slopes. The height of the mast sensor is adjusted relative to this plane so that cutting and filling is accomplished simply by driving the tractor over the field.

Tractor and Grading Equipment

The tractor is a standard agricultural tractor. Land grader is also standard equipment, may be as simple as a land plane or a bucket type which loads and carries earth. The size of the land plane or the bucket dimensions and its capacity determine the HP of the tractor required (Table 11.4). It is not advisable to use under power tractors for this operation. The hydraulic and control systems must be modified so that the whole system operates with the electronic controller.

Table 11.4: Selection of Tractor and Scraper Size

Tractor Size (Min/Max, HP)	*Scraper Width (Min/Max, m)*
30-50	2.0
50-100	2.0-3.0
100-125	3.0-3.5
125-150	3.5-4.0
>150	4.0-5.5

Min= Minimum, Max=Maximum.

Levelling the Field

During the actual levelling operation, following sequence of steps are used:

- ☆ Set the average elevation value of the field in the control box.
- ☆ The laser-controlled bucket is positioned at a point that represents the mean height of the field, normally around the centre of the field. The cutting blade is set slightly above ground level (1.0-2.0 cm).
- ☆ The tractor is then driven in a circular direction from the high areas to the lower areas in the field.
- ☆ When the whole field has been covered in this circular manner, a final levelling pass is made in long runs from the high end to the lower end of the field.
- ☆ The field is then re-surveyed to make sure that the desired level of precision has been attained. Repeat the process, if necessary.

Time Requirement for Levelling

Approximate time taken for laser levelling in hours per ha is as follows:

- ☆ 15 h and above for undulating land where elevation difference is > 60 cm.
- ☆ 12-15 h for rough scrub land which has never been farmed
- ☆ 5-8 h for land which has been farmed and tilled but has never been laser leveled
- ☆ 2 -4 h for land laser leveled anytime during the past 2-3 years

Land Smoothing and Maintenance Levelling

Levelling operations leaves irregular surfaces due to dumping of earth. Thus, use of large land planes or floats is recommended to remove minor irregularities. The process is the final operation in land levelling and is known as land smoothing. It helps to obtain a smooth surface. Even if a field is perfectly leveled, the levels are disturbed due to tillage and/or irrigation. These activities result in settling of the fill areas while cut areas are fluffed up. If such a thing is anticipated then it may be advisable to delay smoothing until this initial settlement has taken place. Moreover, the levels of surface-irrigated fields undergo gradual changes due to agricultural operations and the action of irrigation water. Therefore, periodical maintenance levelling should be undertaken to sustain high irrigation efficiency and crop yields. The levelers and floats are used to maintain the smoothness of leveled fields. It is now emerging that repeat levelling may be undertaken every 3 to 5 years depending upon the field conditions (Ambast *et al.*, 2014). It may be noted that repeat levelling costs much less than the original levelling.

Precision of Land Levelling Operation

Precision of land levelling is measured using any of the following two indices.

Coefficient of Land Uniformity

An index of the effectiveness with which land levelling operation has been carried out is defined as follows (Tyagi and Singh, 1979):

$$LUC = 1 - [\Sigma \mid (LD - LE) \mid / \Sigma LD] \qquad (11.9)$$

Here LUC is the coefficient of land uniformity, LD is the desired elevation at a grid point, LE is the existing elevation at that grid point. Summation is made over the total number of grid points used in the field. The summation term between the two lines means that all values irrespective of the signs are to be added. The LUC may range from 0.0 (very poor graded land) to 1.0 (A precisely graded land without any deviations from the desired level).

Levelling Index

Agarwal and Goel (1981) used another but similar index to define the precision of land levelling. According to them, levelling index LI, is written as:

$$LI = \frac{\Sigma[(LD - LE)]}{n} \qquad (11.10)$$

All the parameters have the same meaning as for eq. (11.9) and n is the number of grid points. It has the units of cm. The lower limit of LI is zero, which indicates perfect levelling of the field. Increasing values of LI reflect poor quality of land levelling.

Agarwal and Khanna (1983) showed that application and distribution efficiencies were higher when levelling index was less in comparison to when it was high (Table 11.5). There was not much change in storage efficiency although numerically it was higher with higher levelling index.

Table 11.5: Effect of Levelling Index on Application, Storage and Distribution Efficiencies

Levelling Index (cm)	*Application Efficiency (Per cent)*	*Storage Efficiency (Per cent)*	*Distribution Efficiency (Per cent)*
2.0	67.6	68.0	88.7
2.5	63.6	67.5	80.6
3.0	60.0	68.1	77.3
3.7	52.6	72.2	67.9

Studies conducted at other places also documented the benefits of precision of land levelling on irrigation efficiencies (Khepar *et al.*, 1982). In general, it emerged that all the application, storage and distribution are high with low than high LIs. Jat *et al.* (2006) reported that efficiencies were higher in a field having LI of 2.9 cm as compared to a field having LI of 13.0 cm. The results reported by Tyagi (1984) reveal that

- ✰ The application efficiency is as high as 90 per cent at a LI of 0.75 cm and it decreased with increasing LI. Lower application efficiencies were associated with higher application depths.

- For the same depth of water applied, the distribution efficiencies decreased with increasing LI although improved with increasing application depth per irrigation for the same LI.
- The non-uniformity in levelling was reflected in a water productivity value of 93.1 kg/ha-cm at LI of 0.75 cm to 59.1 kg/ha-cm at LI of 6.75 cm. It also emerged that to apply a desired system application depth of 5 cm, required to achieve optimum productivity and income, the levelling quality had to be such that the LI is equal to or less than 3 cm (Table 11.6).

Table 11.6: Effect of LI on Average Depth of Water Applied

Levelling Index	*Average Depth (cm)*
0 1.5	4.2
1.5-3.0	5.1
3.0-4.5	7.0
4.5-6.0	8.1
6.0-7.5	9.5

Field Experiences

As already pointed out, laser levelling provides much higher precision of land levelling than the conventional methods. A large number of studies have been directed to gather evidences of increased yields, water saving and other benefits of laser land levelling over the conventional methods. A compilation of these studies have been made by Ambast *et al.* (2014). These benefits are illustrated in Table 11.7 for rice-wheat cropping and Table 11.8 for various crops. A pictorial view of the performance of wheat crop is given as Figure 11.9 showing the difference in crop performance under conventional and laser land levelling.

Table 11.7: Comparison of Conventional and Laser Guided Levelling

Parameter	*Rice*	*Wheat*
Saving in water (per cent)	18.6	21.1
Increase in yield (per cent)	8.3	11.0
Per cent increase in water productivity (kg/m^3 of applied water)	34.6	37.6
Per cent increase in water productivity (Rs./m^3 of applied water)	4.7	5.9
Increase in income (Rs./ha)	5,040	5,870
Total (Rs./ha)		10,910

Cost of Land Levelling

Practically speaking, the cost of land levelling varies depending upon the kind of soil, initial undulations, field size, type of machinery used and the capability of the person carrying out the operation. Land levelling cost is not proportional to the differences in the initial and final levels as it may increase almost exponentially with increasing precision of land levelling. The cost may vary from a few hundred rupees to several thousands of rupees. Therefore, it may be prudent to determine the cost for individual projects to plan and economize on cost.

Table 11.8: Crops Yield and Water Saving with Laser Land Over Conventional Levelling

Crop	*Grain Yield (t/ha)*		*Water Saving Over Without Laser Leveled Field (Per cent)*
	Laser Leveled Field	*Without Laser Levelling*	
Paddy	6.79	6.50	38
Wheat	4.75	4.55	20
Sugarcane	112.00	98.75	24
Summer green gram	0.50	0.38	20
Potato	10.00	9.00	25
Onion	10.00	9.00	20
Sunflower	2.25	2.00	20

Figure 11.9: The Relative Performance of Wheat in a Conventional (Left) and Laser Levelled Field (Right).

One of the ways to calculate the cost of land levelling is to work out the volume of expected earthwork. The earthwork multiplied by the cost of per unit of earth moved will provide the cost of land levelling. The other way to determine the cost is to assess the per hour cost of tractor and related levelling machinery using the principles of fixed and operational costs. For custom hiring the profit margin of the vendor needs to be added to the cost so arrived at (Ambast *et al.*, 2014). Students will learn these principles in some other course.

Ambast (2006) calculated the cost of land levelling using various equipments in conventional along with laser land levelling. Apparently, to achieve the same precision, the cost is minimal in laser than any of the conventional equipments.

It is attributed to the fact that to achieve the same precision it takes more time in conventional than laser land levelling (Table 11.9).

Table 11.9: Cost of Precision Levelling (LI < 1.5 cm) by different Equipments

Implement	*Operation Time (h/ha)*	*Cost (Rs./ha)*
Levelling blade	26.4	5280
Levelling float	15.0	3000
Levelling float bed	18.5	3700
Super leveller	19.5	3900
Laser land leveller	5.0	2500

In a study conducted in Egypt (https:/wocatpedia.net/wiki/Land_levelling) it has been revealed that cost increases with precision of land levelling and so the benefits (Table 11.10).

Table 11.10: Relative Cost and Profits with Various Methods of Land Levelling

Method of Levelling	*Relative Cost*	*Relative Net Profit*
Manual	100	100
Hydraulically operated machines	83.3	103
Laser land levelling	200	191.4

Most of the work for laser land levelling now-a-days is done on an hourly basis at commercial rates on custom-hiring basis. Although the costs do not distinguish between soil types and conditions, the degree of finish, amount of topsoil saved, type and cost of equipment or the experience of the equipment operators, but all conditions being similar, an inexperienced operator may take more time to finish the job or the job may not be finished properly to the desired level. Some farmers have reported that their fields have been leveled as table top. It has resulted in more time to apply the water per unit land. Apparently, the improper machines have been used by the vendors. It might have saved some investments for the vendor but such practices may kill a very good technology that has become quite popular amongst the farming community.

Chapter 12

Groundwater Aquifers and Pumps

What is Groundwater?

Groundwater is broadly defined as the underground water present in the zone of saturation of variable thickness and depth, below the ground. Cracks and pores in the existing rocks and unconsolidated crystal layers, make up a large underground reservoir, where part of precipitation is stored. The groundwater is utilized through wells and tube wells.

Practically, all groundwater originates as surface water, being a part of the hydrological cycle. The major components of this cycle are precipitation, evaporation, transpiration, infiltration, surface runoff (overland flow and interflow), and subsurface runoff (vadose-water flow and groundwater flow). Water enters the aquifers from the land surface or from surface water bodies through the vadose zone, and then it travels slowly within the aquifer for varying distances until it finally returns to the land surface by natural flow, plants, or humans. Sometimes the groundwater is brought to the surface by some natural source like springs, and sometimes these waters are tapped by artificial means by constructing wells, infiltration galleries *etc.*

Groundwater is found in aquifers (water-bearing geologic formations), which act as conduits for water transmission and as underground reservoirs for water storage. The zone of saturation is technically called 'aquifer'. Aquifers are significantly porous and permeable to supply water to wells and springs (Raghunath, 1987). On the other hand, water stored in ponds, lakes, rivers, streams, seas/oceans and other surface reservoirs is called surface water. Not all underground water is groundwater, rather only free water or gravitational water (the water that moves freely under the force of gravity into wells) constitutes the groundwater. Therefore, a precise and practical definition of groundwater is (Bouwer, 1978): "Groundwater is that portion of the water beneath the earth's surface, which can be collected through wells, tunnels, or drainage galleries, or which flows naturally to the earth's surface via seeps or springs". Depths to groundwater may range from 1 m or less to 1000 m or more. There are also places where groundwater does not exist at all.

Occurrence of Groundwater

All the materials of variable porosity near the upper portion of the earth's crust are places of the potential storage for groundwater (Kumar, 2014). This storage of groundwater is sometimes referred to as the Groundwater Reservoir. The possibility of occurrence of groundwater mainly depends upon two properties of the underground soil, they are;

- Porosity
- Permeability

Porosity is the quantitative measurement of the voids present in the soil; which is the ratio of the volume of voids and the total volume given as a percentage. Permeability is defined as the ability of a rock or unconsolidated sediment to transmit or pass water through itself and is generally measured in terms of coefficient of permeability.

A permeable stratum or geological formation of permeable material, which is capable of yielding appreciable quantities of groundwater under gravity, is known as an aquifer. An aquifer may be an entire formation, a group of formations, or part of a formation. Geologic formations may be either consolidated or semi-consolidated or unconsolidated. Consolidated geologic formations, also known as hard rocks, are formed by cementation, compaction and recrystallization, their grains being tightly held together. Examples of consolidated geologic formations are igneous and metamorphic rocks such as granite, basalt and schist, and indurated sedimentary rocks such as sandstone, shale and limestone. Semi-consolidated geologic formations are sedimentary rocks wherein the induration process is incomplete and the primary porosity (inter-granular porosity) is preserved to a varying degree. Among the semi-consolidated sedimentary rocks, sandstone is considered most productive because in an early stage of cementation, its primary porosity is very high like sand. Groundwater in these consolidated formation flows through fracture networks and/or pore space.On the other hand, unconsolidated geologic formations comprise of non-indurated colluvial, alluvial, aeolian (wind-borne sediments), lacustrine, marine (coastal) and glacial deposits. These formations/deposits consist of sand, silt, clay, gravel and pebbles. Groundwater flows through spaces between the grains. Geologic processes can likewise erode and metamorphose unconsolidated sediments.

An aquiclude is porous and may contain groundwater but cannot transmit it fast enough to be of consequence as a water supply. That is why, it may be considered an aquifer in one area but may be considered an aquiclude in another because of differences in demand and availability of alternate supplies. Clay is an example of aquiclude. An Aquitard is defined as a geologic formation that can store some water as well as transmit water at a relatively low rate compared to aquifers. Although an aquitard may not yield water economically, it can hold appreciable amounts of water. Sandy clay is an ideal example of an aquitard. An aquifuge is defined as a geologic formation that can neither store nor transmit water *e.g.* solid granite. Aquifuge is essentially a non-leaky confining layer, whereas aquitards and aquicludes are leaky confining layers. In practice, even aquiclude is considered as a non-leaky confining layer because leakage through aquicludes is so small that it can be considered practically insignificant.

Types of Aquifers

The types of aquifers are: 1) Groundwater or unconfined aquifer or non-artesian aquifer, 2) Confined aquifer and 3) Perched aquifer

Groundwater or Unconfined Aquifer

As already defined, a permeable stratum or a geological formation of a permeable material, which is capable to yield appreciable quantities of groundwater under gravity is known as an aquifer. An unconfined aquifer, also known as water table aquifer, is the one which is not confined by an upper impermeable layer (Figure 12.1). The upper surface of the zone of saturation is known as water table, which is at atmospheric pressure. If a well is constructed in these aquifers the level of the water table is the level of water in the well.

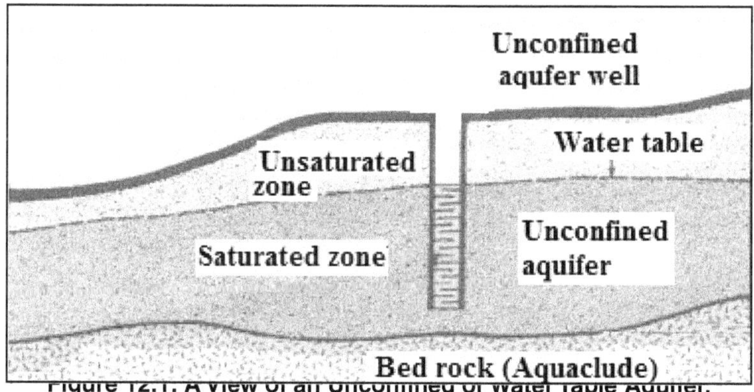

Figure 12.1: A View of an Unconfined or Water Table Aquifer.

Confined Aquifers or Artesian Aquifers

When an aquifer is confined on both sides by impervious rock formations *i.e.* aquicludes, and is also broadly inclined so as to expose the aquifer somewhere to the catchment area or recharge area at a higher level for the creation of sufficient hydraulic head, it is called a confined aquifer or an artesian aquifer (Figure 12.2). Water in these aquifers is under pressure above the atmospheric pressure. If a well is put in these aquifers, water will rise to a level above the water table of the upper confining layer because of the pressure under which the water is held. The imaginary level to which water will rise in wells located in an artesian aquifer is known as the piezometric level. A well that yields water by artesian pressure at the ground surface is a "flowing" artesian well.

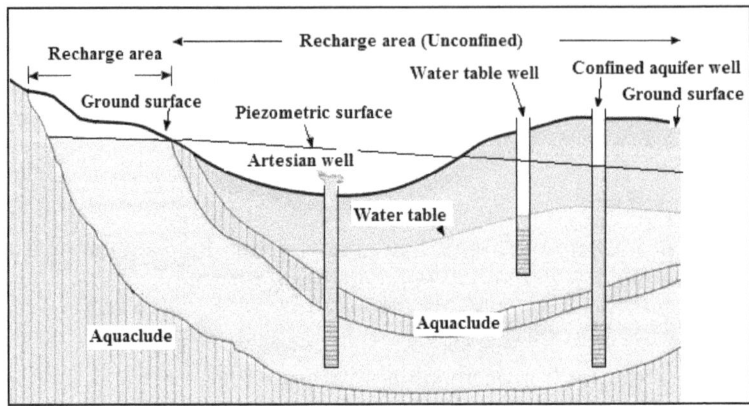

Figure 12.2: A Schematic Diagram of a Confined Aquifer.

Perched Aquifer

Perched aquifer occurs whenever a groundwater body is separated from the main groundwater by a relatively impermeable stratum of small areal extent (Figure 12.3). The upper surface of this perched zone of saturation is a perched water table, either temporarily or permanently. It is underlain by a negative confining bed which stops or retards the down-ward movement of water under the force of gravity. Many islands have perched water tables, which are important sources of fresh water.

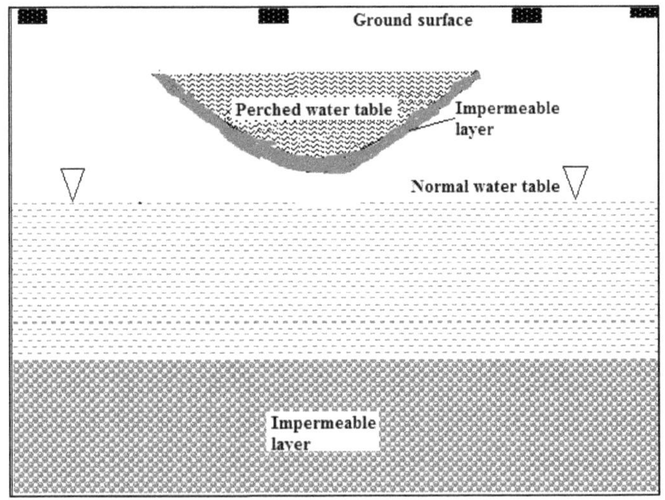

Figure 12.3: Schematic Diagram of Perched Water Table.

Types of Wells and Tube Wells

Water well is a hole usually vertical, excavated in the earth for bringing groundwater to the surface. The wells are usually classified in a number of ways.

On the basis of construction, wells and tube wells have been classified as dug wells, driven wells and tube wells.

Dug Well

These are shallow wells, usually constructed in soft ground and soils having sand and gravel. The open wells are usually unconfined wells. The walls of these wells may be unlined or constructed with preset RC blocks, bricks or stone masonry. These are comparatively bigger diameters wells, diameter may vary from 1 to 9 m. These usually are low discharge wells say about 5 l/s or less. The depth of these wells may be limited to about 20 m. Cost of these wells is low. Hence they are popular in rural areas and small towns. If higher discharge is needed, a borehole is made in the center of the well. In that case it is known as dug-cum-bore well.

Driven Well

It is also a shallow well. A casing pipe with a sharp point at the bottom is driven into a water bearing stratum. The lower portion of the casing is perforated. The perforated portion of the pipe is covered with a fine mesh to retain sand particles so that these do not enter inside the well. The discharge of these wells is very small.

Tube Wells

Tube wells are used when more quantity of water is required. The tube wells are constructed using blind pipes and strainer pipes to tap water from one or more than one water bearing aquifers. The tube wells are usually 3 cm to 15 cm diameter. The depth of tube well can be in the range of 30 m to 500 m or more. Depending upon the depth tapped the tube wells are categorized as shallow and deep wells. Generally deep wells are drilled to an aquifer below an impervious layer like clay. Depending on geological formation of the site, tube wells are designed to yield required quantity of water that can be sustained.

On the basis of the aquifer being tapped by the well tube wells are categorized as confined tube well (penetrating into the confined layer) and unconfined wells (pumping from unconfined aquifer) as described in a previous section of this chapter.

On the basis of length of penetration in the confined or unconfined aquifer, the wells are classified as:

- ✰ Fully penetrating wells
- ✰ Partially penetrating wells
- ✰ Non-penetrating wells (Cavity wells)

As the name suggest, fully penetrating well penetrate the whole aquifer (Figure 12.4 right), while a partially penetrating well penetrates the aquifer partially (Figure 12.4 left). The hydraulics of flow in the two kinds of tube wells is quite different as shown by the arrows in each case. A non-penetrating well on the other hands does not pierce the aquifer but ends just at its top (Figure 12.5). It is also called a cavity well and is discussed in the following section.

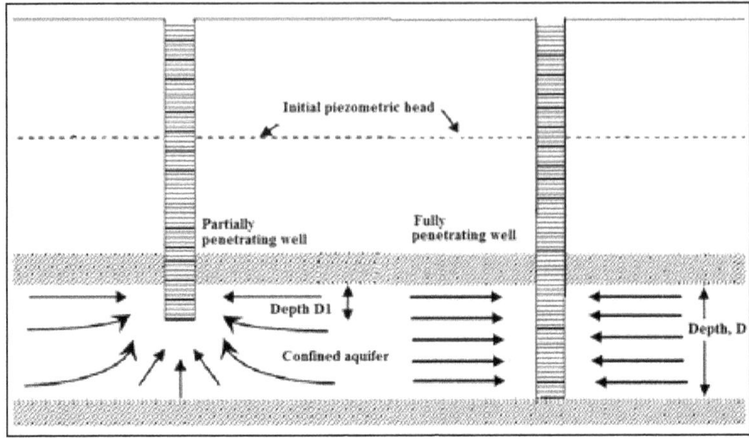

Figure 12.4: Partially (Left) and Fully (Right) Penetrating Wells.

Cavity Type Tube Well

A cavity well essentially comprises of a blind pipe in its entire length. The bottom is extended up to the aquifer where sufficient water is available. For the stability of the cavity, the roof of the cavity must be strong. Therefore, a pre-requisite

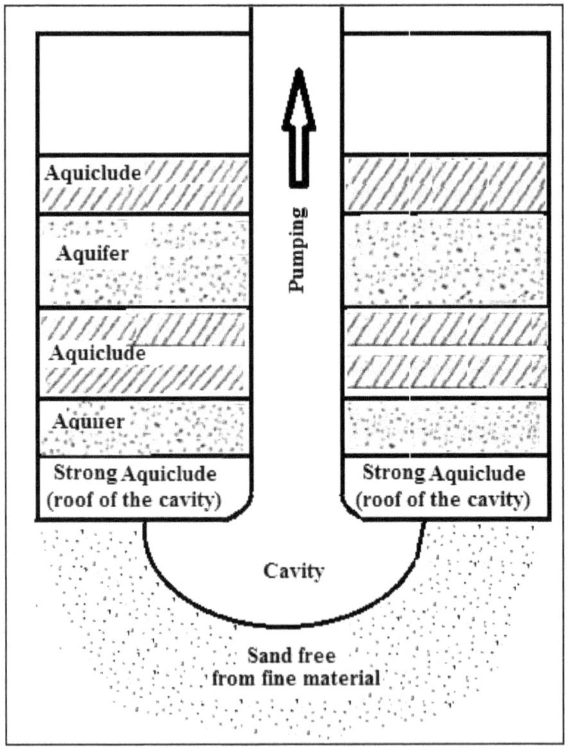

Figure 12.5: A Non-Penetrating or Cavity Well.

for the construction of cavity wells is presence of strong aquiclude or clay layer. After the construction is complete, the well is developed by pumping at a faster rate than envisaged during normal pumping of water. In the initial stages of development, fine sand will come with the water. It will form a hollow or cavity at the bottom (Figure 12.5). The bottom of the cavity for some thickness is thus made free of finer particles. The size of the cavity determines the yield of a well. The hydraulics of a strainer and cavity tube well is quite different, being radial in the former and spherical in the later.

Screen or Strainer Type Tube Well

The name of the well is derived from the strainer used in these kinds of wells. A strainer is nothing but a slotted pipe with fine wire mesh wrapped around and soldered or welded to it. It acts as a filtering device, serves as the intake portion of wells constructed in unconsolidated or semi-consolidated aquifers, and serves structurally to support the aquifer material. Well screens are manufactured from a variety of materials and range from crude handmade to highly efficient and long-life models made on machines.

The strainer type tube wells are more commonly used wells. These wells tap water from one or more water bearing formations. Strainers are installed in one or all water bearing formations to allow entry of sand free water. The perforated pipe (strainer) extends only for the aquifer portions of the formation while for other portion a non-perforated plain pipe is used. The bottom of the tube well is plugged. The plug is kept a little above the bottom to prevent failure of the plug due to the weight of the well. The grain-size distribution of aquifer material in such situations is such that a properly selected well screen allows finer particles to enter the well, which are removed during well development. After the well is fully developed, coarser particles are retained outside the well screen that forms a permeable envelope around the well screen. It is known as a 'natural gravel pack' and the well is called a naturally developed well. If the formation contains fine sand or silt, even strainer type of wells may require gravel packing around the strainer pipe.

Slotted Type or Gravel Packed Tube Well

If the water bearing strata is very deep, then slotted pipes are used as screen. In this case slotted wrought iron pipe pass through the main water bearing formation. This pipe is surrounded by gravel packing. Thus, a gravel-packed well is the well having an artificially placed gravel envelope around the well screen. Salient advantages of the artificial gravel pack are:

- It stabilizes the aquifer tapped by the well
- It avoids/minimizes sand pumping
- It allows to use a large screen slot with a maximum open area
- It provides a zone of high permeability surrounding the well screen, which increases the well radius and well yield.

In an artificial gravel pack well, the size of the screen openings is decided on the basis of size of the gravel used in the gravel pack.

Shallow and Deep Tube Wells

Tube wells are also characterized on the basis of depth as shallow tube wells and deep tube wells. Although no hard and fast rule is applied in this categorization, usually tube well of low discharge (20 l/s) with an average depth of 60 m are categorized as shallow wells. Cavity tube wells and strainer tube wells with coir strainers generally fall in this category. Shallow stainer wells mostly tap the unconfined aquifer. Deep tube wells on the other hand are wells of high capacity, tapping one or more than one aquifer with depth usually ranging from 60-500 m or more. These are either strainer or gravel-pack wells

Other Categories

Besides the classifications described in the previous paragraphs, structures to exploit groundwater have been classified in different ways. Some of these are as follows.

Hard Rock Bore Wells

Tube wells bored in hard rock formations are classified as hard rock bore wells.

Multi-Strainer Wells

A multiple-well system is a group of closely installed shallow tube wells, usually connected to a common header pipe or manifold and pumped by suction lift of a centrifugal pump. These tube wells are adapted under the following conditions:

- ✰ Where the water table is at shallow depth
- ✰ Where the installation of medium and deep tube wells is not economical due to hydraulic characteristics of the aquifers being poor
- ✰ Where salts are present in the deeper layers of the water-bearing formation and it is proposed to skim the fresh water
- ✰ To solve the problem of water logging
- ✰ Where the conditions are such that a constant level of water is required to be maintained at some depth from ground surface.

Radial Collector Wells

A radial collector well system comprises of a series of horizontal wells discharging water into a central caisson. They are located at or close to rivers and other surface-water bodies.

Skimming Wells

These are usually partially penetrating wells designed to skim a fresh water layer floating on saline water.

Besides these main structures, other structures to draw groundwater are springs, infiltration wells and infiltration galleries.

Springs

The natural outflow of groundwater at the earth's surface is said to form a spring. A pervious layer sandwiched between two impervious layers give rise to a natural spring. These are generally capable of supplying small amounts of water and are not considered as a good source of supply.

Infiltration Wells

Infiltration wells are the shallow wells constructed in series along the bank of a river to collect the river water seeping through their bottoms.

Infiltration Galleries

Infiltration galleries are the horizontal tunnels. These are constructed at shallow depths such as 3 to 5 meters along the banks of rivers, through water bearing strata. They are sometimes called horizontal wells. These galleries are generally constructed of masonry walls with roof slabs, and extract water from the aquifer by various porous drain pipes.

Mechanisms of Water Lifting

When the source of water is at a lower level than the area to be irrigated and when free gravity flow is not available to drain surface or subsurface water, the water needs to be lifted. For this purpose water lifting devices are used. Water may be lifted by the application of any one (or a combination thereof) of the following six mechanical principles, which are mostly independent (FAO, 1986):

Direct lift: it involves physical lifting of water such as in a container.

Displacement: Since the water is effectively incompressible, it can be 'pushed' or displaced.

Velocity head creation: When water is propelled to a high speed, the momentum can be used either to create a flow or to create a pressure.

Buoyancy of a gas: Air or other gases bubbled through water will cause movement of columns of water due to difference in specific gravity.

Impulse: Water hammer phenomenon creates impulse due to which a small portion of the water supply is lifted to a considerably high level.

Gravity: Water flows downward under the influence of gravity.

Types of Water Lifting Devices

Several types of water lifting devices are in use in irrigation. These devices may be classified according to the principle of lifting used in their operation. However, there are other kinds of classification systems as well. The most commonly used classification criterion is according to power source used in water lifting. Thus, the water lifting devices are classified as manual, animal and power operated devices.

Manually Operated Devices

Humans have limited physical power output, which may be in the range of 0.08 to 0.1 HP. This power can be used to lift water from shallow depths for irrigation. The common human powered devices are:

Swing Basket

The device consists of a basket made from the cheap materials like woven bamboo strips, leather, or iron sheet to which four ropes are attached. Two persons hold the basket facing towards each other, dip the basket in water source and by swinging, the basket is lifted and tilted in water course from where the water flows to the fields (Figure 12.6). The device is useful up to a lift of 0.15 m. The discharge may vary from 3500 to 5000 l/hr depending upon the lift.

Figure 12.6: Use of Swing Basket to Drain a Mustard Field.

Counterpoise Lift

It is also known as *dhenkli* or *picattach*. It is generally used for lifting of water from unlined wells, stream or pond for irrigating small fields. It consists of a lever rod supported at a suitable point on a vertical post about which it can swing in vertical direction. Water at the rate of about 2000 1/hr can be lifted from depths of 2 to 3 m.

Don

The principle of operation of don is similar to counterpoise lift. The don consists of a trough made from wooden log or iron sheet; closed at one end and open at the other. The open end of the trough is connected to a hinged pole with a counter weight through rope. For operation, the trough is lowered by exerting pressure on it by pulling the rope and also by foot of the operator till the closed end is submerged in water. Upon releasing pressure, the trough comes to its original position due to action of counterweight along with water. Water can be lifted from this device from a depth of 0.8 to 1.2 m.

Archimedean Screw

It consists of a helical screw mounted on spindle, which is rotated inside a wooden or metallic cylinder. One end of the cylinder remains submerged in water and is placed in an inclined position at an angle of about 30°. It is used for lifting of water from a depths of 0.6 to 1.2 m and may give discharge of about 1600 1/hr.

Paddle Wheel

It is also known as *Chakram* and is mostly used in coastal regions for irrigating paddy fields. It consists of small paddles mounted radially to a horizontal shaft, which moves in close fitting concave trough, thereby pushing water ahead of them. The number of blades depends on the size of wheel, which may be 8 for 1.2 m and up to 24 for 3 to 3.6 m diameters. The wheel having 12 blades may lift about 18000 l/hr from a depth of 0.45 to 0.6 m.

Animal Powered Devices

Animal power is abundantly available in India. Besides other field operations and processing works animal power is used to lift the water. A pair of bullocks may develop approximately 0.80 HP, which may help to lift water from a depth of 30 m or even more, the rate of discharge reducing with increasing lift. Some of the devices operated by animal power are as under.

Rope-and-Bucket Lift

Also known as *Mote, Charsa* or *Pur*, it is used to lift water from lined wells up to a depth of 30 m. The device consists of a bucket or bag made of GI sheet or leather, and pulley arrangement. A rope is attached to the bucket-or bag, which passes over a pulley and finally fixed to the yoke of bullocks. The bullocks walk down on an earthen ramp sloped at an angle of 5-10 degrees to lift the water. About 9000 1itres of water can be lifted per hour with two pairs of bullocks from a depth of 15 m.

Self-Emptying Bucket

The arrangement is similar to rope and bucket lift device. The system consists of a leather container shaped like a funnel which has a spout on the lower end and the upper portion resembles a conical cylinder. The container is open from both ends having the capacity in the range of 100 to 150 liters. The device is suitable when the lift does not exceed 9 m. The discharge at this lift is about 8000 l/hr.

Two Bucket Lift

In this device two buckets are raised and lowered alternately. The bullocks move in a circular path. With the help of central rotating lever, rope and pulley arrangement the buckets move up and down. The carrying capacity of each bucket may be up to 70 liters. The buckets are provided with hinged flap at the bottom, which acts as a valve. Guide rods in the well control the movement of buckets. The buckets are automatically filled and emptied during the operation. The device is suitable for lifts up to 5 m at which discharge may be about 14000 l/hr.

Persian Wheel

It is also known as *Rahat*. It is used to lift water from depths up to 20 m. The efficiency of the device is considerably reduced if the lift exceeds 7.5 m. The device consists of a endless chain of buckets made of GI sheet having capacity in the range of 8-15 liters. The chain of bucket is mounted on a drum and is submerged in the water to sufficient depth. The drum is connected to a toothed wheel held in vertical plane by a long shaft, which is usually kept below the ground level. The vertical toothed wheel is geared with a large toothed horizontal wheel connected to a horizontal beam. This beam is yoked to a pair of animals. The animals move in rotary mode that rotates the buckets carrying water through the gear system. The water is released when the buckets reach the top. Average discharge of Persian wheel operated by one pair of bullocks is about 10,000 l/hr from a depth of about 9 m.

Chain Pump

The chain pump is used to lift water from shallow wells and works most satisfactorily when the lift is about 6 m. The pump consists of an endless chain on which discs are mounted at an interval of about 25 cm. The endless chain usually passes over two drums. The upper drum is above the top of well to which axle and handle is attached for operation. The chain with disc passes through a pipe, which is about 10 cm in diameter and extends downward from the top of well to about 0.6 to 0.9 m below the surface of water. The discs on rotation of chain entrap the water and carry it to the top, which is discharged into the trough. The pump can be operated either manually or by animals having a system like that of a Persian wheel.

Hydraulic Ram

Hydraulic ram has been used for many decades to lift water. Its popularity stems from two main reasons: i) it need no external source of power as the kinetic energy of the moving water gives the power it needs and ii) The machine is extremely simple with just two moving parts. As such, no repair and maintenance is required except for changing of washers in the valves, whenever required. The ram can operate 365 days in a year without any trouble.

The basic idea behind a ram pump is simple. The pump uses the momentum of a relatively large amount of moving water to pump a relatively small amount of water uphill (USDA, 2007). The essential components of the system are check

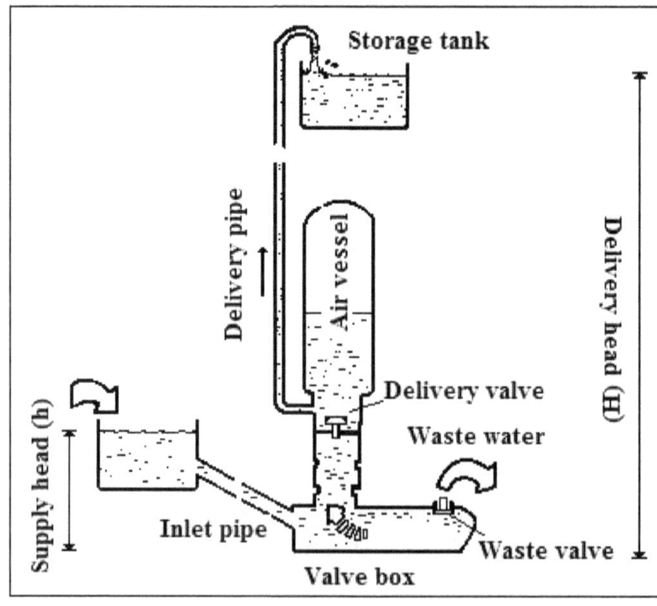

Figure 12.7: Working Principle of a Ram.

dam, supply pipe, hydraulic ram comprising of air vessel, waste valve and delivery valve, discharge pipe and storage tank (Figure 12.7).

To use a ram pump, a source of water must be available that should be situated above the pump. For this purpose, a check dam may be constructed on flowing water in a stream to create low head. Once the water flow is initiated, the water enters the chamber having two flap valves and an air vessel. The delivery valve is fitted to the air vessel. As the water flowing from the supply tank enters the chamber, it gets filled up making the waste valve to move upwards. A stage comes, when the waste valve suddenly closes. This sudden closure of valve creates high pressure inside the chamber. This sudden increase in pressure opens the delivery valve. Thus, the water from the chamber enters the air vessel and compresses the air inside the vessel. The compressed air exerts a force on water inside the air vessel. A small quantity of water is raised to a greater height. As the water in the chamber loses momentum, the waste valve gradually opens in downward direction and flow of water from the supply tank starts flowing to the chamber. This cycle is repeated.

Advantages

- No moving parts
- No power requirement
- Inexpensive
- Quiet pumping continuously over a long period
- Pollution free or "Green" pump
- Simple construction and easy to install
- Only initial cost and very low or negligible maintenance cost.

Disadvantages

- It can pump only one tenth of the water volume received, remaining being wasted through waste valve
- It must have a continuous source of supply at a minimum height of not less than 90 cm.
- It cannot pump viscous fluids to a greater height. Usually used for pumping drinking or potable water.

Ram Efficiency

If the energy efficiency of the ram is assumed to be 100 per cent, one can expect that only 20 per cent of the supplied water will be available if the source is 2 m above the ram and the water is lifted to 10 m above the ram. Actual water delivered is further reduced by the energy efficiency factor. If the energy efficiency is 70 per cent, the water delivered will be only 14 per cent (70 per cent × 20 per cent). Very high ratios of delivery to supply head usually result in lowered energy efficiency. Let Q be the discharge rate of water into the chamber, q is the discharge rate of water raised per second, h is the height of water in supply tank above the chamber, H is the height of water raised from the chamber. Thus, knowing the energy supplied by the supply tank to the ram and energy delivered by the ram, the efficiency of the ram is given as:

$$\eta = Q \times h / qH \qquad (12.1)$$

The efficiency η, specified above is called as D' Aubuisson's efficiency.

Rankine argued that the height of water raised is not H, but (H-h) and therefore, he defined the ram efficiency as:

$$\text{Rankine efficiency, } \eta = q \times (H\text{-}h)/(Q\text{-}q) \times h \qquad (12.2)$$

Suppliers of rams provide tables giving expected volume ratios based on actual tests.

Mechanically-Powered Lifting Devices

Mechanically powered water lifting devices are usually termed as pumps, which are operated with the help of auxiliary power sources such as engine or electric motor. A pump is a hydraulic machine which converts mechanical energy into hydraulic energy or pressure energy. The pumps are capable of lifting large quantity of water to higher heads and are usually employed for irrigation. Basically there are four principles involved in pumping water:

- Atmospheric pressure
- Centrifugal force
- Positive displacement and
- Movement of column of fluid caused by difference in specific gravity.

Pumps are usually classified on the basis of operation, which may employ one or more of the above principles. These are broadly classified into (Figure 12.8):

- Displacement pumps: Rotary and reciprocating
- Dynamic such as centrifugal pumps further categorized as volute, diffuser, turbine and propeller
- Airlift pumps.

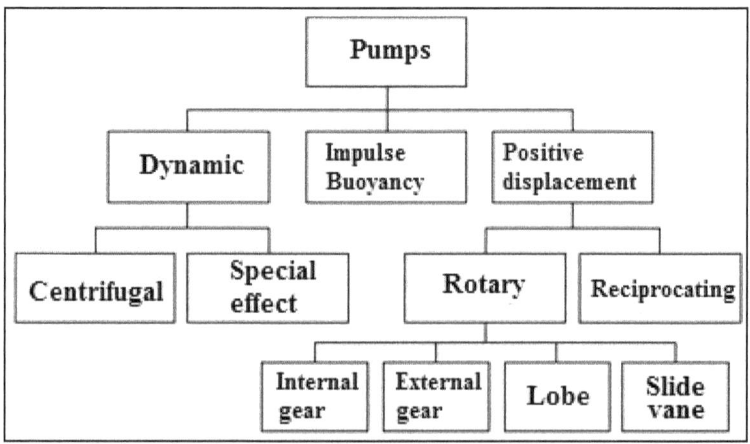

Figure 12.8: Classification of Pumps.

Positive Displacement Versus Dynamic Pumps

A practical difference between dynamic and positive displacement pumps is the ability of the dynamic pumps to operate under closed valve conditions. Positive displacement pumps physically displace the fluid; hence closing a valve downstream of a positive displacement pump will result in a continual build-up in pressure resulting in mechanical failure of either pipeline or pump. Dynamic pumps differ in that they can be safely operated under closed valve conditions (for short periods of time). Other major differences are listed as follows:

Positive displacement pumps	Dynamic pumps
More complex, consists of several moving parts	Simple in construction
Speed is limited by the higher inertia of the moving parts and tho fluid	Can operate at high speed and hence compact
Suitable for fairly low volumes of flow at high pressures	Suitable for large volumes of discharge at moderate pressures in a single stage
Higher maintenance cost	Lower maintenance requirements
Fluctuating flow	Delivery is smooth and continuous
Can be used for less viscous fluid otherwise valves may pose problems	Can be easily used for viscous or muddy fluids
Needs more area	Needs less area being compact
Can be used to compress air	Cannot be used for air
Low head pumps have low efficiencies	Low head pumps have high efficiencies
Priming is not required in this pump	Priming is required

Reciprocating Pumps

Reciprocating pump is a positive displacement pump. The water is drawn or forced into a finite space, and sealed in it by mechanical means and then forced out to flow. This cycle is repeated to lift the water. Simply stated, the principle can be stated as: ´reciprocating pump operates on the principle of pushing the liquid by a piston that executes a reciprocating motion in a closed fitting cylinder´. The reciprocating pumps are available in various designs and models, and can be operated manually, with animal power and/or auxiliary power sources.

Reciprocating pump are characterized as single acting or double acting. On the basis of number of cylinders these are also categorized as single cylinder, double cylinder and triple cylinder pumps. Pumps even with 4 or 5 cylinders are in the market.

In a rotary positive displacement pump, the motion is rotary rather than reciprocating. The rotary positive displacement pumps are classified as gear pumps, lobe pumps and vane pumps depending upon the kind of rotary motion.

Major components of a reciprocating pump are:

Piston or Plunger: A piston or plunger that reciprocates in a closely fitted cylinder.

Crank and Connecting Rod: Crank and connecting rod mechanism is operated by a power source. Power source gives rotary motion to crank. With the help of connecting rod this rotary motion is translated to reciprocating motion that moves the piston in the cylinder.

Suction Pipe: One end of suction pipe remains dip in the liquid and other end attached to the inlet of the cylinder.

Delivery Pipe: One end of delivery pipe attached with delivery part and other end at discharge point.

Suction and Delivery Valves: Suction and delivery valves are non-return values provided at the suction and delivery ends respectively.

Air Vessels: These are mainly provided to reduce flow fluctuations. These also reduce acceleration head and friction head.

Air Vessels

Air vessels are closed chamber containing compressed air at the top and liquid at the bottom (See Figure 12.10). Air vessels can be fitted on one or both sides of the pump (on suction and delivery side). They are located close to the cylinder. Air vessel on the delivery side is connected to the reciprocating pump through the opening at the base through which liquid enters into it during delivery stroke of the pump and compressed. It flows out during suction stroke of the pump as the compressed air expands in the air vessel. The uses of an air vessel are:

- It provides uniform discharge from pump.
- The chances of cavitation or separation are considerably reduced.
- A considerable amount of work is saved as a result of reduced frictional resistance.
- The pump can run at higher speed to provide higher discharge.

Single Acting Reciprocating Pumps

A single acting reciprocating pump is shown in Figure 12.9. The motion of piston in the cylinder creates partial vacuum. Because of the low pressure, water will rise from the well through suction tube and fill the cylinder by forcing the suction valve to open. This operation is known as suction stroke. In this stroke crank rotates from 0° to 180°. Delivery valve remains close during this stroke. When the crank rotates from 180° to 360° piston moves inwardly. Now piston exerts pressure on the liquid due to which suction valve closes while the delivery valve opens. The liquid is then forced up through delivery pipe. This stroke is known as delivery stroke. Now the pump has completed one cycle. The same cycle is repeated as the crank rotates. It may be noted that discharge in a single acting reciprocating pump will be intermittent. The discharge rate can be varied by changing either the frequency of reciprocation or stroke length of the piston.

The theoretical discharge of a single acting reciprocating pump can be given as:

$$Q = ALN/60 \qquad (12.3)$$

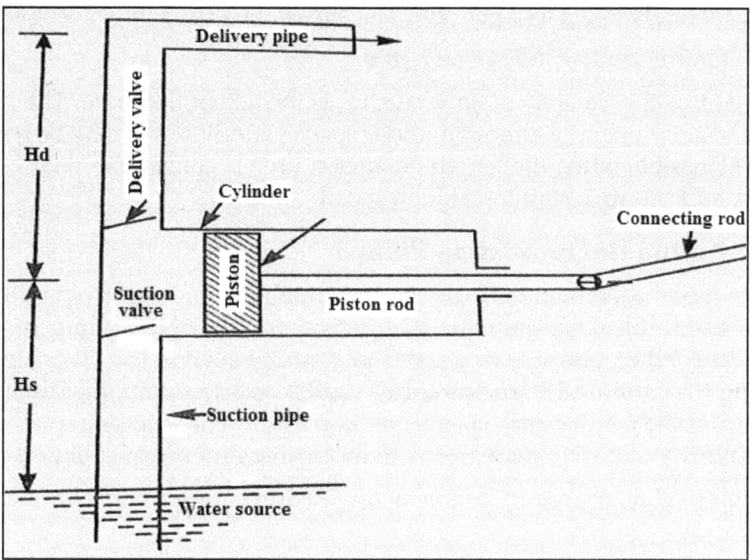

Figure 12.9: A Single Acting Reciprocating Pump.

Such that Q is the discharge in m^3/s, A is the cross sectional area of the cylinder, m^2, L is the stroke length, m and N is the speed of rotation, rpm. The constant 60 is the conversion factor that converts rpm into revolution per second (rps).

Theoretical work done by a single acting pump is given as:

$$= w\,Q\,(Hs+Hd) \quad (12.4)$$

Such that w is the specific weight and is given as ρg, ρ is the density of water and g acceleration due to gravity, m/s^2, Hs is the suction head and Hd the delivery head. Thus, power can also be given as:

$$\text{Power } P = w\,(ALN)\,(Hs+Hd)/60 \quad (12.5)$$

However, due to the leakage and frictional losses, actual power input will be more than the theoretical power.

$$\text{Actual power } P_a = w\,(ALN)\,(Hs+Hd)/(60\,\eta) \quad (12.6)$$

Here η is the efficiency of the pump.

Slip

Normally, the actual discharge is always less than the theoretical discharge. The coefficient of discharge (Cd) is accordingly written as ratio of the actual discharge to the theoretical discharge (Q_A/Q_T), such that Q_A is the actual discharge. The value of coefficient of discharge is around 95-98 per cent for pumps in good conditions. The slip of the pump is defined as

$$\text{Slip} = Q_T - Q_A$$

It is generally taken as 2 per cent. The slip of the pump in per cent is given as:

$$\text{Slip (per cent)} = [(Q_T - Q_A)/Q_T]100 \qquad (12.7)$$

Here Q_T is the theoretical flow rate; Q_A is the actual flow rate. The slip of the pump is usually positive although under special conditions it may be negative as well. It is possible when the length of suction pipe is considerably long, delivery head low and pump is running at high speed.

Double Acting Reciprocating Pumps

To reduce the fluctuations in discharge, a double acting pump or multi-cylinder pump is used. The diagrammatic sketch of a double acting pump is shown in Figure 12.10. When one side of the piston is under suction the other side will be delivering the fluid under pressure. In a double acting pump, the discharge will be nearly double the discharge per revolution as compared to single acting pump. The complexity of design increases with increasing number of cylinders.

Theoretical discharge of double acting reciprocating pumps is calculated as follows:

Discharge = Water discharge in forward stroke + water discharge in reverse stroke

$$Q = ALN/60 + (A-Ap)\,LN/60 \qquad (12.8)$$

Here Ap is the area of the piston rod.

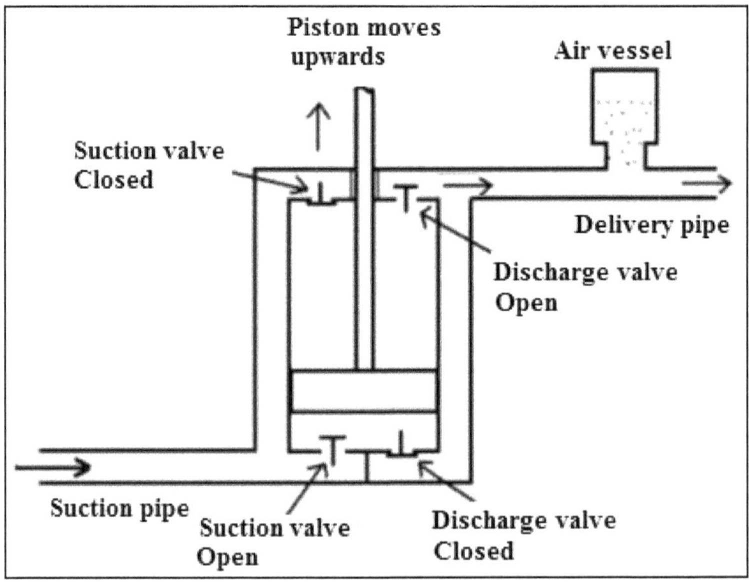

Figure 12.10: A Double Acting Reciprocating Pump.

If D is the diameter of the piston and d that of the piston rod, then

$$A-Ap = \pi D^2/4 + \pi(D^2 - d^2)/4 = \pi D^2/4 + \pi D^2/4 - \pi d^2/4$$

$$= \pi\,(2D^2 - d^2)/4 \qquad (12.9)$$

If d is neglected in comparison to D, then

$$A\text{-}Ap = 2\pi D^2/4 \tag{12.10}$$

Or

$$Q = 2ALN/60 \tag{12.11}$$

Example 12.1

A single acting reciprocating pump discharges 0.023 m^3/s of water when running at 60 rpm. Stroke length is 50 cm and the diameter of the piston as 25 cm. If the total lift is 15 m determine the theoretical discharge, slip and percentage slip, coefficient of discharge and power required for running the pump. Assume any condition to solve the problem.

$Q_T = 0.5 \times 60 \times (\pi/4)\,(0.25)^2/60 = 0.5 \times (\pi/4)\,(0.25)^2 = 0.0245\ m^3/s$

Slip = 0.0245 – 0.023 = 0.0015

Per cent slip = (0.0015/0.0245) × 100 = 6.12 per cent

Coefficient of discharge = 0.023/0.0245 = 0.938

Power input = $\rho.g.Hs.Q_T$ = 1000 × 9.81 × 15 × 0.0245 = 3601 W = 3.61 kW

Here it is assumed that there is no leakage and frictional losses. In practice, actual power input will be more than the theoretical power. The calculated value needs to be divided by the pump efficiency in fraction, η. Assuming $\eta = 0.7$

Power required to run the pump = 3.61/0.7 = 5.16 kW.

Rotary Pumps

In these pumps, the reciprocating motion is substituted by the rotary motion. The rotary motion is achieved by cams or by gears. There are two cams or gears which fit with each other. They rotate in opposite directions. The water enters through the suction pipe and it is trapped between cams or teeth of gears and casing then thrown with force into the discharge pipe. This type of pump is useful for moderate heads and small discharges not greater than 40 l/s.

Dynamic or Variable Displacement Pumps

Variable displacement pumps also known as dynamic pumps have been categorized as follows.

- ☆ Centrifugal Pump
- ☆ Turbine Pump
  - ❖ Deep well turbine pump
  - ❖ Submersible pump
- ☆ Propeller Pump
- ☆ Jet Pump
- ☆ Air Lift Pump

Centrifugal Pump

A centrifugal pump is also known as a Rotodynamic pump or dynamic pressure pump. It works on the principle of centrifugal force. It is commonly used to move liquids through a piped system. Advantages of centrifugal pumps are:

- Low cost
- High efficiency
- Uniform and continuous discharge
- Easy installation and maintenance
- Can run at high speeds without the risk of separation of flow

Centrifugal pumps may be classified into the following types.

1. According to casing design or energy conversion these are: Volute type or diffuser type pumps
2. According to number of impellers or stages these are grouped as: Single stage pump or multistage or multi impeller pump
3. According to number of entrances to the impeller these are: Single suction pump or double suction pump
4. According to disposition of shaft (axis of rotation) these are categorized as: Horizontal shaft pump or vertical shaft pump
5. According to liquid handled pumps are grouped as: Open impeller pump, semi open impeller or closed impeller types
6. According to specific speed pumps are: Low specific speed or radial flow impeller pump (further grouped as relatively low, medium and high specific speed), medium specific speed or mixed flow impeller pump, high specific speed or axial flow type or propeller pump.
7. Based on the working head, centrifugal pumps are classified as follows: Low head if $H < 15$ m, medium head if $15 < H < 40$ m and High head if $H > 40$ m
8. On the basis of method of drive: Direct connected- mono-blocks, geared and belt or chain driven
9. According to flow direction in the pump: radial, axial and mixed flow types

Some and not all the classifications are discussed in the following paragraphs.

Volute and Diffuser Types

Volute Pump

In this case, a spiral casing is provided around the impeller. The water which leaves the vanes is directed to flow in the volute chamber circumferentially. The area of the volute chamber gradually increases in the direction of flow resulting in reducing velocity and increasing pressure. As the water reaches the delivery pipe a considerable part of kinetic energy is converted into pressure energy. Since, the eddies are not completely avoided, some loss of energy takes place due to the continually increasing quantity of water through the volute chamber.

Diffuser Pump

In the case of a diffuser pump an additional component, the guide wheel containing a series of guide vanes are provided. The diffuser blades which provide gradually enlarging passages surround the impeller periphery. They serve to augment the process of pressure built-up that is normally achieved in the volute casing. Diffuser pumps are also called turbine pumps in view of their resemblance to a reaction turbine.

Single and Multi-Stage

Single-Stage: Single-stage centrifugal pumps have only one impeller fitted to the pump shaft. It is usually a low-lift (low-head) pump.

Multi-Stage: Each impeller and matching casing is called a stage. The number of stages necessary for a particular installation is determined by the height the water must be raised from the surface of the water source. When the pressure is more than that can be practicably or economically furnished by a single stage, additional stages are used. A pump with more than one stage is called a multi-stage pump. The impellers are surrounded by guide vanes and the water is led through a by-pass channel from the outlet of one stage to the entrance of the next until it is finally discharged into a wide chamber from where it is pushed on to the delivery pipe. Multi-stage pumps commonly used are the turbine and submersible types. Vertical turbine pumps can have as many as 25 stages.

Single and Double Suction Pump

Single Suction: The impeller of single suction pump has only one water inflow end. In a single suction pump, also known as end-suction pump, when fluid in the impeller is pushed towards the peripheral direction; low pressure area is formed in the center of impeller. With the impact of total static pressure difference between fluid level in the storage tank and the impeller center, the fluid is sucked into the impeller center.

Double Suction: Double suction pump has two ends; benefit of which is to decrease the axial force and increase the flow quantity. In fact, double suction pump impeller consists of two back-to-back impellers, from which the water flow into a volute. One double suction pump works just like two same diameter end-suction pumps work at the same time, so its flow quantity can be doubled by using the same outer diameter impeller. Meanwhile, the entry end and exit end of double suction pump are in the same direction, being perpendicular to the pump shaft, which is helpful for the layout and installation of pumps and water pipes. Double suction pump is widely used for the fluid delivery in drainage and irrigation of farmland, waterworks, pump stations, power stations, fire fighting and shipping, *etc.*

Horizontal and Vertical Shaft Pump

The primary difference between these pump types is the shape and the position of the shaft. In a horizontal centrifugal pump, the shaft is normally in a horizontal position. It is sometimes overhung or placed between bearing design. While in case of a vertical pump, the shaft is in vertical position; it is always an overhang and of radial-split case type design.

Open, Semi-Open and Closed Impellers

Open Impeller: An open impeller has vanes that are attached to a center hub and mounted directly on to a shaft. There is no wall surrounding the vanes. It makes open impellers weaker than closed or semi-closed valves. On the other hand, open impellers are easier to clean and repair. Open impellers are mostly used in smaller pumps and pumps that handle suspended solids.

Semi-open Impeller: Semi-open or semi-closed impellers have a back wall that adds strength to the impeller. Reduced efficiency is a common problem, but the ability to pass solids is an important trade-off.

Closed Impeller: Closed impellers have a back and front wall around the vanes, to increase strength. Closed impellers are used primarily in larger pumps. These types of impellers are commonly used in clear liquid applications. They don't do well with solids and are difficult to clean when clogged.

Low, Medium and High Head

Low-Head: The impeller is usually surrounded by a volute and there are no guide vanes. The shaft is generally horizontal and water may enter the impeller from one or both sides depending on the quantity of water to be delivered. These are capable to produce heads up to 15 m.

Medium-Head: They are capable of generating heads as high as 40 m. They are usually provided with guide vanes. Water may enter from one or both sides depending on the quantity of water to be pumped.

High-Head: They can generate heads exceeding 40 m. They are generally multi-stage pumps because an ordinary single impeller cannot build up such a high pressure. High-head centrifugal pumps may be horizontal or vertical; the latter is mostly used in deep wells.

Specific Speed

Specific speed of a centrifugal pump is defined as the speed of a geometrically similar pump when delivering 1 L/s against a head of 1 meter. The term is used for comparing numerically different centrifugal pumps. This is a variable obtained from the service data which has great practical significance for the design and choice of pumps. The most commonly adopted expression for the specific speed, N_s is:

$$N_s = N\sqrt{Q}/H^{3/4} \qquad (12.12)$$

Here, N is speed of the pump (rpm), Q is discharge of the pump (L/s), and H is the total dynamic head (m). It may be noted that the specific speed, dimensionless, increases with N and Q and decreases with H.

The values of Q and H used to calculate specific speed are corresponding to the maximum efficiency of the pump at its normal working speed. Furthermore, for a multi-stage pump, the value of H is obtained by dividing the actual head developed with the number of stages. Similarly, the value of Q for a double-suction pump is taken as half the actual discharge delivered by the pump.

Types of Centrifugal Pumps Based on Specific Speed

The specific speed provides a sound basis for the technical classification of centrifugal pumps. It is the only characteristic index of a pump when several impellers can be used for the same head and capacity. The performance and dimensional proportions of pumps having same specific speed will be the same even though their outside diameters and actual operating speeds may vary. Based on the specific speed (Modi and Seth, 1998), centrifugal pumps are classified into five classes (Table 12.1).

Table 12.1: Types of Centrifugal Pumps Based on the Specific Speed

Sl.No.	Type of Pump	Specific Speed (N_s)
1	Low-speed radial flow	300–900
2	Medium-speed radial flow	900–1500
3	High-speed radial flow	1500–2400
4	Medium speed mixed flow or screw type	2400–5000
5	High speed axial flow or propeller type	3400–15000

Components and Working of Centrifugal Pump

Centrifugal pump has two main components: an impeller and a stationary casing, housing, or volute (Figure 12.11). Besides these two major components, it has some other components such as pump inlet, pump outlet, suction valve, delivery valve, priming device, foot valve with a screen and/or pressure gauge. The impeller attached to a rotating shaft, consists of a number of blades, called vanes. The vanes are arranged in a regular pattern around the shaft. The impeller is made of bronze, stainless steel, cast iron, polycarbonate, and a variety of other materials. As the impeller rotates, fluid is sucked in through the eye of the casing. Depending upon the direction of the fluid path through the pump in relation to the shaft, the pumps are characterized as radial, axial and mixed flow types.

Radial Flow Pumps

As the fluid flows from the eye to the periphery of the blades, energy is added to the fluid by the rotating blades, and both pressure and absolute velocity are increased. The vanes discharge the water flowing radially outward i.e. perpendicular to the rotating shaft into a diffuser or volute. The volute is shaped to slow down the flow of water. This decrease of the flow further increases the pressure by converting the velocity head to pressure head. From here water exits into the downstream piping system. Centrifugal pumps are radial-flow machines that operate most efficiently for applications requiring high flow rate at relatively low heads. Although centrifugal pumps are most often associated with the radial flow type, however, the term "centrifugal pump" is used to describe all impeller type rotodynamic pumps including the radial, axial and mixed flow variations.

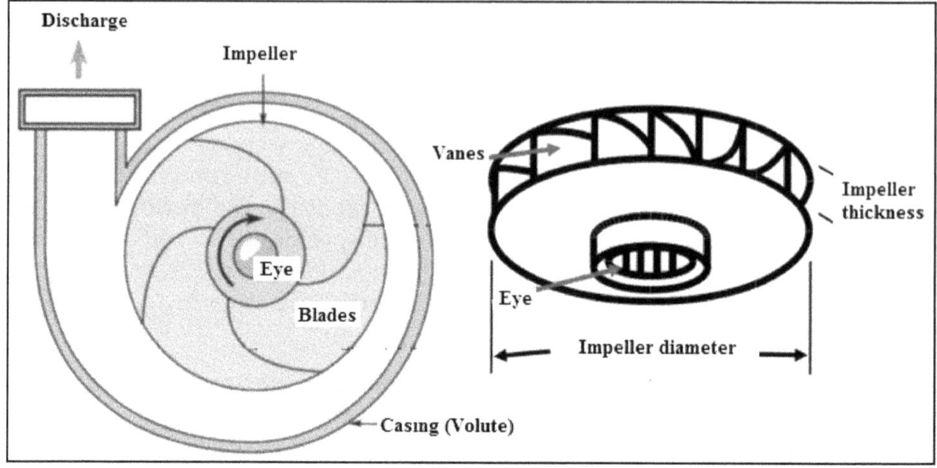

Figure 12.11: A Schematic View of the Centrifugal Pump (Left) and a Close Up of the Impeller (Right).

Axial Flow Pumps

Many applications such as those associated with drainage and irrigation require high flow rate at low head. Under such conditions, axial-flow pumps are commonly used. Axial-flow pump, often called propeller pump consists of a propeller confined within a cylindrical casing. As against the radial flow pumps, in axial flow pumps fluid enters and exits along the same direction (parallel to the rotating shaft). The fluid is not accelerated but instead "lifted" by the action of the impeller.

Mixed Flow Pumps

Mixed flow pumps, as the name suggests, function as a compromise between radial and axial flow pumps, the fluid experiences both radial acceleration and lift and exits the impeller somewhere between 0-90° from the axial direction. As a consequence mixed flow pumps operate at higher pressures than axial flow pumps while delivering higher discharges than radial flow pumps. The exit angle of the flow dictates the pressure head-discharge characteristic in relation to radial and mixed flow.

Priming

For the same speed, centrifugal force exerted on air is much-much less than in the case of liquid. As a result, pump cannot expel all the air to go out of casing and suck water. Moreover, centrifugal pump does not suck/pull the liquid but pushes it; the pump will not work, if the chamber is not full of water. It is why priming is compulsory in the case of centrifugal pump. Priming is the process in which the impeller of a centrifugal pump is fully submerged with water from outside without any air trap inside. Priming should always be done before starting the pump unless arrangements are made to keep the impeller submerged by using foot valve or a self-priming pump is used. Two commonly used accessories in the pump to help in priming of the pump are reflux valve or a foot valve.

Reflux Valve: Its main purpose is to retain water in the suction pipe section between the pump and the reflux valve, to help priming. The reflux valve is used when pumps are directly connected to the suction pipe. When pumping to overhead tanks or to convey water long distances, the reflux valve is also used in the delivery side so as to permit flow only in the forward direction. It not only checks the return flow but also prevents water hammer.

Foot Valve with a Screen (optional): The suction side of the pump consists of foot valve at the bottom of suction line. The foot valve is a one-way valve which helps to prevent the trash from entering the suction side and at the same time retains water in the suction line. As a result, priming of the pump from outside may not be required.

Self Priming Centrifugal Pump

Most centrifugal pumps are not self-priming. That means, the pump casing must be filled with liquid before the pump is started. Otherwise the pump will not function. Self-priming pumps on the other hand are capable of evacuating air from the pump suction line without any external auxiliary devices. The self-priming pumps are provided with a separation chamber. This retains some of the fluid. When the pump is started, the fluid is pumped and the entrained air bubbles are also pumped into the separation chamber by the impeller action. The air escapes through the pump discharge nozzle but the fluid drops back down. The process continues until the suction line is fully evacuated of the air. The low efficiency of these pumps and the large dimensions of the separating chamber are major issues with these pumps. For these reasons this method is adopted only for small pumps. More frequently used types of self-priming pumps are side channel and water ring pumps. On the other hand these pumps can handle air and gases entrained in water. The pump also does not need a foot valve. The pump has high suction lift characteristics and can suck water from much lower depths. Self-priming pumps are available both in stationary and portable models. Centrifugal pumps with an internal suction stage such as water jet pumps or side channel pumps are also classified as self-priming pumps.

Pump Characteristic Curves

The pump performance curves are of great importance to the engineer who is responsible for the selection of pumps for a particular flow system. Selection of the pump is usually made on the basis of characteristic curves namely

- ☆ Head capacity curve which shows how much water a given pump will deliver at a given head
- ☆ Overall efficiency curve represents the relationship between the pump efficiency and the discharge

- ☆ Brake horse power curve, which gives the relation between the discharge and horse power
- ☆ Net positive suction head required plotted over the capacity range (not shown in Figure 12.12)

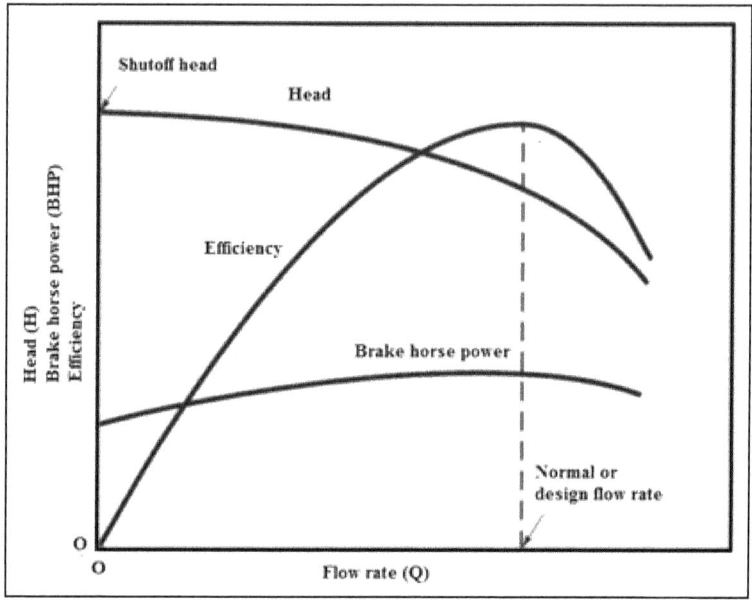

Figure 12.12: Characteristic Curves of a Centrifugal Pump.

The typical characteristic curves show the total dynamic head, brake horsepower, efficiency, and net positive Suction head all plotted over the capacity range of the pump at a constant impeller speed. (Figure 12.12). Sometimes, the curve relating the net positive suction head required against the capacity range may not be included in the performance curves. Some of the characteristics and important terms related to performance curves of the pumps are given as follows.

Rise Head Curve: The head curve rises with decreasing flow rate.

Shut-off Head: It indicates the head developed by the pump at zero discharge. It represents the rise in pressure head across the pump with the discharge valve closed.

Brake Horse Power: As the discharge is increased from zero the brake horsepower increases, with a subsequent fall as the maximum discharge is approached.

Efficiency: The efficiency is a function of the flow rate and reaches a maximum value at some particular value of the flow rate. It is commonly referred to as the normal or design flow rate or capacity for the pump.

Best Efficiency Points (BEP): The points on the various curves corresponding to the maximum efficiency.

Knowledge of pump characteristics enables one to select a pump which fits the operating conditions and thus attain a relatively high efficiency with low operating cost. For a centrifugal pump used to pump water from an open water source, the total head is calculated to select a suitable pump. In the case of wells, the head capacity curve of the well is matched with the pump head-efficiency-horse power curves to select the pump (Figure 12.13). It is because the point where the pump operates on its curve is dependent upon the characteristics of the system in which it is operating, commonly called the system head curve or the relationship between flow and hydraulic losses in a system (https:/beeindia.gov.in/sites/default/files/3Ch6.pdf). It is completely independent of the pump characteristics. Since friction losses vary as a square of the flow rate, the system curve is parabolic in shape. The system curve is drawn considering the friction losses through the system including all bends and valves and represented in a graphic form (Figure 12.13). The positive static head involved in this case must be taken into account. Although it does not affect the shape of the system curve or its steepness, but it does dictate the head of the system curve at zero flow rate. The actual pump operating head and flow rate are determined by the intersection of the pump and system curves. It is important to select a pump such that the intersection of the pump and system curves is near the BEP.

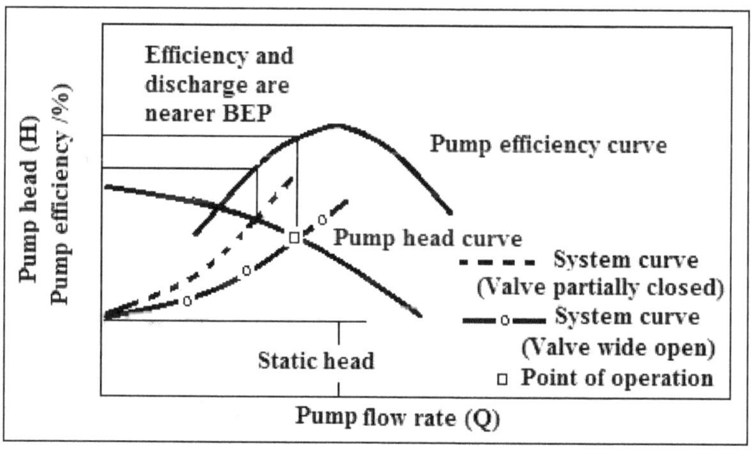

Figure 12.13: Matching of System Curve with the Pump Characteristic Curve.

Capacity of the Tube well

There are many components that determine the capacity of a tube well. Besides the aquifer capacity to yield water, it depends upon the type of well, intended use, well depth and intake design. On the other hand, proper design of an irrigation system requires that the pumping system be precisely matched to the irrigation distribution system. When an irrigation system is designed or modified to use an existing pumping system, it is necessary to know the capacity of the existing pump. The capacity of the tube well for irrigation is decided on the basis of irrigation scheduling or the area irrigated is matched to the capacity of the tube well. It further depends upon the total area to be irrigated and the irrigation interval. Without going into this issue in detail at this stage, following two examples illustrate the procedure to determine the tube well capacity.

Example 12.2

A centrifugal pump delivering 850 liters per minute is used to irrigate an area of 1.6 ha. What is the depth of irrigation per day if the pump is operated for 8 hours a day?

Water delivered per day (m^3) = $(850 \times 60 \times 8)/1000 = 408\ m^3$

Depth of irrigation (mm) = $408 \times 1000/1.6 \times 10000 = 25.5$ mm

Example 12.3

A tube well is proposed to be operated for 15 hours each day during the irrigation season. What should be its capacity, if it is proposed to command an area of 5 ha? Take the irrigation interval as 20 days and depth of irrigation as 5 cm?

If Q is the capacity of the tube well, then

The quantity of water pumped in a day = $Q \times 15 = 15\ Q\ m^3$

Area to be irrigated per day = $5/20 = 0.25$ ha

Water required to irrigate 0.25 ha = $10000 \times 0.25 \times 5/100 = 125\ m^3$

Therefore, $Q \times 15 = 125$

$Q = 8.33\ m^3/hr = 2.33\ l/s$

Example 12.4

A 12-hectare field is to be irrigated with a sprinkler system. The root zone depth is 0.9 m (D) and the field capacity of the soil is 28 per cent while the permanent wilting point is 17 per cent by weight. The soil bulk density (Db) is 1.36 g/cm^3 and the water application efficiency is 70 per cent. The soil is to be irrigated when 50 per cent of the available water has depleted. The peak evapotranspiration is 5.0 mm/d and the system is run for 10 hours a day.

Field capacity = 28 per cent

Permanent wilting point = 17 per cent

Available moisture, Am = 28 - 17 = 11 per cent

Root zone depth = 0.9 m

Bulk density = 1.36 g/cm^3

Depth of available moisture, = Am. Db. D = $0.11 \times 1.36 \times 900 = 135$ mm

Allowing for 50 per cent depletion of available moisture before irrigation

Depth of readily available moisture = 0.5×135 mm = 67.5 mm

Net irrigation depth = Depth of the readily available moisture = 67.5 mm

Gross irrigation = Net irrigation depth/Application efficiency = 67.5/0.7 = **96.4 mm**

Irrigation interval =Net irrigation depth/Peak ET = 67.5/5 = 13.5 days ~ 13 days

In design, irrigation interval = irrigation period = 13 days

Total area to be irrigated = 12 ha

Area to be irrigated per day = Total area/irrigation period = 12 ha/13 days = 1 ha/d

System Capacity, Qc can be obtained by directly incorporating the values in the following equation:

$$Qc = \frac{Ad}{F\,H\,Ea} m^3/s$$

Here A is the area = 12 ha = 120,000 m^2, Net irrigation depth, d = 67.5 mm = 0.0675 m,

Irrigation period, F = 13 days, Number of hours of operation, H = 10 hr/d and Irrigation efficiency, Ea = 0.7

Qc = (120,000 × 0.0675)/(13 × 10 × 0.7) = **89.01 m ^{3}/hr**

Recall: 1 m ^{3}= 1000 L and 1 hr = 3600 s

Therefore Qc = {89.01 × 10 ^{3} L}/3600 = 24.73 = 25 L/s

The pump must have capacity equal to or greater than 25 L/s.

Calculation of Total Head

The energy imparted by a pump to the liquid increases the pressure which is reflected in the increased head. Total head generated by a pump is called 'pumping head' which is also known as 'manometric head' or 'total dynamic head' (TDH). Total dynamic head can be written as:

$$TDH = \Delta Elev + Hf + P/\gamma + V^2/2g \qquad (12.13)$$

Here ΔElev is the elevation difference between the point from where water is being lifted and the point at which water is being delivered. Hf is the total friction head loss from the entrance to the exit point including minor losses due to fittings *etc.* Pressure P and velocity V are usually measured at the outlet. If the water is discharging at the atmosphere pressure, P is zero. But, if the water is discharging in open air, there may be some velocity head at the outlet. It should be considered in the calculations of the TDH. Note that one external parameter which does not appear in Eq. (12.13) is the operating head that is needed to operate sprinkler, gated pipe or drip irrigation systems. Therefore, to compute the total head which a pump must provide to operate an irrigation device, it is necessary to add the required operating head. The procedure to calculate the various parameters in eq. (12.13) is illustrated in Figure 12.14.

Friction Head Losses

Frictional resistance offered to the liquid flowing through suction and delivery pipes resulting in head loss are known as friction head losses. In several instances,

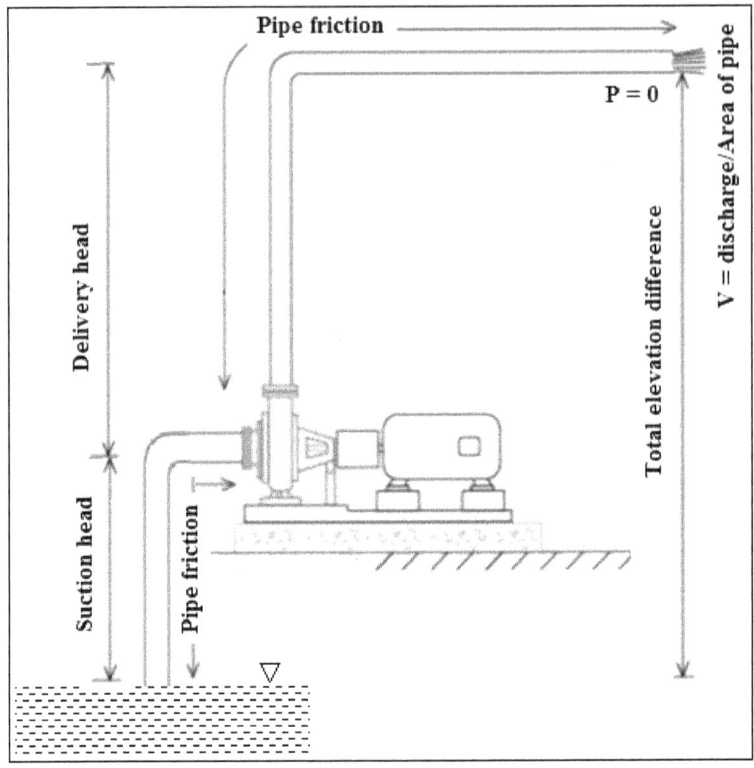

Figure 12.14: Calculation of Total Dynamic Head.

these losses constitute significant components of the total pumping head. Friction head losses are classified as: (a) major friction head losses, and (b) minor friction head losses. Major friction head losses are the friction head losses due to the roughness of the inner surface of pipe-network, and the viscosity and density of the flowing fluid. Minor friction head losses are the head losses due to pipe fittings, bends, entry and exit, sudden expansion and contraction, valves, screen, foot valve and so on. The sum of major and minor friction head losses constitutes the 'total friction head losses'.

The major friction head losses (h_f) in the suction and delivery pipes of a pumping system are calculated by the Darcy-Weisbach equation, which is given as:

$$Hf = (f\,L\,V^2)/(2gd) \tag{12.14}$$

Where, f is the friction factor (For the values of f see chapter 14 or consult any textbook for Moody's diagram), L is the length of the pipe, V is the velocity of flow, d is the inside diameter of the pipe, and g is acceleration due to gravity. Besides, Darcy-Weisbach equation, Hazen-William and Blacius equations are also used for this purpose.

Minor friction head losses are calculated with the help of standard tables (Table 12.2). Herein depending upon the kind and size of the fixture, the equivalent length of pipe is determined and added to the total length of the pipe to calculate the friction head losses. In general, the value of major friction head losses is much higher than that of minor friction head losses. Therefore, minor friction head losses are often neglected or a fraction of major head losses is taken as minor head losses depending upon the number and kind of fittings. However, if the value of minor friction head losses is significantly large, these must not be neglected while calculating the total pumping head.

Table 12.2: Friction Loss Equivalent Length (PVC) - Feet of Straight Pipe

Fitting	*Nominal Pipe Size (inches)*												
	1/2	*3/4*	*1*	*1 1/4*	*1 1/2*	*2*	*2 1/2*	*3*	*4*	*6*	*8*	*10*	*12*
90° Elbow	1.5	2.0	2.5	3.8	4.0	5.7	6.9	7.9	12.0	18.0	22.0	26	32
45° Elbow	0.8	1.1	1.4	1.8	2.1	2.6	3.1	4.0	5.1	8.0	10.6	13.5	15.5
Gate valve	0.3	0.4	0.6	0.8	1.0	1.5	2.0	3.0					
Tee Flow - Run	1.0	1.4	1.7	2.3	2.7	4.3	5.1	6.2	8.3	12.5	16.5	17.5	20.0
Tee Flow - Branch	4.0	5.0	6.0	7.0	8.0	12.0	15.0	16.0	22.0	32.7	49.0	57.0	67.0
Male/Female adapter	1.0	1.5	2.0	2.8	3.5	4.5	5.5	6.5	9.0	14			

Water Power or Water Horse Power

Water power is the theoretical power required to pump awter. It is the power required by a pump in lifting a given quantity of water in a unit time assuming no losses of power in the pump. Let, Q be the discharge of pump and H the total head against which the pump is working, then the water power generated by the pump is given as:

$$WP - \gamma Q H \text{ Newton-m/s} \quad (12.15)$$

Where, WP is the water power (W), γ is the specific weight of the water (N/m^3), Q is the discharge of the pump (m^3/s), and H is the total head or total dynamic head (m). Since 1 Newton-m/s is equal to 1 W, the water power in eq. (12.15) is in watts. Since γ (N/m^2) is also written as $\rho.g$ where ρ is the density of the liquid (Water ~ 1000 kg/m^3) and g is acceleration due to gravity (9.81 m/s^2), we can write

$$WP = 1000 \times 9.81\, Q H = 9810\, QH \quad (12.16)$$

Since 1 watt is also equal to 0.00134 horse power, we have

$$WHP = 1000 \times 9.81\, Q H = 9810\, QH \times 0.00134 = QH/0.076 \quad (12.17)$$

Writing Q in l/s, we have:

$$WHP = QH/76 \sim QH/75 \quad (12.18)$$

Shaft Horse Power: It is the power required at the pump shaft and is given as:

$$\text{Shaft Horse Power} = \text{Water Horse Power}/\text{Pump Efficiency} \quad (12.19)$$

Brake Horse Power (BHP): It is the actual horse power required to be supplied by the engine or electric motor for driving the pump.

For direct driven pump, BHP = Shaft Horse Power (12.20)

With belt or indirect drives, BHP = Water Horse Power/(Pump Efficiency × Drive Efficiency)

Horse Power Input to Electric Motor (IHP) = BHP/Motor Efficiency (12.21)

Kilowatt Input to electric motor or energy consumption in kilowatt hours is given as:

Energy in Kilowatts = IHP × 0.746 or BHP × 0.746/Motor Efficiency (12.22)

Since energy consumed is given on KWH, it can be determined by multiplying the energy in KW by the hours of pumping as Energy (KW) × Hours of pumping.

Cost of operation for electric motor =
KWH × Cost per Kilowatt Hour (12.23)

The pump efficiency of most of the pumps generally ranges from 60 to 70 per cent and the drive efficiency of motor is about 80 per cent. The overall efficiency of the system may be approximately 50 to 55 per cent.

Example 12.5

A pump delivers 21000 liters of water to a height of 45 m in every five minutes. If the efficiency of the pump is 75 per cent, what is the horsepower required to drive the pump?

Water delivered per second = 21000/(5 × 60) = 70 l/s

WHP = 70 × 45/75 = 42

Assuming that the pump is direct driven, we have

SHP = BHP = 42/0.75 = 56 HP

Example 12.6

A pump lifts 100 m^3/hr against a total head of 20 m. Calculate water power of the pump? If the pump has an efficiency of 75 per cent, what size of prime mover is required to operate the pump? If a direct drive electric motor with an efficiency of 80 per cent is used to operate the pump, compute the cost of electric energy in the month of April. Assume that the pump is operated for 10 hours per day for the whole month. Assume the cost of electricity charges as Rs 6.00 per unit.

Discharge of the pump (Q) = 100/60 × 60 = 0.028 m^3/s, Total head (H) = 20 m

Water Power, WP = $\gamma \times Q \times H$ = 9810 × 0.028 × 20 = 5493.6 W or 5.5 kW

Check: Water Power, WP = Q×H/75 = (0.028 × 20/75) × 0.746 = 5.6 kW (Nearly same)

Since the efficiency of the pump is 75 per cent, the shaft power is given as

Shaft power = 5.5/0.75 = 7.33 kW

Since direct drive is used to operate the pump, the shaft power determines the required size of the prime mover. Input power is given as:

Input power = 7.33/0.80 = 9.16 kW

Total energy consumption per month = Input power × Hours of Pump Operation × Number of days in a month

= 9.16 × 10 × 30= 2748 kW-h

Since 1 kW-h is equal to 1 unit, therefore, the cost of electric energy = Rs. (2748 × 6) =

Rs. 16488 or 16500.

Example 12.7

A centrifugal pump is required to lift water at the rate of 100 l/s. Calculate the brake horse power of the engine from the following data when the water is directly supplied to the field channel.

- ✰ Suction head = 6 m
- ✰ Coefficient of friction = 0.01
- ✰ Efficiency of pump = 65 per cent
- ✰ Water is directly supplied to the field channel (Delivery head may be taken as zero)
- ✰ Diameter of pipe = 15 cm
- ✰ f of the Darcy's-Weisbach equation is= 0.01

$Q = 100$ l/s or $0.1\ m^3/s$

Delivery head, Hd = 0 (As water is directly supplied to field)

H = 6 m, f = 0.01, d = 15 cm = 0.15 m

The length of the pipe where frictional effect may occur is taken equal to the suction head, so

$Hf = f \times l \times Q^2/\ [((\pi/4)\ d^2)^2 \times 2g \times d)] = 0.01 \times 6 \times (0.1)^2/\{(12.1) \times (0.15)^5\}$ (Note: Q = V/A)

Hf = 0.65 m

Neglecting the minor head losses, we have

Total head, H = Hs + Hd + Hf = 6 + 0 + 0.65 = 6.65 m

Efficiency, η = 65 per cent = 0.65

Horse Power (H.P) = $QH/75\ \eta = 100 \times 6.65/75 \times 0.65 = 17.70 = 13.64$ or 15 HP

Example 12.8

Calculate the cost of pumping 4 million liters of water from a well with a centrifugal pump from the following data.

Suction head = 3 m; Delivery head = 7 m; Friction head = 1.5 m

Output of the pump = 120000 l/hr; Pump efficiency = 70 per cent

Motor efficiency = 85 per cent ; Cost of electricity = Rs. 2 per unit.

Total head = 3 +7+1.5 +0.1 × 1.5 = 11.65 m (Assuming minor friction losses as 10 per cent of major friction losses)

Water Horse Power = (1200000/3600) × 11.65/75 = 5.17

Brake horse power = WHP/pump efficiency = 5.17/0.7 = 7.4 ~ 7.5

Time required to pump 4 million liters of water = $4 \times 10^6/4 \times 10^4$ = 100 hr

Therefore units consumed = 7.5 × 100/0.85 = 882.4 kwH

The cost of pumping = 882.4 × 2 = Rs. 1764.8 ~ Rs. 1765

Affinity Laws

The affinity laws express the effect of changes in certain application variables on the operational performance of pumps (Michael *et al.*, 2008). The affinity law variables which affect pump performance are:

- ☆ Pump speed in revolutions per minute (RPM)
- ☆ Impeller diameter

Change in the Pump Speed (RPM)

When the impeller diameter of a centrifugal pump is held constant, the effect of changing the speed (RPM) of the pump is in accordance with the following relations:

Capacity: $Q1/Q2 = N1/N2$ (12.24)

Head: $H1/H2 = (N1/N2)^2$ (12.25)

BHP: $BHP1/BHP2 = (N1/N2)^3$ (12.26)

Here Q is the capacity, H is the head, BHP is the brake horsepower and N is the pump speed, RPM. Here subset number 1 shows performance at the initial speed and subset number 2 shows performance at the new speed.

It may be seen that changing the speed (RPM) of a pump affects the flow, head and input brake horsepower of the pump in different proportions. Changing the speed affects the flow through the pump by a proportion equal to the increase or decrease in speed. The pump head is changed by the square of the proportion of speed change, while the brake horsepower is changed by the cube of the proportion of speed change.

Change in the Impeller Diameter

When the speed (RPM) of a centrifugal pump is held constant, the effect of changing the impeller diameter (D) is as follows:

Capacity: $Q1/Q2 = D1/D2$ (12.27)

Head: $H1/H2 = (D1/D2)^2$ (12.28)

BHP: $BHP1/BHP2 = (D1/D2)^3$ (12.29)

Here all variables are as discussed previously and D is the impeller diameter.

It may be seen that changing the diameter of a pump impeller affects the flow through the pump by a proportion equal to the increase or decrease in diameter. The pump head is changed by the square of the proportion of diameter change, while the brake horsepower is changed by the cube of the proportion of diameter change.

Example 12.9

It is proposed to increase the volume flow capacity of an existing system by 10 per cent by changing the pump speed. What would be the increase in power requirement?

The flow capacity of the system can be increased by increasing the speed of the pump (RPM) by 10 per cent (eq. 12.24).

Using the affinity law (Eq. 12.26), we have

$BHP1/BHP2 = (N1/N2)^3 = (1/1.1)^3$

$BHP1/BHP2 = 0.75$

$BHP2 = BHP1/0.75 = 4/3\ BHP1 = 1.33\ BHP1$

Or the power requirement increases by 33 per cent.

Efficienty of Centrifugal Pumps

The overall efficiency (η_o) of a pump is defined as the ratio of the power output from the pump to the power input to the pump from the prime mover.

$$\eta_o = \gamma Q H/\text{Shaft power in Watt} \quad (12.30)$$

$$\eta_o = \gamma Q H/(1000 \times \text{Shaft power})\ \text{in (kW)} \quad (12.31)$$

Here Q is the discharge of the pump (volume of liquid actually delivered per second by the pump), and γ is the specific weight of the liquid and H is the total head delivered by the pump. The eq. (12.31) can be derived as discussed in the following paragraphs.

The efficiency described by eq. (12.31) comprises of manometric efficiency, volumetric efficiency and mechanical efficiency such that overall efficiency is written as the product of these three efficiencies *i.e.*

$$\eta_o = \eta_m \times \eta_v \times \eta_{me} \quad (12.32)$$

Such that η_m is the manometric efficiency, η_v is the volumetric efficiency and η_{me} is the mechanical efficiency.

Manometric Efficiency

Also known as hydraulic efficiency, manometric efficiency (η_m) is defined as the ratio of the manometric head developed by the pump to the head imparted by the impeller to the liquid. That is,

$$\eta_m = H/(H+\Delta H) = H/H_{th} \quad (12.33)$$

Here H is the actual head develop by the pump and H_{th} is the theoretical head developed by the impeller given as $H+\Delta H$.

Volumetric Efficiency

Practically, there is some power loss due to leakage of the fluid between the back surface of the impeller hub plate and the casing, or through other pump components. This leakage reduces the overall efficiency and is termed as the volumetric loss (η_v). If the leakage loss that occurs in the centrifugal pump is represented by ΔQ, then the volumetric efficiency is given as:

$$\eta_v = Q/(Q+\Delta Q) \qquad (12.34)$$

Where, Q is the actual discharge of the pump.

Thus, volumetric efficiency is defined as the ratio of the quantity of liquid discharged per second from the pump to the quantity of liquid passing per second through the impeller.

Mechanical Efficiency

Mechanical efficiency (η_{me}) is defined as the ratio of the power actually imparted by the impeller to the power supplied to the shaft by the prime mover. That is

$$\eta_{me} = P_h/Ps \qquad (12.35)$$

$$P_h = (\gamma\,(Q+\Delta Q)\,H_{th}/1000) \qquad (12.36)$$

Ps = Shaft horse power

Overall Efficiency

Combining all the efficiencies as per eq. (12.32)

$$\eta_o = H/H_{th} \times Q/(Q+\Delta Q) \times (\gamma\,(Q+\Delta Q)\,H_{th}/1000)/\text{Shaft horse power} \qquad (12.37)$$

Simplifying

$$\eta_o = (H \times Q \times \gamma)/(1000 \times \text{Shaft horse power}) \qquad (12.38)$$

It is the same equation as the overall efficiency given in the beginning of this section.

Installation of Centrifugal Pumps

A major issue in any installation of centrifugal pumps is the suction head. If a pump can create perfect vacuum, the water can be theoretically sucked up to 10.34 m at sea level (Figure 12.15). Since the atmospheric pressure is a function of the altitude of the site, the theoretical suction lifts decreases with increasing altitude. The suction lift of an ordinary centrifugal pump in practice is limited to 5-6 m or much less if the suction pipe is unusually long. Therefore, the length of the suction pipe should be restricted to a minimum. A larger diameter pipe also adds to the performance of the pump but cannot be too large due to cost considerations. If the suction is more than the practical suction lifts, pumps may be installed in sumps (pits) so that these pump can be placed within close proximity of the water table.

To have an efficient operation of a centrifugal pump, its proper installation at an appropriate location with a suitable foundation is a must. The pump base-plate should be leveled. The type of installation depends on the nature of water source (surface water bodies or groundwater), type of well (open well or tube well), extent

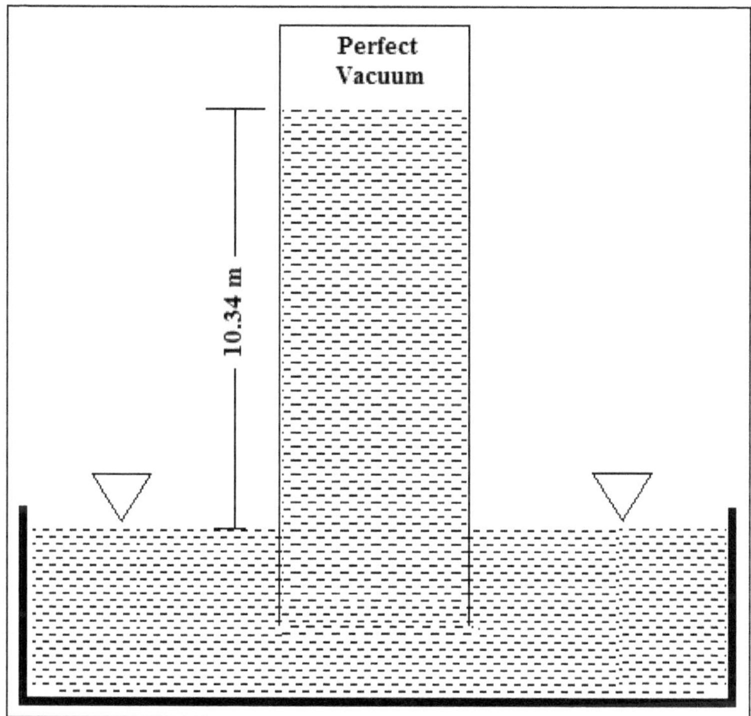

Figure 12.15: Theoretical Suction Lift at Sea Level.

of lining in case of tube wells, seasonal variation in the static water level, and kind of prime mover (electric motor or diesel engine) used in operating the pump. The pump and prime mover should be appropriately aligned. Significant causes of poor pump performance and maintenance problems are cavitation and air-lock. Suction piping should be adequately sized and properly designed to avoid cavitation. If the delivery pipeline is long, it is important to have a check valve (also known as clack valve, non-return valve, reflux valve, retention valve or one-way valve) at the pump outlet. In fact both a check valve and a manual stop valve (usually gate valve) should be installed in the discharge piping with the check valve placed between the pump and the stop valve. It will protect the pump from reverse flow and excessive back pressure. It helps to control water hammer and cavitation (See section on water hammer and cavitation of this chapter).

Turbine Pump

Turbine pumps are used in tube wells or in open wells or sumps where the water level is below the practical suction lift limit of the centrifugal pumps. The vertical turbine pump consists of a shaft on the bottom end of which is the turbine and on the top a driving motor, placed above the ground level (Figure 12.16 left). Turbine comprises of a suction bell through which water enters and passes into the pump bowl. Most turbines have more than one impeller, being either the volute or mixed flow type. Each impeller and bowl is called as one stage. Adding stages

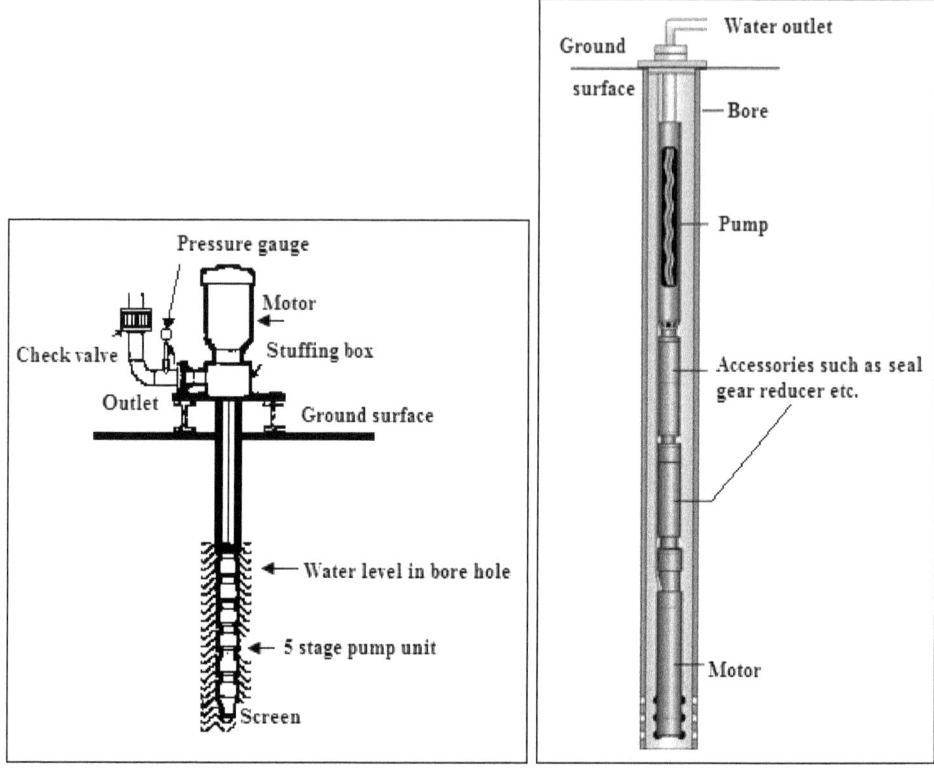

Figure 12.16: A Schematic Diagram of a Vertical Turbine Pump (Left) and a Submersible Pump (Right).

to a turbine pump adds head to which water can be lifted. The pump unit remains submerged in water and the pumping unit is located below the drawdown level of the water source. The outlet is provided at the top of the pump. The long vertical shaft supported on bearings is either water or oil lubricated. From a sanitary point of view, lubrication of pump bearings by water is always preferable, since lubricating oil may leak and contaminate the water. Although the turbine pump also operates on centrifugal principle, it differs from volute type as stationary guide vanes guide upward the water thrown by the impeller to the periphery. The gradual enlarging vanes guide the water to the casing, thus converting kinetic energy into potential energy. Therefore, the turbine pump generates heads several times that of a volute pump. Turbine pumps are most effective on tube wells requiring high heads and suction lifts are also high. A well-maintained pump provides trouble free service for several years.

Submersible Pumps

These are special kind of centrifugal pumps where long vertical shaft connecting the motor and pump unit (as in the case of turbine pump) is replaced by a short shaft. The prime mover and pump become closely coupled and both are submerged in water (Figure 12.16 right). Thus, a centrifugal pump driven by a closely coupled electric motor constructed for submerged operation as a single unit is called a submersible pump. All submersibles are multi-stage pumps mounted on a single shaft, and all rotating at the same speed. Each impeller passes the water to the eye of the next impeller through a diffuser. The diffuser is shaped to slow down the flow of water and convert velocity to pressure. Each impeller and matching diffuser is called a stage. The water inlet into the submersible pump is between the pump and the motor. Submersible pumps are suitable for tube wells having a bore of 100 mm or more. The electrical wiring to the submersible motor must be water proof. The electrical control should be properly grounded to minimize the possibility of shorting and thus damaging the entire unit. The pump and motor assembly are supported by the discharge pipe; therefore, the pipe should be of such size that there is no possibility of breakage. For the same output submersible pumps consume less power besides requiring less space. These pumps are ideal for lifting water from deep tube wells for irrigation, domestic and industrial applications.

Jet Pump

A Jet pump is a diffuser pump used to lift water from shallow as well as deep wells. When the pump is in operation, the output of the diffuser is split with half to three-fourths of the water being sent back down the well through a pressure pipe. At the end of the pressure pipe, water is accelerated through a cone-shaped nozzle, which is combined with a venturi. The jet nozzle and venturi are known as ejectors/ejector kits. The venturi has two parts: the venturi throat, which is the pinched section of the suction tube; and above that is the venturi itself which is the part where the tube widens and connects to the suction pipe. The venturi speeds up the water causing a pressure drop which sucks in more water through the intake at the very base of the unit. The increased suction in this manner enables a centrifugal pump to have a higher effective suction lift. But, the amount of water delivered becomes less as the distance from the pump to the water increases. In that case more water has to be recirculated to operate the ejector. Although a deep well ejector works in the same way as the shallow well ejector, the difference between a shallow well jet pump and a deep well jet pump is the location of the ejector. On a shallow well, the ejector kit is located in the pump housing in front of the impeller. On the deep well, ejector is located in the well below the water level.

Air Lift Pump

When compressed air is forced through the air pipe, a mixture of air and water is formed and rises up in the form of bubbles (Figure 12.17). Thus, the pressure of the water in the educator pipe becomes less than the pressure of water in the casing pipe. This pressure difference forces the water to rise through the educator pipe to discharge through the outlet. The advantages of the air lift pump are: higher lift capacity at less cost, lesser chances of clogging, consistent flow rate and a better control on flow rates.

Water Hammer and Cavitation

Water Hammer is a pressure surge or wave that occurs when there is a sudden momentum change of a fluid (the motion of a fluid is abruptly forced to stop or

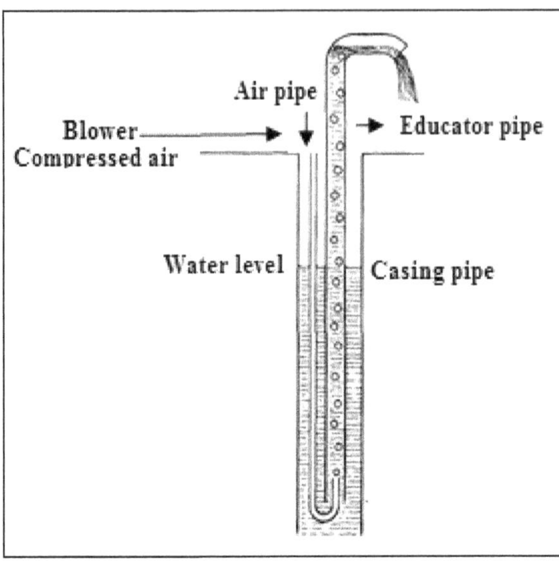

Figure 12.17: A Schematic Diagram of Air Lift Pump.

change direction) within an enclosed space. The most recognizable sign, to indicate that a water hammer occurs, is a loud banging in the pipes, as if they are being hit with a hammer. Vibrations in the pipe are also an indicator of water hammer. It commonly occurs in pipelines when a valve is closed suddenly at the end of a pipeline where the velocity of the fluid is high (Figure 12.18). It will result in a change in fluid momentum. The pressure wave created will propagate within the pipeline. When this occurs, a loud hammer noise is produced and vibrations are sent through the pipe. Another instance that produces a water hammer effect is pump and turbine failure. When a pump fails, the sudden halt in flow will produce the momentum change causing the water hammer effect. In addition, severe water hammer can also occur when the flow reverses which causes the instantaneous closure of the foot valve. The pressure wave produced from such events can cause

Figure 12.18: Schematic Illustration of the Changes Due to Sudden Stoppage of Flow.

significant damage to pipe systems. The large increase in pressure (Figure 12.18) can cause pipes to crack and in some cases pipes may burst. It also causes cavitation within the pipelines and if severe enough, it can cause the pipeline to implode. A check valve on the discharge line (at the pump outlet) can protect the pump from any such back surge down the pipeline.

To illustrate the field practicability, water hammer in sprinkler irrigation systems is usually caused by

- ☆ Valve closure
- ☆ Uncontrolled flow velocities in empty pipes
- ☆ Trapped air in long runs of pipe
- ☆ Reverse flow when pumps stop

Plastic pipe systems are particularly susceptible to water hammer damage. While it is impossible to eliminate water hammer, several strategies can be adopted to minimize the adverse effects.

- ☆ Use slightly oversized valves, especially valves designed to minimize pressure surging
- ☆ For residential or small commercial systems use pipes that are rated at least twice the design static pressure, and limit the velocities to 9 ft/s or less
- ☆ Install automatic air release values at high points in pipelines
- ☆ Use silent check valves instead of conventional check valves. Silent check valves close while the water is still flowing forward and prevents reverse flow
- ☆ Use water hammer arresters. These are devices that use the energy from the pressure wave to compress air. The most inexpensive water hammer arrester is a vertical pipe, slightly larger than the supply line, with a pressure release valve at the upper end. The pressure release valve is set at a pressure less than the bursting strength of the supply line, but more than the operating pressure of the pipe system.

Water hammer can also be caused by cavitation. Cavitation is defined as the phenomenon of formation of vapor bubbles of a flowing liquid in a region where the pressure of the liquid falls below its vapour pressure. It results in the sudden collapsing of the vapour bubbles in a region of higher pressure. When the vapour bubbles collapse, a very high pressure is created. The metallic surfaces at these places are subjected to high pressure, which cause pitting action on the surface resulting in the formation of cavities. Besides, cavitation results in considerable noise and vibrations and reduced efficiency of the pumps. Common causes of cavitation are:

Drop in Pressure at the Suction Nozzle due to Low NPSHa

If the fluid at the pump suction is not available at sufficiently above the vapor pressure of the liquid at operating conditions, then vaporization of liquid will result in formation of gas bubbles that may result in cavitation.

Increase in the Temperature of the Pumped Liquid

An increase in liquid temperature at the pump suction point increases the vapor pressure of the liquid. Thus, there are chances for the operating pressure to fall below this vapor pressure limit, leading to formation of bubbles and cavitation.

Increase in the Fluid Velocity at Pump Suction

An increase in fluid velocity at the pump suction can typically be caused by higher liquid flow rates than the design. As per Bernoulli's principle, higher liquid velocity means lower pressure head. Frictional pressure drop in the pump suction also rises with the rise in the flow rate, making low pressure and cavitation at pump suction to occur.

Reduction of Flow at Pump Suction

Certain minimum flow is required to keep the centrifugal pumps from running dry, as may be indicated by the pump performance curves. If liquid flow falls below this limit, possibility of developing vapor in pumps and cavitation increases.

Undesirable Flow Conditions

Sharp elbows, valves, other fittings and obstructions cause more frictional pressure loss in the pump suction, thus increasing the possibility of low pump suction pressure leading to cavitation.

Improper Pump Selection

Every centrifugal pump has a certain requirement of positive suction head ($NPSH_R$). If the pump is not selected properly, $NPSH_A$ might fall below this $NPSH_R$ limit, causing cavitation. The minimum value of $NPSH_a$ that is needed to prevent cavitation in the pump is the value of $NPSH_a$ at the pump inlet (suction) in excess of the vapor pressure. $NPSH_R$ is determined experimentally by pump manufacturers and reported as a function of pump flow rate.

Following methods are adopted by the operators and designers to prevent cavitation.

Operator

- Reduction of geodetic suction lift or increase in suction head so as to always operate with $NPSH_a \geq NPSH_R$, the value of $NPSH_R$ recommended by the manufacturer
- Valves, bends, curves are avoided wherever possible or the higher radii fixtures are used
- Short suction line with largest possible cross section
- The temperature of the fluid is controlled
- Injection of high pressure air into the flowing liquid at low pressure zones
- Reduction in speed

Designer

- ☆ Design the system so that there is no separation of flow and avoiding sharp corners or curvatures of flow in the system that may produce vertices or eddies
- ☆ Use of cavitation resistant materials such as starlite, bronze, nickel *etc.* (Note: No material is cavitation proof but can resist)
- ☆ High degree of surface finish helps to reduce the occurrence of cavitation as compared to a low surface finish.
- ☆ Use of impellers with double curvature blades drawn well forward into the suction orifice
- ☆ Reducing the thickness of the impeller blades

Another problem that may be encountered during the operation of a centrifugal pump is that of aeration. It happens when the air bubbles enter pump at low pressure side. These bubbles expand in partial vacuum. When the mixture of fluid and air bubbles travel to high pressure side, bubbles collapse forming micro-jets causing rapid erosion in the system.

Failure: Casuses, Prevention and Corrective Actions

Whenever a pump is operated, the operator may face certain problems. Any problems or abnormal events can be diagnosed by referring to the following general guidelines. Accordingly, the operator may check and take appropriate remedial action(s).

Pump in Operation but No Liquid Delivered

- ☆ Pump not primed
- ☆ Inlet pressure too low
- ☆ Strainer or foot valve on the suction line clogged
- ☆ End of suction line is not submerged in water
- ☆ Total system head is higher than the pump total head at zero capacity

Pump Delivers Less than the Rated Capacity

- ☆ Air leak in suction line or pump seal
- ☆ Inlet pressure too low
- ☆ Suction line strainer partially clogged or having insufficient inlet area
- ☆ System total head is higher than the designed head
- ☆ Impeller partially clogged
- ☆ Impeller installed in reverse direction or impeller rotates in wrong direction
- ☆ Suction or discharge valves partially closed
- ☆ Impeller speed too low

Loss of Prime While Pump is Operating

- ✰ Water level falls below the suction line intake
- ✰ Air leak develops in pump or seal
- ✰ Air leak develops in suction line
- ✰ Water vaporizes in suction line

Pump is Noisy

- ✰ Cavitation, the pump is starving
- ✰ Misalignment of pump and driver
- ✰ Foreign material inside pump
- ✰ Bent shaft
- ✰ Impeller might be touching the casing
- ✰ Pump has come loose from its base

Pump Takes Too Much Power

- ✰ Impeller speed too high
- ✰ Shaft packing too light
- ✰ Misalignment of the pump and motor
- ✰ Impeller touching the casing and dragging
- ✰ System total head too low causing the pump to deliver too much liquid
- ✰ Impeller diameter too large
- ✰ Impeller installed in the wrong direction

Motor Runs Back/Reverse Rotation

- ✰ Phase drop out, a "humming" noise can be heard when single phasing occurs
- ✰ Phase rotation

Besides what has been stated, guidelines for the general maintenance of pumps to ensure maximum working efficiency are listed as under.

- ✰ The suction lift should be periodically checked and it should be within the permissible limits
- ✰ The gland packing in the pump should be checked and replaced if necessary. The water should drip through the packing at a rate of 15 to 30 drops minute
- ✰ Periodical inspection of impeller of the pump for wear and tear is necessary
- ✰ The RPM of the prime mover should be at the rated value
- ✰ The alignment of the pump and motor shaft should be regularly checked

Chapter 13

Open Channel Flow

Kinds of Flow

Flow in channels or conduit (Pipes) is of interest in science, engineering, and everyday life. Open-channel flows are those whose boundaries are not entirely a solid and rigid material; a part of the boundary of such flows may be fluid or nothing at all. Open-channel flow is a flow of liquid in a conduit with a free surface. That is a surface on which pressure is equal to local atmospheric pressure. Flows in closed conduits like pipes or air ducts, on the other hand, are entirely in contact with rigid boundaries. Most closed conduits in engineering applications are either circular or rectangular in cross section. The pipe flow does not have a free water surface. Irrigation water is generally conveyed in either open channel (canal network) or closed conduits (Underground pipeline systems or drip and sprinkler irrigation systems). Major differences between the two kinds of flow are given in Table 13.1.

Table 13.1: Comparison of Open Channel Flow and Pipe Flow

Open Channel Flow (OCF)	*Pipe Flow (PF)*
OCF must have a free surface	No free surface in pipe flow
A free surface is subject to atmospheric pressure	No direct atmospheric pressure, hydraulic pressure only
The driving force is mainly the component of gravity along the flow direction	The driving force is mainly the pressure force along the flow direction
Hydraulic grade line (HGL) is coincident with the free surface	HGL is (usually) above the conduit
Flow area is determined by the geometry of the channel plus the level of free surface, which is likely to change along the flow direction as well as with time	Flow area is fixed by the pipe dimensions

Open Channel Flow (OCF)	*Pipe Flow (PF)*
The cross section may be of any form circular to irregular forms such as that of natural streams, which may change along the flow direction as well as with time	The cross section of a pipe is usually circular
Relative roughness changes with the level of free surface	The relative roughness is a fixed quantity
The depth of flow, discharge and the slopes of channel bottom and of the free surface are interdependent	No such dependence

Important examples of open-channel flows are the flows in rivers, irrigation canal network including irrigation channels, even sheets of water running across the ground surface after a rain and tidal currents. Even a partially flowing pipe will be a case of open channel flow (Figure 13.1). Irrigation water is often conveyed in open channels, which receive water from natural streams or underground water and convey water to the farm for irrigation. The basic equations that govern the water flow in open channels are continuity equation, Bernoulli equation and Darcy-Weisbach equation (www.nptel.iitm.ac.in/courses/105107059/module1/./lecture1.pdf Opened on 13.02.2018).

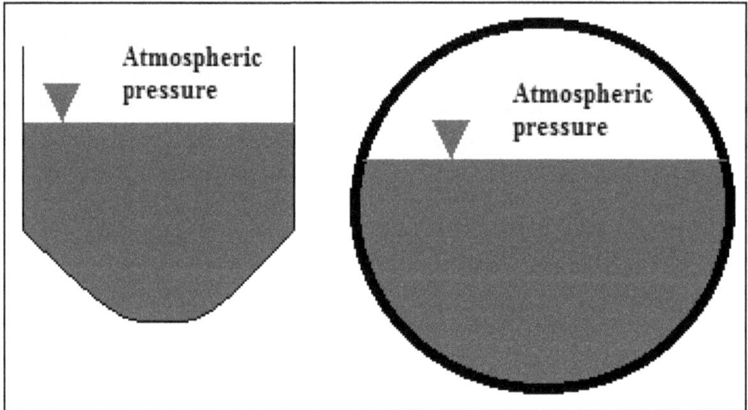

Figure 13.1: Free Surface in Open Channel Flow, Open Channel (Left) and Partially Flowing Pipe (Right).

Types of Open Channels

Prismatic and Non-Prismatic Channels

A channel in which the cross sectional shape, size and the bottom slope are constant over long stretches is termed as prismatic channel. Most of the man-made or artificial channels such as the rectangular, trapezoidal (Figure 13.2), triangular and half-circular in shape are prismatic channels. All natural channels generally have varying cross section and consequently are non-prismatic (James, 1988).

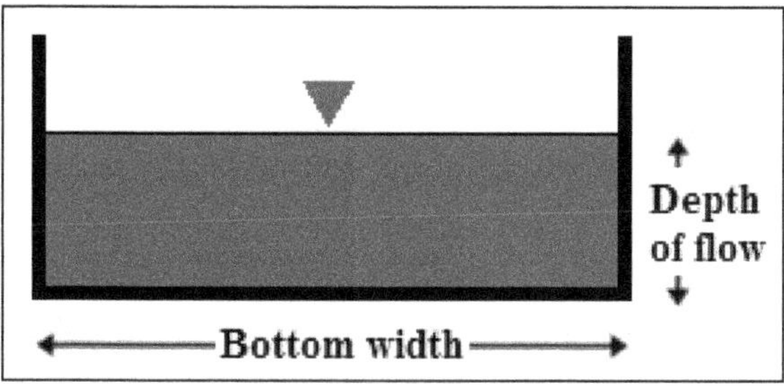

Figure 13.2: A Rectangular Prismatic Channel.

Rigid and Mobile Boundary Channels

Rigid channels are those in which the boundary or cross section is not deformable. The shape and roughness magnitudes are not functions of flow parameters. The lined channels and non-erodible unlined channels are rigid channels. Channels are classified as mobile channels when the boundary of the channel is mobile and flow carries considerable amounts of sediment through suspension and is in contact with the bed. In the mobile channel, depth of flow, bed width, and longitudinal slope of channel may undergo changes with space and time depending on kind of flow.

Unlined and Lined Channels

Depending upon the kind of channel surface, channels have been categorized as unlined and lined. Unlined channels are earthen channels without any lining. Lined channels on the other hand are lined with stones, masonry or even plastic sheet. As already described lined channels are rigid channels. Unlined channels are on the other hand mobile boundary channels.

Advantages of Unlined Channels

Earthen channels are suitable for areas where water is available in abundant quantities. The advantages of earthen channels are:

- ☆ Easy to construct
- ☆ Easy to change layout of field channels
- ☆ Low construction cost, minimal investments
- ☆ It does not require great skill for operation or maintenance
- ☆ Farmers are accustomed to handle these channels as knowledge has been acquired from past practices

Disadvantages

The disadvantages of earthen channels are as follows:

- ☆ High seepage losses
- ☆ High water evaporation and leakage losses

- ☆ Large fraction of water is wasted. In association with other unlined network components, it adds large amount of water to groundwater resulting in water table rise and soil salinization
- ☆ Takes larger areas for the same discharge due to limitations of velocity resulting in land loss
- ☆ There is a tendency for sediment deposition in the channels
- ☆ Weed and other aquatic plant growth can obstruct water flow; regular maintenance is required

Design of Open Channel

Open channel design comprises of finding out the cross-section dimensions of the channel for the amount of water the channel must carry (*i.e.*, capacity) at a given flow velocity, slope, and shape. It also includes determining the discharge capacity, if cross-section and other dimensional components are given.

Terminology

The commonly used terms in the design of open channels are:

Area of Cross Section (A): The flow area, A, is the cross-sectional area of flow normal to the direction of flow. Area of cross section is given by depth of flow times the bottom width in the case of a rectangular channels (Figure 13.2). If b = bottom width of the channel and d = depth of water flow, the area of wetted section of channel, A = b.d.

Wetted Perimeter (P): It is defined as the length of line of intersection of channel wetted surface with a cross-sectional plane normal to the flow direction. It is the sum of the lengths of that part of the channel sides and bottom which are in contact with water. The wetted perimeter for a rectangular channel (Figure 13.2) and with dimensions as in the area of cross-section, P = b+2d.

Hydraulic Radius (R): It is the ratio of area of wetted cross section to wetted perimeter. The hydraulic radius for a rectangular cross-section is given as:

$$R = A/P = bd/b + 2d) = d/(1 + 2d/b) \quad (13.1)$$

Hydraulic Slope (S): In free-surface flow, the hydraulic grade line usually, but not always, coincides with the free surface. If the velocity head, $V^2/(2g)$, in which V is the mean flow velocity for the channel cross section, and g is the acceleration due to gravity, is added to the top of the hydraulic grade line and the resulting points are joined by a line, then this line is called the energy-grade line. This line represents the total head at different sections of a channel. For a steady uniform flow, the channel bottom slope is taken as hydraulic slope. It is the ratio of vertical drop in longitudinal channel section (h) to the channel length (l).

$$S = h/l \quad (13.2)$$

The area of cross section, wetted perimeters and hydraulic radius for other commonly used cross-sections are described in the following sections.

Trapezoidal

The cross sectional area of flow in the trapezoid area, A is given as

$$A = d(b + B)/2 \tag{13.3}$$

$A = (d/2)(b + b + 2zd)$, because $B = b + 2zd$ (Figure 13.3 left).

Simplifying, the area is: $A = bd + zd^2$. (13.4)

Here z is the side slope, H:V = z:1

The wetted perimeter is: $P = b + 2l$. By the Pythagoras Theorem: $l^2 = d^2 + (dz)^2$, or $l = [d^2 + (dz)^2]^{1/2}$, so the wetted perimeter is:

$$P = b + 2d(1 + z^2)^{1/2} \tag{13.5}$$

and the hydraulic radius for a trapezoid is:

$$R = (bd + zd^2)/[b + 2d(1 + z^2)^{1/2}] \tag{13.6}$$

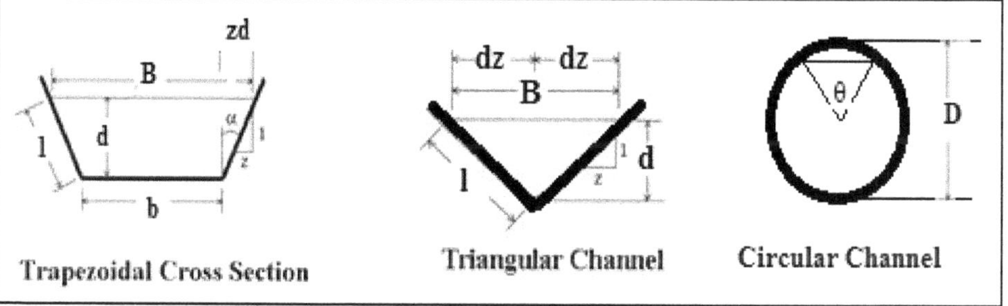

Figure 13.3: Trapezoidal (Left), Triangular (Middle) and Circular Cross-Sections of Open Channels.

Triangular

A triangular open channel cross section where the two sides are sloped at the same angle is shown in Figure 13.3 (middle). Fewer parameters are needed for the triangular area than for the trapezoidal area. The parameters, as shown in the diagram are: B, the surface width of the liquid; l, the sloped length of the triangle side; d, the liquid depth measured from the vertex of the triangle; and the side slope specification, H:V; z : 1.

The triangular area is: $A = Bd/2$, but the figure shows that $B = 2dz$, so the triangle area becomes:

$$A = d^2z \tag{13.7}$$

The wetted perimeter is: $P = 2l$ with $l^2 = d^2 + (dz)^2$. This simplifies to $P = 2[d^2(1 + z^2)]^{1/2}$

The hydraulic radius is thus

$$R = A/P = d^2z/\{2[d^2(1 + z^2)]^{1/2}\} \tag{13.8}$$

Circular

The hydraulic radius for a circular pipe flowing full is easy to calculate. The cross-sectional area is $A = \pi D^2/4$ and the wetted perimeter is $P = \pi D$. Substituting into the equation R = A/P and simplifying the expression gives: R = D/4. Here D is the diameter of the pipe.

If the pipe is not flowing full and the arc makes an angle θ at the center (Figure 13.3 right), then

$$\text{Top width} = D \sin \theta/2 \quad (13.9)$$

$$\text{Area, } A = 1/8(\theta - \sin \theta)D^2 \quad (13.10)$$

$$\text{Wetted perimeter, } P = \theta D/2 \quad (13.11)$$

$$\text{Hydraulic radius} = D/4(1 - \sin \theta/\theta) \quad (13.12)$$

Parabola

For a parabolic cross section,

$$A = 2/3\, t\, d$$

$$\text{Wetted perimeter, } P = t + 8\, d^2/3t \quad (13.13)$$

$$R = t.\, 2d/(1.5\, t^2 + 4d^2) \quad (13.14)$$

Here t is the top width. The approximate relation d =1.5 R is commonly used in the design of parabolic channels.

Freeboard

The term freeboard refers to the vertical distance between the highest water level anticipated in channel and the top of the retaining banks. It means providing extra depth above the designed depth, to take care of the following:

- Margin of safety when actual discharge exceeds the designed discharge capacity
- To take care of deviations between design and construction
- Non- uniform settlement of land after construction
- Operational flexibility
- Accommodating transient flow conditions (design is based on uniform flow)
- Increase in hydraulic roughness due to weed growth *etc.*

There is no universally accepted rule for the determination of free board. Free boards varying from less than 15 per cent to 25 per cent of the design depth are commonly used.

Discharge Capacity of Channel

For an irrigation channel, required channel capacity is estimated by the following equation.

$$Q = (Dr \times A)/(H \times E) \quad (13.15)$$

Here, Q is the channel capacity, cumecs, Dr is the design daily irrigation requirement, m/d, A is the irrigated area supplied by canal or ditch, m^2, H is hours per day that water is delivered and E is the irrigation efficiency including conveyance efficiency of canal or ditch, fraction.

Changing the units in normally used terms, we have:

$$Q = 16667 \times (Dr \times A)/(H \times E) \qquad (13.16)$$

Such that Q is the channel capacity, l/min, Dr is the design daily irrigation requirement, mm/d, A is the irrigated area supplied by canal or ditch, ha, H is hours per day that water is delivered and E is the irrigation efficiency including conveyance efficiency of canal or ditch, per cent.

Economical Section of a Channel

A channel section is said to be economical when the cost of construction of the channel for the same flow is minimum. Since the cost of construction of a channel depends on the depth of excavation and construction for lining, the cost of construction is minimum when the channel passes maximum discharge for a given cross sectional area. It is also evident from the continuity equation and uniform flow formulae that for a given value of slope and surface roughness, the velocity of flow is maximum when hydraulic radius is maximum. The hydraulic radius is maximum for given area if wetted perimeter is minimum. Hence the wetted perimeter, for a given discharge should be minimum to keep the cost down or minimum. This condition is utilized to determine the dimensions of economical sections of different forms of channels (Michael, 2010). Most economical section is also called the best section or hydraulic efficient section as the discharge passing through the most economical section of a channel for the given cross-sectional area (A), slope of the bed (S) and a roughness coefficient (n), is maximum.

The conditions for the most economical section of channel are:

A rectangular channel section is the most economical when either the depth of flow is equal to half the bottom width ($d = b/2$ or $b = 2d$) or hydraulic radius is equal to half the depth of flow ($R = d/2$).

A trapezoidal section is the most economical if b and d follow the relation:

$$b = 2d \tan \theta/2 \qquad (13.17)$$

Where θ is angle formed by side slope. If the trapezoid is half hexagon then

$$P = 2\sqrt{3}\, d \qquad (13.18)$$

$R = d/2$ or hydraulic radius is half the depth of flow. (13.19)

It may not always be possible to design for the most efficient cross section. As a compromise between hydraulic efficiency and other considerations, the ratios between bed width (b) and water depth (d) for trapezoidal channels are approximated (Table 13.2; Smedema and Rycroft, 1983).

A triangular channel section is the most economical when each of its sloping side makes an angle of 45° with vertical or is half square described on a diagonal and having equal sloping sides.

Table 13.2: Recommended Bed Width, b and Depth, d Ratios for Open Channels

Drain depth	Recommended b/d ratios for trapezoidal channels
Small channel, d=0.75 m	1 for clay to 2 for sand
Medium channel, d=0.75 to 1.5 m	2 to 3
Large channel, d=1.5m	3 to 4

Amongst the various shapes of the open channel the semi-circle shape is the most hydraulically efficient cross sectional shape. Since, the construction of semi-circle cross section for earthen unlined channel is difficult, trapezoidal section is commonly used for open channels.

Channel Dimensions

The channel dimensions are obtained using uniform flow formula, which is given by

$$Q = AV \tag{13.20}$$

Here, V is the flow velocity, m/s, A is the cross-sectional area of canal perpendicular to flow, m^2, Q is the capacity of the channel, m^3/s.

Velocity of flow is computed by the Manning's formula or the Chezy's formula. Herein, it is proposed to describe the use of the Manning´s formula only.

$$V = (1/n)\,(R^{2/3})\,(S^{1/2}) \tag{13.21}$$

Here, n is the Manning's roughness coefficient, R is the hydraulic radius of flow = A/P, m, P is the wetted perimeter, m, S is the channel bed slope, m/m. From the longitudinal section of the ground along the proposed alignment, the average slope of the ground shall be determined. This would be the maximum average slope which can be provided to the channel (ISI, 1974). For eq. (13.21) to be dimensionally homogeneous, n in MKS system has the unit of $T/L^{1/3}$. Manning's roughness coefficient, n, is selected from Table 13.3 or Table 13.4 or any other such tables reported in the literature (See Table 17.4).

Table 13.3: Limiting Velocities for Clear and Turbid Water from Straight Channels after Aging (*Source*: Fortier and Scobey, 1926, Schwab *et al.*, 1993)

Material	*Maximum permissible velocities (m/s)*	*n*
Fine sand	0.5	0.020
Vertical Sandy loam	0.58	0.020
Silt loam	0.67	0.020
Firm loam	0.83	0.020
Stiff clay	1.25	0.025
Fine gravel	0.83	0.020
Coarse gravel	1.33	0.025
Gravel	1.2	
Disintegrated rock	1.5	

Material	Maximum permissible velocities (m/s)	n
Hard rock	4.0	
Brick masonry with cement pointing	2.5	
Brick masonry with cement plaster	4.0	
Concrete	6.0	

Besides what is stated, the channel design should ensure:

The flow velocities remain within a range of acceptable minimum and maximum values so that neither scouring nor sedimentation occurs in the channels (Kumar and Singh, 2005). The minimum permissible velocity is the lowest velocity that will prevent sedimentation and vegetative growth (some estimate give 0.6 – 0.9 m/s for sedimentation and 0.75 m/s for vegetation). The maximum velocities to avoid scouring of the channel bed are given in Table 13.3 and Table 13.4.

Table 13.4: Maximum Allowable Velocities for Non-vegetative Channels having different Channel Materials

Channel Materials	Manning's n	Maximum Allowable Velocities (m/s)		
		Clear Water	Water with Colloidal Silt	Water with Sand Gravel/ Fragment
Fine sand, colloids	0.020	0.46	0.76	0.46
Sandy loam, non-colloids	0.020	0.53	0.76	0.61
Silt loam, non-colloids	0.020	0.61	0.91	0.61
Alluvial silts, non-colloids	0.020	0.61	1.07	0.61
Ordinary firm loam	0.020	0.76	1.07	0.69
Stiff clay, very colloidal	0.025	1.14	1.52	0.91
Alluvial silt, colloidal	0.025	1.14	1.52	0.91
Fine gravel	0.025	0.75	1.52	1.14
Coarse gravel, non-colloidal	0.025	1.22	1.83	1.98
Cobbles and shingles	0.035	1.52	1.68	1.98
Shales and hard pan		1.83	1.83	1.52

The side slopes depend primarily on the engineering properties of the material through which the channel is excavated (Chow, 1955). The commonly used side slopes for different kinds of soil vary from nearly vertical to 3:1 (Table 13.5).

Table 13.5: Suitable Side Slopes for Channels Built in Various Types of Materials (Chow, 1959)

Material	Side Slope
Rock	Nearly vertical
Muck and peat soils	1/4 : 1
Stiff clay or earth with concrete lining	1/2 : 1 to 1 : 1

Material	*Side Slope*
Earth with stone lining for large channels	1 : 1
Firm clay or earth for small ditches	1 1/2 : 1
Loose, sandy earth	2 : 1
Sandy loam or porous clay	3 : 1

Example 13.1

Compute the mean velocity and discharge for a depth of flow of 0.30 m for a lined trapezoidal channel of 0.6 m wide and side slope of 1.5:1 (H:V). The Manning's roughness (n) is 0.012 and the bed slope is 0.0003.

Area of cross section, $A = bd + zd^2 = 0.60.\ 0.30 + 1.5\ (0.30)^2 = 0.315\ m^2$

Wetted perimeter, $P = 0.60 + 2(0.3)\ \sqrt{[(1.5)^2 + 1]} = 0.60 + 1.08 = 1.68$ m

Hydraulic radius $(R) = A/P = 0.315/1.68 = 0.187$

$V = (1/n)\ (R^{2/3})\ (s^{1/2}) = (1/0.012)\ (0.187)^{2/3}\ (0.0003)^{1/2} = 0.473$ m/s

Discharge $(Q) = A \times V = 0.315 \times 0.473 = 0.149\ m^3/s$

Example 13.2

Calculate the best hydraulic rectangular cross-section to convey a discharge $Q=10\ m^3/s$ with $n= 0.02$ and $S= 0.0009$.

For the best rectangular hydraulic cross-section,

$R = d/2$

$A = 2\ d^2$

$Q = A\ (1/n)\ (R^{2/3})\ (s^{1/2})$

$10 = 2d^2\ (1/0.02)\ (d/2)^{2/3}\ (0.0009)^{1/2}$

$d^{8/3} = 5.29$

$d = 1.87$

$b = 2d = 3.74$

Example 13.3

A trapezoidal ditch with a bottom width of 1.2 m and side slopes of 1V:4H is constructed on a slope of 0.010 m/m and has a depth of flow of 0.5 m. If n is 0.020, calculate the capacity of the channel.

$A = db+zd^2 = 0.5\ (1.2 + (4 \times 0.5)) = 1.6\ m^2$

$P = b+2d(1+z^2)^{1/2} = 1.2 + (2 \times 0.5)(1 + 16)^{1/2} = 5.32$ m

$R = A/P = 1.65/5.32 = 0.30$ m

If Manning's n is taken as 0.020 then

$Q = A\ (1/n)\ R^{2/3}\ S^{1/2}$

$Q = (1.6)\ (1/0.020)\ (0.30)^{2/3}/(0.01)^{1/2} = 0.22\ m^3/s$

Example 13.4

Design an open ditch to carry a discharge of 0.26 m^3/s. The soil is silt loam and maximum permissible slope of the channel bed is 0.1 per cent. Assume n as 0.02.

Discharge rate Q = 0.26 m^3/s

Side slopes for silt loam is taken as= 1.5:1

tan θ = 1/1.5=0.666 or θ=33.66 or θ/2= 16.83

Assuming d= 0.5 m, b=2 d tan θ/2 =2 × 0.5 × 0.3025=.30 m ~ 30 cm

$$\text{Length of side} = \sqrt{(1.5\times0.5)^2+(0.5)^2} = 0.901\,\text{m}$$

P= 2 × 0.90+0.30= 2.10 m

$$A = \frac{(3\times0.5+0.30)+0.30}{2}\times0.5 = 0.53\,\text{m}^2$$

$$S = \frac{0.1}{100} = 0.001\,\frac{\text{m}}{\text{m}}$$

$$R = \frac{A}{P} = \frac{0.53}{2.1} = 0.252\,\text{m}$$

$$V = \frac{(0.252)^{2/3}\times(0.001)^{1/2}}{0.04} = 0.315\,\text{m/s}$$

The velocity is within permissible limit.

Q= 0.53 × 0.315 = 0.167 m^3/s

Since the discharge to be handled is less than the design discharge, assumed dimensions will not suffice. Therefore, one more iteration should be made by increasing d by 0.1 m.

Assuming d= 0.6 m, b=2 d tan θ/2 =2 × 0.6 × 0.3025=.363 m ~ 36 cm

$$\text{Length of side} = \sqrt{(1.5\times0.6)^2+(0.6)^2} = 1.12\,\text{m}$$

P= 2 × 1.12+0.36= 2.6 m

$$A = \frac{(3\times0.6+0.36)+0.36}{2}\times0.6 = 0.756\,\text{m}^2$$

$$R = \frac{0.756}{2.6} = 0.29\,\text{m}$$

$$S = \frac{0.1}{100} = 0.001\,\frac{\text{m}}{\text{m}}$$

$$V = \frac{(0.29)^{2/3} \times (0.001)^{1/2}}{0.04} = 0.346\,\text{m/s}$$

The velocity is within permissible limit.

$Q = 0.756 \times 0.346 = 0.261\ m^3/s$

The capacity of the channel is higher than the design capacity and hence the dimensions are OK in this respect.

Example 13.5

Calculate the dimensions of the channel, if the land slope is 0.1 per cent and side slopes are H:V = 1:1. Assume Manning´s n =0.030. Assume that the drain is to carry a discharge of 4.95 cumecs.

Q = 4.95 cumecs

If we assume b = 2d

$A = bd + zd^2 = bd + d^2 = 3\ d^2$

$P = b + 2d\ (z^2+1)^{1/2} = b + 2.83\ d = 4.83\ d$

$R = A/P = 3\ d/4.83 = 0.62\ d$

Applying the Manning´s equation

$V = 1/n\ (R^{2/3} \times S^{½})$

For n = 0.030

$V = 1/0.030 \times (0.62d)^{2/3} \times (0.001)^{½}$

$= 33.33 \times 0.488\ d^{2/3} \times 0.0316 = 0.514\ d^{2/3}$

$Q = A \times V = 3\ d^2 \times 0.514\ d^{2/3} = 1.542 \times d^{8/3}$

$4.95 = 1.542 \times d^{8/3}$ or $d^{8/3} = 3.21 = 1.55$ m

With 20 per cent free board, d = 1.55 + 0.31 = 1.86 m

b = 3.72 m

Top width, $T = b + 2 \times z \times d = 3.72 + 2 \times 1.86 = 7.44$ m

Example 13.6

A trapezoidal channel has a base width b = 6 m and side slopes 1H:1V. The channel bottom slope is S = 0.0002 and the Manning roughness coefficient is n = 0.014. Compute the depth of uniform flow if $Q = 12.1\ m^3/s$

$A = b \times d + 2 \times (d^2/2) = d\ (b + d)$

$P = b + 2\sqrt{2}\ d = 6 + 2\sqrt{2}d$

$S = 0.0002$, $n = 0.014$, $Q = 12.1\ m^3/s$

$Q = (A/n)\ (R^{2/3} S^{1/2})$

$AR^{2/3} = 11.98$ (Incorporating S, Q and n)

$11.98 = d (6+ d)[(d (6+ d)/(6 + 2 \sqrt{2}d)]^{2/3}$

By trial and error d= 1.5 m

Example 13.7

A trapezoidal channel with bottom width of 6 m, side slopes of 3H:1V carries a flow of 50 m^3/s on a channel slope, S of 0.0015. The uniform flow depth in the channel is 1.3 m with n = 0.025. This channel is to be excavated in stiff clay. Check whether the channel will be susceptible to erosion or not.

d =1.3 m

b = 6 m

$A= (b+zd)\ d = 6+ 3 \times 1.3 \times 1.3 = 12.87\ m^2$

$V = Q/A = 20/12.87 =1.554\ m/s$

Which is higher than the permissible velocity (V = 1.25 m/s, Table 13.3)

Suggestion: Increase width, b, to reduce velocity:

b = 8.4 m, d = 1.3 m Corresponding area of flow $A = 8.4+3 \times 1.3 = 15.99$ m

V = 20/15.99 = 1.251 m/s which is equal to the permissible velocity

On Farm Structures for Water Conveyance

To make the best use of water being conveyed through the irrigation channel, regulation and control of water is achieved through water control structures. Some of the flow control structures used to regulate water flow at the farm are presented in this section.

Structures for Diversions and Channel Crossings

Water from the farm irrigation channel is diverted to branch channels or into field channels by means of junction boxes, gates and siphons. These structures are used as diversion structures or application structures *i.e.* to apply water from irrigation channels to the fields.

Check Gate

The check gate is a structure used to maintain or increase water level in an open channel. Check is placed in an irrigation channel to form an adjustable dam to control or raise the elevation of the water surface upstream by at least about 8 to 12 cm above ground surface. It allows the use of siphon tubes or turnouts to efficiently divert the water from field channels to the fields. Checks may be permanent or temporary, portable or stationary. They may check the entire flow or allow a portion of it to pass.

Divisor Box

These are used to divide the flow into two or more channels. The divisor boxes are constructed/placed in the water course. The box may have movable dividers for variable divisions or may have rigid construction for fixed divisions of flow.

Turnout

Turnouts or outlets are used in the conveyance system to deliver water from a small channel directly to field basins, borders, and distribution laterals. They usually include a structure with a control gate. Turnouts can be concrete structures or pipe structures. A turnout may have a fixed opening in the side and equipped with the device to control the area of opening. A structure is sometimes used upstream and downstream for stability and to control erosion. Other turnout structures use gates, siphons and pipes. Turnouts are needed to insure an equitable supply to each field and/or farm.

Siphon Tube

Siphon tubes are curved tubes made of plastic, rubber or aluminum. These pipes are laid over the bank of delivery channels to deliver water to the borders or the furrows (Figure 13.4). The siphon tubes are completely filled with water by dipping in the channel. One end of the tube is pulled (siphoned) over the bank of the delivery channel, and turned into the borders or the furrows. During this process, it is ensured that tube remains full of water. The water begins to flow due to siphon action, if there is sufficient operating head, and the tube is properly positioned and is full of water (primed). For a free-flowing tube, the effective operational head is difference in elevation between the water surface at the tubes entrance and the center of its outlet end (Figure 13.4 top). For a submerged tube, the effective operational head is difference in elevation between the water surface at the tube entrance and the tube outlet (Figure 13.4 bottom). The discharge of a siphon tube depends on its diameter, length, inside roughness, the number and degree of bends, the operating head, and the difference in elevation between the water surface at the upstream and downstream ends of the tube. Siphon tubes are available in diameters ranging between 13 and 150 mm and length of 1.2 to 3.0 m.

Flume

Flumes are constructed to carry irrigation water across streams, canals, gullies, ravines or other natural depressions. They may be open channels or pipes which are often supported by pillars or may be fixed to bridges. Open channels are made of concrete or wooden having rectangular or trapezoidal shapes. Alternatively steel, concrete or vitrified clay pipes are also used. However, when using pipes, one should ensure the these are positioned below the water surface at the upstream end so as to ensure that they flow full. The supporting structure may be made of timber, steel or concrete. Manning's equation is used to estimate the discharge of the flumes. Flumes constructed in specially shaped and stabilized channel section may also be used to measure flow.

Culvert

A culvert is any structure not classified as a bridge that provides an opening under a roadway, and other type of access or utility (CDoT, 2004). A culvert is designed and used under the following situations:

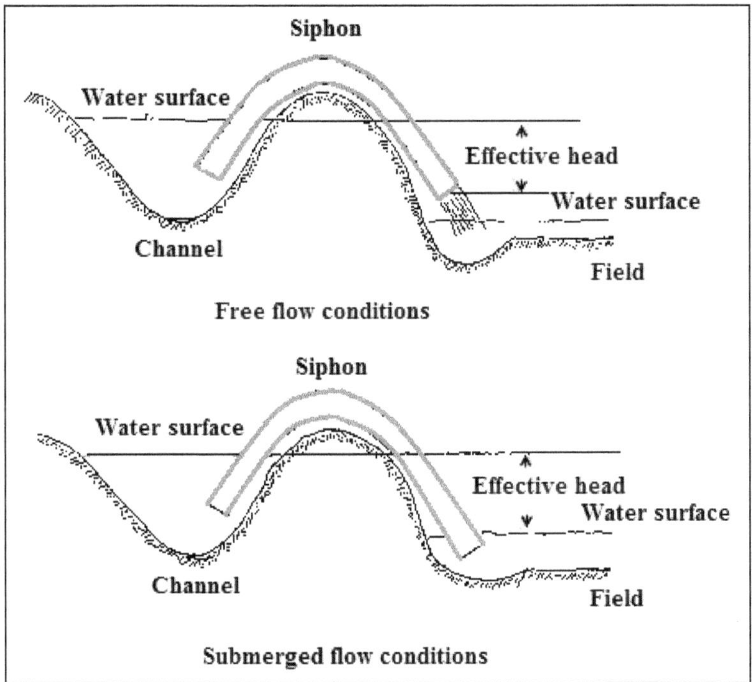

Figure 13.4: Siphon System Free Flow (Top) and Submerged Flow (Bottom).

- ☆ A structure that is usually designed hydraulically to take advantage of submergence to increase hydraulic capacity
- ☆ A structure used to convey irrigation water or surface runoff through embankments
- ☆ A structure, as distinguished from bridges, that is usually covered with embankment and is composed of structural material around the entire perimeter, although some are supported on spread footings with the streambed serving as the bottom of the culvert
- ☆ A structure that is 20 feet or less in centerline span width between extreme ends of openings for multiple boxes. However, a structure designed hydraulically as a culvert is treated as culvert regardless of its span

Culvert is most suitable structure at a channel crossing when the road fill is sufficiently high and the channel bed lies on the field surface on either side. Commonly used types of culverts are:

1. Pipe culvert (single or multiple)
2. Pipe arch (single or multiple)
3. Box culvert (single or multiple)
4. Arch culvert
5. Bridge culvert

About 45 cm of soil cover above the culvert pipe is desirable (Michael, 2010). The entrance and outlet conditions of the culvert structure directly impact its hydraulic capacity. A careful approach to culvert design is essential. For a pipe culvert, the design problem is primarily the determination of the type of flow that will occur under a given head and tail water conditions. Pipe flow (conduit controlling capacity) will occur under most conditions when the slope of the culvert is less than the natural slope and entrance capacity is not limiting.

Inverted Siphon

The inverted siphon (Figure 13.5) is constructed when a channel has to cross a wide depression or where the road surface lies close to the field surface. It has an inlet and an outlet tank connected together at their bottom by a pipe. A check gate is used at the inlet end to control the water surface level in the upstream channel. In some cases, the tank of an inverted siphon also acts as a stilling basin. In that case, the bottom of the tanks is kept about 15-20 cm below the bottom of the outlet pipe to collect the silt that will be deposited due to slow down of the water velocity.

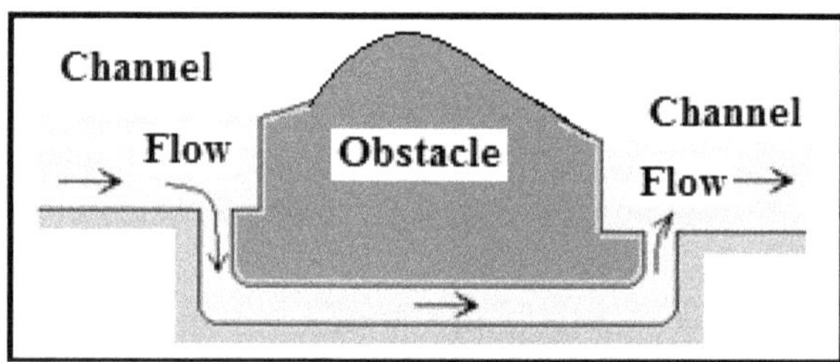

Figure 13.5: Inverted Siphon.

Drop

A drop structure is used to convey water in the channel when the channel has to deliver water from a higher to a lower elevation. In such cases, drops are used to control the flow velocities and to dissipate the energy of the flowing water in a protected section. The major function of the drops are:

- To control the water velocities upstream from the drop to prevent erosion
- To covey the flow to a lower level
- To dissipate the excess energy
- To control downstream erosion.

Chute Spillway

Chutes are lined, high -velocity open channels used to convey water from steep slopes. Chute structures are constructed with concrete, bricks or cement. They have an inlet, a steep-sloped section of lined canal where the elevation change occurs, a stilling pool or other energy dissipation structures, and an outlet section.

Pipe Drop Spillway

Pipe drop structure is used where a channel has to cross an embankment. In such cases water can be safely discharged from a higher to a lower level by providing a pipe drop. This type of structure allows the discharge of water through a pipeline, without disturbing the existing bunds or embankment. The components of structures are gated pipe, and stilling basin with end sill. Stilling basin is used to dissipate the energy of water flow.

Seepage in Unlined Channels

Seepage loss in unlined canals and farm ditches is a major problem of canal irrigation systems. In extremely sandy or gravelly ditches, a major fraction of the water is lost through seepage. If these losses can be reduced by improving water conveyance facilities, it can substantially increase the water availability to crops. Besides, it will permit irrigation of additional land, prevent water logging, increase in channel capacity, reduction in maintenance cost and more importantly enable to use available water sustainably. Minimizing the seepage losses is assumes greater significance in regions of water scarcity. One of the means to reduce seepage losses is through channel lining (Figure 13.6).

Figure 13.6: A View of a Lined Field Irrigation System.

Lining of Channels

The major advantages of earthen channel lining are as follows:

- ✰ Reduced seepage losses resulting in saving of water
- ✰ Prevention of water logging by reducing seepage contribution to water table
- ✰ Reduction in cross-section area of channels. As a result of increase in permissible velocity one can save land in channels. Lined channels also allow to avail steeper slopes in design wherever available, and thus, minimize excavation costs

- ☆ Improved discharging capacity of existing channels
- ☆ Improved operational efficiency since possibility of breaches are remote owing to sound structural stability of lined sections
- ☆ Prevention or minimization of the menace of aquatic weeds
- ☆ Reduced maintenance cost
- ☆ Long economic life
- ☆ Insures cross section stability from scour, low flow conditions *etc.*
- ☆ Considerable economy in the acquisition of cultivable land due to relatively narrow section of the canal
- ☆ Better environmental and aesthetic look

Disadvantages of Lining

Some of disadvantages are:

- ☆ Higher initial investment; for the same capacity a lined channel may be 3 to 4 times costlier than an unlined channel
- ☆ Costly repairs, whenever required
- ☆ Longer construction period

Materials for Lining Canals and Field Channels

Several lining materials are used to control seepage from canals and field channels. The commonly used channel lining materials are:

a) Hard Surface Lining Materials

1. Cement concrete or pre-cast concrete
2. Stone masonry
3. Brick tile or concrete tile
4. Asphaltic concrete

b) Earth Type Lining Materials

1. Compacted earth
2. Soil cement
3. Bentonite - clay soil mixture

c) Synthetic Sheet/Film

1. Rubber or synthetic materials
2. Low density polyethylene sheet

Selection of Lining Material

The following points are normally considered while selecting the method of lining and/or the lining materials.

- ☆ Availability and cost of the material at the site

- ☆ Labour available for lining at a reasonable cost
- ☆ Degree of seepage control desired
- ☆ Velocity of flow in the channel
- ☆ Useful life of the lining material
- ☆ Maintenance cost.

Other Alternatives

Alternative to lining of channels are:

- ☆ Portable surface pipe system (Figure 13.7)
- ☆ Underground pipeline systems

The underground pipeline system is discussed in chapter 14.

Figure 13.7: Water Conveyance through Portable Surface Pipe System.

Chapter 14

Underground Pipe Conveyance of Irrigation Water

Introduction

Water is one of the most valuable natural resources for agriculture. On the other hand, large volumes of water are lost in the conveyance network through seepage and leakage especially in surface irrigation systems comprising of main canal, branch canals, distributaries and minors and water courses. Besides, huge losses occur in on-farm unlined field channels. Ill-maintained field channels add to the woes resulting in poor on-farm water use efficiencies. Therefore, lining of the field channels has been recommended to minimize seepage/leakage losses. Although, a perfect lining can prevent most of the seepage/leakage losses, but the channel lining deteriorates with time. If not maintained properly, after few years losses in these channels may turn out to be of similar proportion as that of unlined channels. Therefore, current emphasis is on underground pipeline (UGPL) to deliver water from tube wells or canal water courses to the site of irrigation. To encourage UGPL, governments defray a part of the cost as subsidy. This chapter describes various components, and construction aspects of the underground or buried pipeline distribution system. It can make a major contribution in improving water management in on-farm surface irrigation systems. It will not only reduce seepage/leakage from the distribution system and but will also provide users with a more flexible irrigation supply.

Description

An UGPL water distribution system consists of buried pipes for conveying water to different points on the farm. It also includes allied structures required for the efficient functioning of the system.

Suitability

The pipeline distribution system is adopted under the following conditions.

- ☆ Limited water availability to serve a large command
- ☆ Steep topography or undulating terrain where open channels are expensive and difficult to construct
- ☆ Light textured soils where heavy seepage losses are anticipated
- ☆ Farmer´s willingness to implement modern technologies having sufficient funds for initial capital investment.

Advantages

Major advantages of underground conveyance of irrigation water are:

- ☆ No or minimal loss of water due to evaporation and seepage
- ☆ No or minimum loss of culturable command area
- ☆ Minimum interference with farming operations
- ☆ Flexible and better operational facilities
- ☆ Lower maintenance cost
- ☆ Higher project life span
- ☆ Requires lesser number of cross masonry and cross drainage works.

Limitations

- ☆ Higher initial investments
- ☆ Requires skilled manpower for installation
- ☆ Unsuitable if water has high silt content
- ☆ Difficult to identify leakage during normal operations.

Types of Pipeline System

Three types of irrigation water conveyance pipeline systems are commonly used.

- ☆ Fully portable system or surface pipes (See Figure 13.7)
- ☆ Combination of buried and surface pipes
- ☆ Fully buried system

In the first system water is supplied from source and applied to the field from open end of pipeline or using gated outlets. In the second system, buried permanent pipelines are used to transmit water from source to risers. These risers supply water to surface pipes to deliver to the fields. In the third system, water is delivered in the same manner as in the second case but here riser/alfalfa valves obviate the need for surface pipes. Water is released on the portion of the field to be irrigated from risers. This chapter discusses a fully buried system.

Classification

On the basis of pressure in the pipelines, UGPL systems are classified as low pressure systems (less than 10 m), medium pressure systems (10 m to 20 m) and high pressure system (more than 20 m). The low pressure systems are used for water conveyance while the medium pressure ones are used with drip systems and high pressure ones with sprinkler system (Murthy and Jha, 2018).

Pipes Types

Several kinds of pipes are used in constructing UGPL systems. Some of these are:

- Non-reinforced cement concrete (CC) pipes
- Reinforced concrete pipes (On-site made or factory-made)
- Vitrified clay pipes
- Asbestos cement pipes
- Plastic or PVC pipes
- Ductile iron (DI) pipes for 0.5 m to 1.0 m diameter
- Mild steel (MS) pipes for more than 1.0 m diameter

The CC pipes are generally used under low pressure up to a head of 6 m or 0.6 kg/cm^2 of water (Michael, 1978). Although CC pipes are cheaper but their use is limited due to low heads. Even at low heads, much skill and supervision is required to make good quality pipes.

At higher operating heads reinforced cement concrete (RCC) pipes or unplasticized polyvinyl chloride (UPVC) or polyvinyl chloride (PVC) pipes are used. If pressure in the main pipeline is up to 10 m, then RCC pipes of NP2 class can be used. For the design pressures in the range from 5 m to 25 m, PVC Class 2.5 pipes can be used conveniently.

Asbestos cement pipes are usually more costly than CC and RCC pipes but can be easily installed, have a long service life and can be adapted to a wide range of water pressures. These are made in 4 different grades to withstand pressure of 350 kN/m^2 to 1400 kN/m^2 as per IS 1592 -1989.

Most commonly used pipes are the reinforced concrete pipes and PVC pipes especially in the command areas of tube wells. Now-a-days, PVC pipes are preferred as these are easily available, are easy to handle, easy to install and easy to ensure leak proof joints. Besides, fittings of the same or similar materials are available that makes it easy to assemble the system quickly and efficiently. Moreover, for the same diameter, PVC pipes can convey more water than concrete pipes because of lesser internal friction of the PVC pipes. The only prevention required is to see that all sections of PVC pipes are buried under the ground so as to avoid UV degradation of pipes. To achieve this objective under undulating topography, one may have to lay pipe deeper. However, HDPE pipes can be used even above ground level.

Dicharge Capacity

The discharge capacity of the pipeline can be determined using the Darcy–

Weisbach formula. By re-writing the eq. (12.14) in the following form, one can calculate the velocity and hence the discharge.

$$V = (Hdg/2f\,L)^{1/2} \quad (14.1)$$

Here V is the velocity of flow of water through the pipe, cm/s, H is the available head causing flow, cm, d is the diameter of pipe, cm, g is the acceleration due to gravity, cm/sec^2, L is the length of pipe, cm and f is the Darcy's roughness coefficient. Recommended maximum velocities in low pressure pipelines are in the range of 1.3 to 1.5 m/s. High velocities help to reduce the diameter of pipe and hence cost but may result in higher frictional losses and higher cost of protection against water hammer. Minimum flow velocities should be around 0.5 m/s to prevent sedimentation of fine sand.

The values of f for laminar flow in circular pipes can be calculated by the following equations.

$$f = 64/Re \quad (14.2)$$

Here Re is the Reynold's number in the range of 500 to 2300.

For turbulent pipe flow, the equation is given as:

$$1/\sqrt{f} = -1.8 \log [\{(e/d)/3.7\}1.11 + 6.9/Re] \quad (14.3)$$

The values of Re in eq. (14.3) vary from 100 to 10^6. The parameter e/d is the relative roughness of a pipe, *i.e.* it is roughness divided by its internal diameter. Alternatively, the Darcy friction factor is related to Manning's n through the following relationship:

$$n = R/6\,(\sqrt{f}/8g) \quad (14.4)$$

Here n is the Manning's friction factor and R is the hydraulic radius.

For concrete pipes, the values can also be selected directly from Table 14.1. For other materials, the tabulated values are available in the literature.

Table 14.1: Darcy-Weisbach Coefficients for Concrete Pipes

Diameter (cm)	*Velocity (cm/s)*				
	30	*60*	*90*	*120*	*150*
15	0.0102	0.0094	0.0090	0.0086	0.0084
22.5	0.0080	0.0074	0.0070	0.0067	0.0064
30	0.0067	0.0062	0.0059	0.0056	0.0054
45	0.0056	0.0051	0.0048	0.0046	0.0044
60	0.0050	0.0046	0.0044	0.0042	0.0040
90	0.0044	0.0040	0.0038	0.0036	0.0034
120	0.0039	0.0037	0.0035	0.0033	0.0030

For the given diameter, discharge capacity of the pipe can then be determined by the relation:

$$Q = AV \quad (14.5)$$

Such that Q is the pipe capacity, m^3/s, A is the area of cross section of the pipe, m^2 and V is the velocity, m/s. Thus, V calculated with the help of eq. (14.1) need to be divided by 100 for use in eq. (14.2).

Similarly, one can also use the Hazen-Williams equation for the same purpose.

$$V = 0.85\, C_h\, R^{0.63}\, S^{0.54} \qquad (14.6)$$

Here, v is the average velocity of flow, m/s, C_h is the Hazen-Williams coefficient (dimensionless), R is the hydraulic radius of flow conduit, m and S is the ratio of h_L/L: energy loss/length of conduit (m/m). The values of C_h to be used to solve eq. (14.6) are given in Table 14.2.

Table 14.2: Hazen-Williams Coefficients for Various Pipe Materials

Type of Pipe	*C_h*	
	Average for New Clean Pipe	*Design Value*
Steel, ductile iron or cast iron with centrifugally applied cement or bituminous lining	150	140
Plastic, copper, brass, glass	140	130
Steel, cast iron, uncoated	130	100
Concrete	120	100
Corrugated steel	60	60

Exmple 14.1

What is the velocity of flow of water in a concrete pipe of 15.4 cm diameter if the energy loss is given as 6 m of head over a length of 300 m? Compute the volume flow rate at this velocity.

The nearest value of f is selected from Table 14.1 as 0.0084

Using eq. (14.1)

$V = (Hdg/2f\,L)^{1/2} = [600 \times 15.4 \times 981/2 \times 0.0084 \times 30000]^{1/2} = 134.10$ cm/s or 1.34 m/s

The area of the pipe is $(\pi/4)\,(15.4/100)^2 = 0.0186\ m^2 \sim 0.019\ m^2$

Using eq. (14.5)

$Q = VA = 1.34 \times 0.019 = 0.0254\ m^3/s$

Exmple 14.2

What will be the velocity of flow of water in the above example if the pipe is made of plastic? Compute the volume of flow rate at this velocity. Use the Hazen-Williams equation for the solution.

$S = h_L/L = 6/300 = 0.02$

$R = d/4 = (0.154)/4 = 0.0385m$

$C_h = 130$ (Value of C_h is taken from Table 14.2)

Then, using eq. (14.3)

$V = 0.85C_hR^{0.63}s^{0.54} = 0.85(130)(0.0385)^{0.63}(0.02)^{0.54} = 1.717$ m/s

$Q = A\ V = (0.019 \times 1.717) = 0.033$ m^3/s

Structures for Underground Pipeline System

For the efficient performance of the underground system, several kinds of structures are used. These are:

- ☆ Pump stand (inlet structure)
- ☆ Gate stand (Diversion structure or distribution stand)
- ☆ Air vents
- ☆ Outlet (riser, alfalfa valve, hydrants and gated pipes)
- ☆ End plugs

Pump Stand

The pump stand, located at the inlet end of underground pipeline system, is to convey the flow of the water into the pipe system. The size of the pipe used to construct pump stand is larger than the diameter of pipeline. Besides the flow, it also performs several other functions. For example, it helps to dissipate high velocity stream and release of entrapped air before water enters the pipeline. The larger diameter also allows access for repairs. The height of the pump stand is designed in such a way that it provides needed pressures at all the pipe outlets. Depending upon the source of water or the need for pumping, pump stands have been categorized as:

- ☆ Gravity inlets
- ☆ Pumped inlets

Gravity Inlets

A gravity inlet is used only when water surface elevation of the water source is sufficient to allow gravity flow into the pipeline and to provide adequate pressure at every point of pipeline and outlets. The low head UGPL system directly connected with water source is generally used to deliver water from canal networks. If the irrigation water contains appreciable quantities of silt, a silt trap to remove the suspended material is built into the inlet structure so that water passes through it before entering the inlet. If the silt load is relatively less, then stand itself can functions as silt trap (Figure 14.1). In that case, it has an extra-large diameter to ensure low velocity of water and its bottom is set about 60 cm below the bottom of the pipeline. A debris screen is fixed to the inlet to keep the trash out of pipeline. The top of the structure is provided with a cover to prevent accidents and to keep the trash out from blowing into it. For small field scale systems, inlet consists of simple masonry tanks of required height and about one meter square. A concrete bed 10 cm thick is prepared at the bottom for stands up to 3 m height.

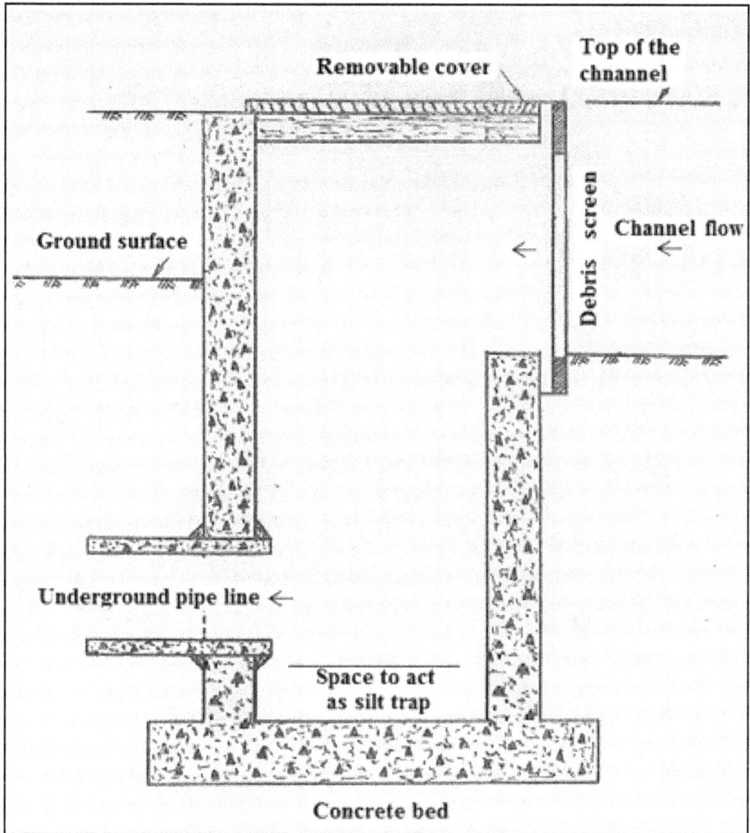

Figure 14.1: A Gravity Inlet for Underground Pipeline (Note the space for silt trap).

The total height of the pump stand is calculated as follow:

Total height of inlet stand = 0.60 m below the bottom of pipe + Diameter of pipe + Depth of pipeline below the ground + Maximum depth of flow in the canal + Freeboard (14.7)

Pumped Inlets

If the water surface elevation of the source is insufficient to allow gravity flow into the pipeline and to provide adequate pressure at every point of pipeline and outlets, a pumped inlet is used. In most tube well based irrigation systems, this kind of inlet is used. There is not much difference in the design of a gravity and pumped inlet (Figure 14.2). Reinforced cement concrete (RCC) rings having an internal diameter of 1.2 m or 0.9 m can be used to construct pump stand. The rings with base plate are also available (monolithic) that makes the process of laying the bottom piece easier.

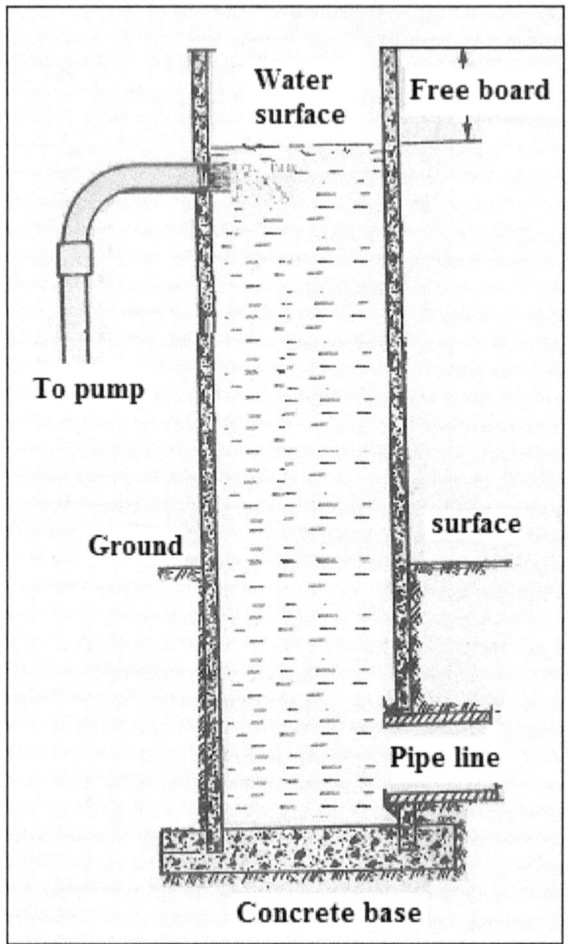

Figure 14.2: Pump Stands for Underground Pipeline System.

Gate Stands

Gate stands are installed to divert and control the flow into branch lines. As such, these are installed where branch lines take off from the main line. The gate stands also prevent high pressure and act as surge chamber. Each outlet of a gate stand is equipped with slide gate or gate valve so as to release water through a particular gate valve. Figure 14.3 shows branching off water from main pipeline in three directions (1, 2 and 3) through the gate stand. The gate stand is constructed in a similar manner as the pump stand.

Air Vent

If air is entrained in a pipeline at any location, it can cause discontinuous flow, resulting in reduction of conveyance capacity and creating high pressure induced surges or water hammer. Air vents are vertical pipe structures to release

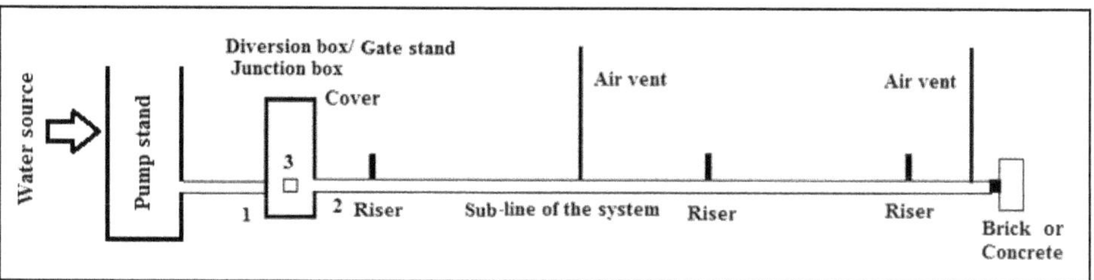

Figure 14.3: Gate Stand and Other Structures for Underground Pipeline System.

air entrapped in the pipeline and to prevent vacuum. The area of vent pipe should not be less than half the area of pipeline. In no case the diameter of the small pipe should be less than 5 cm. The first air vent is installed near (3 to 5 times the diameter of the pipeline) to the pump stand (Raghuvanshi, 2016). If the pipe system has significant changes in relief, vent stands should be placed at summit points in the pipeline or where there are changes in grade in a downward direction of more than ten degrees. For a long pipe, air vents should also be installed at every 150 m. They are also required immediately upstream from gates where closure of gates would make such points the downstream end of laterals or line. The air vents are also provided at the end of the pipeline. The height of the air vent is kept in such a

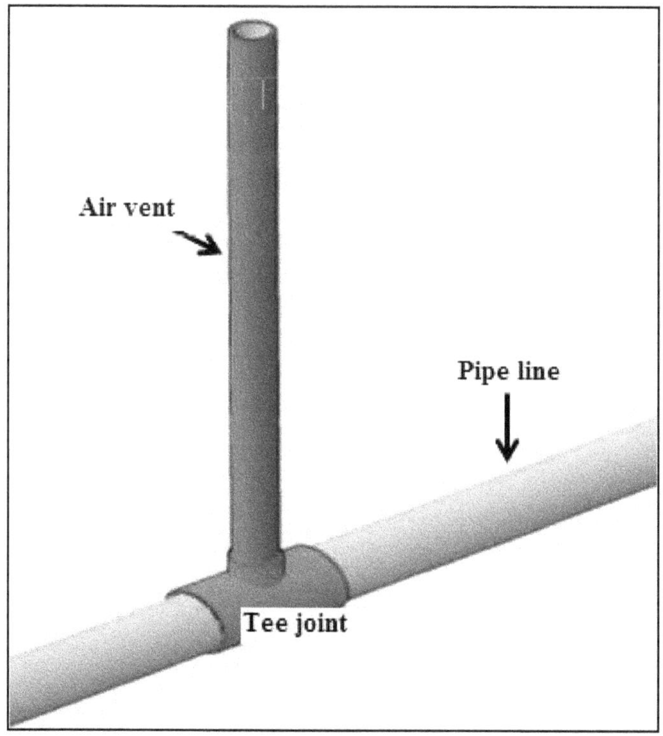

Figure 14.4: An Air Vent on the Underground Pipeline System.

way that the top of the pipe is at least 25-60 cm (free board) above the maximum-design hydraulic grade line. A typical vent stand is shown in Figure 14.4 and there locations on a branch line in Figure 14.3. Note the simplicity in the installation of air vents in a PVC system. A simple TEE joint is used to install the air vent. It can also be provided in two sizes. In that case, the diameter of pipe up to 60 cm above the soil surface is not less than ½ the diameter of the pipe. Thereafter the diameter can be reduced up to 5 cm as a means to reduce the cost. An additional reducer is all that will be required to install this kind of an air vent. Pipelines designed to operate under pressure may be provided with air relief valves than open air vents.

Outlets

These are established mainly on branch lines (can also be installed on main line) to deliver the water from the pipeline system to the fields. An outlet consists of riser pipe joined vertically to the main pipeline. While designing the pipe distribution network, outlet are provided as per requirement of the farm and to ensure minimum application loss. The diameter of the riser pipe is kept the same as the pipeline system where the entire flow of the pipeline is to be released through the valve. The riser pipes are installed on the pipeline by cutting a hole on the upper surface of the pipe (Figure 14.5). The riser pipe is saddled in the opening and permanently joined. At the end of the riser pipe near the ground level a valve (an alfalfa or alpha-alpha valve) is fitted (Figure 14.5). The valve is used to control the flow of water. Two variants of the valves used in underground pipeline using plastic pipes is shown in Figure 14.6. Hydrants and gate pipes are connected with riser pipe to distribute water to furrow or a border or a basin.

Total available head at any outlet = Head at the pump stand + Difference in elevation at inlet and the outlet - Total head loss (14.8)

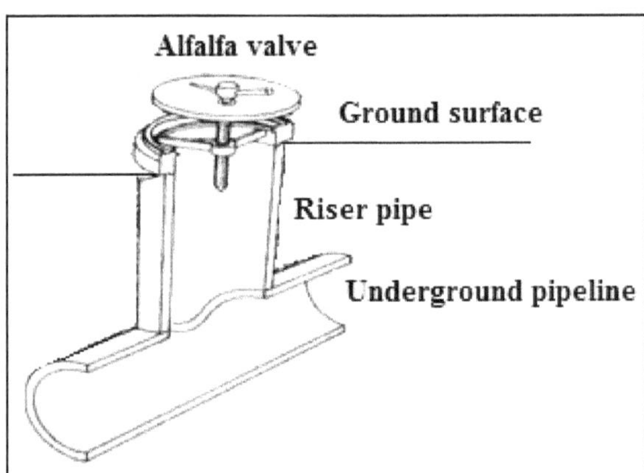

Figure 14.5: A View of the Riser and Alfalfa Valve at the Outlet.

Hydrants

Hydrants are portable devices placed over the riser valve outlets. They are moved from one valve outlet to another to serve the portion of the field which is being irrigated at a particular time. These are used to connect portable gated pipes to the pipeline. If the water pressure in pipeline is low, then hydrants can also be used for connecting the suction hose of a pump to the water supply, to develop high pressure.

Figure 14.6: Different Kinds of Valves Used in Underground Pipeline Systems of Plastic Pipes.

End Plug

The end plugs are provided at all closing points of the pipeline. The plug should protrude inside the pipe for about 10 to 15 cm (Figure 14.7). The plugs absorb the pressure developed at the end of the line on account of water hammer. The end of the plug should be supported sufficiently to withstand the operating pressure that may develop due to sudden opening or closing of valves. As such, the plugs are backed by concrete/masonry blocks to provide sufficient strength.

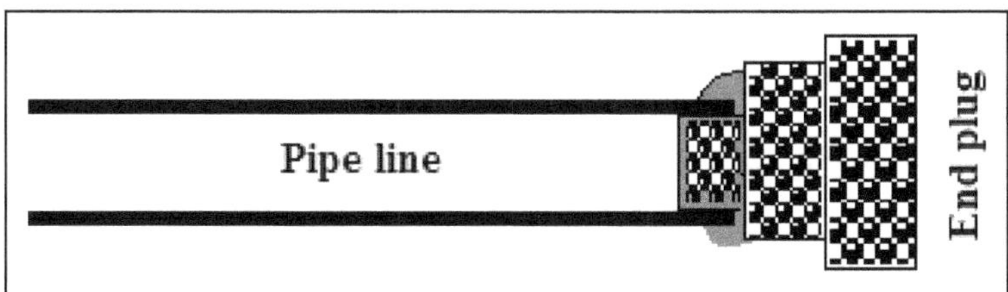

Figure 14.7: Concrete End Plug and Wall.

Installation of Underground Pipelines

For a trouble free operation of UGPL system one should rigorously follow the sound engineering practices. Following steps may add up to the construction of an efficiently running system.

- ☆ Selection of pipes, depth and grade as per the hydraulic design. No cracked pipe should be used
- ☆ Digging the trenches to proper depth and grade with sufficient working space
- ☆ The bottom of the trench is compacted and smoothened for uniform foundation
- ☆ The bed at each joint is scraped to a small pit so that sand filling and jointing work can be conveniently accomplished
- ☆ using rope for lowering the pipe and for bringing the pipes end to end
- ☆ Sealing the joints
- ☆ Use of moist soil for the back-fill after 12 hours of sealing the joints.

Some of the activities are briefly discussed in the following sections.

Grade and Alignment

Pipelines should be straight and of uniform gradient from reach to reach or between stands or anchors. This can be controlled by proper excavation of the trench. The alignment and grade for the bottom of the trench should be properly established prior to the excavation of the trench.

Excavation

Trenches should be reasonably straight, with the bottom reasonably free of undulations and humps. All trenches should be excavated with vertical sides and a well-graded bottom. All excavated material should be deposited on one side of the trench and the pipe placed on the other side. Curves are allowed with a maximum deflection of five degrees in mortar joints and a maximum of 3 degrees in rubber gasket joints.

The width of the trench should be enough to permit laying of pipe correctly and for proper bonding or finishing of the joints. There should be a minimum clearance of 15 cm from the outside of the pipe to the sides of the trench.

Where trenches are excavated in clay soils or those that show evidence of shrinking or swelling, the trench should be excavated 10-15 cm below grade and brought up to grade with a compacted layer of loam or coarser textured soil. In all clay or expansive soils, the pipe should be deep enough so that the soil mass surrounding it will not be subject to changes in moisture and temperature.

When water is encountered in the trench, it should be removed by draining or pumping. Should water get into the trench before the pipe is laid, postpone the laying of pipe until the trench is dried sufficiently to provide a firm foundation for the pipe. The other option could be to remove the mud and softer material followed by re-establishing the grade by backfilling with some suitable compacted material.

Dimensions

All pipes should be placed deep enough below the land surface to protect it from hazards imposed by movement of heavy machinery during farm operations or soil cracking. The minimum depth of cover should be 45 cm for pipe sizes up to and including 30 cm diameter, and 60 cm for pipelines greater than 30 cm diameter in cultivated coarse textured soils. In heavy soils such as clay, the depth of cover should be increased by about 15 cm. The soil texture in this case is for the soil at the pipe depth, rather than that of the surface soil.

Foundation

Foundations for pipelines should be firm but should yield slightly and uniformly when pipe is bedded to establish line and grade. Sometimes a firm foundation is not encountered, due to soft, spongy or other such materials present in the profile. All such unstable material under the pipe and for a width of not less than one diameter on each side of the pipe should be removed. The space so created should be backfilled with suitable material, such as coarse sand. It should be properly compacted to provide adequate support for the pipe.

Placement

At the time of laying the pipe, the prepared trench should be reasonably dry. Necessary facilities should be provided to lower and properly place the sections of pipe in the trench without damage. Immediately before placing each section of pipe in final position for jointing, the bedding for the pipe should be made by scraping away or tamping under the body of the pipe and not by wedging or blocking. The interior of the pipe should be kept free from dirt and other foreign material while the pipe laying work is in progress, and left clean at the completion of the work. Any pipe which is not in true alignment, or which shows any undue settlement after laying or is damaged, should be taken out and re-laid.

Backfill

Backfill should be placed in uniform layers. Each layer should be carefully and uniformly compacted. If the fill is not compacted, a mound over the top of the trench should be made as to allow for settlement. There should be an initial backfill of soil around the pipe covering the pipe to a depth of at least 15 cm for the full width of the trench. Necessary care should be exercised so that joints are protected from drying out. If the soil used in the initial backfill is not thoroughly moist, the joints should be covered with a layer of jute bags which are kept moist.

Joints

All joints shall be constructed to withstand the design maximum working pressure for the pipeline without leakage and shall leave the inside of the line free of any obstruction that may reduce its capacity below design requirements. Manufacturers' recommendations for joining pipe shall be used when not in conflict with specified requirements of the organization. The common types of joints are:

- Collar joints
- Tongue and groove joints
- Bell mouth-socket and spigot joints

Collar Joints

Collar joints are commonly used to join the concrete pipes having plain ends. The pipes are butted together and joined with a concentric concrete collar. Before the pipe is laid a collar is slipped on one of its ends. The butted pipe ends are caulked with mortar or jute rope dipped in hot bitumen. If rope and bitumen are used, the pipes must be pressed together before placing the collar. A jack is commonly used to squeeze the pipes. Four or five pipes can be squeezed at one time. The collars are then slipped over so that half the collar width covers each side of the joint. The gap between the pipe and collar is then filled with jute and tampered with an appropriate tool. Once the gap is sealed satisfactorily the collar is lined with mortar and beveled at an angle of 45° to the pipe. The mortar used in this operation is prepared by using 1 part of cement to 2 parts of sand.

Bell Jointed Pipes

Pipes with bell end are laid with their sockets (bell ends) facing upwards. Jute or hemp rope after dipping in a cement paste is wrapped around the plane end of the pipe. After applying about 1 cm thick mortar (1:2 cement : sand) on the spigot and bell mouth of the two pipes to be joined. The spigot end of the pipe is pushed into the bell taking care that the packing is not pressed through the pipe. A chisel is used to ram the rope so that it tightly sets into the bell. Finally the joint is finished with the mortar with a bead on the outside.

Tongue and Groove Joints

Pipes with tongue and groove joints are laid with the groove ends facing upstream. The two ends are cleaned and wetted with a brush. Sufficient mortar is applied to the lower side of the joint. The groove end of the pipe filled with mortar is then tipped over so as not to dislodge the mortar but forming a snug tight joint. Excess mortar will be squeeze out of the joint on both sides. It is used to form a mortar band on the joint. Additional mortar is used to finish the band.

Joining Plastic Pipes

Solvent cement for use with PVC pipe and fittings shall meet the requirements of ASTM 02564. To make good joints, the following procedure should be adhered to.

Joints

- ☆ The surfaces to be joined must be softened and made semi-fluid
- ☆ To get a leak proof joint, sufficient cement must be applied so as to fill gaps between the pipe and the fitting. The pipe and the fitting must be assembled while the surfaces are still fluid and wet.
- ☆ Joint strength develops with drying of the applied cement. In the tight part of the joint, the surfaces tend to fuse together. In the loose part, the cement bonds both the surfaces. If the cement coatings on the pipe and fittings are wet and fluid when assembly takes place they will tend to flow together and become one cement layer.

Openings in Pipelines

All openings cut into the pipe for outlets and connections should be cut to fit closely and should be strongly cemented to the pipe with mortar.

Testing

Before backfilling, the pipeline should be thoroughly and completely tested for pressure, strength and leakage. The line shall be slowly filled with water. The velocity of water input shall not exceed 0.3 m/s (1 ft/s). Adequate provision shall be made for air release while filling taking care to bleed all entrapped air in the process. The pressure shall be slowly built-up to the designed working pressure. Entire length of the pipeline should be checked while the maximum working pressure is maintained. If leaks are discovered they shall be repaired and the line retested.

According to the Mridha (1993), tests on eight schemes revealed that leakage problems are encountered as below:

- Maximum seepage was observed when handmade vertical moulded pipes were used. These pipes contained more voids in the pipe wall. Besides, pipe strength as well as its durability was affected
- Use of faulty materials such as cracked pipes and pipes cracking when exposed to sun while a leakage was being repaired
- Inadequate curing and poor joints.

Cost of Plastic Pipes and Related Materials

Although, it may be difficult to assign a reasonable value to the likely investments required in laying an underground pipeline system, yet a good guess can be made for a specific site, if cost of various components are added together (Table 14.3). Depending upon the area and other requirements, the bill of quantities (BoQ) may be prepared for the items in Table 14.3. The BoQ multiplied by the unit cost of each item when added together will give a reasonable first assessment of the cost.

Table 14.3: Cost Components of an Underground Pipeline System

Cost Component	*Remarks or Units*
Detailed Project Report	For large projects only (5 per cent of the total cost)
Earth work in trenches (Lx Bx D)	m^3
Cost of pipes	Rs. per m (See Table 14.4 for plastic pipes)
Cost of other materials	Rs. per number (See Table 14.4 for plastic pipes)
Cost of collars	Rs. per number
Cost of bricks, cement and sand	Rs. per number, bags and m^3
Lowering of pipelines in trenches	Lump sum
Joining of pipes	Labour rate
Riser with riser valves	Rs. per m
Pump stand	Rs. per system

Cost Component	*Remarks or Units*
Gate stand	Rs. per number
Air vents	Rs. per m
End plug	Rs. per number
Carriage charges	Actual cost to be assessed
Contingencies	10 per cent of the total cost

Currently plastic pipes are commonly used to construct UGPL systems at the farms (CWC, 2017). The costs of HDPE and PVC pipes and few other accessories used in the system and approved by Government of Haryana are given in Table 14.4 (Director of Agriculture, Haryana, 2016).

Table 14.4: Cost (Rs.) of Plastic Pipes and Related Accessories

Sl. No.	*Component Size (mm)*	*Size (mm)*							
	HDPE Pipe System								
		75	90	110	125	140	160	180	200
1	Pipe 6 m (2.5 kg/m²)	455	600	875	1110	1380	1720	2150	2660
2	Pipe 6 m (3.2 kg/m²)	545	790	1130	I350	1660	2150	2590	3210
3	PCN (I pc)	250	325	410	502	601	671	801	911
4	Tee (I pc)	116	150	188	207	243	261	308	340
5	Bend (I pc)	98	120	152	190	207	224	279	330
6	End cap (I pc)	72	85	iI0	120	135	150	175	200
7	Quick coupler (I pc)	110	125	150	185	220	250	285	320
8	Air release valve (1 pc)	235	235	235	235	235	235	235	235
	PVC Pipe System								
1	Pipe 6 m (2.5 kg/m²)	-	390	570	700	900	1150	1380	1690
2	Pipe 6 m (4 kg/m²)	400	580	830	1040	1320	1670	2100	2520

Chapter 15

Farm Drainage

Introduction

From management point of view, waterlogged lands are classified into 2 categories with lands having shallow water table further categorized in 2 sub-categories:

Waterlogged Lands

Surface water stagnation	
Shallow water table	Good quality groundwater
	Poor quality groundwater

Surface stagnation is either due to excess rainfall or irrigation. In the monsoon climatic conditions prevalent in India, surface stagnation of water is a serious problem especially in the humid and sub-humid areas. Even arid and semi-arid regions experience surface stagnation because one day maximum storms can be quite high. Flood irrigation often result in short-term water stagnation. Problem of shallow water table is quite widespread in irrigation commands, coastal areas and areas having saline groundwater. Shallow water table in general accentuates the problem of surface stagnation as rainwater cannot be absorbed in adequate quantities under this situation.

Salinity problem is often associated with waterlogged areas having shallow water table and where groundwater is of saline/sodic in nature. Leaching, an essential pre-requisite to reclaim or manage salt affected soils, often fails under the condition of shallow water table. Thus, depending upon the kind of problems and soil conditions, either surface or subsurface drainage or a combination of both are needed to minimize adverse effects of water logging and soil salinity. This chapter briefly describes some of the appropriate surface and subsurface drainage technologies that can be adopted under varying agro-climatic conditions.

Irrigation and Drainage Systems

For optimum crop productivity, both irrigation and drainage systems are essentially required. Whereas in irrigation system water is diverted from a river to the command for irrigation purpose, a drainage system collects the excess water in the command and takes back to the river system. Thus, together these systems save the crop from inadequate and/or excess water, both being harmful to the crop. The size of the irrigation network reduces with distance from the diversion point but the size of the drainage channels increase with distance and is largest at the outlet. It is because the amount of water collected increases with distance. Since both the systems are complimentary in nature, all fields on a farm must have an irrigation channel at the top and a drainage channel at the bottom (Figure 15.1).

Definition

Agricultural drainage is defined ´as the removal and disposal of excess water from agricultural lands´. Since agricultural drainage besides removal of water is also necessary to remove salts from the soil, definition of drainage has been refined by the International Commission on Irrigation and Drainage (ICID, 1979). According to ICID (1979), drainage is defined as: ´Land drainage is the removal of excess surface and subsurface water from the land to enhance crop growth, including the removal of soluble salts from the soil´. The above definition is most commonly applied in practice although several other definitions are in use.

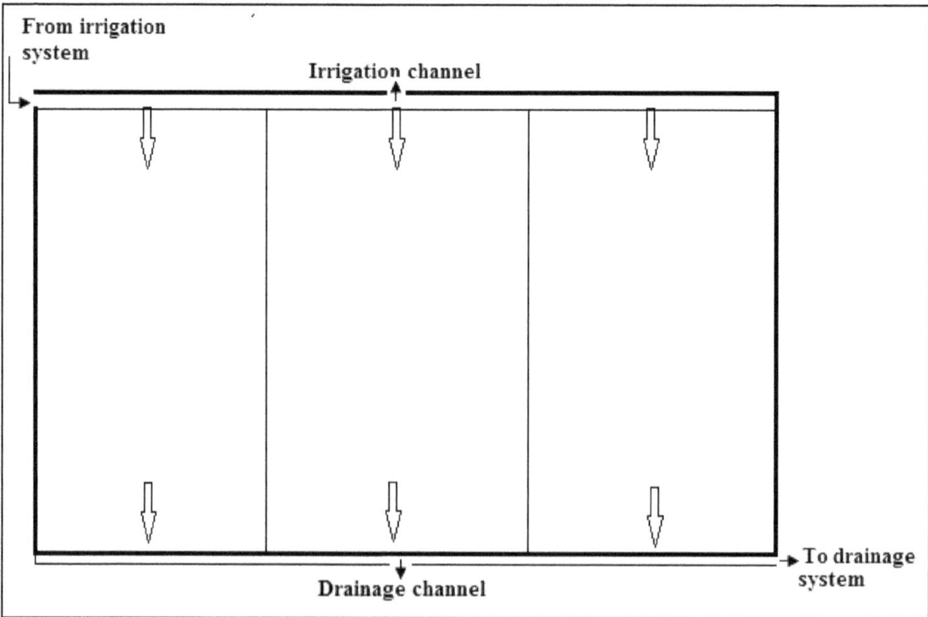

Figure 15.1: Schematic View of the Layout of Irrigation and Drainage Systems at the Farm.

Nomenclature

Drainage is mainly classified into two groups:

- ✰ Surface drainage
- ✰ Subsurface drainage

In the first case, removal of water is sought from the surface of the land mainly to remove the surface stagnation. On the other hand, in subsurface drainage, removal or control of groundwater is the prime objective. In many cases control of salinity may also be sought using water as the transporting agent. Subsurface drainage has further been classified on the basis of materials/structures used, purpose and/or orientation of the drains with respect to ground surface. On this basis subsurface drainage is classified as:

Subsurface Drainage (On the basis of materials used or their orientation)	
Horizontal drainage (System is laid parallel to the ground surface, horizontal)	Pipe drainage (concrete tiles, plastic pipes or pipes of any other suitable materials are used); earlier known as tile drainage because tiles made of clay or concrete were used
	Pipe less or mole drainage
Vertical drainage (System is laid normal to the ground surface, vertical)	Also known as tube well drainage
Bio-drainage (biological material, an exotic tree, is used to drain water)	It is also categorized as vertical drainage since plants are normal to the land surface
Subsurface Drainage (On the basis of purpose)	
Relief drainage	To control water table
Interceptor drainage (Intercepts the seepage water from the canal network or from the sloping land before it appears on the land surface)	Open or pipe interceptors (drains)
	Bio-interceptor

Some other classification may also be prevalent but for the purpose of this book, we restrict to the nomenclature as discussed above.

Benefits of Land Drainage

The positive impacts of land drainage are reflected in the improved physico-chemical environment of the root zone. Following benefits accrue from effective land drainage.

- ✰ Improved soil aeration resulting in an optimum mix of air and water in the root zone
- ✰ Dry soils (drained) are warmer with relatively higher temperature in the root zone, because specific heat of dry soils is lower than the water
- ✰ Soil structure is maintained as tillage operations are carried out under optimum moisture conditions
- ✰ Length of the growing season is more because soils dry earlier making early cultivation possible

- ☆ Quality of crops is better with timely harvest of crops. Timely harvest also allows ample time for field preparation for the next crop
- ☆ Lesser runoff because of greater storage capacity of the soil to absorb water
- ☆ No or reduced production of toxic substances which otherwise may retard plant growth
- ☆ Higher cropping intensity can be realized
- ☆ Improved biological quality of soils with profuse development of micro-organisms
- ☆ Facilitates leaching of salts resulting in reduced electrical conductivity of the soil profile
- ☆ Enhanced availability of plant nutrients

Surface Drainage

Surface drainage is defined as 'the removal of excess water stagnating over the ground surface through natural or constructed channels with adequate capacity outlets´. ICID (1982) defined surface drainage as ´the diversion or orderly removal of excess water from the surface of land by means of improved natural or constructed channels, supplemented when necessary by shaping and grading of the land surface to such channels´. Flat or nearly flat areas, heavy soils with low permeability and lands in humid or tropical and subtropical regions where high intensity storms are common, are subjected to surface inundation. The problem may or may not be of local origin as in lowlands, it may be due to the ingress of surface runoff from uplands. Overflow from rivers or natural channels sometimes contribute to the drainage problem of an area. The following factors contribute to the problem.

- ☆ Uneven land surface causing hindrance in the natural run-off from the area
- ☆ Inadequate capacity of the drainage channels during critical periods, may be due to under design or due to poor upkeep of the systems
- ☆ Inadequate outlet conditions, may be due to flat topography or due to natural drainage course in high flood at the critical periods
- ☆ Back water flow.

Surface drainage is carried out to meet one or more of the following objectives:

- ☆ In flat areas for timely removal of water so as to allow optimum growth of plant
- ☆ In sloping areas for orderly removal of water such that it does not cause erosion
- ☆ For the collection and disposal of run-off from surface irrigation.

Surface drainage is not a panacea to all drainage problems of a region since it is incapable to lower the shallow water table. Rate of build-up of the water table to some extent is minimized through efficient surface drainage.

Investigations for Surface Drainage

Practically speaking there is not much difference between the investigations for surface and subsurface drainage. However, the investigations for surface drainage are less intensive. One can save considerable investments in drainage investigation, if the initial investigations reveal that surface drainage is likely to suffice to improve land drainage. Following investigations are usually required to plan and design a surface drainage system.

- ☆ Topographic surveys
- ☆ Weather analysis
- ☆ Soil surveys
- ☆ Limited geologic investigations
- ☆ Land use and cropping pattern
- ☆ Water sources survey
- ☆ Profile and cross-sections of existing streams and ditches
- ☆ Stage-frequency investigations for high water in the outlets.

Since most of these investigations are also required in subsurface drainage, each one of these issues would be taken up under the investigations for subsurface drainage.

Tolerance of Crops to Surface Stagnation

Not all crops suffer equally from inundation or root submergence, the damage varying from one crop to another. Grasses, forages crops and forest plantations are usually more tolerant to inundation as compared to cereals or other crops. Yield reduction in all cases increases with increasing degree and duration of submergence. The damage also depends upon the stage of growth at which submergence occurs, generally crops being more sensitive to submergence occurring at initial than latter growth stages. In the Indian context, only limited data and that too only for few crops are available in this regard. Analysis of these data sets reveal that 50 per cent yield reduction may occur in about 4-6 days of water stagnation for pigeon pea and about 8-10 days in pearl millet and cowpea. In a semi-reclaimed alkali land, one day water stagnation reduced the yield of *berseem* (seed) by only 2 per cent whereas wheat and mustard yields reduced by 8 per cent (Gupta, 2014). A yield reduction of about 40 per cent has been observed in the case of wheat following 6 days of water stagnation (Figure 15.2). According to Bureau of Indian standard, water stagnation must be removed within a day from vegetable crops but sugarcane is able to tolerate water stagnation for about 7 days. Rice of course can tolerate water stagnation even for longer periods. Some varieties of the same crop are able to tolerate water stagnation for a longer period than other varieties.

Drainage Coefficient

One of the major requirements to design a collection and disposal system is to determine the drainage coefficient. It is defined as the amount of water to be removed from a given area within 24 hr. Commonly the unit is depth of water per

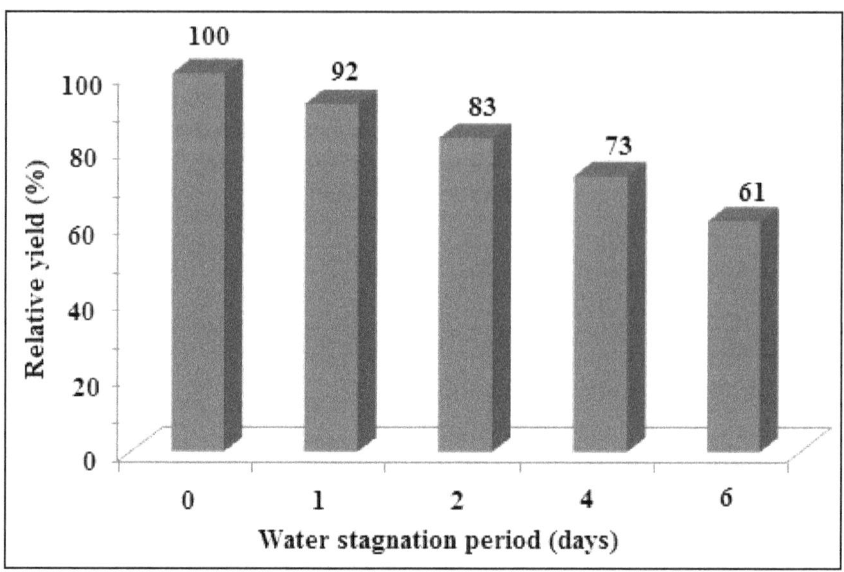

Figure 15.2: Effect of Duration of Water Stagnation on Relative Yield of Wheat in a Semi-Reclaimed Alkali Land.

day but many a times, the drainage coefficient is also expressed in flow rate per unit area. It is not often possible to determine drainage coefficient with any degree of accuracy although it is the key factor in establishing the needed capacity of the drainage system.

Several methods are used to evaluate the drainage coefficient. The use of a particular method mainly depends on the availability of required data. The simplest method of assessing surface drainage coefficient for farm drainage consists of the following steps (Finkel, 1983):

- Evaluate maximum 24 hr rainfall depth which may occur with a probability of 10 to 20 per cent. Frequency analysis procedures are commonly employed for this purpose
- Evaluate the basic infiltration rate of the soil and estimate the potential evapotranspiration in the area
- Estimate the tolerance of crop to surface inundation. For example let us say it is 'n' days. Multiply the infiltration rate and expected potential evapotranspiration by the number of days, n
- Subtract the value calculated in step (iii) from the value in step (i). The resulting value when divided by number of days, n will give the drainage coefficient in depth of water/day.

These steps are described in the form of a mathematical equation written in the following form.

$$q = (1/n)\,[\,R - n(E + I)] \qquad (15.1)$$

Here q is the drainage rate in mm/day, R is the rainfall in mm, E and I are potential evapotranspiration and infiltration rate in mm/day and n is number of days. When n is greater than one, it may be useful to increase the duration of rainfall in step 1 also from 1day maximum to n days maximum. Since some depth of water say d mm is allowed to stand in the rice fields after the drainage period, necessary correction in eq. (15.1) can be made by subtracting this depth. For the rice crop, eq. (15.1) is modified as:

$$q = (1/n)\,[\,R - n(E + I) - d\,] \qquad (15.2)$$

Normally, empirical equations are used to determine the drainage volume for disposal through the collectors and main drains. The commonly used formula can all be modified and described by following equation.

$$Q = C \times (A/n)^{m} \qquad (15.3)$$

Here Q is the discharge in cumecs, A is the catchment area in km², and C is a coefficient, which depends upon rainfall and n and m are constants. The values of C, n and m are given in Table 15.1.

Table 15.1: Values of C, n and m of Eq. 15.3 for different States to Design Main/Collector Drains

State	*Values/Guideline*		
	C	*n*	*m*
Punjab and Haryana	3.5 for 500 mm rainfall to 35 for rainfall of 1000 mm	6 up to 260 km²	0.5
		4 up to 261-650 km²	0.5
		2 > 651 km²	0.5
Uttar Pradesh and Andhra Pradesh	0.11	1	1
Madhya Pradesh	0.22 for A>79 km² 0.44 for A<13 km²	1	1
Tamil Nadu	0.44 for A= 13 km²	1	1
	0.6 for A> 52 km²	1	0.67
Odisha	0.3 for discharge of 10-12 cusecs	1	0.83

Note: 1 cumecs = 35.3 cusecs.

A number of workers have attempted to estimate drainage coefficients for different agro-climatic regions of India. Most comprehensive analysis made by Gupta *et al.* (1971) revealed that drainage coefficients in India varies from a low value of 2 mm/day to about 40 mm/day depending upon the crop (rice, grain or vegetable), soil and the climatic conditions.

Components of a Surface Drainage System

A surface drainage system essentially consists of the following three parts:

- Collection system

- ☆ Disposal system
- ☆ Outlet

Collection system mainly concerns with on-farm drainage with disposal and outlets forming a part of the overall system. A major component of the disposal system and the outlets are designed, executed and maintained by the State Governments. On-farm collection system is left to the farming community.

Methods of Field Drainage

Surface drainage of agricultural lands can be accomplished through land forming, open ditches and terracing. Land forming is defined as the process of changing the natural topography so as to control the movement of water on to or from the land surface. In the drainage context, it includes land smoothing, land grading or levelling and shallow field drains which are crossable with farm machinery. Levees commonly used for flood control, may or may not form a part of surface drainage system although these are essential to control the run-off that might enter the area from adjoining lands.

Land Smoothing/Land Grading

Flat lands where there are only minor differences in field levels, land smoothing is carried out to achieve good surface drainage. In this process general contours of the ground surface remain unaltered but only small ridges are removed and depressions filled-up to achieve a smooth surface. For best results, it is desirable to use the equipment three or more times by going across the field lengthwise, crosswise and diagonally. Land grading or land levelling are synonymously used in surface irrigation and drainage. Land grading is more descriptive than land levelling where a sloping plane surface than a level surface is designed. Land grading is a onetime operation in which pre-decided design slope is achieved. The design slope should meet the requirements of both the surface irrigation and drainage. For drainage, the grade in the direction of drains can be in the range of 0.05 to 0.5 per cent but a preferable slope is 0.1 per cent. For more details readers may refer chapter 11 of this book.

Layout of Drainage Systems

A number of surface drainage systems are in vogue. Some of them are bedding system, parallel field drain system, the random drain system and the parallel open ditch system.

Bedding System

Bedding system is probably the oldest method of surface drainage. It is suited to soils with poor hydraulic conductivity with slopes not exceeding 1.5 per cent. In this system, the level surface is converted into a sloping area by ploughing several times in such a way so as to always throw the back furrow in the channel (Figure 15.3). The beds are always kept parallel to the greatest slope so that dead furrow act as field drains. The bed width depends upon factors such as land slope, the land use, soil permeability, farm machinery to be used and the nature of the farming operations. Based on hydraulic conductivity, the bed width varies from 8-12 m (K≤0.05) to 20- 30 m (0.10<K≤ 0.2). The length of the beds may range from 100 to 300 m. The height of the crown from the bottom of the furrow should not exceed 45 cm for pastures and 115 cm for arable lands. Field laterals are provided perpendicular to the field drains which collects the water and carries it to the main drain. The method is most useful for pasture lands. Some of the limitations of bedding system are:

- The flow to the furrows may be obstructed when crop are cultivated in rows parallel to the dead furrows
- The slope of the furrow may not always be sufficient to take care of the overland flow
- Since the system requires movement of the top soil from the side of the beds to the middle, fertility gradients are established at least in the initial years
- The system may pose hindrance in the mechanized farming
- Maintenance costs are also high as the beds require regular maintenance.

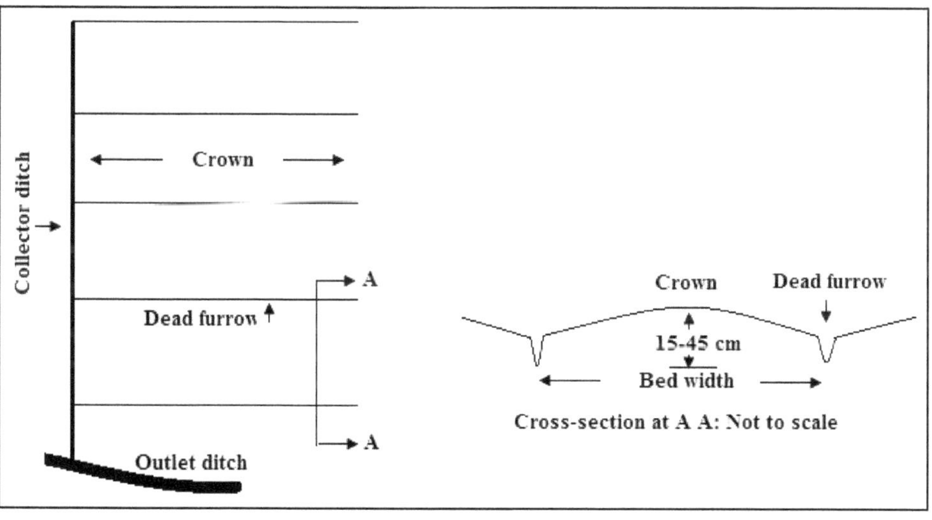

Figure 15.3: A Schematic View of the Bedding System of Land Drainage.

Parallel Field Drain System

The parallel system is used for flat surfaces having only few noticeable ridges or depressions. The channels in this system are spaced farther apart than a bedding system and have greater capacity than the dead furrows (Figure 15.4 left). The design and layout is similar to that of bedding system but it is not necessary that drains are spaced equally or they drain on both sides. The depth of the drain should be a minimum of 20 cm and have a minimum cross section of 0.45 m^2. The side slopes are 8:1 or flatter. The field drains are perpendicular to the field laterals. The system is most suited to fields where agricultural operations are mechanized. Land smoothing of field surface between laterals is required on a regular basis so as to keep the land well graded. It will also improve the efficiency of the drainage system.

Random System

This layout is used when drainage problem is encountered in scattered depressions (Figure 15.4 right). The system is especially used in areas where depressions are deep and it is technically or economically infeasible to fill up the depressions. Field or row drains and field laterals in this case meander from one depression to another to drain the same to an outlet ditch. Topography of the land dictates the alignment of lateral ditches but these should follow a route that requires minimum cut. The slope of the channels can be as flat as 10:1 when depth of drains is more than 60 cm or 4:1 if fields are farmed parallel to the drains. The recommended minimum cross section area is 0.45 m^2. While aligning the channels, it should be ensured that there is minimum interference in the farming operations. The system is quite economical and gives more returns on investments. The maintenance cost is relatively more than other systems of drainage.

Parallel Lateral Open Ditch System

The parallel lateral open ditch system besides providing surface drainage also helps to control the groundwater table. The layout of this system is similar to the field drain system but the ditches are deeper in this case. When financial or other

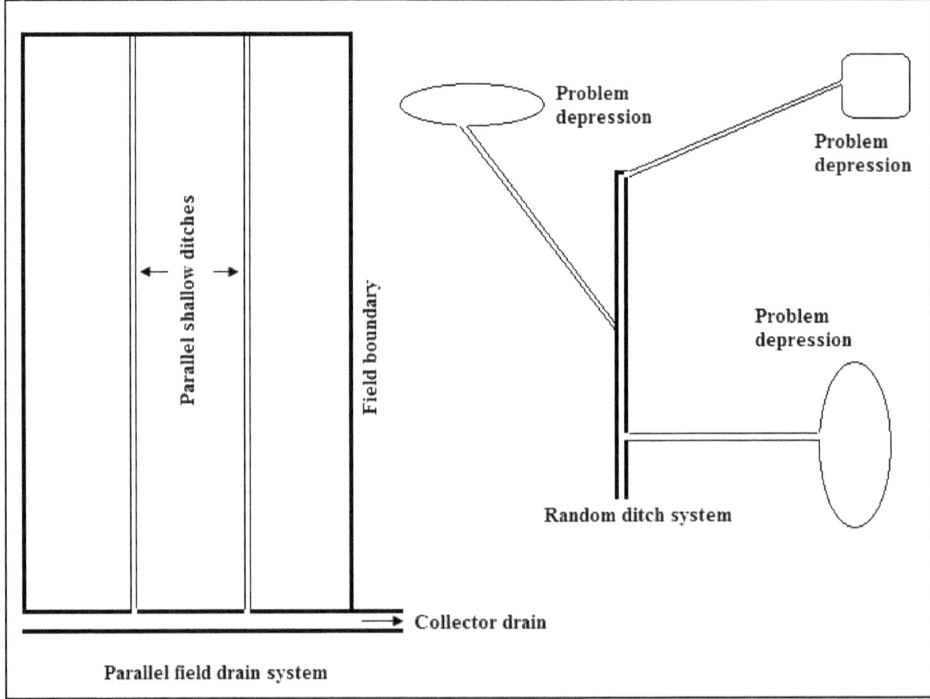

Figure 15.4: Parallel Field Drains (Left) and Random Ditch System (Right) of Surface Drainage.

constraints do not allow going for tile (pipe) drainage, this system of drainage is used till finances are generated to construct a pipe drainage system.

Design of Surface Drainage

The design of various components of the drainage system is governed by the drainage parameters collected during the drainage investigation. The design essentially comprises of hydrologic design and hydraulic design. Hydrologic design quantifies the excess water to be drained and the rate at which it has to be drained. For on-farm drainage it can be accomplished using eqs. (15.1) and (15.2). For collector drains either eq. (15.3) is used or peak flow is estimated using rational formula (See Chapter 29) or curve number technique. The latter methods are also used to design main drains. For hydraulic design, Manning's equation is used. For detailed discussions on use of Manning's equation to design open channels, readers may refer chapters 13 of this book.

Subsurface Drainage

Crop yields are usually low when the water table is near or within the root zone. It is illustrated in Figure 15.5 for the soybean crop, which reveals that yield reduced by 45 per cent when the water table was at 40 cm compared to 80 cm. In order to develop a sustainable highly productive agriculture, all agricultural lands with shallow water table need some kind of subsurface drainage. It is defined as: 'Removal of excess groundwater to control the groundwater table or to remove excess salts in time to prevent damage to crops or to prevent build-up of salts in the root zone using water as the medium of salt transport'. If conduits are used then it is known as pipe drainage. If no conduits are used then it is known as pipe less or mole drainage. Both the systems are part of subsurface drainage system. The challenge

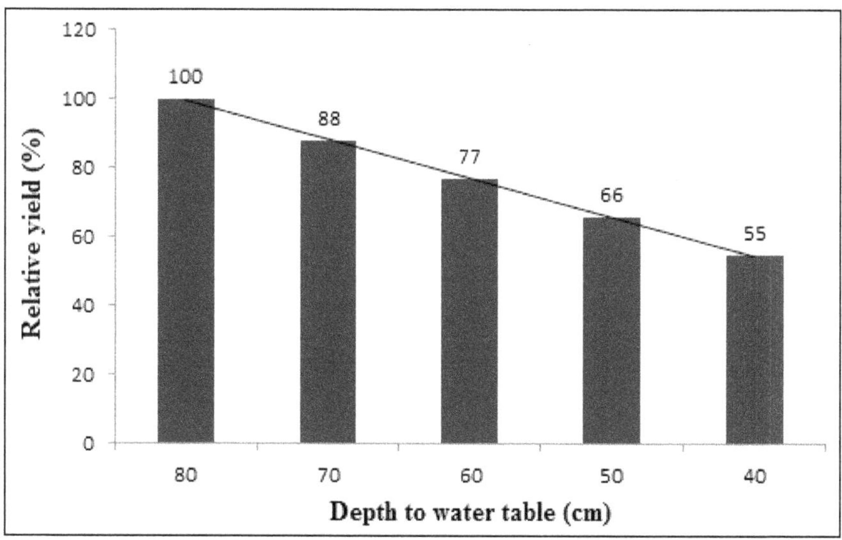

Figure 15.5: Relative Yield of Soybean as Affected by Depth to Water Table in Chambal Command, Rajasthan.

is to correctly evaluate the problem, design and install an economical drainage system and to maintain the system so that maximum benefits accrue from the land.

A horizontal drainage system consists of the laterals, the collectors and the main drains. Lateral drains (Figure 15.6) aligned parallel to each other cover the entire field to primarily control the groundwater table as they accept the excess groundwater from the soil above the drain level. The water entering into the laterals flows into the collectors which carry it to the main drain. The main drains carry the water to the outlet from where it is disposed out of the area either by gravity or by pumping from a sump.

Drainage Investigations

In order to determine the feasibility of the drainage and to design and install an efficient and economical drainage system, investigation of site conditions is essential. The following list present various components of a systematic survey and investigation plan to collect desired information with reasonable accuracy.

- Reconnaissance Survey
- Topographic Survey
- Soil and Geologic Surveys
    - Soil texture
    - Depth to impermeable layer
    - Cross -section of the profile
    - Particle size distribution, structure and macro pores
    - Engineering properties
    - Salinity and alkalinity
- Photogrammetric Survey
- Groundwater Surveys
- Water Transmission Characteristics
    - Infiltration rate
    - Hydraulic conductivity
    - Drainable pore space

The reconnaissance survey begins with a visit to the project area to collect qualitative information on depth to water table, intensity of drainage problem, crops grown, water management practices being adopted, physical boundaries, surface water bodies, sources of irrigation water and possible drainage outlets. Reconnaissance survey provides information on the necessity of drainage, enables a tentative layout of main drains and the outlet, and helps in taking a decision on drainage scheme to be adopted.

A topographical map with all characteristic features natural or manmade on an approximate scale of 1:3000 or 1:4000 is an essential pre-requisite. Contours at an interval of about 20 cm should be drawn to understand localized slopes in the project area. The main line may be located in the direction of general slope, while

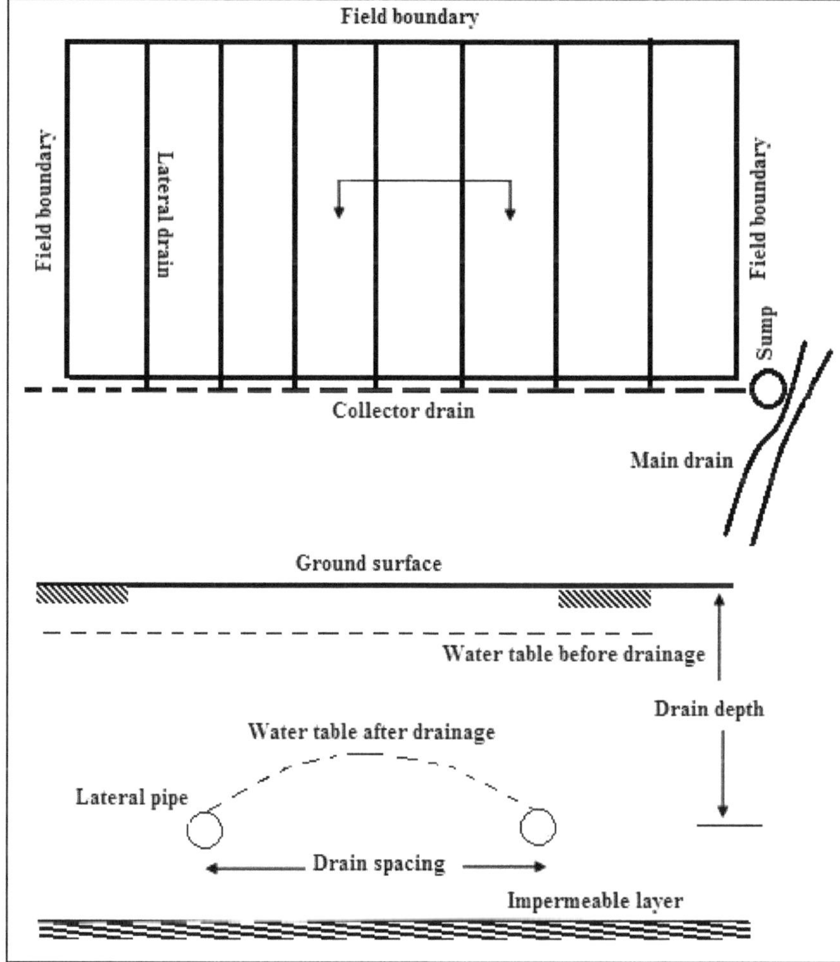

Figure 15.6: Plan and Section of Subsurface Drainage System Showing Various Components.

the laterals may be located across the general slope. The lowest point in the area should be selected for the construction of sump/outlet.

A soil sampling plan is prepared in which most of the determinations required are integrated to arrive at a cost effective soil investigation plan. While only shallow bores are made at certain points within the grid, deep bores are made at other locations to investigate profile characteristics and depth to impermeable layer. Samples are also drawn for soil texture analysis, particle size analysis and to assess salinity and alkalinity problems. Groundwater samples are collected at a grid of 30 to 40 m. The chemical properties such as Electrical Conductivity (EC), Sodium Adsorption Ratio (SAR) and RSC (Residual Sodium Carbonate) of the groundwater are determined to assess the chemical composition and nature of the groundwater.

The groundwater composition foretells the threat of salinization or alkalization during and after the reclamation. It also helps to decide the disposal options or reuse possibility of drainage effluent.

Aerial photographs, remotely sensed data and Geographical Information System (GIS) have been used extensively to study salinity and water logging. Because of the extreme variability in their occurrence, it has not been possible to develop a standard methodology so far. Nevertheless, these techniques are now increasingly applied mainly because the surface water logging can clearly be identified.

Groundwater surveys are the most important part of drainage investigations. The grid designed during the soil survey is used to install observation wells and piezometers to collect data on groundwater condition of the area. The water table behavior is studied on the basis of water level data of pre and post- monsoon period. The groundwater hydrographs give a fairly good idea of the time at which water is shallow and an insight to the causes of rise in the water table. The isobath maps provide justification to the project. Water level contour map provides information on the direction of groundwater flow. These maps also help to assess whether interceptor drain is necessary and to decide the directions of main and lateral lines.

The infiltration characteristics and hydraulic conductivity of the soils in the project area are measured through field tests. The infiltration rate is determined using double ring infiltrometers. The horizontal hydraulic conductivity is determined using 'Auger Hole Method' for below the water table condition and 'Inversed Auger Hole Method' for above the water table condition. Drainable pore space is not usually required for designs based on steady state theory commonly used under field conditions. The representative values of drainable porosity of different textures of soil can be taken as 3-5 per cent (volume basis) for heavy clays, 6-9 per cent for medium textured soils and 10-15 per cent for sandy soils.

The amount of investigations required would mainly depend upon the investigator's knowledge of the area, manpower and the funds at disposal, amount of data already in hand or available with different agencies operating in the area, size of the project and data available for similar projects executed earlier. Projects covering large areas would require more intensive investigations.

Design of Subsurface Drainage System

Following main items are considered in the design of a subsurface drainage system.

- Drainage criteria
- Layout
- Subsurface drainage coefficient
- Depth of laterals
- Spacing of field laterals
- Grade and size of the laterals and collectors
- Drainage materials
- Construction

Water logging is a common problem in the irrigation command. To control the water table at a reasonable depth is often the main criteria to design the drainage system. Current thinking is that shallow water table of good quality is not a curse and its adverse impact can be resolved through management. Lowering the water table to facilitate leaching for salinity control is now the major drainage design criterion. Following five criteria are commonly adopted in the design of subsurface drainage.

- Steady state
- Non-steady state
- Fluctuating water table
- Salinity control criterion
- Trafficability criterion

Although fluctuating water table criterion is most realistic as it represents the field conditions, yet steady state design criterion is commonly used because of its simplicity in use. In this case it is assumed that water table will stay most of the time at a pre/decided depth (steady state) such that drainage coefficient is equal to average rate of recharge. In the non-steady case, it is assumed that water following an irrigation or rain event may rise and should be lowered to a pre-decided level within a pre-decided interval. Under most critical conditions, water table may reach to the soil surface. It should be lowered to 30 cm within a period of 2-3 days following its rise to the soil surface. Salinity control criterion often determines some of the parameters to be used in the steady, non-steady and fluctuating water table theories. In the trafficability criterion, relation between soil moisture content of the soil surface versus depth to water table and soil moisture content and soil capacity to bear the load are used to assess the trafficability of the machinery in the fields. The field should permit free movement of the machinery during harvesting or pre-sowing tillage activities.

Types of Systems

There are two kinds of subsurface drainage. These are relief drains and interceptor drains. Relief drains are installed to remove excess groundwater percolating through the soil or to control a shallow water table. Depending upon the topography and nature of the subsurface drainage problem, several layout or patterns are used. The layout is adjusted to utilize the natural slope and to minimize earthwork against slope. Moreover, the layout should be such that it needs minimum length of main and collector drains to reduce the cost of the system. The layout of subsurface drainage systems is classified into four general types namely: parallel, herringbone, double main and random.

Parallel System

In the parallel field drainage system laterals are perpendicular to the main drain (Figure 15.5 top). Variations of this system are often used with other patterns. Parallel system is often preferred because it provides intensive drainage of a given field or an area.

Herringbone System

The herringbone field drainage system consists of laterals that enter the main drain at an angle, generally from both sides (Figure 15.7). If site conditions permit, this system can also be used in place of the parallel system. It can also be used where the main is located on the major slope and the lateral grade is obtained by angling the laterals upslope. This pattern may be used with other patterns in laying out a composite system in small or irregular areas.

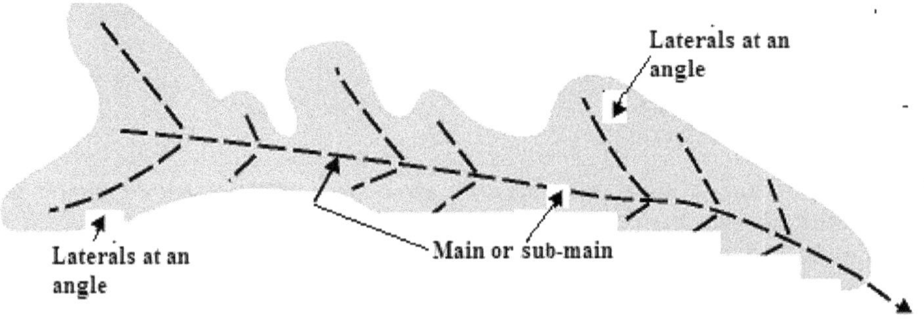

Figure 15.7: Herringbone Layout of Subsurface Drainage.

Random System

A random system of drains is used where the topography is undulating or rolling and contains a number of scattered isolated wet areas.

Double Main System

The double main system is a modification of the herringbone or parallel system and is used when a depression, mostly a natural water course, divides the field to be drained.

Interceptor Drainage

The interceptor drainage intercepts the subsurface flow before it re-surfaces on the earth surface. Normally it is used on long slopes and on shallow permeable surface soils overlying relatively impermeable soils. It is also used to intercept the seepage from water distribution networks to prevent development of water logging in the areas adjoining these systems.

Drainage Coefficient

To design a drainage system it is necessary to determine the drainage coefficient accurately because it control the drain spacing, the size of the field laterals, and the capacity of the collectors and pumping station (if any). Excess rain or irrigation water or net groundwater inflow that causes rise in the water table to a level shallower than the permissible limits, must be drained out. Although elaborate procedures are used in evaluating drainage coefficients, the drainage coefficients in the range of 2-5 mm/day have been used in India. General guidelines for selecting drainage coefficient based on the climatic conditions are available (Table 15.2).

Table 15.2: Guidelines on Drainage Coefficient for Subsurface Drainage

Climatic Conditions	*Range (mm/d)*	*Optimum Value (mm)*
Arid	1-2	1
Semi-arid	1-3	2
Sub-humid	2-5	3

Source: Gupta *et al.* (2002).

Depth of Laterals

The depth of subsurface drains is decided on the basis of soil characteristics, type and value of crops, outlet conditions, and the presence or absence of impervious soil layers in the upper portion of the soil profile. Important aspects to be considered in determining the depth of drains are:

- ☆ The desirable depth to water table for proper aeration in the root zone
- ☆ Since depth and spacing are inter-related, the depth of drain placement should be such that the cost per hectare as given by the combination of depth-spacing is minimum
- ☆ The soil conditions such as unstable or hard soils and quick sand conditions are avoided
- ☆ Availability of machinery
- ☆ The outlet conditions and available depth of outlet drains, topography, minimum pipe slopes.

Subsurface drains should not be placed under an impermeable layer, and they should be located within the most permeable layer. The drain depth is also controlled to a greater extent on the construction method, the available machinery and drain specifications. Generally, the greater the depth of a field drain, the wider the spacing can be between drains. Drains should be deep enough to provide protection against tillage operations, equipment loading, and frost. Main and sub-mains must be deep enough to provide the specified depth for outlets of lateral drains. Also, the maximum depth at which drains can be laid to withstand trench loading varies with the width of the trench and the crushing strength of the drain to be used. All subsurface drains in mineral soils should have at least 60 cm of cover for protection against overloading from heavy machinery (Table 15.3). Subsurface drains in organic soils should have at least 75 cm of cover. General guidelines on drain depth are given in Table 15.4, indicating that one can go for shallow lateral drains, if it is possible to have gravity outlets.

Table 15.3: Minimum Soil Cover Required for Pipe Drains

Soil Type	*Minimum Soil C(cm)*
Mineral	60
Deep peat and muck	120
Organic	75

Table 15.4: Guidelines on Drain Depth

Outlet Conditions	*Depth of the Drains*	*Optimum Depth (m)*
Gravity	0.9-1.2	1.1
Pumped	1.2-1.8	1.5

Grade

The design grade must be sufficient to allow the water to flow to the outlet. It must ensure that the pipe capacity is sufficient to drain the area. The minimum grade is a function of the pipe size. Guidelines on minimum grade for pipelines are reported in Table 15.5. In areas where sedimentation is not a hazard, the minimum grade shall be based on site conditions and a velocity not less than 15 cm/s. If the soil sedimentation is anticipated, a velocity of not less than 45 cm/s (self cleaning velocity) shall be used to establish the minimum grades, if site conditions permit. Otherwise, provisions shall be made for preventing sedimentation by applying suitable envelope materials.

Table 15.5: Minimum Grade for Pipe Drains

Drain Size (mm)	*Grade (per cent)*
100	0.10
125	0.07
150	0.05

Spacing of Lateral Drains

The water table between two parallel drains is usually curved, its elevation being highest midway between the drains (Figure 15.8). The factors which influence the height of water table are:

- ☆ Precipitation and other sources of recharge
- ☆ Evaporation and other sources of discharge
- ☆ Soil properties
- ☆ Depth and spacing of the drains
- ☆ Cross-sectional area of the drains
- ☆ Water level in the drains

Several equations have been developed interrelating the above factors. These equations are used to calculate the lateral drain spacing. Most commonly used is the steady state equation proposed by Hooghoudt (1940). For details of this and other equations readers may refer the book on Drainage Engineering by Gupta (2019).

The applied spacing in India ranges from less than 20 m to more than 100 m, higher spacing being for lighter and lower for heavy texture soils. Typically, drain lines are not spaced closer than 30 m for agriculture because of the associated costs and other associated problems with the close spacing (Table 15.6). For very high

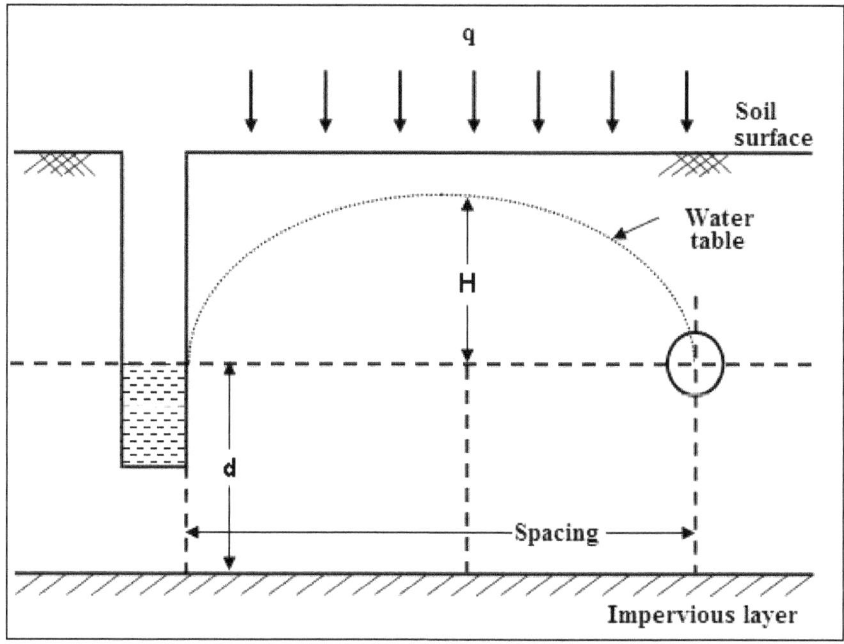

Figure 15.8: Definition Sketch of Subsurface Drainage Under Steady State Conditions. Note the post drainage water table being shallowest at mid-way between the drains.

value crops and for special use areas such as athletic fields, closer spacing may be justified.

Table 15.6: Guidelines on Lateral Drain Spacing

Soil Texture	*Spacing of Drains (m)*
Light textured	100-150
Medium textured	50-100
Heavy textured including Vertisols	30 × -50

× Can use lesser spacing, if financial analysis reveals that additional cost is recoverable within 3 years

Size of Lateral Drain Pipe

Wessling's equations for uniform flow in smooth pipes and corrugated pipes derived from Manning equation can be used to calculate the size of the lateral drain pipes. The equations are:

For corrugated PVC pipes

$$Q = q \times A = 2.0 \times 10^6 \times d^{2.67} \times S^{0.5} \text{ if } d \leq 100 \text{ mm} \quad (15.4)$$

$$Q = q \times A = 2.5 \times 10^6 \times d^{2.67} \times S^{0.5} \text{ if } d > 100 \text{ mm} \quad (15.5)$$

The corresponding equations for rigid smooth PVC pipes derived on the basis of Darcy-Weisbach equations are:

$$Q = 4.6 \times 10^6 \times d^{2.714} S^{0.572} \text{ for } d \leq 100 \text{ mm} \quad (15.6)$$

$$Q = 5.79 \times 10^6 \times d^{2.714} S^{0.572} \text{ for } d > 100 \text{ mm} \quad (15.7)$$

Here, Q is the discharge of the pipe, m^3/sec, d is the internal diameter of pipe, m and S is the hydraulic gradient, m/m. Also, A is the drainable area (m^2) equal to length of the pipe multiplied by drain spacing and q is the drainage coefficient (m/d). The computed pipe diameters are to be rounded off to the next higher commercially available size. For clay and cement concrete tiles the prescribed minimum diameter is 10 cm, while for PVC and PE pipes, it is 7.5 cm.

Drainage Materials

Pipes

Several kinds of tiles/pipe materials are used in subsurface drainage. Based on the flexibility of pipes, these have been categorized into two groups:

Rigid Conduits: The materials include clay, concrete, asbestos and locally available materials such as bricks, stoneware and sewage pipes/tiles. Metal pipes are not used except some pieces at emergent locations such as outlets or road crossings *etc.*

Flexible Conduits: These include plain or corrugated plastic pipes made of polyvinyl chloride (PVC) and high density polyethylene (HDPE).

It is now customary to use flexible conduits since it is easy to construct a drainage system using mechanical equipment.

Envelope

Soils in which drains are prone to mineral clogging are commonly referred to as problem soils because the soil particles tend to migrate into the drain. In practice, all very fine sandy or silty soils with low clay content are probable problem soils and require some kind of envelope. The drain envelope material is used on or around a subsurface drain for one or more of the following reasons:

- To stabilize the soil structure of the surrounding soil material, more specifically a filter envelope
- To improve flow conditions in the immediate vicinity of the drain, more specifically a hydraulic envelope
- To provide structural bedding for the drain, also referred to as bedding.

Drain envelopes have been divided into mineral or granular (river run or crushed gravel), organic (rice husk, saw dust, coconut fiber *etc.*) and synthetic or manmade or fabric (geo-textile, woven socks *etc.*) envelopes. Granular materials commonly used so far are going out of practice in favour of synthetic envelopes because of extensive use of PVC pipes as drain material. Since the performance of organic envelopes is not well tested, therefore, under critical conditions, it is advisable to go for other kinds of envelope materials.

Outlet

The execution of drainage system begins with the construction of the outlet which could either be a gravity or a pumped outlet. If the natural slope permits, the disposal of drainage water from the collector or main drain into an open water body through gravity is desirable. The gravity outlets are cheaper to construct and do not involve operational cost. In flat or gently sloping areas, where the gradient is generally insufficient for free fall, disposal of water through gravity into the open watercourse is difficult. Under these conditions, a sump is constructed in between the subsurface collector drain and the open channel. The sump provides space for storage of water flowing from the collector drain, before it is pumped into the open watercourse.

To construct a drainage sump, a circular pit of diameter about 1 m more than the diameter of the proposed sump is dug up to the groundwater level. At the bottom of the circular pit an iron ring is placed to serve as cutting edge over which a RCC kerb is constructed to support the masonry wall to a height 1m above the ground. The wall is plastered from inside and outside and is provided with holes of 1 cm diameter at regular intervals. The sump is sunk into the ground by removing the soil from the bottom. Simultaneously, the wall of the sump is also raised such that the sump is 2 to 2.5 m deeper than the depth at which collector drain has been laid. In waterlogged sandy soils, quick sand conditions may interfere with the installation of sump. Such situations can be tackled by simultaneously dewatering the area all around the sump while sinking the structure. On completion, a gravel layer of about 0.5 m thick is placed at the bottom of the sump to serve as a filter for the groundwater flowing into the sump. Holes are provided in the side walls to connect the collector drains. A sump of 3.5 m to 4 m inner diameter and of 4 m depth below the ground could serve an area of about 40 ha. With mechanization and availability of pre-cast reinforced concrete rings, the construction of a sump is a lot easier now. It is a preferred alternative to manual construction.

Installation of Subsurface Drainage System

The starting point in planning a subsurface drainage system is normally the location of the outlet. Based on the design, a layout plan is prepared to suit the field boundaries. The topography of the project site receives due weight while deciding the layout plan. To install the system, the collector and lateral drain lines are staked and the location for outlet and junction boxes are identified in the field. Commonly, three methods, namely, manual, mechanical and semi-mechanical methods are used to construct drainage systems in India. Irrespective of the method used, general features of system construction should be observed.

Digging should begin from the outlet end. it is easier to empty the drains, if there is rain or water accumulates in the drains for some other reason. The width of the trench should be kept as minimum as possible but should be enough so that no obstructions are encountered while installing the pipes at the desired drain depth. The bottom width should be the outside diameter of the drain pipe plus 130 mm. Minimum length of the tile line is twice the spacing. Maximum length should not exceed 300 m if conventional flushing methods are likely to be used to clean the drains. If the length exceeds 300 m, a manhole should be provided at every 300 m. Alignments should be in straight sections. Avoid 90° bends. Bends, if necessary, should be provided as smooth curves. Stakes to govern alignment should be provided at 30 m intervals in straight section and 15 to 18 m on curves. As far as possible, maintain the desired grade during the digging operation itself. If a trench is accidentally excavated below grade, the section should be backfilled, tamped and shaped to grade. For slope control, line and gauge method is commonly used. In mechanical laying, machines are usually equipped with laser control to set the desired grade.

A number of engineering structures are designed and used during construction for disposal of the drainage effluent, to maintain the system, maintain the infrastructure of the project area, to transport the surface runoff and to minimize or optimize the drainage flow. Some of the structures besides the outlet are: manhole/ junction box, drops, surface water inlets, relief and breathers, crossings: surface drain, canal/distributary/minor/watercourse/road, drain bridges, closing devices and flow management devices. All structures may not be required in all the projects.

Vertical Drainage

In many irrigated areas, pumped wells system for drainage (Vertical drainage) is the most feasible approach. It is especially true for regions where drainage outlets for pipe drainage systems are not easily available. Vertical drainage is defined as ´the removal of groundwater by means of a borehole (or well) either under pressure (artesian conditions) or through pumping´. The vertical drainage system practically includes one or more deep/shallow tube well(s). The network of open or closed drains to dispose of the water also constitutes a part of the system. The group of tube wells having sufficient overlap of their individual cones of depressions controls the groundwater table or artesian groundwater conditions. The overlapping of cone of depressions ensures uniform lowering of the water table at all the points. Lowering of water table in many parts of India ascribed to overexploitation through tube wells attests to its feasibility in lowering the water table. The pumped water can be used for irrigation either alone or in conjunction with fresh water, provided the proper salt balance is maintained. A comparative statement given in Table 15.7 describes the site characters where vertical drainage may prove to be successful

Bio-Drainage

Biological drainage or bio-drainage as already indicated before is defined as ´the application of a biological material to absorb, utilize, transport and evapo-transpire water stagnating on the land surface or in the root zone to drain the land´. Quick growing plants are employed as bio-drain materials as these are capable of absorbing and transpiring huge quantities of water. Since water is transpired straight into the atmosphere, there are no disposal problems. In essence, bio-drainage involves three processes (Figure 15.9):

- ✰ The uptake of water from the soil by the roots (absorption)
- ✰ Transportation of water from root to plant parts
- ✰ Transpiration of water by the plants through the tree leaves.

Table 15.7: Favourable and Non-favourable Site Features for Vertical Drainage

Favourable Features	*Non-favourable Features*
Good quality aquifers are available	Non-availability of good aquifers; low transmissivity does not allow construction of tube wells of sufficient capacity
Groundwater is of good quality	Groundwater of high EC, SAR or RSC resulting in problems for its disposal. Tube well components are more prone to wear and tear/corrosion due to salinity resulting in higher maintenance cost
Cost of tube well construction is less and electrical power is available	Lack of fresh water without which possibility of conjunctive use of drainage water may pose problems
The system is constructed and operated by the stakeholders	High operational and maintenance costs when systems are installed under government set-up
Upward seepage pressure from a confined aquifer is cause of drainage problem and the cost of horizontal subsurface drainage is prohibitive	
The areas to be reclaimed are only waterlogged with no/minimal salinity	
Potential of using drainage water for irrigation with minimum disposal related problems	
One window system available for agricultural activities and groundwater development	

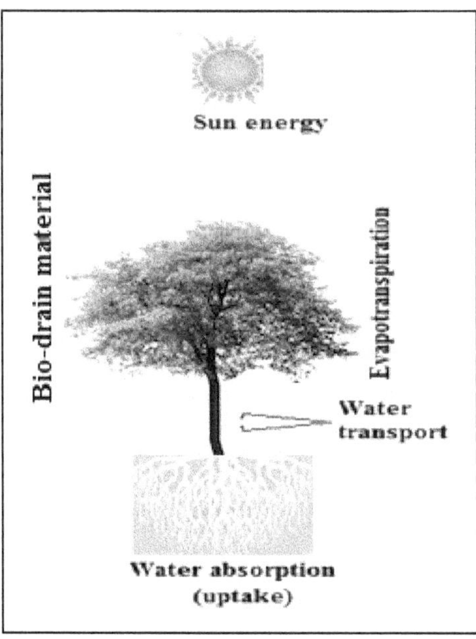

Figure 15.9: Processes Involved in Bio-drainage.

Apparently, a good bio-drainage friendly plant must be able to absorb, transport and transpire water under stresses of water logging and soil salinity (Gupta, 2019). Some of the plants that can be used as bio-drain material are: *Eucalyptus, Casuarina glauca, Acacia ampliceps, Terminalia arjuna* (*Arjun*), *Pongamia pinnata* (*Papri*) and *Syzygium cumini* (*Jamun*). *Eucalyptus* with its quick growing characteristics fits well into any bio-drainage scheme. A major problem with this system arises due to the fact that bio-drainage is not capable of taking care of salt balance since plants do not absorb salts. Therefore, these systems may be used as a preventive measure for water logging and salinity control.

Section III

Irrigation Water Management

Chapter 16

Classifcation of Irrigation System

Definition

According to Cambridge University dictionary, meaning of irrigation is to supply land with water so that crops and plants will grow. According to https:/ www.vocabulary.com/dictionary/, irrigation comes from the Latin word meaning "moist" or "wet," but it means the purposeful wetting of something. Irrigation is also defined as the application of controlled amounts of water to plants at needed intervals. Scientifically, irrigation is defined as the science of artificial application of water to the land, in accordance with the 'crop water requirement' throughout the 'crop period' for full-fledged nourishment of the crops (Garg, 1996).

Irrigation is very ancient practice and can be traced to the beginning of human civilization and advent of agriculture. Importance of irrigation in agriculture is very well documented in Indian Hindu Literature. In Narada *Smriti,* it is mentioned that "no grain is ever produced without water, but too much water tends to spoil the grain; an inundation is as injurious to crop growth as the dearth of water." Clearly it emphasizes on application of water as per 'crop water requirements' neither more nor less. Irrigation is designed to permit farming in arid regions and to offset drought in semi-arid regions. Even in areas where total seasonal rainfall is adequate on an average, it may be poorly distributed during the year and might vary from one year to another. Thus, even in humid and sub-humid regions, the plants must receive additional water at critical junctures.

The necessity of irrigation mainly arises from:

- ✰ Rainfall is less than the water requirement of the plants
- ✰ Rainfall is sufficient, but the spatial distribution of rainfall is not as per requirement of the crops

- ☆ Rainfall is sufficient and the spatial distribution is also good, but the temporal distribution is not as per requirement
- ☆ Advanced scientific development resulting in essentiality of irrigation to adopt high tech agriculture such as high yielding varieties, inorganic fertilizers *etc.*

The irrigation essentially serves the following purposes:

- ☆ To supply moisture for plant growth
- ☆ Transportation of fertilizers into the root zone
- ☆ To leach or dilute salts present in the soil
- ☆ To help in field preparation, dust control *etc.*
- ☆ Other benefits of irrigation include cooling of the soil and atmosphere to create more favourable environment for crop growth and frost control

The advantages and disadvantages of irrigation are listed below.

Advantages

- ☆ Irrigation plays an important role in increasing and stabilizing food production against weather based fluctuations. Thus, not only agricultural productivity is high, the risk of crop failure is reduced
- ☆ Irrigation helps the country to be self-sufficient in meeting the food requirement of its people
- ☆ Allows for multiple cropping during an year and/or helps to grow cash crops
- ☆ It creates employment opportunities
- ☆ It makes people prosperous and improves socio-economic conditions of farmers. The living standard of the people is improved
- ☆ Higher productivity results in steady supply of food at lower prices (supply -demand principle of economics)
- ☆ Modify soil environment and improves micro-climate
- ☆ Irrigation improves the groundwater storage as water lost due to seepage adds to the groundwater recharge
- ☆ With watering facility, the land value appreciates.

Disadvantages

There are no potential disadvantages in the principle of irrigation as such. The disadvantages mentioned herein have been observed in such areas where systems have not been appropriately designed, operated or maintained. Under these conditions, irrigation practices have often lead to few problems and even failed to achieve targeted benefits. Some of the problems encountered are:

- ☆ Excessive irrigation may result in lower crop yield

- ☆ Excessive irrigation may cause leaching of pesticide, insecticide, nitrogen and nitrates that may pollute the groundwater and may also transport them to surface water systems.
- ☆ Poorly designed and maintained canal networks may result in excessive seepage and thereby rise in water table. It may lead to development of water logging
- ☆ In irrigated areas inappropriate provision of drainage may result in water logging and soil salinity (Figure 16.1)
- ☆ In lift irrigation systems, excessive groundwater pumping may result in overexploitation of groundwater and consequently lowering of groundwater levels. Such a situation may damage aquifer structure, increase pumping costs and increase the risk of land subsidence (Table 16.1)
- ☆ Alignment of irrigation networks often obstructs the natural drainage causing prolonged submergence of standing crops and even villages
- ☆ At micro-level, colder and damper climate may bring in new insects and diseases
- ☆ Poorly managed irrigation systems may prove to be a source of many water borne diseases and environmental damage.

Table 16.1: Rate of Groundwater Table Decline in 4 Blocks of 3 Districts in Haryana

Block (District)	*Rate of Decline (m/annum)*	*Period*
Babain (Kurukshetra)	1.0	1995-2005
Panipat (Panipat)	1.1	1999-2010
Samalkha (Panipat)	0.96	1999-2010
Ganaur (Sonipat)	0.34	1979-2008

Sources of Water

Source of water for all kinds of water resources, be it lakes, tanks, canals or groundwater, is the natural precipitation in the form of rain, snow, hail and sleet. Surface and groundwater resources commonly considered as the main sources of irrigation water, are all derived from the natural precipitation (A part of the hydrological cycle).

Figure 16.1: Shallow Water Table in Sharda *Sahayak* Canal Command, Uttar Pradesh (Left) and Soil Salinity in Western Yamuna Canal in Haryana.

Surface Water

Water present on the surface of the earth in the form of oceans, rivers, lakes, ponds and streams is called surface water. The amount of surface water available depends largely upon rainfall. Water for irrigation in canal network is diverted mainly from rivers by buildings dams or weirs. From there, water is conveyed to the farm fields where it is directly applied to the field or stored in farm irrigation structures like ponds or tanks. On the basis of source of water, surface irrigation systems are categorized as:

1. Canal irrigation
2. Tank irrigation
3. Irrigation with harvested rainwater in farm ponds and through check dams or other means

Groundwater

A part of the water after any rainfall event infiltrates into the soil a part of which joins the groundwater table. Wells are generally used to extract the groundwater. The extraction of groundwater is mainly achieved through: dug wells, dug-cum-bore wells, tube wells, strainer wells, cavity wells, radial collector wells and infiltration galleries (See chapter 12).

Irrespective of the source of water, the quantity, the quality, and the reliability of water availability at appropriate time are crucial for the efficient utilization of the water resource for irrigation.

Development and Utilisation of Water Resources

With a geographical area of about 329 million ha (M ha), India accounts for 2.4 per cent of the world surface area. On an average over space and time, annual rainfall including snow over the Indian sub-continent is estimated at 1200 mm. Thus, the main water resource of India consisting of this precipitation is estimated at 4000 km^3 per annum (400 M ha-m per annum). It includes net inflow from outside the national boundaries. After meeting interception, evaporation and soil moisture requirement, this resource gets distributed in surface and groundwater resources. Rainfall in itself is the most important source of water for rainfed agriculture and rain forests.

Surface Water Resources

From surface water resource point of view, India has been divided into 20 river basins. These comprise of 12 major basins each having a catchment area exceeding 20,000 km^2 and 8 composite river basins combining suitably all the other remaining medium and small river systems. The total water potential of these basins has been revised over time. The revised value of 1869 km^3 of surface water resource by Central Water Commission is now taken as reliable. The largest potential of water is available in Ganga/Brahmaputra/Barak basins making a total of 1110 km^3 followed by Godavari and by Tapi west flowing rivers each having an average

Table 16.2: River Basin-Wise Water and per Capita Availability

Sl. No.	River Basin	Average Annual Surface Water Availability (km³)	Replenishable Groundwater Resource (km³)	Utilizable Surface Water Resource (km³)	Per capita Water Availability in the Year 2000 (m³)
1	Indus	73.31	31.23	46.0	1482
2	Ganga-Brahmaputra-Barak		209.85		
	(a) Ganga	525.02		250.0	1239
	(b) Brahmaputra and Barak	585.6		24.0	14057
3	Godavari	110.54	37.50	76.3	1734
4	Krishna	78.12	26.65	58.0	1088
5	Cauvery	21.36	10.15	19.0	619
6	Pennar East Flowing Rivers	6.32	5.10	6.9	550
7	(a) Between Mahanadi and Pennar	22.52	14.17	13.1	808
8	(b) Between Pennar and Kanyakumari	16.46	18.11	16.5	311
9	Mahanadi	66.88	17.72	50.0	2131
10	Brahmani and Baitarni	28.48	6.70	18.3	2463
11	Subarnrekha	12.37	5.13	6.8	1118
12	Sabarmati	3.81	2.98	1.9	307
13	Mahi	11.02	3.12	3.1	888
14	West Flowing rivers of Kutchh, Saurashtra including Luni	15.1	11.90	15.0	579
15	Narmada	45.64	12.90	34.5	2628
16	Tapi West Flowing Rivers	14.88	7.36	14.5	853
17	(a) Tapi to Tadri	87.41	12.38	11.9	2870
18	(b) Tadri to Kanyakumari	113.53		24.3	2950
19	Area of Inland drainage in Rajasthan desert	Neg.		-	
20	Minor Rivers basins draining into Bangladesh and Myanmar	31.00	0.40	-	12500
21	Andaman, Nicobar and Lakshadweep		0.34	-	
	Total	**1869.35**	**432.98**	**690.1**	**1869**

annual potential of more than 100 km³ (CWC, 2010, Publication Division, 2011). As per the figures reported in Bhagirath (July-September 2001), the total water resource and per capita water availability in various river basins varies widely (Table 16.2). Lowest per capita water available is in Sabarmati River basin assessed at 307 m³ per capita. Highest per capita water is in the Brahmaputra and Barak basin estimated at more than 14,000 m^3. Due to extreme variability in precipitation, which disallows assured storage of all the water, due to non-availability of storage space in hills and plains, evaporation losses and water going to the sea and outside India, it is anticipated that utilizable surface water resources would be 690 km^3 which will be utilized by the year 2025 (Sharma and Paul, 1998, Table 16.2). An upward revision in the utilizable flows may be expected on account of advancement in technology and on account of large-scale inter-basin transfer of water.

Groundwater Resources

The groundwater resource has been categorized as static and dynamic. The static groundwater resource also sometimes known as "fossil" water resource is defined as the amount of groundwater available in the permeable portion of the aquifer below the zone of water level fluctuation. The dynamic resource is defined as the amount of groundwater available in the zone of water level fluctuation. The dynamic resource is replenished annually or periodically by precipitation, irrigation return flow, canal seepage, tank seepage, influent seepage, *etc.* On the basis of the depth of groundwater and the productivity of deeper aquifers, the water available in the aquifer zones below the zone of water level fluctuation (static) has been assessed at 10812 km^3. As per the National Water Policy, development of groundwater resources is to be limited to utilization of replenishable component of naturally occurring groundwater available in subsurface domain.

Replenishable groundwater resource is mostly derived from the precipitation. Out of 4000 km^3, 2150 km^3 of rain water percolates into the ground out of which only 500 km^3 joins the groundwater or is available for exploitation (CGWB, 1995). While a part of this water regenerates into streams, there is net addition of water through streams and irrigation which is presently estimated at 260 km^3. Deep percolation adds another 110 km^3. The state wise groundwater potential estimated by Central Ground Water Board places the groundwater resource of India at 433 km^3 as on March 2004 (Table 16.3). Basin wise estimates reveal that Ganga basin has the maximum groundwater potential being more than 38 per cent of the total groundwater potential. It is followed by Brahmani and Baitarni at 59 km^3. Amongst the states, Uttar Pradesh, which mostly lies in the Ganga basin has the highest potential. Goa followed by Himachal Pradesh on the other hand, has the least potential amongst the states. After deducting the provision for domestic and industrial purposes, the potential available for irrigation is 362 km^3. The utilizable groundwater for irrigation is thus assessed at 325 km^3 (Table 16.4). It is likely to increase to 350 km^3 by 2025. About 50-60 km^3 of water goes into storage, which results in rising water table and is lost through evaporation. In terms of area, the irrigation potential of the groundwater has been revised and placed at 64.05 M ha (Gupta *et al.*, 2000).

Utilizable Flows and Irrigation Potential

To usher in an era of self-sufficiency in food production, expansion of irrigation was thought to be an essential pre-requisite. In pursuance of this policy a large number of major, medium and minor irrigation projects were initiated to harness surface water. Irrigation potential created increased from 22.6 M ha in 1951 to 108.21

Table 16.3: State-wise Replenishable Groundwater Resource

Sl.No.	States	Annual Replenishable Groundwater Recharge (km^3)	Groundwater Development (per cent)*
1.	Andhra Pradesh	36.50	37
2.	Arunachal Pradesh	2.56	0
3.	Assam	27.23	14
4.	Bihar	29.19	44
5.	Chhattisgarh	14.93	35
6.	Goa	0.29	28
7.	Gujarat	15.81	67
8.	Haryana	9.31	133
9.	Himachal Pradesh	0.43	71
10.	Jammu and Kashmir	2.70	21
11.	Jharkhand	5.58	32
12.	Karnataka	15.93	64
13.	Kerala	6.84	47
14.	Madhya Pradesh	37.19	57
15.	Maharashtra	32.96	53
16.	Manipur	0.38	1
17.	Meghalaya	1.15	0
18.	Mizoram	0.04	3
19.	Nagaland	0.36	6
20.	Odisha	23.09	28
21.	Punjab	23.78	172
22.	Rajasthan	11.56	137
23.	Sikkim	0.08	26
24.	Tamil Nadu	23.07	77
25.	Tripura	2.19	7
26.	Uttarakhand	2.27	57
27.	Uttar Pradesh	76.35	74
28.	West Bengal	30.36	40
	Total States	432.13	
29.	Total UTs	0.90	
	Grand Total	433.03	Average 62 per cent

*as in 2011.

M ha by March 2010, *i.e.* at the end of the first three years of 11th plan (CWC, 2010). As against this, the potential utilized is only 87.39 M ha showing a gap of 20.82 M ha between the potential created and utilized. It is expected that in 2025 AD, about 770 km^3 of water will be available for irrigation and the rest would be for domestic and industrial uses. The ultimate irrigation potential (UIP) of the country stands at about 140 M ha. The share of minor irrigation is higher by 22.96 M ha as compared to that of Major and Medium irrigation (CWC, 2013). Groundwater contribution is more than 79 per cent of the total ultimate potential through minor irrigation.

Table 16.4: Utilizable Groundwater Resource of India

Sl.No.	Distribution	Value (km³/year)
1	Total replenishable groundwater resource	433
2	Provision for domestic, industrial and other uses	71
3	Available groundwater resource for irrigation	362
4	Utilizable groundwater resource for irrigation (90 per cent of the value at sr. no. 3)	325
5	Total utilizable groundwater resource (Sum of sr. no. 2 and 4)	396

Uttar Pradesh and Bihar are the two largest states in term of irrigation potential due to Major and Medium Irrigation Projects. These two states along with Madhya Pradesh, Andhra Pradesh and Maharashtra account for about 54 per cent of the total ultimate potential of Major and Medium irrigation in the country. The largest UIP for Minor Irrigation (Groundwater) exists in Uttar Pradesh. Andhra Pradesh and Madhya Pradesh are two major states in which potential of Minor Irrigation (Surface Water) is much higher than the remaining states. The net sown area in 2025 AD is estimated to be 155 M ha. With anticipated cropping intensity of 136 per cent, the gross cropped area will be about 210 M ha. As already stated, the available water resources even at that time will only be sufficient to provide irrigation to an area of about 140 M ha (Table 16.5). Therefore, with limited resources, it is expected that only 66.7 per cent of the gross sown area of 210 M ha will be under irrigation in 2025. The remaining area will continue to depend upon the rainfall.

Table 16.5: Ultimate Irrigation Potential (M ha) as per Source of Water Availability in India

Sl.No.	Sector	Irrigation Commission (2nd) (1972)	Existing
1	Major and medium Irrigation	58.47	58.47
2	Minor irrigation	55.00	81.43
	Surface water	15.00	17.38
	Groundwater	40.00	64.05
3	Total	113.47	139.90

Current and Projected Water Requirement

India's population is increasing at a relatively fast rate. To feed the burgeoning population, green revolution played a very significant role as the food grains production increased from around 50 MT in the fifties to about 203 MT in the year 1999-2000. Although it hovered around that value for some years, it has again got the momentum and is likely to touch 273.38 MT during 2016-17. Contrarily, projections for our population reveal that for a high growth scenario, population is likely to stabilize at 1581 million in 2050. It means, we not only need to sustain the green revolution but add to the existing productivity. To achieve this, water resource allocation to irrigation cannot be curtailed. On the other hand, increasing demand of other sectors cannot be ignored, more so because of their paying capacity. The likely demand of water for various sectors at a low and high growth scenario as projected in the National Planning Documents is reproduced in Table 16.6. It shows that water allocation for irrigation in per cent term will decrease over the years.

Table 16.6: Annual Water Requirement for Various Uses

Item of use	*1997-98*	*2010*			*2025*			*2050*		
		Low	*High*	*Per cent*	*Low*	*High*	*Per cent*	*Low*	*High*	*Per cent*
Irrigation	524	543	557	78	561	611	72	628	807	68
Domestic	30	42	43	6	55	62	7	90	111	9
Industries	30	37	37	5	66	67	8	81	81	7
Power	9	18	19	3	31	33	4	63	70	6
Inland navigation	-	7	7	1	10	10	1	15	15	1
Environment/Ecology	0	5	5	1	10	10	1	20	20	2
Evaporation losses	36	42	42	6	50	50	6	76	76	7
Total	**629**	**694**	**710**	**100**	**784**	**843**	**100**	**973**	**1180**	**100**

Besides the increasing population, a number of other factors will impact the water requirement. Increasing urbanization, 55.2 per cent of the population is projected to live in the cities by 2050, will greatly impact the water demand. Intensive and extensive cultivation, changes in eating habits, life style changes and increasing emphasis on travel/tourism and environment will lead to a steep increase in water demand. The per capita average annual freshwater availability has reduced from 5177 m^3 in 1951 to about 1869 m^3 in 2001 and is estimated to further come down to 1341 m^3 in 2025 and 1140 m^3 in 2050 (Figure 16.2). As per the international norms, if per-capita water availability is less than 1700 m^3 per annum then the country is categorized as water stressed and if is less than 1000 m^3 per capita per annum then the country is classified as water scarce. Although a large population of India is already in water stressed zone, on an average, whole population will be water stressed by 2025. This change in demand scenario will call upon agriculture, which is currently consuming 80 per cent of the developed water resource, to release fresh water for other sectors of economy. It is in spite of the fact that its own water demand is likely to multiply. Therefore, agriculture needs to look at other unconventional sources particularly the so-called wastewaters released after first use by other sectors.

Spatial and Temporal Distribution of Resource

Annual average rainfall in India is above the global annual average of about 1100 mm yet it fluctuates widely both in space and time. Rainfall is as low as 150 mm at places like Jaisalmer while it is as high as 11,690 mm at Mousinram near Cherrapunji in Meghalaya, being the wettest station in the world. Rainfall distribution in different rainfall zones reveal that more than 50 per cent of the resource is generated in the

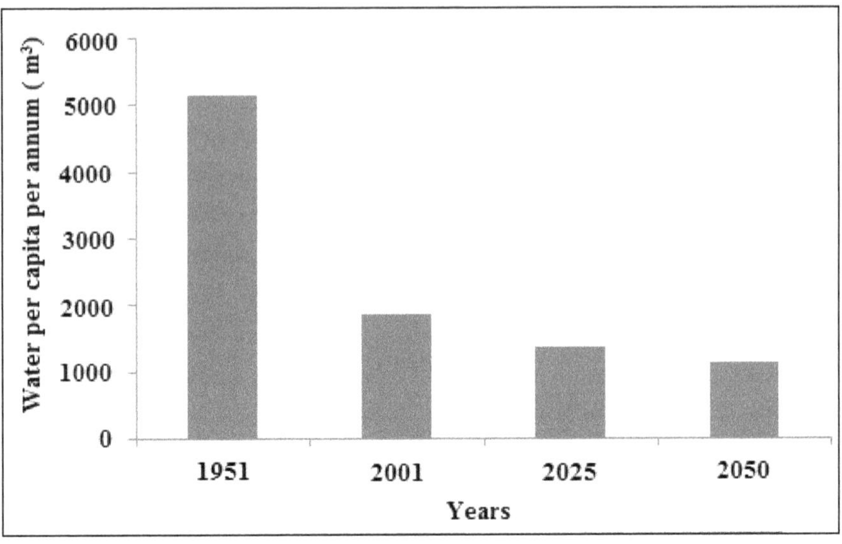

Figure 16.2: Decreasing per Capita Water Resource Over the Years in India.

zone where rainfall ranges from 1000-2500 mm (Table 16.7). The area receiving less than 1000 mm of rainfall, constituting about 48 per cent of the geographical area, contributes only 24.6 per cent to the water resource. Besides 75 per cent or about 3000 km^3 of this resource is generated during June to September. Only about 1000 km^3 is generated during the rest of the year. Another characteristic of the rainfall is that in most cases, it is only a few heavy storms, which account for most of the annual rainfall. Absence of heavy storms generally means reduced rainfall. Moreover, the variations as revealed by the standard deviation increase with decreasing rainfall. Thus, large spatial and temporal variability in the water resource presents several managerial and investment problems that require storage, transport and distribution to areas of deficient water supply.

Table 16.7: Distribution of Geographical Area in different Rainfall Zones in India

Rainfall Zone (mm)	*Geographical Area (M ha)*	*Precipitation Availability (km^3)*
100-500	52.07	156.2
500-750	40.26	251.6
750-1000	65.86	576.3
1000-2500	137.24	2058.6
> 2500	32.57	957.3
Total	328.00	4000.0

Classification of Irrigation Projects

Irrigation sector has always been the largest user of water and would continue to be so. The irrigation projects vary from gigantic structures which collect, convey, and deliver water to the fields to small scale on-farm systems which pump and convey water to the field for irrigation. A number of approaches have been used to categorize the irrigation projects.

Based on Culturalable Command Area (CCA)

On this basis projects have been categorized in 3 groups.

Major Irrigation Projects: The area covered under irrigation in major irrigation projects is more than 10000 ha (CCA>10,000 ha). These projects comprise of huge storage reservoirs, flow diversion structures and a large network of canals. These are often multi-purpose projects covering aspects like flood control and generation of hydro-power.

Medium Irrigation Projects: Medium irrigation projects are the projects having CCA less than 10,000 ha but more than 2,000 ha. These are also multi-purpose surface water projects. Medium size storage, diversion and distribution structures are the main components of such projects.

Minor Irrigation Projects: Projects having CCA less than or equal to 2,000 ha are termed as minor irrigation project. The main sources of water are tanks, small reservoirs and groundwater. These irrigation projects could independently serve an area and/or may exist within the command area of a major or medium irrigation project.

Based on Storage

Sometimes irrigation systems are also categorized on the basis of storage.

- Storage irrigation
- Direct irrigation

In the storage irrigation system, part of the excess water of a river during monsoon season is stored in a reservoir formed at the upstream of a dam constructed across a river or stream (Figure 16.3 left). This stored water is then used for irrigation and other multi-purpose uses.

In the direct irrigation project water is directly diverted from the river into the canal by constructing a diversion structure across the river like a weir or a barrage. These may be designed to have some storage to take care of flow variations (Figure 16.3 right). The structures also help to raise the river water level to enable the water to flow into the off-take channels by gravity. The flow in the channel is usually controlled by a gated structure and this in combination with the diversion structure is sometimes called the headwork.

Another form of direct irrigation is the inundation irrigation which may be called river-canal irrigation. In this type of irrigation there is no irrigation work across the river to control the level of water in the river. Inundation canal off-taking from a river is a seasonal canal which conveys water as and when available in the river.

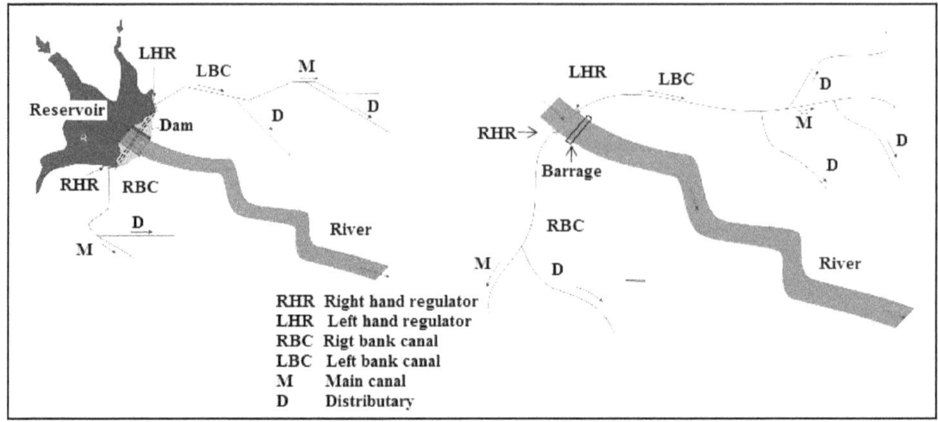

Figure 16.3: Storage Irrigation Scheme (Left) and Diversion Irrigation Scheme (Right).

Based on the Way of Water Application

On this basis, irrigation schemes are classified into two types.

Gravity/Flow Irrigation Scheme: This is the type of irrigation system in which water is stored at a higher elevation so as to enable water to reach the fields by gravity flow. Such irrigation schemes consist of head works across river to store the water and canal network to distribute the water. The gravity irrigation scheme is further classified as:

Perennial Irrigation Scheme: In these schemes water is assured to the command area throughout the crop period so as to meet the irrigation water requirement of the crops.

Non-perennial Irrigation (Restricted Irrigation): In these schemes canal water is generally made available in non-monsoon period from the storage.

Lift Irrigation Scheme: Irrigation systems in which water is pumped from lower elevations to higher elevations (fields are at higher level) are categorized as lift irrigation schemes. It may be the canal water that is pumped to supply water to the fields located at higher elevation or the groundwater based system where groundwater is pumped for irrigation. A major and medium irrigation project can have one or more than one lift irrigation scheme to take care of the un-commanded areas within the command of the project.

Classification based on Productive Potential

The design practice in irrigation engineering is to design irrigation systems in such a manner that water supply covers the full crop water requirement, either completely by irrigation or in addition to rainfall. However, in many cases it does not happen. On this basis irrigation projects are categorized as:

- ✰ Protective irrigation
- ✰ Productive irrigation

Most large scale irrigation systems in India are essentially based on the concept of protective irrigation. These systems are designed and operated on the principle that the available water is spread thinly over a large area, in an equitable manner. Irrigation water is only a supplementary source of water. The idea is to reach as many farmers as possible, and to protect them against crop failure and famine. The primary objective of protective irrigation thus has an explicit social dimension.

In productive irrigation, available water is distributed to ensure high productivity *i.e.* each stakeholder is provided water in amounts equal to the full crop requirement. It is believed that it will ensure high productivity in the project as a whole as compared to protective irrigation.

Commonly Used Terms in Irrigation

Command Area (CA): It is defined as the area that can be irrigated by a canal system, the CA is further be classified as GCA and CCA.

Culturable Command Area (CCA): It is the proportion of the Gross Command Area which is culturable and cultivable. This is the actually irrigated area within the GCA. However, the entire CCA may or may not be put under cultivation during any crop season due to several reasons. For that season, area is termed as culturable cultivated area while the remaining area not cultivated during a crop season is termed as culturable uncultivated area.

Cumecs: An acronym for water flow rate in cubic meter per second. $1\ m^3/s = 1000\ l/s$.

Cusecs: An acronym for water flow rate in cubic feet per second.

Gross Command Area (GCA): It is defined as total area that can be irrigated by a canal system on the perception that unlimited quantity of water is available. It is the total area that may theoretically be served by the irrigation system. But this may include inhabited areas, roads, ponds, uncultivable areas *etc.*, which are not to be irrigated.

Intensity of Irrigation: It is defined as the percentage of the irrigation proposed to be irrigated annually. Usually the areas irrigated during each crop season (*Rabi, Kharif, etc.*) are expressed as a percentage of the CCA which represents the intensity of irrigation for the crop season. The yearly intensity of irrigation is obtained by adding the intensities of irrigation for all crop seasons.

Irrigation Potential Created (IPC): This term refers to the total gross area proposed to be irrigated under different crops during a year by a scheme. The area proposed to be irrigated under more than one crop during the same year is counted as many times as the number of crops grown and irrigated.

Irrigation Potential Utilized (IPU): This term is defined as the gross area actually irrigated during the reference year out of the gross proposed area to be irrigated by the scheme.

Irrigation System: A set of components which includes the water source (canal or groundwater), water distribution network, control components (*e.g.*, valves and controllers), emission devices (*e.g.*, sprinklers and emitters) and possibly other general irrigation equipment for irrigation of field crops.

Project Efficiency: Overall efficiency of irrigation water use in a project setting that accounts for all water uses and losses, such as crop ET, environmental control, salinity control, deep percolation, runoff, ditch and canal leakage, phreatophyte use, wetlands use, operational spills, and open water evaporation. Efficiency is also defined as the ratio of the water beneficially used to the total water delivered into the project area.

Ultimate Irrigation Potential (UIP): This term refers to the gross area that could be theoretically irrigated if all available land and water resources would be used for irrigation.

Chapter 17

Measurement of Irrigation Water

Introduction

The science and practice of water measurement is known as hydrometry. The measurement of irrigation water in the field is generally carried out using current meters, calibrated or rated channel cross sections, flumes or standardized weirs. It may be done by manual or automatic recorders. This chapter discusses some of the irrigation water measurement equipment and procedures of measurements.

Importance of Water Measurement

Irrigation water is an essential input for crop production. Over- exploitation of natural water resources has resulted in depletion of groundwater table in high productivity areas. Poor irrigation efficiencies resulting from high seepage losses from the conveyance system, poor land development and mismanagement of the irrigation water resources is being debated. Systematic water measurements at various scales, proper recording and interpretation will help in:

- ✰ Accurate accounting of the precious resource
- ✰ Allocation of equitable share of water between competitive users
- ✰ Help on-farm water users to design conveyance systems, achieve the best use of the irrigation water with minimal negative environmental impacts
- ✰ Higher yields by avoiding under-irrigation or over irrigation of the crops
- ✰ First step in volumetric supply and pricing of water
- ✰ Evaluation of seepage losses in channels
- ✰ Assess cost-benefits of proposed canal and ditch improvement programmes required in irrigation water management
- ✰ Design of channels in studies related to testing of wells, runoff measurement and planning of soil conservation structures according to anticipated flow rates.

Open Channel and Pipe Flow

A major category of flow is open channel flow, which occurs when there is a free liquid surface open to atmospheric pressure. In open channels, the only force that can cause flow is the force of gravity on the fluid. As a result, with steady uniform flow under free discharge conditions, a progressive fall or decrease in the water surface elevation always occurs as the flow moves downstream. The flow types include tunnels, partially filled pipes, canals, streams and rivers. The term pipe flow is often used to refer to flow in any closed conduit under pressure. The closed conduit is often circular, but may also be square or rectangular, such as a heating duct. Flow occurs in a pipeline when a pressure or head difference exists between ends. The discharge in pipe flow mainly depends upon (1) the amount of pressure or head difference that exists from the inlet to the outlet; (2) the friction or resistance to flow caused by pipe length, pipe roughness, bends, restrictions, changes in conduit shape and size, and the nature of the fluid flowing; and (3) the cross-sectional area of the pipe (Also see Chapter 13).

Units of Measurement

Several units are used to measure irrigation water and runoff. While water is usually measured in volumetric units, flows are usually described in terms of rate *i.e.* volume of water per unit time. In metric unit's volume is measured in terms of liter (l), cubic meter (m^3) or cubic cm (cm^3) or hectare centimeter (ha-cm) or hectare meter (ha-m). Flow rates are measured in liter per second (l/s) and cubic meter per second (m^3/s commonly written as cumecs). Since dimensions of many structures are given in British units, British units of volume or flow rates notably cubic feet and cubic feet per second (cusecs) are commonly used in irrigation measurements. Conversion tables are then used to convert the British units into metric units.

Methods of Water Measurement in Open Channels

Several devices are commonly used to measure irrigation water and runoff *etc.* (Miller, 1984, Bos 1976). They are grouped into several categories although principle used in some of the devices may be similar:

1. Volumetric measures
2. Stage-discharge relations
3. Velocity-area methods
   a. Float method
   b. Current meters
4. Slope-hydraulic radius-area methods
5. Measuring structures
   a. Weirs
   b. Orifices
   c. Flumes
6. Tracer Methods

7. Coordinate method
8. Modern instruments

Although slope-hydraulic radius method is also a velocity-area method, yet it is separately listed because principally the method is substantially different. Besides, tracer methods may have limited applications but are included to have a fair and complete view of the methods used for water measurement.

Volume Measurement or Timed Gravimetric Method

The flow rate is calculated by collecting the entire content of the flow stream in a container of sufficient capacity over a fixed duration. This method is suited to situation where small discharges are measured say of shallow tube wells or small runoff plots. The discharge rate is calculated by noting the time taken to fill the container or for the water surface to rise to a predetermined level. Rate of flow is then calculated by the following formula:

$$\text{Discharge (l/s)} = \text{Volume of the water (l)/Time (s)} \quad (17.1)$$

Stage-Discharge Relations

Since, there is a strong relation between channel stage and discharge, stream gauges are used to measure stage in the canals/rivers to workout discharge. The whole facility varies from a simple staff gauge to a structure in which instruments used to measure, store, and transmit the stream-stage information are housed. One common approach is to have a stilling well in the canal/river bank or attached to a bridge pier. Water enters and leaves the stilling well through underwater pipes allowing the water surface in the stilling well to be at the same elevation as the water surface in the canal/river. The stage is then measured inside the stilling well either visually or using a float or a pressure, optic, or acoustic sensor. Staff gauges are commonly used for visual measurement of the stage. The stage at a stream gauge is always measured with respect to a constant reference elevation, known as a datum.

Calibration charts are usually prepared to define the stage-discharge relationship by measuring discharge at a wide range of stages (Figure 17.1). These charts are then used to measure the flow rates once the stage of the channel is known.

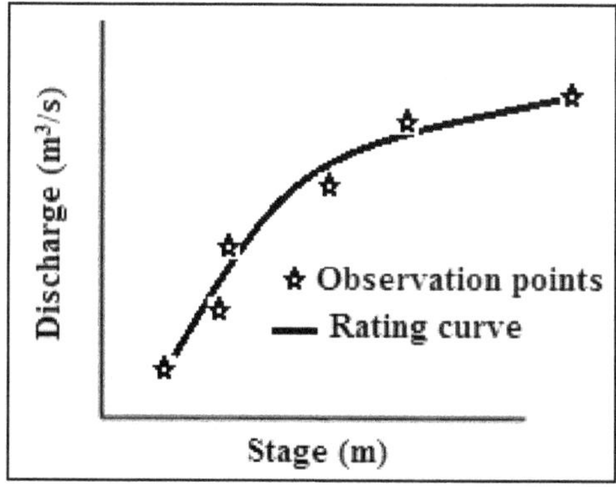

Figure 17.1: A Typical Stage-Discharge Relation of a Channel.

Use of Existing Structures

Existing structures such as culverts or bridge openings are used as control sections to have an estimate of the flows. For rectangular culverts, an approximate value can be calculated from the general formula for flow through a rectangular weir.

$$Q = c\,W\,H^{3/2} \qquad (17.2)$$

Here, Q is the flow, m^3/s, W is the width of opening, m, H is the depth of flow, m and c is the coefficient of discharge which depends on the geometry of the culvert. A typical value is 0.6. More precise values can be selected/derived from tables/formulae reported in USDA-SCS (1962).

Velocity and Discharge Measurement

Velocity-area methods are commonly used to measure water discharge in open channels. The two most commonly used methods are the float method and current meters. In these methods, velocity of water flow passing a point in an open channel is measured. The flow rate is then estimated by multiplying the average velocity of water by the cross-sectional area of the flow perpendicular to the flow direction. It can be described by the following equation:

$$\text{Discharge, } Q = \text{Area} \times \text{Velocity} = A \times V \qquad (17.3)$$

Here, Q is the discharge rate, cm^3/s or m^3/s, A is the cross-sectional area, cm^2 or m^2 and V is the velocity of water, cm/s or m/s

Float Method

Float such as a wooden object or a long necked bottle is placed in steady, uniformly flowing reaches of channels to measure the flow velocity. Following procedure is used.

Select a straight section of the channel about 30 m long with a fairly uniform cross-section. Find out the average cross-sectional area (A). Several measurements may be required to arrive at a reasonably good estimate. Stretch a string across initial starting point to the end of the flow measurement point at right angle to the direction of flow. The float used to measure the velocity is placed slightly above the upstream of the initial starting point of the trial section. The time is noted as soon as the float reaches the starting point. Similarly, the time is noted when the float reaches the lower end. Time taken by the float to travel the known distance is noted and used to calculate the flow velocity (V). The discharge is then calculated with the help of eq. (17.3) and multiplied by a factor C to have the actual discharge.

The factor C is used to take care of the flow velocity at the surface, which will be higher than the mean velocity of flow. In small field channels, most text books mention that a coefficient of 0.85 would suffice if a wooden float or long necked bottle is used to measure the velocity. But normally the value depends upon the depth of the channel (Table 17.1). Sometimes, float is calibrated in a representative channel to find out the value of C. In other cases, float is even designed to be carried along at the average velocity.

Table 17.1: Coefficient to Convert Surface Flow Velocities to Mean Channel Velocities

Average Depth in the Reach (m)	Coefficient, C
0.30	0.66
0.60	0.68
0.90	0.70
1.2	0.72
1.5	0.74
1.8	0.76
2.7	0.77
3.5	0.78
4.5	0.79
>6.0	0.80

Current Meter

The velocity of flow in a stream or river can be measured directly with the help of a current meter. The current meter is a small instrument containing a revolving wheel or vane that is turned by the flowing water. The conical cup type meter revolves about a vertical axis while the propeller type about a horizontal axis. In each case numbers of revolutions in a given time are counted, either on a digital counter or as clicks heard in earphones worn by an operator. The velocity is then determined using calibration charts having a plot against number of revolutions.

In the conical cup type category, Price AA current meter developed by USGS has a wheel of six metal cups (Figure 17.2). It has the following general features:

- ✰ Vanes to keep the front of the meter headed into the current
- ✰ Either a cable or a rod for handling the meter
- ✰ Weights for sinking the meter when suspended on a cable
- ✰ An electric device for signalling and/or counting the number of revolutions
- ✰ Connections from the current meter to a 12-volt battery-powered headphone

The Price AA meter can be attached to a wading rod for discharge measurements in shallow waters or can be mounted just above a weight suspended from a cable and reel system to measure discharge in fast flowing or deep waters (Government of Rajasthan, 2016). Pygmy meters are similar to Price meters in that both contain a cup-type wheel mounted on a vertical shaft. The pygmy cup wheel is 2" in diameter compared with 5" for conventional Price meters. It is a two-fifths scale version of the Price AA meter. Thus, the pygmy meter can measure velocities closer to flow boundaries and can be used for shallow water channels.

The propeller type current meters have some advantages over the cut type. They are less sensitive to velocity components being not parallel to the meter axis, they are smaller, and they are more suited for mounting in multiple units to assess

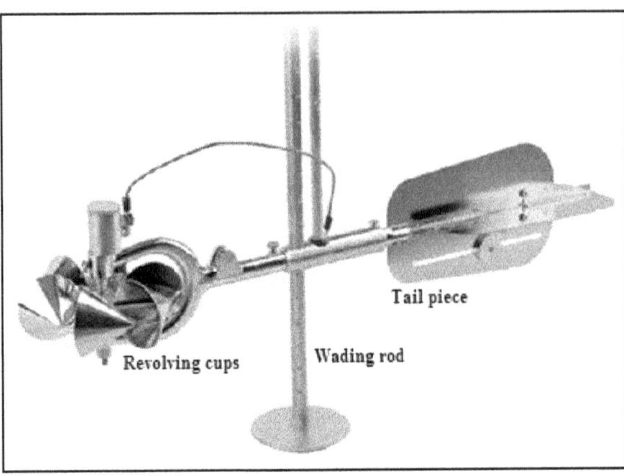

Figure 17.2: A View of the Price Current Meter on a Wading Rod.

the velocity profile in the channels. Several handy new designs of current meters are now in the market to suit specific conditions.

Since a current meter measures the velocity at a single point, several measurements are required to calculate the total flow, if the canal is wide. The procedure is to measure and plot the cross-section on graph paper and to imagine that it is divided into strips of equal or various widths (Figure 17.3). In each section, current meter is suspended to measure the water velocity. Similarly, the area of each section is obtained by measuring the width and depth of that section. The average velocity for each strip is estimated from the mean of the velocity measured at different depths (Table 17.2).

Table 17.2: Measurement of Velocity in Channels

Channel Characteristics	*Depth of Measurement Point*	*Average Velocity, V*
Shallow channels	0.6 d	V0.6, taken as the average velocity
Medium Channels	0.2d and 0.8 d	(V0.2+ V0.8)/2
	0.2d, 0.6d and 0.8d	(V0.2+2V0.6+ V0.8)/4

d is the depth of channel, m. All measurements are from the water surface, m

The discharge in each subsection is computed by multiplying the area of each subsection by the measured velocity of the respective subsection. The total discharge is then computed by summing the discharge of each section. The calculations procedure is illustrated through the solved example 17.1.

Example 17.1

The areas of each section and section wise velocities in a channel are shown in Figure 17.4 and reported in Table 17.3. Fill-in the blanks of all the rows of columns 6 and 7 and calculate the discharge.

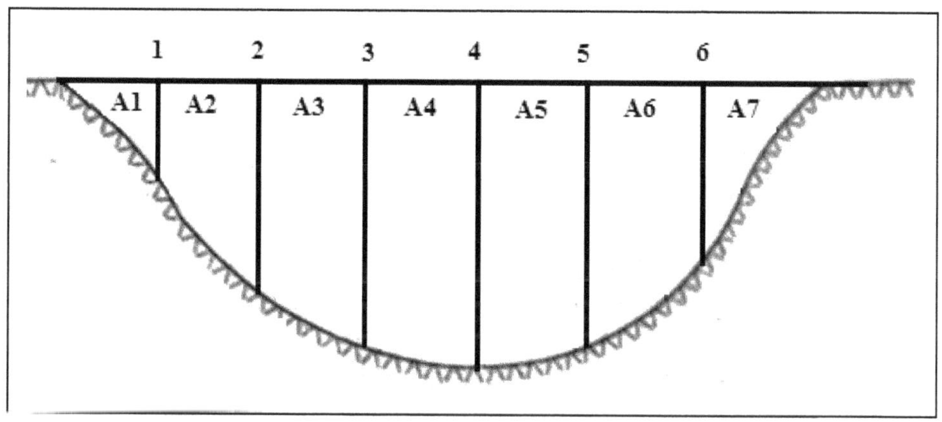

Figure 17.3: A View of Subdivisions of a Large Channel in Sub-section.

The filled/in values are shown in columns 6 and 7 in bold.

Sum of the values in column 7 gives the discharge as 8.09 m^3/s.

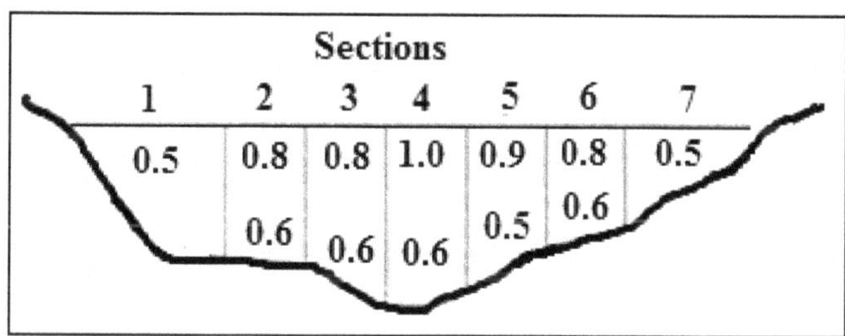

Figure 17.4: Flow Variation in a Stream as Measured Using a Current Meter.

Table 17.3: As given for the Example is Used to Fill-in the Blanks

1	2	3	4	5	6	7
Section	*Area (m^2)*	*Flow Velocities (m/s)*				*Flow (m^3/s) (2 × 6)*
		0.2d	*0.6 d*	*0.8d*	*Mean*	
1	2.4	-	0.5	-	**0.50**	**1.20**
2	1.5	0.8	-	0.6	**0.70**	**1.05**
3	1.6	0.8	-	0.6	**0.70**	**1.12**
4	1.7	1.0	-	0.6	**0.80**	**1.36**
5	2.0	0.9	-	0.5	**0.70**	**1.40**
6	1.8	0.8	-	0.6	**0.70**	**1.26**
7	1.4	-	0.5	-	**0.50**	**0.70**
Summing the values in various rows of column 7, we get Q as						**8.09**

Slope-Hydraulic Radius-Area Methods

Several formulae are used to calculate the velocity and the discharge in the channels. Measurement of water surface slope, cross-sectional area, and wetted perimeter over a length of uniform section channel are needed to calculate the flow rate. Chezy´s and Manning´s are the two important formulae that can be used to estimate the discharge. Herein, more commonly used Manning's equation is adopted to illustrate the procedure.

The Manning's equation (Manning, 1890) is given as:

$$V = (1/n)\,(R^{2/3} S^{1/2}) \qquad (17.4)$$

Here n is roughness coefficient of the channel, R is the hydraulic radius of flow = A/P, m, P is the wetted perimeter, m, S is the channel bed slope, m/m. The roughness coefficients can be selected for a given situation from Tables 13.3 or 13.4 or from Table 17.4. If R and S for a particular reach are known or can be determined, it is possible to calculate the velocity using eq. (17.4). The discharge then can be calculated using eq. (17.3).

Table 17.4: Values of Manning's n for Wet Cross Section Well Covered with Weedy or Grassy Vegetation

Channel Description	*n-value*
(i) Small channel (water depth <0.75 m)	
Sandy soil	0.050
Clayey soil	0.060
(ii) Medium channel (water depth 0.75-1.5m)	0.035
Sandy soil	0.050
Clayey soil	0.020-0.025
(iii) Large channels (water depth >1.5 m)	0.012-0.017
(iv) Concrete lined channel	
(v) Earth Channels	
Straight and uniform	0.023
Winding, sluggish	0.025
Stony bed with weeds on banks	0.035
Small drainage ditches	0.040
Clay or concrete drain tile	0.011
Cast iron	0.012-0.013
Steel	0.015-0.017

Source: For (i) to (iv) Smedema and Rycroft (1983); (v) Michael (1978).

Measurement Devices: Weirs, Orifices and Flumes

Some water measuring devices that use measurement of head, h, to determine discharge, Q, are:

1. Weirs
2. Orifices
3. Flumes

Weirs

A weir is an obstruction or dam built across an open channel over which the liquid flows, often through a specially shaped opening. These simple devices can be used to measure even small flows since the height of the water flowing over the crest of the weir correlates to the flow rate. In its simplest form a weir consists of a weir wall of concrete, timber or metal with a sheet metal weir plate fixed to it. These are built as stationary structures or are portable. According to the shape of the opening, weirs are classified as:

- ✰ Rectangular weir
- ✰ The triangular (or V-notch) weir
- ✰ Trapezoidal (or Cipolletti) weir
- ✰ Compound weirs or a combination of two shapes

These weirs are known as sharp crested weirs *i.e.* a weir having thin-edged crest such that the over-flowing sheet of water has the minimum surface contact with the crest. Amongst the four, rectangular and 90°- V notch weirs are commonly used. The weirs are normally installed at a place where there is a drop in the elevation of the channel bed. It allows the flow through the weir to fall freely to the downstream channel. The discharge through a weir is proportional to the head on the crest and depends on condition of the crest, contraction, velocity of approaching stream and elevation of the water surface downstream of the weir (Kindsvater and Carter,1957). The following equation describes the flow characteristics:

$$Q = CLH^{m} \tag{17.6}$$

Here, Q is the discharge, L/s, L is the length of crest, cm and H is the head over the weir crest, cm and m is an exponent depending on weir opening.

On the basis of shape of the weir crest, weirs are categorized in two groups.

- ✰ Sharp crested weir
- ✰ Broad crested weir

A broad crested weir is normally a flat topped obstruction at the top which extends across the entire channel. In this case the streamlines are parallel to each other over the crest such that the hydrostatic pressure distribution can be assumed along the length of the weir (L). A true broad-crested weir flow occurs when upstream head above the crest is between the limits of about 1/20 and 1/2 the crest length in the direction of flow. For example, a thick wall or a flat wooden log can act like a sharp-crested weir when the approach head is large enough that the flow springs from the upstream corner. If upstream head is small enough relative to the top profile length, the wooden log acts like a broad-crested weir. Sharp crested weirs are used to measure the discharge of smaller rivers and canals; whereas broad crested weirs are used to measure the flow of larger rivers and canals.

Weirs are also classified on the basis of effect of the sides on the emerging nappe.

- ☆ Weir without end contraction (suppressed weir)
- ☆ Weir with end contraction (contracted weir)

When sides of the flow channel act as the ends of a rectangular weir, no side contraction exists, and the nappe does not contract from the width of the channel. End contractions are produced when the crest length or width of the notch is less than the channel width.

Advantages of Weirs

- ☆ Are simple devices and easy to construct
- ☆ Durable but costs less than flumes
- ☆ No obstruction by floating material
- ☆ Relatively easy to install
- ☆ Very accurate when used properly

Disadvantages of Weirs

- ☆ Operate with a relatively high head loss making their use impractical in areas having level land
- ☆ Higher maintenance cost than flumes
- ☆ Accuracy affected by approach velocity and as such cannot be easily combined with other farm structures
- ☆ If gravel, sand or silt *etc.* is deposited above the weir, it may affect the accuracy. Therefore, weirs need periodical cleaning

Commonly Used Terms

The commonly used terms are described as follows (Figure 17.5):

Bottom Contraction: When the crest of the weir is elevated above the floor of the channel, the flow coming along the bottom of the weir pool and up a sufficiently high bulkhead springs up in a curved, underside jet surface or crest contraction. It is related to the vertical distance from the weir crest to the bottom of the weir pond. Since the mean velocity in the headwater, v can be given as Q/b (M + h) (m/s) such that b is crest width, h is the head above the crest and M the height of the weir crest above the headwater bottom. The bottom or vertical contraction will not occur If h/(M+h) tends towards zero (because h is small compared to M) then $v^2/2gh$ also tends towards zero.

End Contraction: End contractions cause the water flow lines to converge through the notch (Figure 17.6). For a weir to be fully contracted, (B-b) must be greater than $4h_{max}$, where h_{max} is the maximum expected head on the weir (USDI, 2001). A partially contracted weir has (B-b) between 0 and $4h_{max}$. Here B is the channel width and b the crest width.

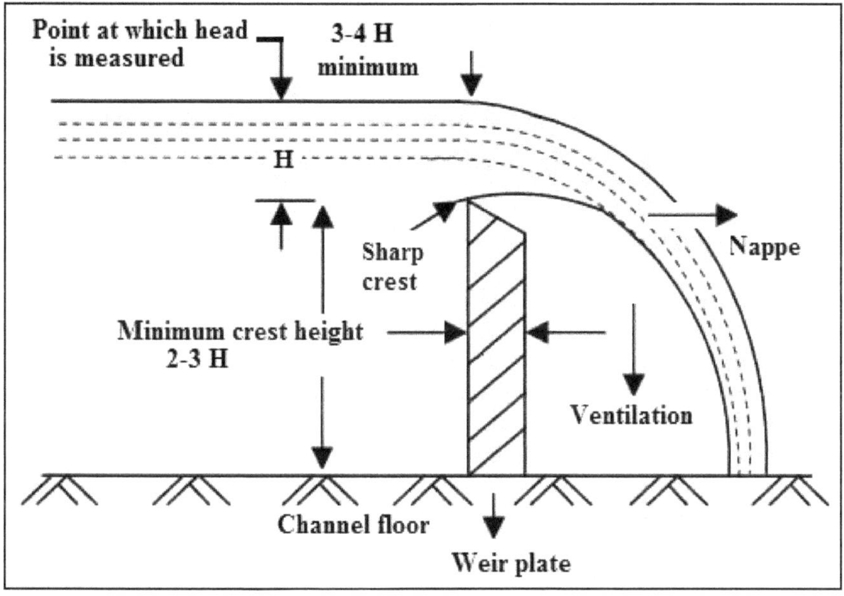

Figure 17.5: General Features of a Rectangular Weir.

Free and Submerged Flow: When the water surface downstream from the weir wall is far enough below the crest so that air has access around the nappe, the flow is said to be free. Downstream water rising near or above the weir crest elevation produces a submerged weir condition (submerged flow).

Head: The depth of water flowing over the weir crest measured at some point in the weir pond.

Nappe: The falling sheet of water springing from the weir plate is the nappe.

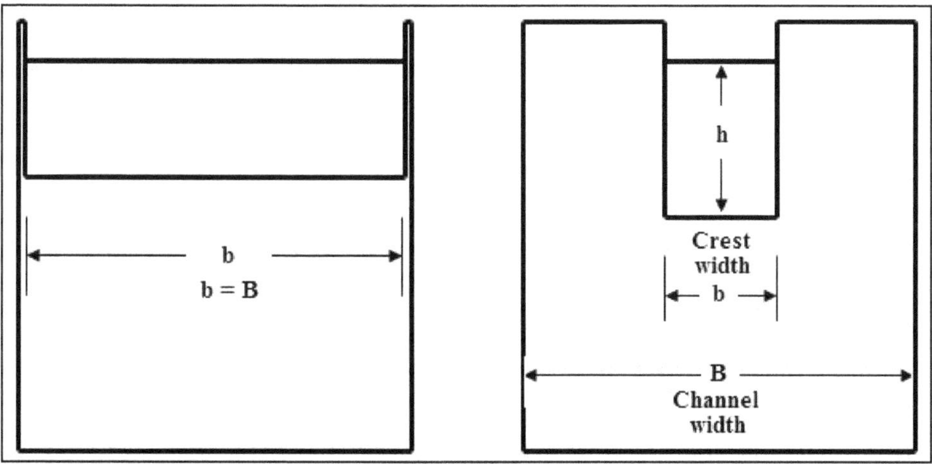

Figure 17.6: A Suppressed (Left) and a Contracted Weir (Right).

Weir Crest: The bottom of the weir notch.

Weir Pond: Weirs act as a dam across the flow of water. As a result, there is a body of water created upstream of the weir. This body of water is the weir pool or pond the portion of the channel immediately upstream of the weir.

Weir Scale or Gauge: The scale fastened on the side of weir or on a stake in the weir pond to measure the head on the weir crest.

Rectangular Weir

A rectangular weir derives its name from rectangular shape of the notch with horizontal crest and vertical sides. These have a sharp crest and are bevelled on the downstream side only. The sides are not bevelled. When sizing a rectangular weir, the minimum length of the crest should be about 30 cm. These are used to measure comparatively large discharges. The discharge through a suppressed rectangular weir is computed by the Francis' formula stated below:

$$Q = 0.0184\ LH^{3/2} \qquad (17.7)$$

Here, Q is the discharge, l/s, L is the length of crest, cm and H is the head over the weir crest, cm. Use of this equation is subject to the condition that $H/P \leq 0.33$ and $H/B \leq 0.33$, such that P is crest height from the bottom of the channel and B is the width of the channel.

Contracted Rectangular Weir

When the length of the crest is less than the width of the approach channel, the water approaches the channel in a converging manner. The contraction may be either on one or on both sides. Following equations are used in such cases.

End contractions on one end:

$$Q = 0.0184\ (L - 0.1H)\ H^{3/2} \qquad (17.8)$$

End contractions on both ends:

$$Q = 0.0184\ (L - 0.2H)\ H^{3/2} \qquad (17.9)$$

Here Q, L and H have the same meaning as for eq. (17.7). Although tables can be used to assess discharge, yet it is easier to calculate the discharge using above equations. Use of the fully contracted rectangular weir equation is subject to the conditions that $H/L \leq 0.33$, $B - L \geq 4\ H_{max}$, and $P \geq 2H_{max}$. L is the weir length, H_{max} is the maximum head over the weir, and H, B, and P are as defined earlier.

Example 17.2

Compute the discharge of a rectangular weir 50 cm long with a head of 12.5 cm. Assume the weir as a suppressed weir. What will be the discharge if the weir is contracted at one of its end and on both the ends.

With no end contraction:

$$Q = 0.0184\ L\ H^{3/2}$$

$$= 0.0184 \times 50 \times 12.5^{3/2} = 40.66\ l/s$$

With one end contraction

$Q = 0.0184 (L-0.1 H) H^{3/2}$

$= 0.0184 (45-0.1 \times 12)123/2 = 39.64$ l/s

With two end contractions:

$Q = 0.0184 (L-0.2 H) H^{3/2}$

$= 0.0184 (45-0.2 \times 12)123/2 = 38.62$ l/s

The discharge reduces from 40.66 l/s to 39.64 l/s for one end contraction to 38.62 l/s for contraction at both the ends.

Triangular or V-notch Weir

The name has been derived from the V shape of the notch. Its crest is the notch where the two sides of the triangle meet. The angle of the notch is commonly 90° (Figure 17.7), but equations and calibration charts are available for other angles, such as 120°, 60°, 45°, and 30°. The V- notch weir is capable to accurately measure small flows, having reasonable accuracy for flows up to 0.3 cumecs. It is highly accurate to measure flows less than 0.03 cumecs. The discharge through a fully contracted 90° V-notch weir with free flow conditions is computed by the following formula:

$$Q = 0.0138 H^{2.48} \quad (17.10)$$

Such that Q is in l/s and H is in cm. In practice, the exponent is taken as 2.5. The conditions for the V notch weir to be fully contracted are (Figure 17.7):

$P \geq 2H_{max}, S \geq 2H_{max}$

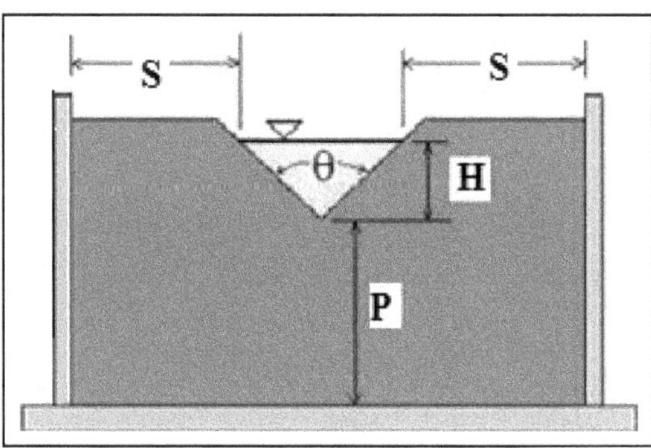

Figure 17.7: Details of 90°-V-notch Weir Showing Typical Dimensions.

Example 17.3

Calculate the minimum and maximum flow rate covered by the range of 6 cm to 40 cm for the head over a V- notch weir that is fully contracted under the free flow conditions.

In order to be fully contracted, P and S must both be greater than $2H_{max}$, that is greater than (2) (40) or greater than 80 cm (Check)

The discharge is given as

$Q = 0.0138.H^{2.5}$

Thus minimum discharge is

$Q_{min} = 0.0138 \times 6^{2.5} = 1.22$ l/s

Ana maximum discharge is

$Q_{max} = 0.0138 \times 40^{2.5} = 139.64$ l/s

Cipolletti Weir (Trapezoidal)

This weir combines the characteristics of the V-notch and the rectangular weir in one (Robinson and Chamberlein, 1960). The Cipolletti weir has a horizontal crest like a rectangular notch and sloping sides like a V-notch (Figure 17.8). The Cipolletti weir is a contracted trapezoidal weir in which each side of the notch has a slope of 1 horizontal to 4 vertical. It is named after its inventor Cesare Cipolletti, an Italian engineer. Its popularity rests largely upon the belief that side slopes of 1 to 4 are just sufficient to correct for the end contractions of the nappe and that the flow is proportional to the length of the weir crest. As such, there is no need to apply the correction for end contractions. The weir has sharp crest and sharp sides, bevelled from the downstream side only. It is commonly used to measure medium flows. It is less accurate than V-notch or rectangular weirs. The discharge can be computed by the following formula:

$$Q = 0.0186\ LH^{3/2} \tag{17.11}$$

There is no particular advantage of this weir over the rectangular weir.

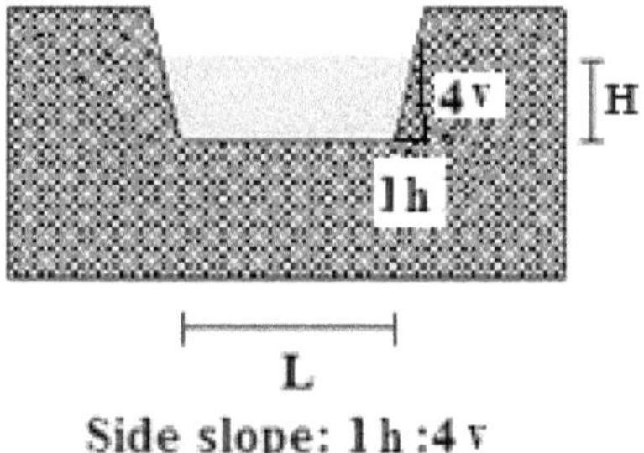

Figure 17.8: A Schematic View of a Cipolletti Weir.

Compound Weirs

Unusual situations may require special weirs. For example, under fluctuating flows, a compound rectangular weir may perform better (Figure 17.9 left). Similarly, a V-notch weir might easily handle the low flows; but occasionally, much larger flows might require a rectangular weir. A compound weir, consisting of a rectangular notch with a V-notch cut into the center of the crest, might be used in such a situation (Figure 17.9 right).

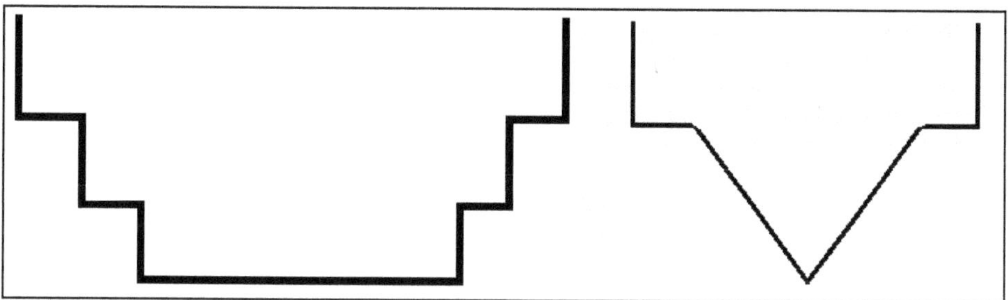

Figure 17.9: Schematic View of a Compound Weir Rectangular (Left) and a Combination of V-notch and Rectangular Weir (Right).

Installation and Operation of Weirs

A properly constructed and installed weir is one of the simplest and most accurate methods to measure water, accuracy under ideal conditions being within 2-3 per cent of the real discharge. On the other hand, improper setting and operation may result in large errors. Following guidelines may be adopted during installation and operation of the weirs.

- ☆ The weir should be set at the lower end of a long sufficiently wide and deep channel so that water flow is even and smooth. Baffles may be put in the weir pond to reduce velocity to about 15 cm/s and equalize the flow.
- ☆ To measure the discharge from tube wells, a portable weir pond made of sheet metal or a built-in weir pond made of brick or stone masonry should be used. The weir plate is fixed at the outlet of the weir pond.
- ☆ The weir wall must be vertical and must extend far enough into the bank.
- ☆ The centre line of the weir should be parallel to the direction of flow.
- ☆ The crest of the weir should be level. It will ensure that the water passing over it will have same depth at all points along the crest.
- ☆ The notch should be of a regular shape with edges being rigid and straight.
- ☆ The weir crest should be above the bottom of the approach channel (weir pond) by at least twice the depth of water flowing over the weir.
- ☆ A minimum distance of about twice the depth of flow should be maintained between the end of the weir crest and the bank of the upstream channel.
- ☆ The crest of the weir should be high enough so that water freely falls below the weir, leaving an air space under the over falling sheet of water.
- ☆ The depth of water flowing over the rectangular weir should not be less than about 5 cm, and not more than about two-thirds the crest width.
- ☆ The scale or gauge used to measure the head should be located at a distance of about three to four times the approximate head (Figure 17.5). It should be far enough to one side so that it is placed in a comparatively still water.
- ☆ The zero of the scale should be exactly at the same level as the crest level of the rectangular or Cipolletti weir or the apex of a V-notch weir.

Orifices

An orifice is a simple opening in a plate made of steel or aluminium and placed across a channel to measure the flow rate. It may be defined as an opening provided in the side or bottom of tank or vessel such that the liquid flows through the entire orifice. Orifices can be of various shapes although the most common shapes are the circular and the rectangular. If the upstream water surface drops below the top of the opening of a rectangular orifice, the opening, in effect, becomes a rectangular weir. The cross sectional area of the orifice is small in relation to the stream cross section. Under free flow conditions, the flow from the orifice discharges freely into air. In submerged flow condition, the downstream water level is above the top of the orifice opening and the flow discharges into water. Free flow orifices plates can be used to measure comparatively small streams. A standard orifice must:

- ☆ Be placed perpendicular to the flow
- ☆ Have sharp edges (no more than 1-2 mm thick)
- ☆ Be contracted from the channel boundary and water surface at least twice the smallest opening dimension, and
- ☆ Have a cross-sectional flow area upstream of the orifice, at least 8 times the orifice cross-sectional area.

Sectional side view of an orifice plate installed in an open channel is shown in Figure 17.10. The orifice shown is fully contracted and submerged orifice (17.10 left), *i.e.* the perimeter of the opening being far enough from the walls of the approach channel and water surface. The jet fully contracts when passing through the opening. The orifice jet is also submerged since downward level of water is above the orifice (Figure 17.10 left). The discharge equation for a standard non-submerged orifice (Figure 17.10 right) is given as follows:

$$Q = 61 \times 10^{-5} A (2gh_1)^{1/2} \tag{17.12}$$

The equation for a submerged orifice shown in Figure 17.11 changes to:

$$Q = 61 \times 10^{-5} A\left(\sqrt{2g(h_1 - h_2}\right) \tag{17.13}$$

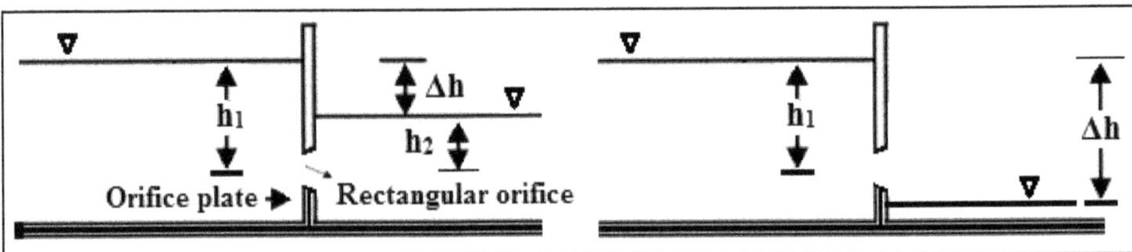

Figure 17.10: A Rectangular Submerged Orifice (Left) and Orifice with Free Flow (Right).

Here, Q is discharge, L/s, A is the cross-sectional area of the orifice, cm^2, g is the acceleration due to gravity which is 981 cm/s^2, h_1 is the depth of water over the centre of the orifice (on the upstream side) in case of free flow orifice or the difference in elevation between the water surface at the upstream and downstream faces of the orifice plate in case of submerged orifice, (h_1-h_2), cm.

If the upstream channel cross-sectional area, A1, is not 8 times larger than the orifice opening area, A, then the approach velocity must be taken into consideration. The upstream velocity head, hv, given by:

$$h_v = v1^2/2g = Q^2/(2g\ A1^2) \tag{17.14}$$

The hv so calculated must be added to the upstream hydraulic head, so that:

$$Q = 61 \times 10^{-5}\ A\left(\sqrt{2g(h_1 + h_v - h_2}\right) \tag{17.15}$$

If the bottom or the side walls of the approach channel are too close to the orifice edges, then the sides and bottom of the orifice jet are not fully contracted. If a portion of the orifice is suppressed (an edge of the orifice does not protrude from, but is even with, the upstream channel floor or walls) the discharge coefficient, Cd (61×10^{-5} in Eq. 17.12), must be increased by 1.5 per cent for every 10 per cent of the orifice perimeter suppressed.

Spiles are small pipes buried in the ditch bank. Siphons on the other hand are small diameter pipes used to convey water over the channel embankment (Figure 17.11). Spiles and siphon tubes act like orifices and therefore, can be used as flow measurement devices if properly calibrated. For short lengths of pipe, the head loss due to pipe friction is small and is proportional to the velocity head. Therefore, it will slightly reduce the discharge coefficient. The discharge in both cases can be estimated using eq. (17.12). The discharge coefficient depending upon the shape and length of the tube and the entrance and exit conditions will vary. Coefficients close to 60×10^{-5} are commonly used.

Advantages and Disadvantages

Submerged orifices conserve head and are therefore, preferred where outfall

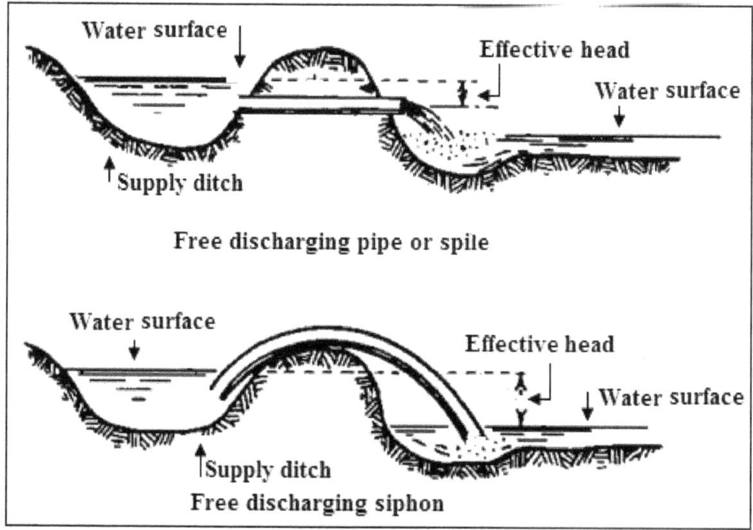

Figure 17.11: Spiles (Pipes) and Siphon Tubes as Measuring Devices.

is insufficient for weirs. Submerged orifices are also used where cost, space limitations, or other site conditions do not permit installing a weir or a flume. The main disadvantage of the submerged orifice is that sand and sediment upstream from the orifice may accumulate to affect measurement accuracy. Severe blockage have been reported in several cases.

Flumes

A flume is a specially shaped flow section that provides a restriction in the channel area and/or a change in channel slope. The flow rate in the channel is determined by measuring the liquid depth at a specified point in the flume. Short-throated flumes commonly used to measure water are:

- Parshall flume
- Cutthroat flume
- H flume

The following sections deal with Parshall and cutthroat flumes. Relatively speaking, cutthroat flumes are simpler to construct than Parshall flumes and can operate under free flow conditions at a higher degree of submergence. H Flumes are commonly used to measure runoff at faraway places because of little maintenance requirement.

Long-Throated Flumes

These types of flumes control the discharge in a throat that is long enough to cause nearly parallel flow lines in the region of flow control. The long-throated ramp flume (also known as a broad-crested weir) are used to measure discharge in flat ditches or canals since free flow conditions can be maintained even at 90 per cent or more submergence. The long throated flume is recommended because of its simple design, easy installation and flexibility in placing them in complex channel shapes with varying flow regimes.

Major advantages and disadvantages of flumes are as follows:

Advantages

- Self-cleaning to a certain degree
- Relatively low head loss
- Accuracy is less affected by approach velocity than weirs
- Lower maintenance cost than weirs

Disadvantages

- High cost
- Difficult to install

Parshall Flume

Parshall flumes are in extensive use in many irrigation projects and at the farms. There use is on the decline because of difficulties in construction (many angles) and being expensive. Yet in many cases, it is preferred because of its self cleaning nature, accuracy and relatively low head loss.

A Parshall flume comprises of three sections (Figure 17.12):

- A converging section
- A constricted section also known as throat
- A diverging section

The floor of the converging section is level, the floor of the throat inclines downward and the floor of the diverging section slopes upward. Size of Parshall flumes is designated by its throat width. Flumes with throat width ranging from 2.5 cm to 1500 cm covering a range of discharges are available. For different sizes, there is considerable overlap in flume discharges. As such, it is possible to pass a given discharge through flumes of different standard sizes. Therefore, besides the anticipated normal and maximum flows, other factors such as throat width, economy, and channel dimensions should be considered. Smallest flume of adequate capacity should be selected unless other factors such as width of the channel require use of a larger size flume.

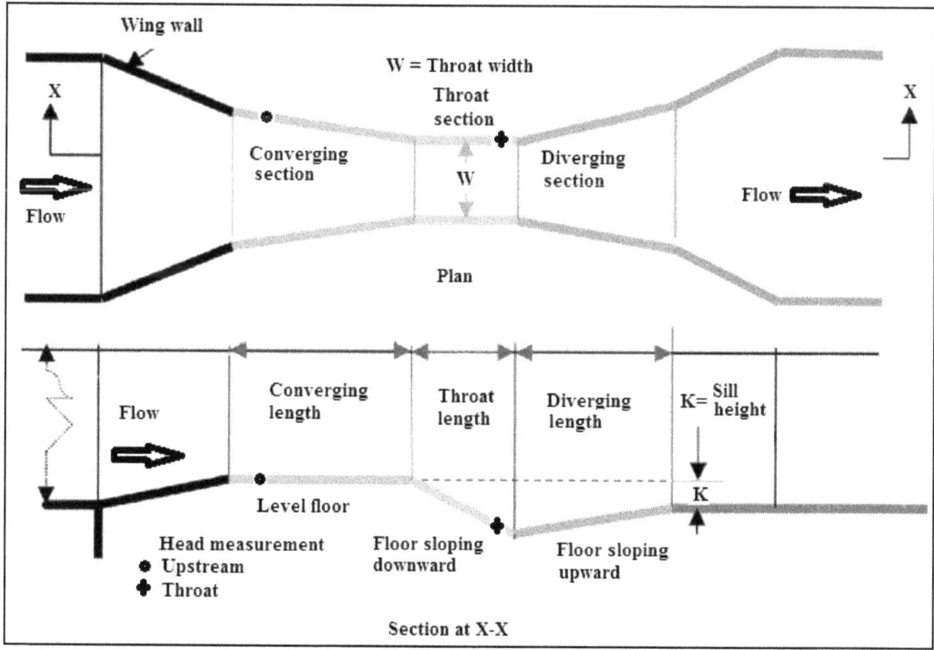

Figure 17.12: Plan View of a Parshall Flume. Note the Measuring point for Free and Submerged Flow conditions in the Converging and Throat sections.

The free-flow discharge equation for a standard Parshall flume of various sizes is given as:

$$Q = C\,h^n \text{ where Q is in } m^3/s \text{ and h is in m} \qquad (17.16)$$

Values of C and n can be selected with the help of Figure 17.13.

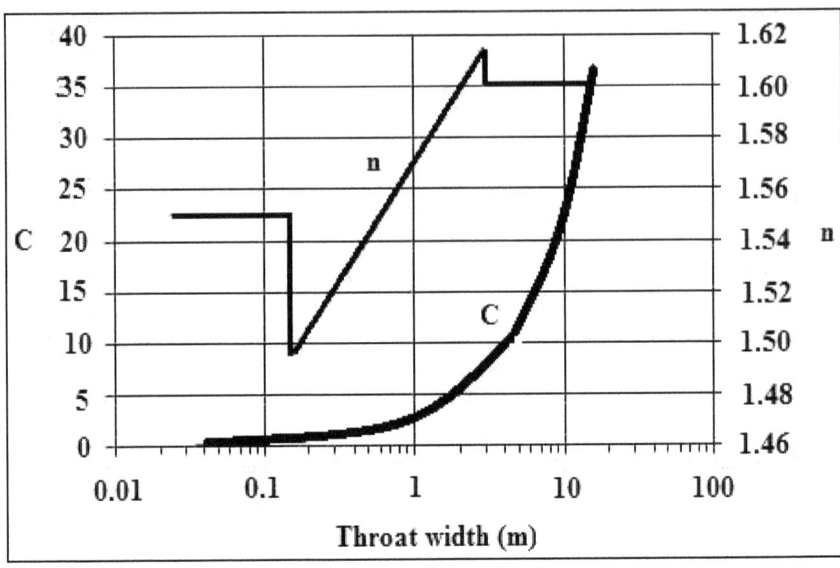

Figure 17.13: The Values of C and n for Use in eq. (17.16).

Submerged Flow

If the free flow discharge is computed using established formulae (say eq. 17.16) then the net submerged flow, Q_{net} is given by:

$$Q_{net} = Q_{free} - Qc \tag{17.17}$$

Such that Q_{free} is the free flow and Qc is the correction to the flow.

To find out whether the flow is free or submerged water heads at two places in the flume are measured (Figure 17.12). Calibration tests have shown that the discharge for a given upstream measuring head is not reduced until the submergence ratio, hb/ha (submergence head to measuring head) expressed in per cent, exceeds the following values:

50 per cent for flumes 1 to 3 inch (2.5 to 7.5 cm)

60 per cent for flumes 6 to 9 inch (15 to 22.5 cm)

70 per cent for flumes 1 to 8 ft (0.3 to 2.5 m)

80 per cent for flumes 8 to 50 ft (2.5 to 15 m)

The method to determine Qc varies with flume size. For small flumes graphical solutions are available. For 1 – 8 ft throat width, the correction for submergence can be determined by using the following equation:

$$Qc = M\,(0.000132\ ha^{2.123}\, e^{9.284\ S}) \tag{17.18}$$

Here Qc is the discharge reduction, ft^3/s, and M is the multiplication factor selected from the Table 17.5. S is the submergence ratio hb/ha.

Table 17.5: Values of Multiplying Factors

Size of Flume (ft)	Multiplying Factor, M
1	1.0
1.5	1.4
2	1.8
3	2.4
4	3.1
5	3.7
6	4.3
7	4.9
8	5.4

For flumes of 10 ft or more width (W), following equations are used.

$$Qc = (W/10) \times X \times ha^2 \tag{17.19}$$

Such that

$$X = (3.364 + 20.19\ S^2 \times \ln S)^2 \tag{17.20}$$

Or

$$X = -308.12 + 1212.85\ (S) - 1597.529\ (S^2) + 704.083\ (S^3) \tag{17.21}$$

Any of the two equations works well for the calculation of X.

For reasonable accuracy following requirements needs to be kept in view.

- Construction should be exactly as per the dimensions for a given size of the flume.
- The crest of the flume (the floor of the converging section where depth measurements are made) must be level both longitudinally and transversely.
- Channel turns, tees, elevation drops or other obstructions should be avoided so that conditions upstream promote laminar flow conditions at the flume inlet.
- The upstream channel slope should not cause excessive velocity at the flume. A slope of almost flat to 3 per cent maximum for very small flumes and 2 per cent maximum for larger flumes is ideal. A 1:4 sloping ramp upstream should be provided for flumes that must be installed above the channel floor.
- Long, narrow, flat or undersized channels downstream can result in a backwater effect at the flume and should be avoided as it may cause submerged flow conditions. A large fall or steep slope immediately downstream of the metering station can eliminate the possibility of submerged flow conditions. The submergence ratio hb/ha in any case should not exceed 0.90

Cutthroat Flume

The Cuthroat flume, developed in the 1960's at the Utah Water Research Laboratory, is able to overcome the limitation of the Parshall flume. It can be installed under flat gradients. The cutthroat flume unlike the Parshall flume has only two sections namely the converging and the diverging sections with no (zero) throat section (Figure 17.14 left). The plan and side view of the flume is shown in Figure 17.14 right. The flume length ranges from 45 cm to 3 m and neck width from 2.5 cm to 1.8 m. The flume has a flat bottom and vertical walls. The flat bottom has made the Cutthroat flume to be the flume of choice for many retrofit applications. It is simple to construct since the Cutthroat flume can be manufactured to fit directly on the channel floor. The flume sits above the channel floor by no more than the thickness of the flume's floor. It works well under both free and submerged flow conditions although it is desirable to operate it under free flow. Under free flow condition, critical depth of flow occurs near the flume neck or flume throat, which is the junction of the converging and diverging sections. The ratio of incoming flow depth at the upstream (ha) to the flume length should be less than 0.4.

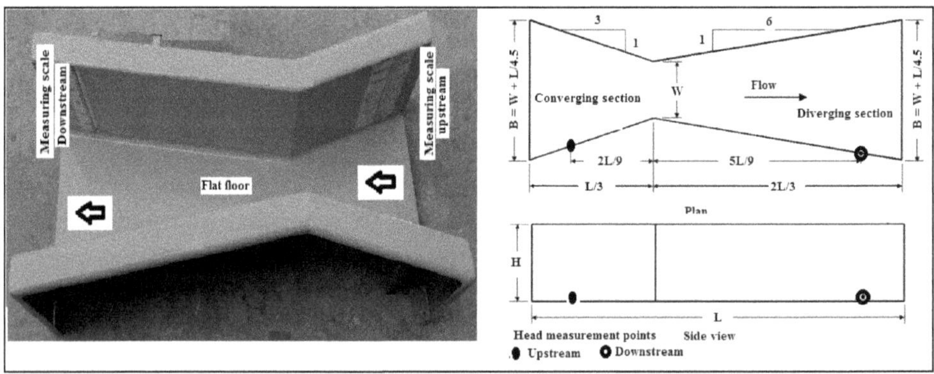

Figure 17.14: A View of the Cutthroat Flume: Pictorial (left), Plan and Side View (Right), Note Zero Length of the Throat Section.

The discharge through a cutthroat flume under free flow conditions is determined from the following equation.

$$Q = C1\ ha^{n1} \tag{17.22}$$

Where, Q is the flow rate, ft^3/s, C1 is the free flow coefficient, which is the value of Q when ha is 1.0 foot that is the slope of the free flow rating curve, ha is the head at the converging section, ft and n1 is the exponent whose value depends on flume length. The value of n1 is constant for all cutthroat flumes of the same length regardless of their throat width, W. The values of c1 and n1 are obtained for any flume length between 1.5 feet to 9 feet by plotting Q and ha on logarithmic paper and reading the intercept and slope.

When the flow depth exceeds the critical depth, the flow becomes submerged. For finding out the discharges, flow depths both at the converging section (ha) and the diverging sections (hb) are measured. The submergence, S is expressed as the ratio, often in per cent, of the downstream depth (hb) to the upstream depth (ha). For a particular size of the cutthroat flume, the value of the transition submergence, St, should be known. If S < St then free flow occurs otherwise the flow is submerged. For example, St for 4″ × 3′, 8″ × 3′ and 12″ × 3′ flumes are 0.580, 0.674 and 0.754 respectively.

A Cutthroat flume is less costly to construct than other types of flumes. In addition, the flat floor allows it to be easily retrofitted into existing channels, reducing installation costs and eliminates the need to either set the flume above the channel floor (increasing the upstream pool and head loss) or modifying the downstream channel. A further advantage of the Cutthroat flume is that the angles of convergence/divergence are the same for all sizes of Cutthroat flumes, thereby allowing intermediate flume sizes of cutthroat flumes to be developed by simply moving the sidewalls in and out as necessary and interpolating from the existing ratings.

Critical Aspects of Flumes Installation, Maintenance and Operation

Following conditions must be ensured during installation of the flume for accurate flow measurements.

- A flume should be located in a straight section of the open channel, without any bend immediately upstream.
- The approaching flow should be well distributed across the channel and should be relatively free of turbulence, eddies and waves. If necessary, to correct the flow patterns, the channel may be deepened widened, or baffles/other spreading devices introduced on the approach channel.
- The channel section immediately upstream from the flume should be free of sediment and debris.
- A solid water tight foundation should be constructed to avoid settlement or heaving and to prevent flow from bypassing the structure that may erode the foundation.
- Flumes should be operated within the prescribed head range and flow limits.
- Generally, a site with a high velocity of approach should not be selected. Even such a condition may not affect the accuracy much, if the water just upstream is smooth with no surface boils and waves.
- Height of the upstream channel should be enough to allow the flow of water without overflow since some increase in depth is expected with the installation of flume.
- The possibility of submergence of the flume due to backwater from downstream obstacles should be avoided. If a flume becomes submerged then either the downstream obstruction should be removed or the flume changed either by raising or narrowing the throat. Minimum channel slope in the downstream section is necessary to maintain critical flow through the throat of the flume and prevent the flume from becoming submerged.

- The flume should be level in both longitudinal and transverse directions. Occasionally a flume is installed with a slight slope, which necessitates adjustment of the zero level on the staff gage so that it is at the same elevation as the flume throat. If the flume is installed in an earthen ditch, the flume bottom should always be placed higher than the ditch bottom. If the flume is installed in a concrete ditch having a flat slope, the flume may become submerged. If this is the case, the flume should also be raised above the bottom of the channel.

Tracer Methods

Three commonly used methods under this head are:

- Dye velocity-area method
- Tracer-dilution Technique
- Radioisotope method

The basic principle of any of these methods is that a substance that easily dissolves in water but does not react with it is introduced into the flow at an appropriate location. The time taken by the substance to travel to other pre-decided point is noted and the concentration of the substance introduced is measured. Both or any one of these parameter is used to calculate the flow.

Dye Velocity-Area Method

In this method, a small strongly coloured dye is poured into the stream quickly with a sharp cut off. Time of flow to travel a known distance to the downstream is measured. As the dye travels downstream in a cloud, the time is measured for the first and last of the dye to reach the downstream station. The average of the two times is used to calculate the average velocity. Discharge is then calculated by the following equation:

$$Q = AL/T \tag{17.23}$$

Where Q is the discharge, m^3/s, A is the average cross-sectional area of reach length, m^2, L is the reach length between introduction and detection points, m, and T is the recorded average time taken by the tracer solution to travel between the stations.

Tracer-Dlution Technique

In turbulent streams, the cloud of the dye is dispersed quickly and cannot be measured. For such situations, other tracers such as chemicals are used. The method then is called dilution method. A solution of the tracer of known strength is added to the stream at a constant measured rate. Samples are taken at a point downstream. The concentration of the sample taken downstream is compared with the concentration of the added tracer. The dilution method does not need measurements of channel geometry or the time elapsed. Since the dilution is a function of the rate of flow, the following material balance equation is used to measure the discharge.

$$Q\,Co + q\,C1 = (Q + q)\,C2 \tag{17.24}$$

Solving for discharge, the above equation results in:

$$Q = q\,(C2 - C1)/(C0 - C2) \tag{17.25}$$

Here, Co is the natural or background concentration of the tracer in the flow, C1 is the concentration of the strong injected tracer solution, C2 is the concentration of tracer after full mixing at the sampling station, including the background concentration of the stream, Q is the discharge being measured and q is the discharge of the strong solution injected into the flow. Thus, the discharge of the channel, Q is measured by determining Co, C1, C2, and the injection rate, q. Only the final plateau value or C2 needs to be recorded rather than a complete record of the passing cloud that is needed with the salt-velocity-area method. Since salt concentration is difficult to measure, the electrical conductivity is measured as it bears a linear relationship with the salt concentration over a useful range. Chemicals such as common salt (NaCl) can be used. Sodium dichromate is another chemical that can be used in dilution tests. Its concentration can be determined by analyzing water samples by chemical analysis or using a flame photometer. Dyes have been used to colour the water and calorimeters and comparators are used to determine the dilution by comparing the samples with standard concentrated or dilute solutions.

Radioisotope Method

In radioisotope method, instead of a dye, a radioisotope is used. In this case, the gamma ray emissions are recorded using Geiger counters or scintillation counters. The flow rate is given as:

$$Q = FL/N \tag{17.26}$$

Here Q is the discharge in m^3/s, F is the counts of radioactivity per unit volume of water per unit time, L is the total radioactivity introduced and N is the total counts.

Since the materials commonly used in this method may pose health hazards, use of this method is avoided. Moreover, use of this method requires permission from public health and other agencies.

Water Meters

Water meters have multi blade propeller made of metal, plastic or rubber that rotates in a vertical plane. The meter is fitted in water pipelines. The propeller is geared to totalize and is provided with a counter that totals the flow in volumetric unit. Water meters are available for a wide range of sizes suiting the pipe sizes.

Dethridge Meter

The device commonly used in Australia is simple and consists of an under-shot water wheel revolved by the moving water through a concrete pipe and is fitted at the outlet (Figure 17.15). The concrete outlet has a special construction to provide only the minimum practicable clearance of the lower half of the wheel at its sides and around its circumference. It gives the direct measurement of the discharge in volumetric units. Head loss is quite small. It is simple and robust in construction and allows ordinary floating debris to pass through without damage or stoppage

Figure 17.15: A View of a Dethridge Meter.

of the wheel. To obtain accurate readings, it may be desirable to calibrate the device *in-situ*. The accuracy of measurement under normal situations exceeds 95 per cent.

Impeller Meters

The impeller meter uses a multi-bladed rubber, plastic, or metal impeller mounted inside the flow pipe. The flowing water causes the impeller to rotate, which in turn totalize the flow through mechanical linkages. The volume of water which has passed the impeller is then indicated on the totalizer. Although these meters register total volume of flow, but can be equipped with a rate meter, if flow rates are measured. The impeller meters are installed at least five pipe diameters downstream and at least one pipe diameter upstream from any bends or obstructions in the pipe. These meters can be installed in any convenient position- vertical, horizontal or inclined. These can also be installed on either pump intake or discharge lines. The pipe must always be flowing full for accurate measurement. If the pipe is not expected to flow full, then either an orifice plate to constrict flow or a U-shaped pipe arrangement should be used to create enough pressure to fill the pipe. Modern impeller meters are highly reliable, accurate (within 2 per cent), easily installed, and economical. These characteristics make the impeller meter one of the most practical methods to measure water flow.

Other Meters

A number of other kinds of meters are now being used to measure water flow.

Bypass, proportional or shunt flow meters measure part of the total flow which is allowed or forced through a small passageway by differential pressure across a fitting and returned to the main flow.

The Magnetic flow meters are based on the principle that a voltage is induced in an electrical conductor moving through a magnetic field.

Deflection meters sometimes called drag or target meters consist of a vane or plate that projects into the flow and a sensing element that measures the deflection caused by the force of the flow against the vane. These meters are.

In variable-area meters, the water flows vertically upward in a conically tapered tube in which area increases with height. The rate of flow is indicated by the height at which a shaped weight attains stable support from the flow in the tapered tube.

Vortex flow meter is based on the principle that obstructions placed in flows generate vortex shedding trails in the flow. A properly shaped obstruction will produce stable vortices that reinforce or interact with each other on each side of the obstruction. Measurement of any of the parameter illustrated for a respective meter is then used to measure the velocity and the discharge.

Coordinate Method

In this method, measurements are taken at the discharging jet of water at the end of the pipe to calculate the rate of discharge. This method can be used whether the pipe is discharging vertically upwards, horizontally, or at some angle to the horizontal. Methods for measurement, when the pipe is discharging horizontally (Figure 17.16) and vertically (Figure 17.17) are outlined here.

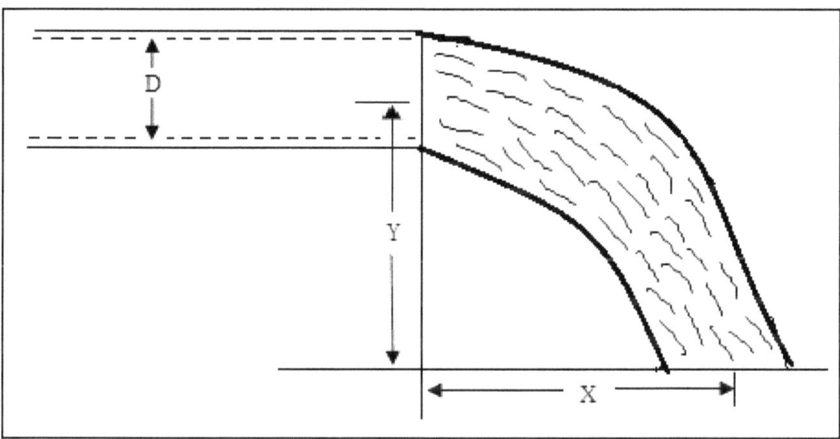

Figure 17.16: Definition Sketch of the Coordinate Method when the Pipe is Discharging Horizontally.

Horizontal Flow

To measure the flow from pipes discharging horizontally, the horizontal distance (X) and vertical distance (Y) from a point at the end of the pipe to a point on the jet (Figure 17.17) are measured. The formula to measure discharge is written as:

$$Q = 0.22\ CA\ X/Y^{1/2} \qquad (17.27)$$

Here Q is discharge, l/s, X and Y are measured distances (Figure 17.17), cm, A is the cross-sectional area of water at the end of pipe, cm^2 and g is acceleration due to gravity, 980 cm/s^2. Vertical distance or fall of jet (Y) is measured at a horizontal distance equal to 8D. Values of the coefficient of discharge C for a pipe flowing completely full at the end are given in Table 17.6 (Hansen *et al.*, 1979). D is the internal diameter of the pipe.

If the pipe is not flowing full, the equation can still be used. In this case, area of flow that is filled with water is taken into account rather than the area of the pipe. Similarly the coordinates are measured from the approximate centre of the jet at the end of the pipe instead from the centre of the pipe as in the case of a full flowing pipe.

Table 17.6: Coefficients, C for the Co-ordinate Method for Pipe Discharging Horizontally

Y/D	*X/D*							
	1.00	*1.50*	*2.00*	*2.50*	*3.00*	*4.00*	*5.00*	*8.00*
0.5	1.44	1.28	1.18	1.13	1.10	1.06	1.03	1.00
1.0	1.37	1.24	1.17	1.12	1.09	1.06	1.03	1.00
2.00		1.11	1.09	1.08	1.07	1.05	1.03	1.00
3.00			1.04	1.04	1.04	1.04	1.03	1.00
4.00			1.01	1.01	1.02	1.03	1.02	1.00
5.00			0.97	0.99	1.00	1.01	1.01	1.00

The method has limited accuracy due to the errors in measuring the coordinates of the jet. Errors up to 10 per cent in the measurements can be expected. For greater accuracy, X/D should not be less than 3.

Vertical Flow

When water is discharging from a vertical pipe in the air (Figure 17.17), the discharge is estimated by measuring the height to which the water rises above the end of the pipe. The proper height that needs to be measured is the height at which the water stays longest time. The discharge is then estimated using the equations given by Lawrence and Braunworth (Bos, 1989).

$$Q = 5.47\, D^{1.25}\, H^{1.35}, \text{ and} \tag{17.28}$$

$$Q = 3.15\, D^{1.99}\, H^{0.53} \tag{17.29}$$

Here D and h are diameter of the pipe and height of water, m and Q is the discharge, m^3/s. Eq. (17.28) is valid for weir flow ($H \leq 0.4D$, Figure (17.18 left) and eq. (17.29) for jet flow ($H \geq 1.4D$, Figure 17.18 right). For heights between 0.4 D and 1.4 D, flow is calculated by both the equations and their average is taken as the value of discharge. The above equations are obtained through actual measurements. The practical range of the pipe diameters is 0.025 m to 0.609 m and that of the head is 0.0025 m to 4.0 m. Pipes should have clear cut edges and a constant diameter over at least a length of 6 D. They should also be vertical for at least a length of 6 D from the top of the pipe.

Modern Measuring Techniques

Currently, a large number of instruments to measure water velocity and the discharge have appeared in the market. Although the list is long, yet some instruments use the following three basic principles to measure water velocity or discharge.

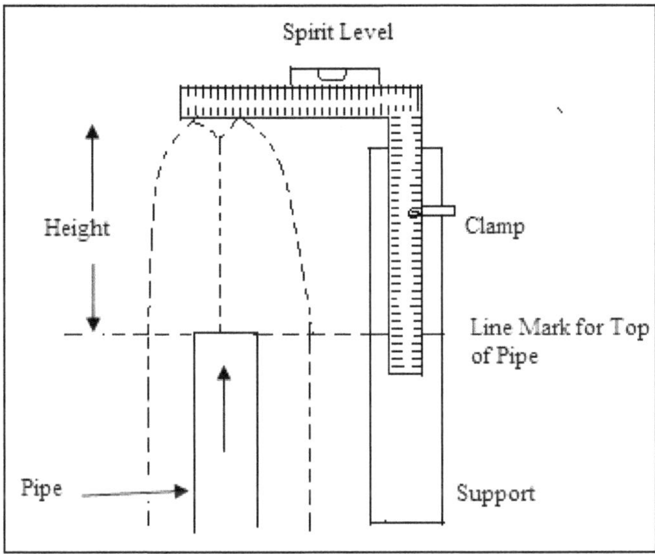

Figure 17.17: Definition Sketch of the Coordinate Method when the Pipe is Discharging Vertically.

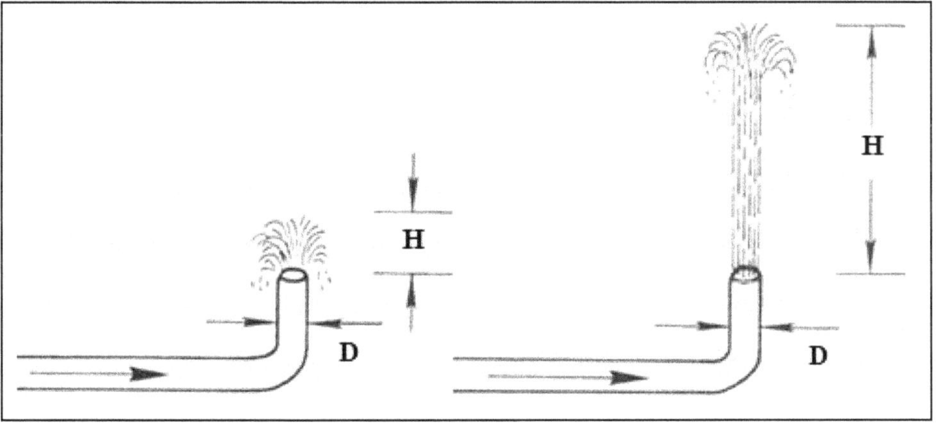

Figure 17.18: Illustration of Conditions for the Validity of Eq. 17.28 and Eq. 17.29.

1. Electro-acoustic (ultrasonic frequencies)
2. Electro-optics
3. The electromagnetic method

The principle of the ultrasonic method is to measure the velocity of flow at certain depth in the channel by simultaneously transmitting sound pulses through the water from transducers located in the banks on either side of the river. The transducers are designed to both transmit and receive sound pulses and are staggered so that angle between the pulse path and the direction of flow is between 30° to 60°. The difference between the times of travel of the pulses in two different directions is directly related to the average velocity of the water at the depth of the transducer. This velocity can then be related to the average velocity of flow of the whole cross section. An area factor to measure discharge can be incorporated in the electronic processor.

The principle of an electro optical instrument is the Doppler Effect (change in frequency or wavelength of a wave) of making particular matter in a coherent light beam and determination of frequency shifts by an optical heterodyning technique. Since only a beam of light enters the flow field, there is no measurable disturbance and therefore, point measurements of velocity are possible. The light beam can be focused to as small an area as few microns. There is no need for prior calibration of the instrument and the response is linear over the entire velocity range of interest. Velocities as small as a friction of cm/s can be measured. Electro-optical velocity instruments are also called, Laser Velocimeter, Laser Doppler Velocimeter and Laser Anemometer.

Faraday was the first person to notice that when the motion of water flowing in a river cuts the vertical components of earth's magnetic field an EMF is induced in the water, which can be picked up by two electrodes. The EMF is directly proportional to the average velocity in the river.

Chapter 18

Surface Irrigation

Introduction

Surface irrigation is the oldest and widely used method of water application on farm lands. The term 'surface irrigation' refers to a broad class of irrigation methods in which water is distributed over the field by overland flow. In this system of water application, the water is applied directly to the soil from a channel located at the upper end of the field. Water is either applied on the entire field (uncontrolled flooding) or part of the field (furrows, basins, border strips) also sometime referred as controlled flooding although it may be a misnomer. Good yield of crops can be obtained from irrigated land only if the water is applied uniformly and in adequate quantities to meet the crops requirement. It is therefore; important to select a method of irrigation which best suits the crop and soil characteristics of the field. The deciding factors that determine the suitability of any surface irrigation method are soil type - its infiltration rate and water storage capacity, land topography, climate, water availability and water quality, crop type, type of technology, previous experiences, required labour inputs *etc.* Moreover the irrigation system for a field must be compatible with the existing farming operations, such as land preparation, cultivation, and harvesting practices. It may sometimes be necessary to use more than one method of irrigation on the same farm because of changes in soil types or crops cultivated.

Flood Method

Flood method of irrigation has been used in India for generations. Since water is applied without any control whatsoever it is often called uncontrolled flooding. It is also known as field to field irrigation. This system is in fact a replica of past where water was made to enter the fields bordering rivers during floods or otherwise. The practice has continued even under irrigation projects. When the irrigation water inundates the fields in this manner, too much water is applied

Figure 18.1: A Flood Irrigated Field in an Irrigation Command, Haryana.

with uneven distribution (Figure 18.1). Sometimes field bunds are used to guide the flow of water, but these are so ill maintained that the practice just replicates the flood irrigation. It is the most inefficient way of water application, a lot of water being wasted in deep percolation losses. Soils especially sloppy lands are prone to erosion. The efficiencies may be as low as 25 per cent but seldom exceeding 40 per cent. The practice discourages the farmers to adopt improved technologies of crop production such as land levelling and others. The only advantage of this method is that it doesn't cost much. Moreover, farmers get ample time to finish other tasks in hand while the fields are being irrigated. The method is commonly used for transplanted rice, low value pastures, lawns and millets *etc.*

Controlled Surface Irrigation Methods

The irrigation applied in a controlled way is more efficient than uncontrolled flooding. Various categories of controlled surface irrigation methods are given in Figure 18.2 while a glimpse of some of these methods is shown in Figure 18.3.

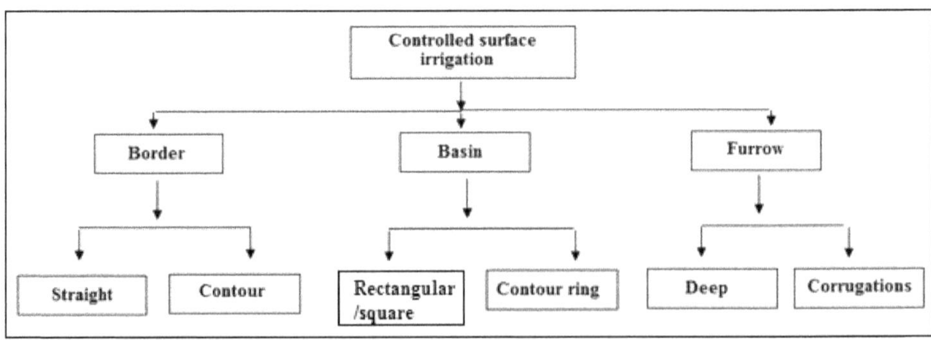

Figure 18.2: Classification of Controlled Surface Irrigation Methods.

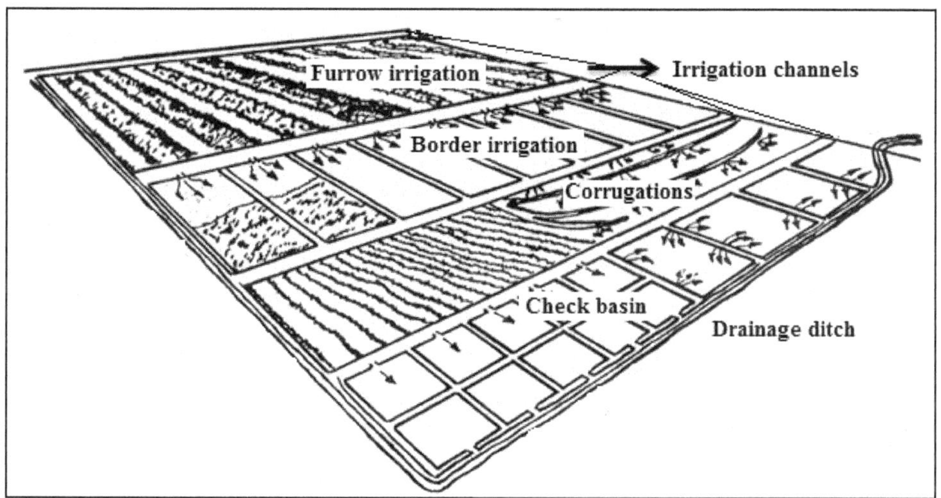

Figure 18.3: Schematic View of some Controlled Surface Irrigation Methods.

Border Irrigation Method

In this method, the fields are divided into a number of long parallel strips, generally 5 to 15 m wide and 75 to 300 m in length separated by small border ridges or low dykes of about 15 cm height (Figure 18.4). Length of the border and slope depend upon the soil type as follows:

Soil Type	*Length of Border (m)*	*Slope (per cent)*
Sandy and sandy loam	60-120	0.25 - 0.6
Medium loam soil	100-180	0.20 - 0.4
Clay loam and clay soil	150-300	0.05 – 0.2

Each strip is irrigated independently as each strip is connected directly to irrigation channel situated at the upstream end of the border strip. Sometimes, siphon tubes are used to transport the water from channel to the border strip. The water spreads and flows down the strip in a sheet confined by the border ridges (Fig 18.4). After sufficient water is applied to one strip, the irrigation stream is turned into another strip. For uniform flow, it is necessary that longitudinal slope must be uniform, and the transverse slope must be zero or negligible (< 0.03 per cent).

The method is most suited to soils having moderately low to moderately high infiltration rates. It is not used in coarse sandy soils that have very high infiltration rates and also in heavy soils having very low infiltration rate. It is suitable to irrigate wide varieties of close growing crops such as wheat, barley, ground nut, fodder crops, pearl millet and *berseem*.

Types of Borders

Straight Border: These borders are formed along the general slope of the field. These are preferred when fields are or can be levelled or be given a gentle slope at an economical cost.

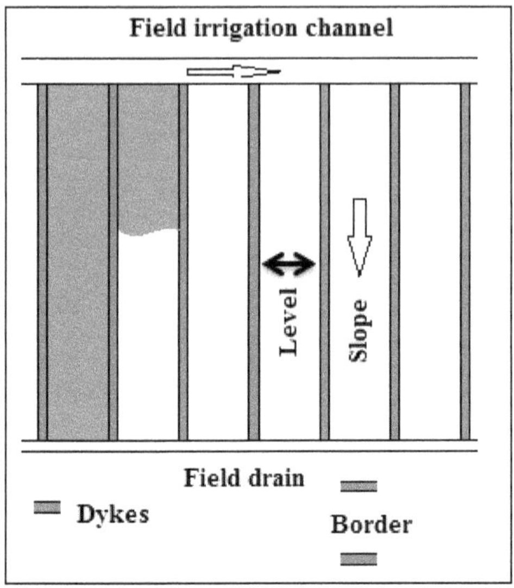

Figure 18.4: Schematic View of a Border Irrigation Method.

Contour Border: These are formed across the general slope of the field and are preferred when land slope exceeds the safe limits.

Based on the management strategy, borders are grouped into closed and open borders.

Closed Borders: These have transverse dykes at the end of the borders. The whole water applied is retained within the border. In the closed border, water saving is achieved through cut-back where water stream is cut-off before the water reaches the tail end.

Open Borders: There is no cross dyke in an open border. In an open border, the water going out of the border is reused for irrigation in the fields downstream. There is no cut-back in this system.

Advantages

- Border ridges can be constructed with simple farm implements like bullock drawn "A" Frame ridger or bund former.
- Labour requirement in irrigation is reduced as compared to conventional check basin method.
- Uniform distribution of water and high water application efficiencies are possible.
- Large irrigation streams can be efficiently used.
- Adequate surface drainage can be provided, if proper outlets are available.

Disadvantages

- Requires relatively large water streams for quick advance of water to minimize deep percolation losses at the upper end of the border strip.

☆ Some wastage of water may occur through deep percolation in coarse textured soils.

Important Terms

Advance Time: It is the time required for a given stream of irrigation water to move from the upper end of the field to any point downstream. A plot of time t on the y-axis and length of the border at which stream has advanced on the x-axis gives a curve that is known as advance curve (Figure 18.5).

Recession Time: As the stream is cut-off, water at the head end begins to disappear and it proceeds downward. Time of water disappearance at a particular point is known as recession time. A plot of time t on the y-axis and length of the border at which water has disappeared on the x-axis gives a curve that is known as recession curve (Figure 18.5).

Cut-off: In a closed border, the flow is cut-off before the water reaches the tail end. It is to avoid water stagnation at the lower end of the closed borders to reduce water losses. Cut off time is the time at which the supply is turned off, measured from the onset of irrigation. Cut-off time is also described as the ratio of the length of the border at which water is cutoff to the total length of the border. The time to cut off the inflow is difficult to determine as it depends upon soil type, stream discharge and even the crop growth stage.

If these 3 parameters are properly managed, it should be possible to attain high efficiency in this method of irrigation.

Infiltration Opportunity Time: Time interval during which water is available to enter the soil is known as infiltration opportunity time. It is the difference between the times water arrives at a point during the advance phase and

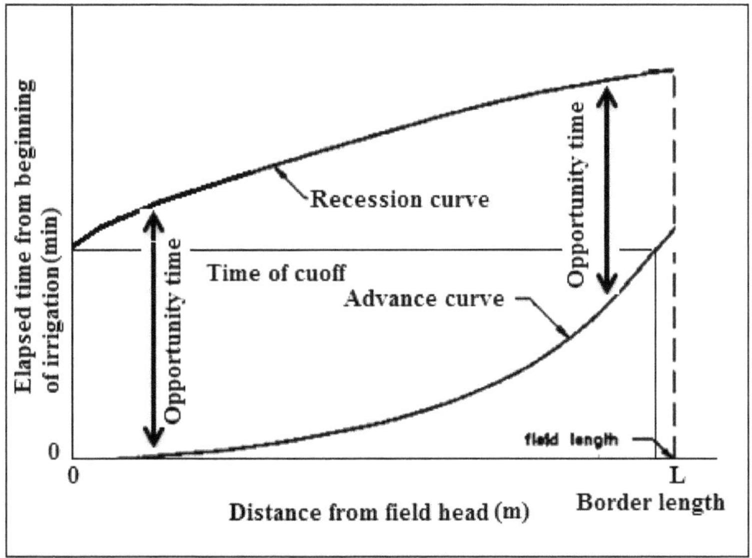

Figure 18.5: Advance and Water Recession Curves in a Closed Border System.

disappears during recession (Figure 18.5). It is the vertical distance between the advance and recession curves. For high irrigation uniformity, an attempt is made to make the opportunity time equal at all the points (Some deviations will anyway remain). In that case, the advance and recession curves will appear parallel on an advance recession graph (Figure 18.5).

Unit Stream Size: It is defined as the size the stream per unit width of the border. It is obtained by dividing the stream size by the width of the border. Unit stream size is commonly used to design the border irrigation. The unit stream size decreases with increasing slope of the border irrigation.

Basin Irrigation

Basins are mainly of two types:

1. Check basin (square or rectangular in shape)
2. Ring basin (circular in shape)

While the former is discussed in this section, the ring basin is discussed in the next section.

In check basin method, fields are divided into square or rectangular plots each having a nearly level surface. The basins are surrounded by dykes or ridges all around the area within which irrigation water is applied (Figure 18.6). The field channels supply water to each basin, during which the basins are filled to desired depth and water is retained until it infiltrates into the soil. Since, it is usually practiced in nearly leveled lands, the depth of wetting is more uniform. The size of the basin varies from 10 m^2 to 25 m^2 depending upon soil type, topography, stream size and crop. The size may be even lesser for growing vegetables and for intensive cultivation and as large as one ha, for growing rice under wet land conditions. For the same steam size, basin area is more in clay than in sand (Table 18.1). The system is most suited to soils having small gentle and uniform land slopes, moderate to slow infiltration rates and permeable soils. It is well adapted to grain and fodder crops in heavy soils and is commonly used to irrigate crops like groundnut, finger millet, sorghum, vegetable crops *etc.* It is not suited to crops which cannot withstand wet or waterlogged conditions for periods longer than one day such as tuber and root crops like potato, cassava beet and carrot. Basins are most commonly used to leach saline soils because a uniform percolation is pre-requisite for uniform leaching.

Table 18.1: Suggested Basin Area (ha) for different Soil Types and Rates of Water Flow

Flow Rate (l/s)	*Soil Type*			
	Sand	*Sandy Loam*	*Clay Loam*	*Clay*
30	0.02	0.06	0.12	0.2
60	0.04	0.12	0.24	0.4
90	0.06	0.18	0.36	0.6
120	0.08	0.24	0.48	0.8

Advantages

- It is possible to conserve rainwater and to reduce soil erosion by retaining a large part of rain within the basins.
- Even small streams can be used to irrigate crops with high efficiencies.
- High water application and distribution efficiency can be achieved.
- Check basins are useful to leach saline soils as more uniform leaching can be achieved.

Limitations

- The ridges may interfere with the movement of farm implements.
- More area occupied by ridges and field channels. Sometimes, it may be as high as 30 per cent of area.
- Surface drainage is impeded.
- Precise land grading and shaping is essentially required
- High labour requirement.
- Not suitable for crops which are sensitive to wet soil conditions.

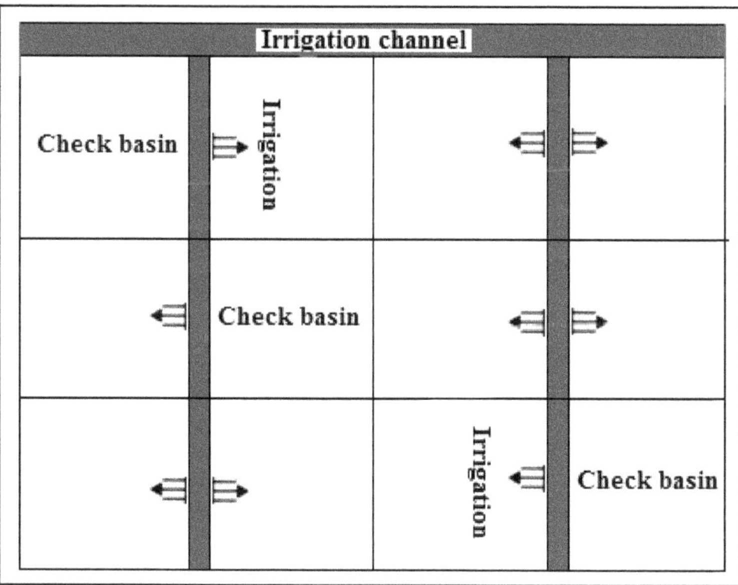

Figure 18.6: Check Basin Method of Irrigation.

Important Terms

Advance Time: It is the time required for a given stream of irrigation water to move from the upper end of the field to any point on the basin along its length. A plot of time t on the y-axis and length of the basin at which stream has advanced on the x-axis gives a curve that is known as advance curve (Figure 18.7).

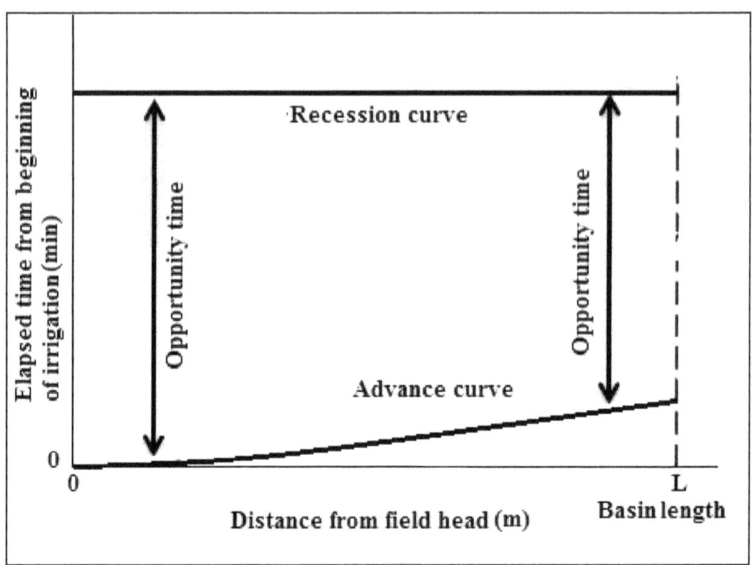

Figure 18.7: Advance and Recession Curves in a Check Basin.

Recession Time: Since irrigation water in this method is filled in the basin uniformly over the whole basin, it is assumed that recession takes place at all the points at the same time (Figure 18.7).

Opportunity Time: It is the time available for water infiltration at a given point of the basin. For uniformity, an attempt should be made that the advance curve is fast enough so that water takes minimum time to reach the basin end. For this purpose an appropriate steam size needs to be used. If advanced curve is properly managed, one can attain a high efficiency in the basin irrigation (Figure 18.7).

The distribution efficiency is controlled by the ratio of the time of advance (Ta) to the net opportunity time (Tn) at the end of the field. Lower the ratio, higher is the distribution efficiency as is evident from Figure 18.7 (Table 18.2). Knowing the infiltration rate curve and irrigation water to be applied, one can calculate the time required for infiltration at the end of the field. Therefore, it should be possible to decide the time of advance to achieve pre-decided distribution efficiency.

Table 18.2: Ratio of Ta to Tn to Achieve Pre-decided Distribution Efficiencies

Ratio of Time of Advance to Net Time of Infiltration at the End of the Field	*Distribution Efficiency (Per cent)*
0.16	95
0.28	90
0.40	85
0.58	80
0.80	75
1.08	70
1.45	65

Ring Basin Method

This method is a modification of check basin method and is suitable for orchard crops and cucurbits (Figure 18.8). In this method a circular bund is constructed around each tree/plant or group of plants/trees to create a basin for irrigation. These basins are suitably connected to irrigation conveyance channels in such a way that either a line of basins is irrigated by flowing water from one basin to another through inter- connections or each basin is irrigated separately (Figure 18.9).

Figure 18.8: Ring Basin Method of Irrigation.

Advantages

- High irrigation application efficiency can be achieved with properly designed system.
- Unskilled labourers can be used.

Disadvantages

- High labour requirement
- Too many bunds may restrict the use of modern machinery in field operations.

Furrow Irrigation Method

Furrow irrigation system is primarily used for row crops that are cultivated either on dykes and sometimes even in furrows especially in saline soils. It is suitable to most soils except sand. Furrows are sloping channels made in the soil with the

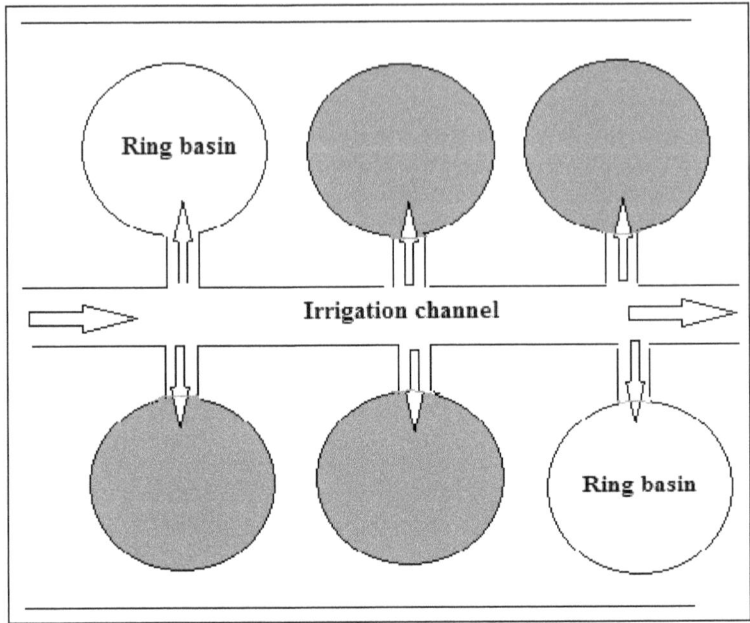

Figure 18.9: Irrigation of Individual Ring Basins.

crops being planted on the ridges. The water is applied only in furrows instead of the whole field. This saves water and at the same time the plant cultivated on the ridges does not come in direct contact with water. It is an added advantage as some plants, like vegetable crops, are very sensitive to water stagnation. Water is applied by small running streams in furrows between the crop rows. Water infiltrates into soil and spreads laterally to wet the area between the furrows (Figure 18.10)

Types of Furrow Irrigation

Based on alignment of furrows, the furrows are characterized as straight furrows and contour furrows.

Based on size and spacing of the furrows, these are characterized as deep furrows and corrugations.

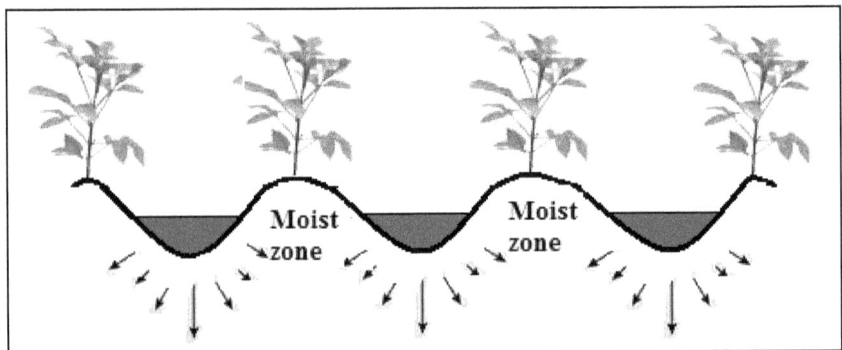

Figure 18.10: A Sectional View Showing Moisture Distribution in Furrow Irrigation.

As the name suggest, straight furrow are constructed on flat land with some slope in the direction of water flow. The contour furrows are constructed on sloping lands with alignment following the contour. Construction of contour furrows is often a difficult task.

The dimension of furrows depend on the crop grown, equipment used and soil type. In general, small plants need small furrows; like vegetables need furrows of 7.5 to 12.5 cm depth while some row crops like orchards need much deeper furrows (Figure 18.11 top). Besides sandy soil usually require deep and narrow furrows while furrows in clay soils are generally shallow but wider, the top width being 5-10 times the depth of water in the furrow (Figure 18.11 bottom).

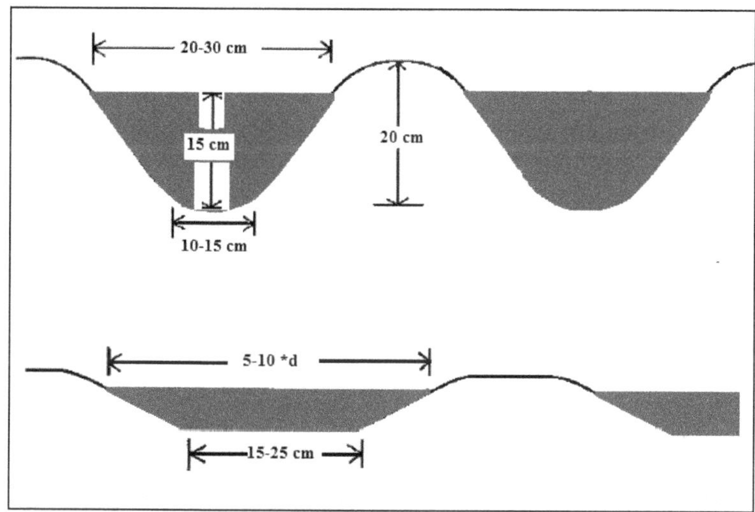

Figure 18.11: Deep and Narrow Furrows in Sandy (Top) and Shallow and Wide Furrows in Clay Soils (Bottom).

Corrugation irrigation consists of running water in small furrows called corrugation which direct the flow down the slope. Corrugations are V-shaped or U-shaped channels about 6 to 10 cm deep, spaced 40 to 75 cm apart. Corrugation method is most suited to broadcast crops such as wheat.

Uniform flat or gentle slopes not exceeding 0.5 per cent are preferred for furrow irrigation to avoid soil erosion. If the land slope is steeper than 0.5 per cent, then furrows can be constructed at an angle to the main slope or even along the contour to keep furrow slopes within the recommended limits. It is desirable to operate the furrow at a non-erosive water velocity. The maximum non-erosive stream size for various furrow grades is reported in Table 18.3 (Booher, 1974). It helps to maintain proper furrow shape as well as reduce sediment moving from the head of the field to the tail of the furrows. Besides what is stated, a minimum gentle slope of 0.05 per cent is desirable to assist drainage following irrigation or excessive rainfall. The length of furrows in sandy soils should be short so that water reaches the downstream end without excessive percolation losses at the head end. The lengths are usually longer in clay soils.

Table 18.3: Relation of Maximum Non-Erosive Flow Rates to Critical Slopes of Furrows

Furrow Slope S (per cent)	*Maximum Flow Rate, Q_{max} (l/s)* $Q_{max} = 0.6/S$
0.1	6.0
0.3	2.0
0.5	1.2
2.0	0.3

Furrow irrigation is most suited to wide spaced row crops including vegetables. It is also suited to maize, sorghum, sugarcane, cotton, tobacco, groundnut and potato. It is also used for irrigating non-cultivated crops like small grains and pastures growing on steep slopes.

Based on management strategies to save water and uniform application of irrigation, furrow irrigation has been classified as follows.

All Furrow Irrigation: It is also called furrow system or uniform furrow system. In this method, water is applied evenly in all the furrows.

Alternate Furrow Irrigation: It is not an irrigation layout but a technique for water saving. Water is applied in alternate furrows. It is further sub-grouped into two classes. In one all the irrigations are performed in the same furrow. Dry furrow remain dry throughout the crop growing season. Root distribution in this kind of irrigation is more towards the wet than the dry furrow. This method is sometimes known as skip furrow irrigation. The skipped furrows are sometimes used to raise intercrops requiring less water. Water saving of about 30-35 per cent can be achieved without much yield reduction.

To overcome the problem of root distribution in skip furrow irrigation, furrows are alternatively irrigated *e.g.* during first irrigation if the even numbers of furrows are irrigated, during next irrigation, the odd number of furrows are irrigated.

Surge Irrigation: Surge irrigation is the application of water in the furrows intermittently in a series of short ON and OFF irrigation cycles. Water is sent down the furrows in pulses. After the initial surge, the furrow is allowed to dry. This wetting and drying cycle continues throughout the whole irrigation event. It has been found that intermittent application of water reduces the infiltration rate over surges thereby the water front advances quickly. Hence, the net irrigation water requirement is much less than in furrow irrigation. This also results in more uniform soil moisture distribution and storage in the crop root zone compared to continuous flow. The irrigation efficiency in this case can be as high as 85 to 90 per cent.

Advantages

- ☆ Water in furrows comes into contact with only one half to one fifth of the land surface thus resulting in low evaporation losses and water saving.
- ☆ Fairly high irrigation application efficiencies among surface irrigation methods.

- ☆ Furrows serve as field drains in areas of heavy rainfall.
- ☆ Compared to check basins, labour requirement for land preparation and irrigation is less.
- ☆ Less wastage of land in field ditches.

Disadvantages

- ☆ Not suitable in coarse textured soils with high infiltration rates.
- ☆ Possibility of intra-furrow soil erosion.
- ☆ Labour intensive in comparison to other methods except check basin.

Relative Comparison of 3 Methods

Based on what has been discussed, a glance at the potential applications of 3 surface irrigation methods can be had from Table 18.4.

Table 18.4: Differences and Similarities of the Three Primary Surface Irrigation Systems

Item	*Furrow*	*Border*	*Basin*
Main slope	< 1 per cent (min. 0.05 per cent)	Up to 2-5 per cent (min. 0.05 per cent)	Usually zero or nearly zero
Soils	Best suited to soils with moderate to low intake rates	Moderately low to moderately high intake rate soils	Medium to fine textured soils
Infiltration	Two dimensional	One dimensional, vertical	One dimensional, vertical
Field size	Large	Large	All sizes
Crops	Best for row crops	Best for close growing crops	All crops but best for crops that can tolerate some water stagnation
Farm mechanization	Adapted to mechanized farming	Easy to apply	Difficult to use
Application efficiency (per cent)	60- 70	60- 70	65 –80
Efficiency and uniformity	Relatively low	High with closed borders	High
Initial cost	Low to medium	Low	High

Subsurface Irrigation Methods

As the name suggest, the water in this system of irrigation is applied from below the ground surface so that it is supplied directly to the root zone of the plants. The main advantages of this irrigation method are reduced evaporation losses and less hindrance to cultivation or other farm mechanization operations. These methods are grouped under the following two groups:

1. Natural sub-surface irrigation method
2. Artificial sub-surface irrigation method

Natural Subsurface Irrigation Method

Under favorable conditions of soils and topography, the water table may be close enough to the root zone of the field of crops. Shallow water table, commonly believed to be a curse, could be a boon to crops if the groundwater is of good quality. Crops are able to draw a major fraction of their water requirement from the shallow water table. Since, the natural presence of the water table may not be able to supply the requisite water throughout the crop growing season, sometimes it needs to be supplemented by surface irrigation or by constructing deep channels in the field. These channels are filled with water to ensure that water table remains at a desired elevation below the root zone.

For this method to be effective, following conditions must be met with.

- The soil in the root zone should be quite permeable.
- There is an impermeable sub-stratum below the water table to prevent deep percolation of water.
- The groundwater should be of good quality and non-saline in nature. Otherwise, upward movement of salts along with water will turn the soil saline.

Artificial Subsurface Irrigation Method

Conceptually, the method attempts to maintain water table at a suitable depth just below the root zone. For this purpose closely spaced perforated pipes are laid in a network pattern below the soil surface. For uniform distribution, pipes are closely spaced, say at about 0.5 m. In order to avoid interference with cultivation, the pipes are buried at least 0.4 m below the ground surface. A pre-requisite for the success of this method is that the soil in the root zone should have high horizontal permeability to permit free lateral movement of water and low vertical permeability to prevent deep percolation of water. This method of irrigation is not very popular because of the high expenses involved, improper distribution of subsurface moisture in many cases, and possibility of clogging of the perforation of the pipes.

Chapter 19

Irrigation Efficiencies

Introduction

All the water withdrawn from a source is unable to reach the root zone of the plants for their use. Part of the water is lost during transit through the canals and on-farm field channels. Only a part of the water applied to the fields remain in the root zone while a part goes as deep percolation loss. The various processes that are operative at various locations in the transit and field are listed as under.

- ☆ Evaporation from the water surface
- ☆ Seepage from canal network and field channels
- ☆ Leakage of water from overtopping of the bunds
- ☆ Tail runoff from canals and fields
- ☆ Deep percolation losses

The above processes not only result in the loss of the precious water resource but create several other problems notably

- ☆ Water logging and soil salinization
- ☆ Leaching of nutrients beyond the root zone resulting in groundwater pollution
- ☆ Aeration problems to the plants
- ☆ Additional time and labour required to drain the fields
- ☆ Poor aesthetic look and environmental problems

Irrigation efficiencies are used to quantify such losses and quantitatively assess the performance of various components of the system. These can then be up-scaled to assess the performance of the complete irrigation system (Israelsen and Hansen, 1962). Let us first consider the various efficiencies used in irrigation projects.

Reservoir Storage Efficiency

Reservoir storage efficiency (Er) is defined as the efficiency with which water is stored in the reservoir. It is expressed as:

$$Er = (Vi - Ve - Vs)/Vi \quad (19.1)$$

$$Er = 1 - (Ve+Vs)/Vi \quad (19.2)$$

At a particular time, Er can also be written as:

$$Er = (Vo + \Delta S)/Vi \quad (19.3)$$

In these equations, Ve is the volume of water evaporated from the reservoir, Vs is the seepage volume, Vi is the volume of inflow to the reservoir, Vo is the volume of out flow from the reservoir and ΔS is the change in reservoir storage. If some amount of water is already present in the reservoir, Vi in the above equations is (Vin + Vi) such that Vin is the initial volume of water in the reservoir. To express the efficiency in per cent, the value so calculated is multiplied by 100.

Example 19.1

A storage reservoir has an inflow of 4200 l/min and outflow of 3500 l/min of water. If the change in storage is 400 m^3during a span of 24 hr, calculate the reservoir storage efficiency.

Inflow = $4000 \times 60 \times 24/1000 = 5760\ m^3$ ($1000\ l = 1\ m^3$)

Outflow = $3500 \times 60 \times 24/1000 = 5040\ m^3$

Using eq. (19.3), we have

Er = (5040+400)/5760 = 94.4 per cent.

Water Conveyance Efficiency

The conveyance efficiency, (Ec) is used to measure the efficiency of water conveyance systems associated with the canal network, water courses and field channels. It is defined as the ratio between the water that reaches a farm or eld and that diverted from the irrigation water source. Mathematically it is represented as follows:

$$Ec = 100\ (Vf/Vd) \quad (19.4)$$

Here, Vf is the volume of water that reaches the farm or field (m^3) and Vd is the volume of water diverted (m^3) from the source.

Conveyance efficiency depends upon the length of the conveyance network, the wetted areas of various components of the network and even the life of the system. Old systems often have low efficiencies. For long networks as in a canal system, Ec can be determined for individual components such as Ecmc, Ecd, Ecm, Ecw and Ecf where Ecmc stands for conveyance efficiency of main canal, Ecd for distributary canal, Ecm for minor, Ecw for water courses and Ecf for on farm field channels. Ec is then given by the product of the efficiencies of individual components. For pipe networks, efficiencies are higher because seepage component is nil and leakage is also minimal. It is why underground pipeline systems are recommended for on-farm conveyance of water.

Application Efficiency

Application efficiency relates to the actual storage of water in the root zone that is used by the plants to in meet their water requirement (Michael, 2008). Application efficiency includes all kinds of application losses. For surface irrigation, it is mainly the deep percolation losses and runoff from the field. However, in case of sprinkler irrigation, it may also include evaporation of droplets in the air, any leaks from sprinkler or drip pipelines, percolation beneath the root zone (can be managed), drift from sprinklers, or runoff from the field (can be managed). Application efficiency, Ea is expressed as:

$$Ea = 100\ (Vs/Vf) \tag{19.5}$$

Such that Vs is the volume of water stored in root zone (m^3) and Vf is the water delivered to the field or farm (m^3).

Application efficiencies of surface irrigation are usually less than pressurized irrigation techniques. Within surface irrigation, flood irrigation is least efficient followed by border, check and furrow irrigation. In later methods of irrigation, efficiencies can be in the range of 60-70 per cent with proper land levelling and designs of systems. Laser land levelling has emerged as one of major intervention to improve water application efficiency of surface irrigation methods.

Example 19.2

A tube well is pumping 10 l/s of water for 8 hours. The water reaching the field head is 250 m^3. If the field capacity moisture content of the soil is 0.3, calculate the water stored in a check basin of 25 × 25 m, if the root zone depth is 0.8 m and the bulk density of soil is 1.4 g/cm^3. Calculate the conveyance and application efficiency.

Water pumped = $(10 \times 60 \times 60 \times 8)/1000 - 288\ m^3$

Water reaching the field head = 250 m^3

Conveyance efficiency = 100 (250/288) = 86.8 per cent (Eq. 19.4)

The water that can be stored in the root zone = $25 \times 25 \times 0.6 \times 0.3 \times 1.4 = 157.5\ m^3$

Assuming the distribution efficiency as 100 per cent

Application efficiency= 100 (157.5/250) = 63 per cent (Eq. 19.5)

Storage and Distribution Efficiency

It may be noted that high application efficiency alone does not mean that irrigation is efficiently applied to the field. It is illustrated in Figure 19.1. It may be noted that in a border length of L m, the application efficiency is 100 per cent if irrigation is terminated at either time T1 or T2 but the storage efficiency is less since a part of the root zone remains devoid of water. On the other hand application efficiency at T3 is less because of deep percolation at the head end but storage efficiency is 100 per cent because whole root zone is now irrigated. Therefore,

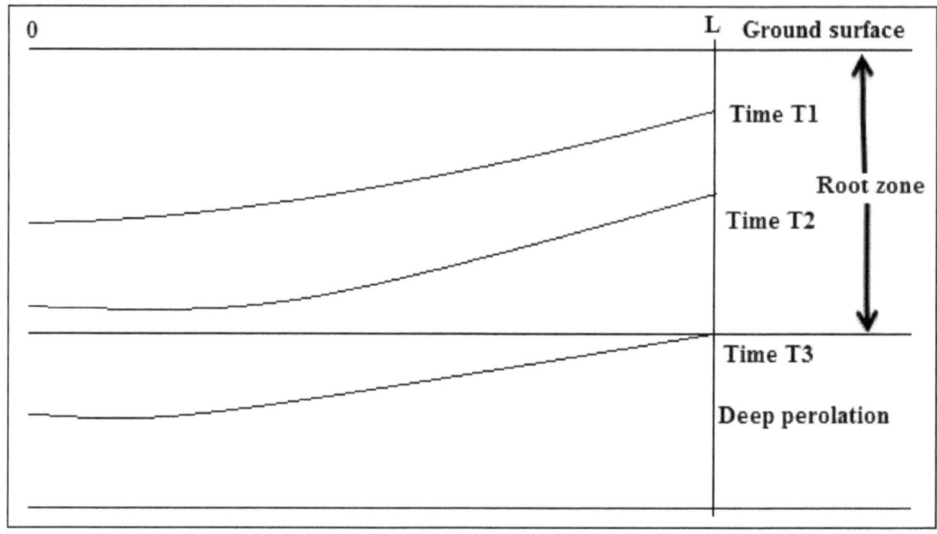

Figure 19.1: Schematic of Application Efficiencies in Border Irrigation at Various Times of irrigation.

application efficiencies must be seen in the context of storage efficiency and distribution efficiency.

Water Storage Efficiency

The water storage efficiency, Es is the ratio of the water stored in the root zone after the irrigation to the need of water in the root zone prior to irrigation. It is also expressed as a fraction or percentage as follows:

$$Es = 100\ (Vs/Vrz) \tag{19.6}$$

Here, Vs is as previously defined and Vrz is the root zone storage capacity (m^3).

The root zone depth and the water-holding capacity of the root zone determine Vrz. The storage efficiency has little utility for sprinkler or drip irrigation because in these methods complete refilling of the root zone is seldom achieved.

Water Distribution Efficiency

Water distribution efficiency, Ed is a measure of water distribution within the field. It is the ratio of the mean of numerical deviations, Y from the average depth of water stored after irrigation and the average depth of water stored after the irrigation (d). It is mathematically expressed as:

$$Ed = \left(1 - \frac{\overline{Y}}{\overline{d}}\right) \times 100 \tag{19.7}$$

here, $\overline{Y}$ is the average numerical deviation of depth of water stored from average depth stored during irrigation and $\overline{d}$ is the average depth of water stored during irrigation.

Low distribution efficiency usually means non-uniform distribution of irrigation water. It may result from improper land levelling or due to variations in soil physical conditions within the field. It may be noted, that one needs to apply excess irrigation to supply water equally in the root zone resulting in additional deep percolation losses lowering the water application efficiency. Water distribution efficiency is similar to the Christiansen's uniformity coefficient used to describe uniformity of water application in sprinkler irrigation.

Example 19.3

A stream of 150 lps was diverted from a canal and 120 lps were delivered to the field. An area of 1.75 ha was irrigated in eight hours. The effective depth of root zone was 1.80 m with an available moisture holding capacity of 22 cm/m depth of soil. The moisture distribution after the irrigation is shown in Figure 19.2. Determine the water conveyance efficiency, water application efficiency, water storage efficiency and water distribution efficiency. It may be assumed that irrigation begins at a moisture extraction level of 50 per cent of the available moisture. The depth of water penetration may also be assumed to be linear. The runoff loss in the field was measured as 450 m^3.

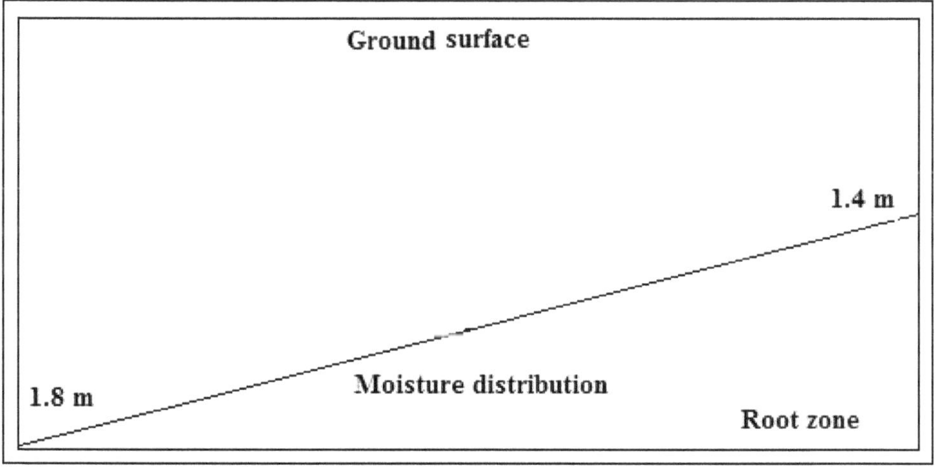

Figure 19.2: Moisture Distribution Pattern in the Root Zone after Irrigation.

Water conveyance efficiency, Ec=100(Vf/Vd) = 100(120/150) =80.0 per cent

Water delivered to the field = (120 × 60 × 60 × 8)/1000 = 3456 m^3

Water stored in the root zone = 3456- 450 = 3006 m^3

Since all the water is stored in the root zone, application efficiency can be written as:

Application efficiency = 100 (3006/3456) = 86.97 per cent

Water required in the root zone before the irrigation begins = 24 × 1.80/2 = 19.8 cm

Volume of water required = 19.8 × 1.75 × 100 × 100/100 = 3465 m^3

Water storage efficiency = 3006/3465 = 86.75 per cent

Water distribution efficiency is given by eq. (19.7). As such the average water depth applied is given as = (1.8+1.4)/2 = 1.6

The deviation at the head end from the mean is = 1.6-1.8 = - 0.2 (Note: Signs are to be neglected while calculating $\overline{Y}$)

The deviation at the lower end from the mean is = 1.6-1.4 = 0.2

Average numerical deviation from depth of penetration = (0.2+0.2)/2 =0.2

Distribution efficiency, Ed = 100(1-0.2/1.6) = 87.5

Water Use Efficiency

Water use efficiency, Eu, is defined as yield of plant product (say kg of grain, Y) per unit of crop water use (mm of water lost by evapotranspiration, ET). Clearly both Y and ET are the outcome of several plants and environmental processes operating over the life of the crop yet the water use efficiency (Eu =Y/ET) is extensively used as a measure of water use efficiency. It is expressed in units of kg/ha-mm.

Since crop ET requirement is met from various sources such as irrigation, rain water and root zone stored water, water use efficiency has been defined and designated differently by different people depending upon the interest of the user. For an irrigation engineer, amount of irrigation water applied is the most important. So water use efficiency has been defined as yield of the crop per unit of irrigation water (IW) applied, Y/IW and is termed as irrigation water use efficiency. For agricultural scientists, it is the ET of the crop that is important irrespective of its source. It is termed as crop water use efficiency and is the ratio of crop yield per unit of water depleted by the crop in the process of evapotranspiration (ET), *i.e.* Y/ET. It is commonly assumed that water stored in the root zone at the time of sowing and harvest is same. However, if measurements of water storage have been made, water depleted from the root zone (It may be positive or negative) is taken into account. The term used then is field water use efficiency or water expense efficiency given as, Y/WR. Here WR is the total water consumed by the crop during its growth period.

Conventionally, it is recommended to continue to use the term irrigation water use efficiency as described in this section. But some people have now questioned and argued that this definition is a misnomer. In fact the water use efficiency, Eum, should be defined as follows:

$$Eum = (Yi - Yr)/IW \qquad (19.8)$$

Here, Eum is modified irrigation water use efficiency, kg/ha-mm, Yi is the yield with irrigation, kg and Yr is yield under rainfed conditions, kg and IW is the amount of irrigation applied, mm.

Crop Productivity

This term is now increasingly used instead of water use efficiency. The argument in favour of this term has emerged from the fact that it is very difficult to compare the water use efficiency across crops. As a result one is unable to decide, whether one should go for either crop A or for crop B? To overcome this, water productivity is defined as the Rs. earned per unit of irrigation water applied.

Crop productivity = Rs. earned per ha/IW = Rs./ha-mm (19.9)

Irrigation Uniformity

Irrigation uniformity describes the evenness of water application over the length and breadth of the field. It is a statistical measure of the distribution of the applied water, which is affected by various factors such as: method of water application, topography, infiltration characteristics, and hydraulic characteristics (pressure, flow rate, *etc.*) of the irrigation system. It is generally expressed by Christiansen's coefficient of uniformity (CU) and distribution uniformity (DU).

Christiansen's Uniformity Coefficient

Uniformity coefficient developed by Christiansen (1942) is stated below:

$$CU = 100\,(1.0 - x/mn) \qquad (19.10)$$

Here "x" is the sum of the deviations of each observation from "m", the mean value of such observations, and "n" is the number of observations. All deviations from the mean are positive numbers *i.e.* any negative number is changed to a positive number. Therefore, eq. (19.10) is commonly written as:

$$CU = 100\left(1 - \frac{\sum |yi - m|}{m.n}\right) \qquad (19.11)$$

Here CU is the coefficient of uniformity, per cent, yi is depth of water at a given point or water measured in each container in sprinkler irrigation, n is the number of points or cans used for measurements (as such i varies from 1 to n), m is mean depth of water applied or collected in all the cans,. Vertical lines states that absolute values of deviations are used. A CU of 84 per cent is desirable.

Example 19.4

Measurements from 15 observations were: 15, 38, 32, 22, 35, 23, 32, 35, 19, 28, 26, 34, 24, 18, 24. Find out the value of CU.

m = (15+ 38+ ....+ 24)/15 = 405/15 = 27

n = 15

$\Sigma |yi - m| = |(15-27)| + |(38-27)| + \ldots |(24-27)| = 90$

Uc = 100 [1.0 – 90/(27 × 15)] = 77.8 per cent

Distribution Uniformity, DU

DU is defined as the ratio of the average application amount received in the least-watered quarter of the field to the average water applied expressed in percentage.

$$DU = 100\,(Lq/Xm) \qquad (19.12)$$

Here Lq is the average low-quarter depth of water infiltrated (or caught) and Xm is the average depth of water infiltrated (or caught). The distribution uniformity gives an indication of the magnitude of the distribution problem.

In sprinkler irrigation, it is also known as pattern efficiency. It is expressed as a percentage by the formula:

PE = 100 × Average of 25 per cent cans with minimum catch/
Average catch in all the cans (19.13)

To ensure that the whole area receives some minimum depth of water, the water applied should be equal to this minimum depth divided by the PE in fraction. For example, if it is desired to have the minimum application of 3 cm and the pattern efficiency is 0.75, then the application depth for the sprinkler should be fixed at 4 cm.

A disadvantage of DU is that it treats under-watering as the critical element but does not tell you how big or severe the dry spot is. On the other side, coefficient of uniformity (CU) treats under-watering and over-watering as the same. Moreover, calculating CU is slightly more complicated than calculating DU.

Acceptable values of uniformity coefficients vary with the type of crop being grown and the specific uniformity equation used. Both the equations result in approximately the same values when uniformity is high. However, DU values are normally much lower than UC values when uniformities are low. For high cash value crops, especially shallow rooted, the uniformities should be high (DU values greater than 80 per cent or UC values greater than 87 per cent). For typical field crops, DU values should be greater than 70 per cent (CU) values greater than 81 per cent). For deep rooted orchard and forage crops, uniformities may be fairly low if chemicals are not injected (DU values above 55 per cent and CU values above 72 per cent). Uniformity coefficients should be high (DU values greater than 80 per cent or CU values greater than 87 per cent) whenever fertilizers or other chemicals are injected into the irrigation systems (Merriam and Keller, 1978). Poor uniformity can deprive some areas of adequate fertilizer while other areas may be over fertilized. If uniformity coefficients are less than these values, system repair, adjustment or modification are required.

In drip irrigation, a term emission uniformity (EU) is used. Emission uniformity is expressed as a percentage, and is a relative index of the variability between emitters in an irrigation block. Emission uniformity is defined as the average discharge of 25 per cent of the sampled emitters with the least discharge, divided by the average discharge of all sampled emitters. Practically it is calculated with the help of eq. (19.12). Several other improved equations to work out emission uniformity in drip irrigation are in use but are being left out being out of course.

Deep Percolation Ratio

It is the ratio of amount or water going as deep percolation to the amount of applied water at the field head. It is expressed as:

$$DPR = 100 \times (Vdp/Vf) \quad (19.14)$$

Here, Vdp is amount of water lost to deep percolation and Vf is the amount of water delivered to the field.

Tail Water Ratio

It is the ratio of amount of runoff to the amount of applied water and is expressed as:

$$TWR = 100 \times (Vro/Vf) \quad (19.15)$$

Here, Vro is the amount of runoff from the field.

Overall Project Efficiency

It is the ratio between the average depth of water stored in the root zone and water diverted from the reservoir. It is mathematically expressed as:

$$Ep = 100 \times (Vs/Vd) \quad (19.16)$$

Here, Eo is the overall efficiency, per cent, Vs is the water stored in the root zone, cm and Vd is the water diverted from the reservoir, cm.

It can also be calculated by multiplying the conveyance efficiency, Ec and application efficiency, Ea each taken in fraction. To express in percentage, resulting value is multiplied by 100.

An overal project efficiency of 50-60 per cent is good; 40 per cent is reasonable, while overall project efficiency of < 30 per cent is poor. Current canal irrigation schemes in India have on an average, efficiency of 38 per cent. These are indicative values and one must strive to achieve a higher value.

Chapter 20

Sprinkler Irrigation

Introduction

Sprinkler irrigation is a system of irrigation in which water is applied in the form of spray somewhat resembling the rain. Although sprinkler irrigation is also referred as "overhead" irrigation, the term may not be a true connotation as in some cases sprinklers (called "under- tree," "low-angle," or "ground") distribute water near the ground. In this system, water is pumped, distributed through a network of pipes and sprayed into the air through spray heads (nozzles). The water at this stage breaks up into small water drops which fall to the ground to wet the land (Figure 20.1). Since sprinklers have a wide range of discharge capacity, the system is adaptable to nearly all irrigable soils.

Historically, a British inventor Sir William Congreve patented a manual sprinkler system in 1812 where perforated pipes along the ceiling were used for fire control. Later, experiments with automatic sprinklers began in and around 1860, but the first automatic sprinkler system was patented by Philip W. Pratt of Abington, MA, in 1872. Since then the sprinkler systems have proliferated around the globe but in India its arrival was delayed for more than a century.

Advantages and Constraints

Sprinkler irrigation systems are used on practically all types of soil, topographic conditions, and on almost all kinds of crops. Its flexibility and efficient water control has permitted a wide range of soils to be classified as irrigable. The main advantages of sprinkler irrigation are as follows (NCPA, 1990):

Advantages

- ☆ High crop yields

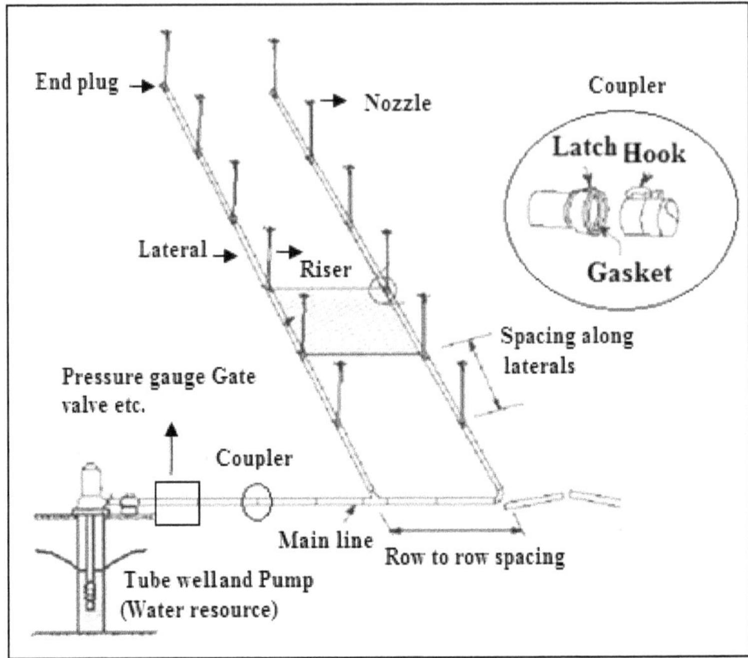

Figure 20.1: Components of a Sprinkler Irrigation System.

- ☆ Water saving as it eliminates channels for conveyance resulting in no conveyance losses and higher water application efficiencies
- ☆ Adaptabilities to use under condition where surface irrigation methods are inefficient or are costly such as undulating area where levelling cost may be high
- ☆ Saves land as no bunds *etc.* are required
- ☆ Suitable for all types of soil except heavy clay
- ☆ Helps in improved and uniform leaching and crop germination in saline soils
- ☆ Suitable for irrigating crops when plant population per unit area is very high; most suitable for oil seeds and other cereal and vegetable crops
- ☆ A convenient method to apply light and frequent irrigation, a must to manage salt affected soils
- ☆ Improved quality of produce especially of fresh vegetables and fruit where colour and quality are quite important
- ☆ Mobility of system to irrigate desired areas as and when required
- ☆ Modifies micro-climate especially in hot summers
- ☆ Areas located at a higher elevation than the source can be irrigated
- ☆ Possible to adopt chemigation

- ☆ Less problem of clogging of sprinkler nozzles due to sediment laden water
- ☆ Most economical way to apply water in areas where labour and water costs are high
- ☆ The same equipment can be used for crop cooling and frost control.

Constraints

Sprinklers, like most physical systems, do have some constraints/disadvantages.

- ☆ High initial cost especially for solid-set systems
- ☆ Require high water pressure (>2.5kg/cm^2) so energy requirement to build-up the necessary pressure is needed
- ☆ Uneven water distribution under windy conditions and may even be difficult to irrigate during high winds
- ☆ High evaporation losses when operated under high temperatures
- ☆ May pose problems when used on highly impermeable or very low permeable soils
- ☆ Damage to foliage of some crops with poor quality irrigation water. Poor quality water can leave undesirable deposits or colouring on the leaves or fruit of the crop
- ☆ Sprinklers can increase the incidence of certain crop diseases due to change in micro-climate
- ☆ Lack of package of practices for most crops as are available for surface irrigation
- ☆ Lack of awareness and lack of social concern to save natural resources.

Therefore, while deciding to use sprinkler system or its variants, all the pros and cons must be economically assessed with other irrigation methods or within types of sprinkler systems.

Potential and Extent of the Area under Sprinklers in India

The most potential soils for sprinkler irrigation are the sandy and undulating riverine areas, un-commanded sandy high areas within canal commands, fringe areas of canal commands, arid/coastal areas where water is either saline or extremely scarce and areas where high value crops are grown on sloppy terraces. Potential area that can be brought under sprinkler irrigation in India has been estimated on the basis of irrigated crops and their extent as per records of 2003. According to these estimates, potential area is assesses at 50 M ha, cereal crop having the highest potential. The crop-wise and state wise distribution of the potential is shown in Figure 20.2 (top and bottom). State wise UP alone accounts for about 27.7 per cent of the total potential, followed by Rajasthan, Punjab, Haryana, MP and Bihar. These positions could substantially change depending upon how the potential is worked out or if one uses the current data on cropping systems to assess the potential. For example; if one excludes the area under cereal crops from the estimate, the total potential itself may only be about 23.52 M ha. Within these figures the potential

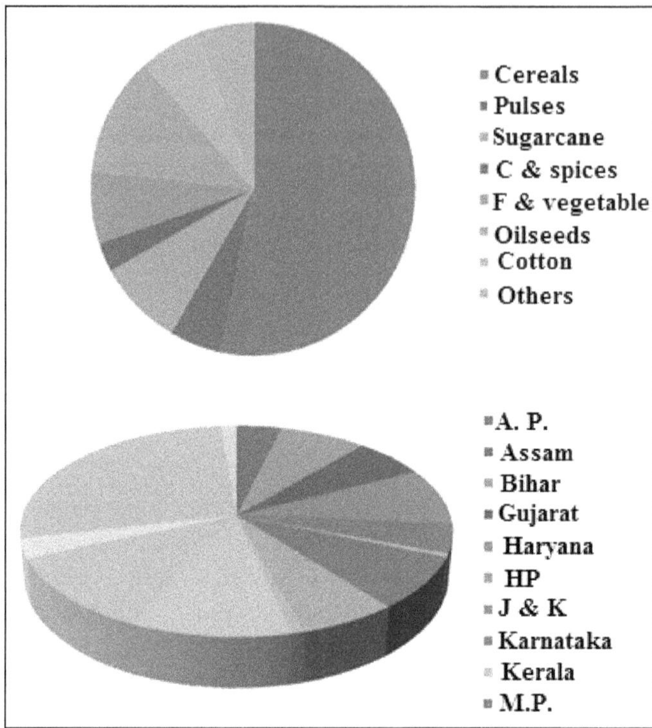

Figure 20.2: Potential Area that can be Brought under Sprinkler Irrigation Crop-wise (Top) and State-wise (Bottom).

area varies widely from 30.6 M ha (Table 20.1) to 4.25 M ha by INCID (INCID, 1998) and to 6.95 M ha by Task Force on Micro Irrigation (TFMI, GoI, 2004). An interesting feature of the data presented in Table 20.1 is that the potential seems to have been underestimated because in some states actual area has exceeded or is nearly equal to the potential. Current experimental studies suggest that sprinkler

Figure 20.3: Application Sprinkler Irrigation in Direct Seeded Rice.

irrigation can also be used efficiently to cultivate paddy crop (Kundu *et al.*, 1998). The potential of sprinkler irrigation in direct seeded rice has now been established and even being adopted by the farming community (Figure 20.3). As such, the extent of the potential area that can be brought under sprinkler irrigation could be vastly different than the earlier figure. It could even touch 74.23 M ha. Irrespective of the estimates, it suffices to say that there is a great scope to extend the area under sprinkler irrigation (MoWR, 1999).

The actual area under sprinkler irrigation has been estimated at 4.38 M ha (Table 20.1). It is only a small fraction of the potential. Since the technical feasibility has been proved on extensive installations throughout the country and the world, the question to be answered is whether such systems or special adaptations are sociologically and economically practical. Given the low canal water rates and electricity tariffs, large scale adoption of sprinkler irrigation may pose many difficulties in convincing the farmers to adopt this practice.

Table 20.1: Potential and Actual Area under MI in different States (Area in '000 ha)

States	*Sprinkler*		
	P	*A*	*Per cent*
Andhra Pradesh	387	328.4	84.90[×]
Bihar	1708	97.4	0.06
Chhattisgarh	189	241.4	127.72[×]
Goa	1	0.9	90.00[×]
Gujarat	1,679	418.2	24.91
Haryana	1992	550.5	27.64
Himachal Pradesh	101	0.7	0.07
Jharkhand	114	9.9	0.09
Karnataka	697	417.0	59.83
Korala	35	6.9	19.71
Madhya Pradesh	5,015	185.8	0.04
Maharashtra	1,598	374.8	23.45
Nagaland	42	5.0	11.90
Odisha	62	82.1	132.42[×]
Punjab	2,819	12.2	0.01
Rajasthan	4,931	1514.5	30.71
Tamil Nadu	158	30.4	19.24
Uttar Pradesh	8,582	21.2	0.00
West Bengal	280	50.6	0.18
Others	188	20.6	10.95
Total	**30,578**	**4376.0**	**14.31**

Source: Raman (2010); P=Potential; A= Actual area. *Source*: (midh.gov.in/AtGlance/MI-AT-A-Glance.docx),[×] Potential might have been underestimated.

Kinds of Sprinkler Systems

Sprinkler systems are classified on the basis of type of the system, portability, field layout, equipment used and type of the system being irrigated (Pillusbury, 1968, Hurd, 1969). On the basis of type of the system, two main categories are the perforated pipes and the rotating overhead system.

Perforated Pipe Systems

Perforated pipes are lightweight portable pipes 5.0-7.5 cm in diameter with holes so spaced so as to discharge water at different angles. The spacing of the small holes varies, depending on the rate of water application. Perforated pipes are suited for soils that absorb water at relatively high rates. The system performs well even at relatively low operating pressure in the range of 0.5 to 2.5 kg/cm^2. The maximum length of the pipeline is governed by variation in pressure that can be allowed for a given set-up. It is around 80 m for a pipe of 5 cm diameter and 171 m for a pipe of 7.5 cm diameter. It ensures an average nominal operating pressure of 1.2 kg/cm^2 if the pressure at the beginning is 1.4 kg/cm^2. It should ensure a fairly uniform water distribution along the line. The width of the wetted strip depending on the pressure varies from approximately 10 m to 16 m (Table 20.2). The move distance is generally 2/3 of the width of the irrigated strip. Besides water application at relatively high rates, perforated pipes have several other limitations. Since the pipes are laid on the ground, plants may grow so close and high that they might impinge on the water jets. Moreover, it may just be impossible to regulate or equalize the pressure along the line, as in the sprinkler heads, except possibly by changing the pipe diameter. Therefore, the system operates inefficiently when used on level land or on a contour line. The system is quite suited to irrigate lawns, gardens and plants of low height.

Table 20.2: Pressure and Width of Irrigated Strips in Perforated Pipe System

Nominal Operating Pressure (kg/cm^2)	*Strip Width (m)*
0.7	10.7
1.0	12.8
1.4	14.6
1.7	15.5
2.1	16.0

Rotating Overhead System

In this system laterals are usually placed on the ground and nozzles are placed on the risers fixed at uniform interval along the length of the lateral pipes (Figure 20.1). Basic sprinkler type commonly used in agricultural fields is the impact-driven rotating sprinklers that rotates 360° or in a pre-decided trajectory in landscape irrigation. The wetted circle and the soil moisture distribution with respect to distance from the riser centre shown in Figure 20.4 reveal that the maximum water application is near the riser. It is why some overlap between the adjacent wetting circles is essential. Overhead sprinklers include fixed installations, centre pivots, traveling guns, and portable sprinklers. Each system has its own characteristics.

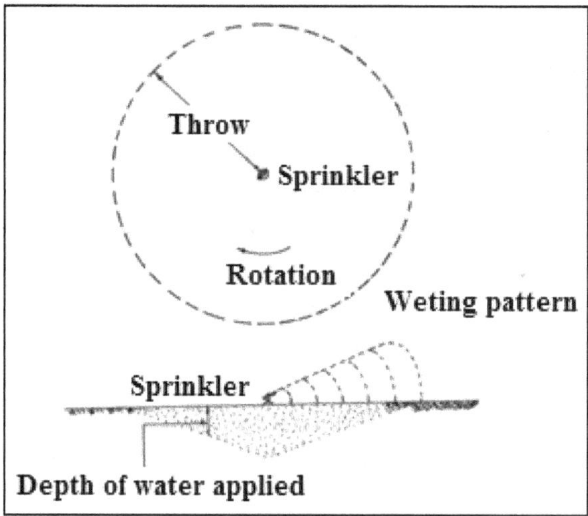

Figure 20.4: Wetted Circle (Top) and Soil Water Depth Distribution (Bottom) being Maximum at the Centre.

Classification on the Basis of Portability

Sprinkler irrigation systems, based on portability, are grouped in the following 5 categories.

Portable (Fully)

A system is fully portable when all of the system components can be moved and fully permanent when none of them can. A fully portable system has portable mains, sub-mains, laterals and even the pumping unit. The system is designed to be moved from one field to another or to different pump site(s) in the same field. The system is moved manually or mechanically. In the latter case, it is known as wheel move system. The former is cheaper but labour intensive while the latter is costly but can save lot of labour.

Semi-portable

It is similar to portable system except that the water source and pumping plant is fixed. The mainline is an extended line so that the system is capable to cover the whole area.

Semi-permanent System

In this system only laterals are portable while main and sub-mains are fixed. The main and sub-mains in this system are usually buried.

Solid Set System

A sprinkler system which remains at a single location during an irrigation season and supplies water by a fixed network of pipes is categorized as a solid-set system. There is no unique field layout for a solid-set system because of the many ways the systems are arranged. A solid set system has enough laterals to cover the whole area. The laterals are positioned in the field early in the crop season. The system is most suitable where crops require frequent irrigations of small depths. Aluminium piping is typically laid along the ground surface. Plastic pipes, usually PVC, are always buried permanently because the quality of these pipes may deteriorate if exposed to ultra-violet (UV) rays. If UV stabilizers and black carbon have been added during manufacturing of LDPE pipes, then these can also be installed on the surface. Solid-set systems are relatively costly but the labour and maintenance costs are minimal. These systems may restrict some cultural operations such as cultivation, spraying, planting and harvesting.

Permanent System

The system is placed permanently in the field with buried mains, sub-mains and laterals. The system is quite suitable for orchard crops of permanent type.

Hand Move and other Travelling Sprinklers

To reduce the required components and minimize interference with other farming operations, sprinkler systems are designed to move the lateral pipelines from set to set. The movement is either manual or tractor-towed or engine controlled. Many names have been given to such sprinklers:

Hand-move

These are further categorized in 2 classes namely tow-line/skid-tow (lateral is pulled across the field) and side-roll (lateral mounted on wheels that roll to move the lateral).

In these systems, sprinklers mounted on moving platforms are connected to the water source by a hose. Lateral move system has a mobile long boom with many sprinklers attached to them. As the machine moves, it collects water from the canal and supplies to the sprinklers connected to the long boom. Automatic moving wheeled systems known as travelling sprinklers are used to irrigate areas such as small farms, sports fields, parks, pastures.

Wheel line irrigation is a series of pipes and wheels with sprinklers attached. Water is pumped through the pipes from a large hose. After enough water has been pumped on to the land, the hose is removed and the wheel line machine is moved to the next section of the field. The hose is attached again to irrigate the next section of the field (Figure 20.5 left). These systems utilize polyethylene tubing wound on a steel drum. As the tubing is wound on the drum powered by the irrigation water or a small gas engine, the sprinkler is pulled across the field. Main characteristics of the wheel line system are:

- ✰ It consists of a lateral mounted on 1.2 to 3.0 m diameter wheels with pipe acting as an axle.
- ✰ This type of system is used mainly on flat ground and is not recommended for slopes greater than 5 per cent
- ✰ Adapted only to low-growing crops

Figure 20.5: A View of the Wheel Line Irrigation (Left) and Centre Pivot System (Right).

- ☆ Medium labor requirements
- ☆ Moderate initial investment
- ☆ Medium operating pressure.

Centre Pivot System

To minimize labour, maintenance and down-time problems with hand-move systems, systems that are continuously on the move have been developed. Centre-pivot, linear move, and big gun systems are typical examples of this concept. In centre pivot irrigation, several segments of pipe (usually galvanized steel or aluminium) are joined together supported by trusses, mounted on wheeled towers with sprinklers positioned along its length while the source of water is stationary. The boom with many sprinklers rotates in a circular pattern about the water source (Figure 20.5 right). New systems have drop sprinkler heads so that water is deposited on the ground between the plants to minimize evaporation and prevent drift of water being blown by the wind. Originally, most centre pivots were water powered but later replaced by hydraulic and electric motor driven systems. Many modern pivots feature GPS devices.

Specialized Systems

Based on the function for which sprinklers have been installed, these have been grouped under the following categories.

- ☆ Landscape and turf irrigation systems
- ☆ Frost protection sprinklers
- ☆ Fire sprinklers

The principle function of these systems is clear from the names. Fire sprinklers are permanently installed. These get activated by heat, smoke or both. Common in hotels, large buildings and industrial settings, these systems inhibit fire from spreading and allow time for people around to escape. Other functional categories are: spray systems, which direct a heavy spray of water in a particular area and deluge systems which use a type of sprinkler head that dumps a large amount of water on an area when activated.

Rain Gun and Long Range Sprinklers

Rain gun is a high pressure sprinkler with a powerful jet system (Figure 20.6 left) often mounted on a mobile machine (Figure 20.6 middle) or may be manually transported (Figure 20.6 right). The speed of the machine determines the application rate. Guns are similar to rotors, except that they operate at very high pressures in the range of 275 to 900 kPa and flows of 3 to 76 L/s (NARC, 1992). The nozzle diameter is in the range of 10 to 50 mm. Depending upon the discharge, low, medium and big volume rain guns are available each having its own features and specifications. The achievable field water efficiencies range from 60 to 70 per cent. These are used for field crops like sugarcane, pulses, oil seeds, cereals, tea, coffee, and vegetables. Rain guns besides irrigation are also used for industrial applications such as dust suppression and logging.

Figure 20.6: A Rain Gun Model (Left), Mounted on Mobile Machine (Middle) and Stationary with Manual Move (Right).

Components of Sprinkler Irrigation System

A sprinkler system usually consists of the following components as shown previously in Figure 20.1.

Pumping Unit

Water is pumped under pressure to the fields, which forces the water through sprinklers or through perforations or nozzles to form a spray. A high speed centrifugal or turbine pump is used. The driving unit is either an electric motor or an internal combustion engine.

Pipes: Mains, Sub-mains and Laterals

Depending upon the functions, pipes are named as mains/sub-mains and laterals. Main line conveys water from the source and distributes it to the sub-mains. The sub-mains convey water to the laterals which in turn supply water to the sprinklers. Aluminium or PVC pipes are used for portable while steel pipes are used for center-pivot laterals. Special aluminium alloys are used to manufacture lightweight pipes that have enough strength and are quite resistant to corrosion.

Asbestos, cement, PVC and wrapped steel pipes are used for buried laterals and main lines.

Couplers

Couplers are used to connect the two pipes quickly and easily. Essentially a coupler should provide:

- ☆ A reusable and flexible connection
- ☆ It should not leak at the joint
- ☆ It should be simple and easy to couple and uncouple
- ☆ It should be light, non-corrosive and durable

In general, a coupler contains at least one rubber gasket. The water pressure holds the gasket tightly in place against the pipe, thereby preventing leakage. Gaskets might deteriorate with time and therefore, needs replacement as soon as leakage is detected. Its life depends on the water quality and local operating conditions. Some type of footing device is used to keep the pipe and coupler from tilting or rolling.

Filters

Sand/media filters are used to remove organic matter and inorganic contaminants from water sources like rivers, tanks and open wells. Hydro-cyclone filters/sand separators are used to remove particles of the size of 75 microns (200 mesh). These equipment are useful for cleaning river water, canal water and tube well water which may contain sand.

Risers

Since the sprinkler needs to be located above the crop to prevent the crop canopy from interfering with the jet from the nozzle, a smaller diameter pipe called a riser is used to conduct water from the lateral to the sprinkler device. Besides, the riser helps to reduce turbulence of the water stream before it reaches the sprinkler.

Fittings, Accessories and Optional Items

The following are some of the important fittings and accessories used in sprinkler system.

- ☆ Water meter is used to measure the volume of water delivered.
- ☆ Flanges, couplings and nipples are used for proper connection to the pump, suction and delivery.
- ☆ Pressure gauge is used to know whether the sprinkler system is working with desired pressure to ensure application uniformity.
- ☆ Bend, tees, reducers, elbows, hydrants, butterfly valves and plugs are necessary to assemble the system.
- ☆ Fertilizer applicator is used to inject soluble chemical fertilizers into the sprinkler system so as to apply it concurrently with water. The equipment

for fertilizer application is relatively cheap and simple and can be fabricated locally (See details in fertigation section).

Sprinkler Head

Sprinkler head is the heart of a sprinkler irrigation system. It is device (usually brass or plastic) with one or more small diameter nozzles that sprinkle the water in the form of rain on the crop canopy. Sprinkler head distributes water uniformly over the field without runoff or excessive deep percolation loss. It is desirable to have nozzles that throw water to great distances using as little pressure as possible. It may help to reduce the operating cost of sprinkler systems.

The sprinkler heads are either fixed or rotating type. Fixed head sprinklers are commonly used to irrigate small lawns and gardens. The rotating type can be adapted for a wide range of application rates and spacing. Most sprinklers will operate satisfactorily over a range of pressures. However, best results are obtained by maintaining pressures at or above the minimum recommended by the manufacturer for a given size of sprinkler and a given number and size of nozzles. They are effective with pressure of about 10 to 70 m head at the sprinkler. Pressures ranging from 16 to 40 m head are considered the most practical for most farmers.

The sprinkler heads are also categorized on the basis of degree of rotation. Full circle sprinkler heads are commonly applied in agriculture and turf/lawn irrigation. The part circle sprinkler has a special mechanism so that a latch engages when the sprinkler head has rotated to the desired angle. It then rotates in opposite direction. Low angle sprinkler head are used where the sprinklers are carried above the crop. The low angle reduces drift and evaporation losses.

Impact sprinkler, a kind of full circle sprinkler is commonly used in agriculture and lawn irrigation. In this case water from the nozzle sprays onto the spoon of the sprinkler arm. As the arm returns to its original position it strikes the sprinkler body. This impact causes the sprinkler body to rotate. The main components of this type of sprinkler head shown in Figure 20.7 are:

- Drive or range nozzle (hits sprinkler arm and throws water out farther)
- Spreader nozzle (optional; Applies more water close to the sprinkler)
- Trajectory angle; the angle varies between 7 and 30 degrees. The low angle is used in orchards or in situations where the spray should be kept close to the ground.

Although new devices are being introduced by the manufacturers yet the three common types of irrigation sprinklers are sprays, rotors (impact or gear drive) and multi-stream rotators. Spray heads and nozzles are the least water efficient of the traditional sprinklers since they are susceptible to wind drift. However, installation is very cost effective for small areas with low-height plant cover. Rotors and multi-stream rotators increase water stream and droplet size. It helps in applying water more efficiently. Installation is very cost effective for medium and large areas. An advantage of multi-streams over rotors is that they are adaptable to small and irregularly shaped areas like parking strips.

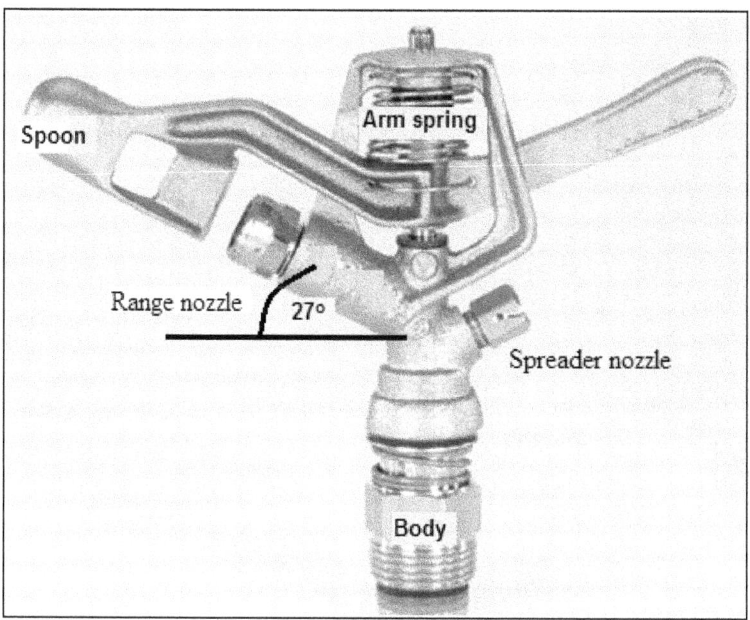

Figure 20.7: Components of an Impact-Driven Sprinkler.

Pop Head Sprinklers

Pop-up or pop-head sprinklers are installed below the ground. A piston that contains the nozzle lifts up from the sprinkler body when the sprinkler is operated but retracts back below ground when not in use. Pop-up sprinklers are available in a variety of heights varying from 2", 3", 4", 6" to about 12". The mechanism and design of a pop-up sprinkler is such that one can adjust its head to regulate the flow of water. The head is easily adjusted by turning the tip of the nozzle and the screw in it. Even the radius of throw can be adjusted. Two major advantages of pop-up sprinklers are in its safety and appearance. Pop-up sprinklers don't cost more than normal sprinklers if the cost of the riser is included.

Selection of Sprinklers

There are hundreds of sprinkler designs and variations and new ones appearing in the market very often. The objective is to provide acceptable uniformity with acceptable hardware and other costs. Sprinklers are selected from the manufacturers' literature considering the site conditions. Low pressure nozzles can reduce pumping costs but tends to have larger drop sizes and high application rates. On the other hand, high pressure sprinklers have high operational costs and high application rates. Thus, medium pressure sprinklers should be chosen if these are able to meet the field requirement. Some guidelines for selection of the rotating head sprinklers are given in Table 20.3.

Table 20.3: Classification of Rotating Head Sprinklers, their Characteristics and Adaptability

Type of Sprinkler	*Gravity Fed under Tree Sprinkler Systems*	*Normal under Tree Sprinkler Systems*	*Permanent Overhead Systems*	*Small Overhead Systems*	*Low Pressure Systems*	*Intermediate Pressure Systems*	*High Pressure Systems*
Pressure range	0.7 - 1.0 kg/cm^2	1 - 2.5 kg/cm^2	3.5 - 4.5 kg/cm^2	2.5 - 4 kg/cm^2	1.5- 2.5 kg/cm^2	2.5 - 5 kg/cm^2	5 -10 kg/cm^2
Sprinkler discharge	0.06 - 0.25 l/s	0.06 - 0.25 l/s	0.2 to 0.6 l/s	0.6 - 2.0 l/s	0.3 - 1 l/s	2 - 10 l/s	10 - 50 l/s
Diameter of nozzles	1 - 6 mm	1.5 - 6 mm	3 - 6 mm	6 - 10 mm	3 - 6 mm	10 - 20 mm	20 - 50 mm
Diameter of coverage	10 - 14 m	6 - 23 m	30 - 45 m	25 - 35 m	20 - 35 m	40 - 80 m	80 - 140 m
Range of sprinkler spacing (square)	—	—	18 - 30 m	9 - 24 m	9 - 18 m	24 - 54 m	54 - 100 m
Recommended speed of sprinkler rotations	—	0.5 - 1 rpm	1 rpm	0.67 - 1 rpm	0.5 - 1 rpm	0.7 rpm	0.5 rpm

Discharge

Discharge of a sprinkler nozzle depends on type of the nozzle, nozzle size and the operating pressure. The sprinkler discharge is related to the operating pressure and nozzle geometry as per the following equation:

$$q = C\,A\,\sqrt{2gh} \quad (20.1)$$

Here, q is the discharge, m^3/s, C is the discharge coefficient, for a good nozzle it is ≈ 0.95-0.96, A is the cross section area of the nozzle, m^2, h is the water pressure at the nozzle, m and g is the acceleration due to gravity, m/s^2 (9.81 m/s^2). The pressure h at the nozzle can be measured using a pitot tube (Figure 20.8)

For C = 0.95 and in terms of diameter of the nozzle, eq. (20.1) can be written as:

$$q = 3.31\,d^2\,\sqrt{h} \quad (20.2)$$

It may be noted that the nozzle diameter increases the discharge relatively more than the increase in pressure as illustrated with the following example.

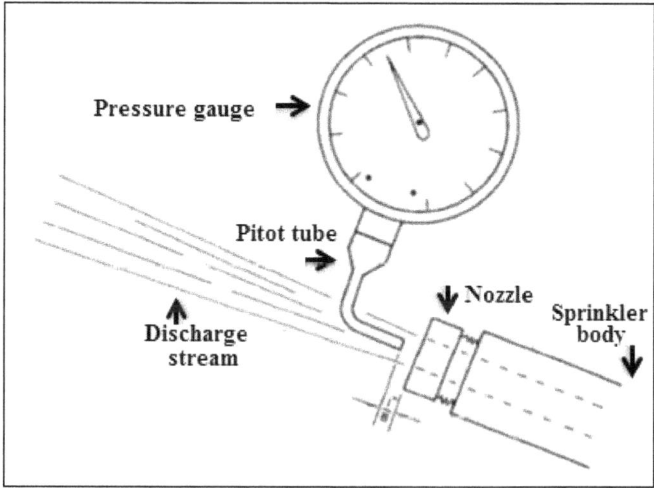

Figure 20.8: Measurement of Pressure using a Pitot Tube.

Example 20.1

A straight bore nozzle with a discharge of 1 m^3/s is used in a sprinkler. Would a 20 per cent increase in nozzle diameter produce more flow than a 20 per cent increase in pressure?

For a 20 per cent increase in diameter, d2 = 1.2 d1 and h2 = h1.

Since initial flow rate q1 = 1 m^3/s, q2 is given as:

$q2 = q1\,\{1.2\,d1/d1)^2 = 1.44 \times q1$.

Thus, a 20 per cent increase in diameter provides a 44 per cent increase in flow rate.

For a 20 per cent increase in pressure, h2 = 1.2 h1 and d2 = d1.

$q2/q1 = [d2^2 (h2)^{1/2}]/[d1^2 (h1)^{1/2}]$

$q2 = q1 \{1.2 h1/h1)^{1/2} = 1.1 \times q1$

Thus, a 20 per cent increase in pressure only changes the flow rate by 10 per cent.

A 20 per cent increase in nozzle diameter produces more flow than a 20 per cent increase in operating pressure.

Sprinkler Spacing and Depth of Water Application

The factors that affect the spacing of the irrigation system are:

- ☆ Type of sprinkler head
- ☆ Water pressure
- ☆ Slope
- ☆ Type of crop/plants
- ☆ Wind conditions
- ☆ Discharge
- ☆ Other local conditions

Three main types of sprinkler spacing patterns are named as square, rectangular and triangular. The square pattern has equal distance running between the four sprinkler positions and it is suitable for irrigating square-shaped areas. The limitation of this pattern is the diagonal distance between sprinklers and it's susceptible to wind effects. To take care of wind effects, one can adopt closer spacing depending on the severity of the wind.

The rectangular sprinkler spacing has sprinkler forming a rectangle with the shorter side of the rectangle across the wind and the longer side along the wind, so as to obtain a good coverage. This pattern can overcome the windy situations and is suitable for areas with well-defined straight boundaries and corners.

In the triangular pattern, sprinklers are arranged in equilateral triangle format so that the distance of sprinkler heads from each other is equal. This pattern permits larger spacing and therefore, requires fewer sprinklers compared to the square pattern. Furthermore, two patterns can also be combined on the same site to achieve optimum sprinkler coverage.

Precipitation Profile

Precipitation profile or water application uniformity is a function of several factors as listed below (Phocaides, 2000):

- ☆ Nozzle pressure
- ☆ Nozzle shape and size
- ☆ Sprinkler head design
- ☆ Presence of straightening vanes
- ☆ Rotation speed

- ☆ Trajectory angle
- ☆ Riser height
- ☆ Wind conditions

If all the factors remain unchanged, the precipitation profile depends upon the operating pressure. The pressure required to operate the system is the sum of maximum water pressure required for normal operation including the friction losses in the piping network from the control station to the distal end of the system, the difference in elevation and the pressure required at the emitter/sprinkler. If operating pressure refers to the pressure measured at the sprinkler, then the sprinkler systems can be classified by the operating pressure as follows (Phocaides, 2000):

1. Low pressure systems, where the pressure required is 200-350 kPa
2. Medium pressure, where the pressure required is 350-500 kPa
3. High pressure, where the pressure required exceeds 500 kPa.

The operating pressure of sprinklers has significant impact on irrigation uniformity and the overall performance of the irrigation system (Figure 20.9). The operating pressure also controls the wetted diameter and the mean water droplet size (Kranz *et al.*, 2005). Under low pressures (less than 276 kPa or 40 psi), the water jet leaving the nozzle does not break up adequately resulting in concentrated water application. Conversely, pressures above 483 kPa (70 psi) breaks the jet excessively (misting). It results in concentrated water application near the sprinklers (King *et al.*, 2006). Since at such pressures, fine mist is created in the sprinkling zone, it results in excessive wind drift and evaporation. The optimum operating pressure of impact sprinklers, with standard straight bore nozzle is 310 kPa to 414 kPa (45 to 60 psi). Armstrong *et al.* (2001) gave the common operating pressure range for overhead impact sprinklers as 240 – 400 kPa.

Overlapping for Uniformity of Water Distribution

The amount of water applied varies with distance from the sprinkler head. In no wind conditions, it is symmetrical around the nozzle head, the depth being maximum at the center. To obtain uniform water application, the wetted circle of the adjacent sprinkler is overlapped so as to add water to the area away from the adjoining sprinkler. The aim is to have the aggregated depth distribution as uniform as possible. For uniform application the overlap, as a thumb rule, should be >50 per cent of the sprinkler wetted diameter although overlapping of 100 per cent has been suggested in many cases.

Effect of Wind

Wind speed in association with sprinkler spacing has significant impact on the uniformity of set-move sprinkler irrigation systems. Besides impacting the wetted diameter, wind also causes loss of water through drift. The problem is pronounced if the wind speed is greater than 8 km/h. The changes in wind speed and direction, however, may tend to increase the cumulative irrigation uniformity calculated over multiple irrigation events.

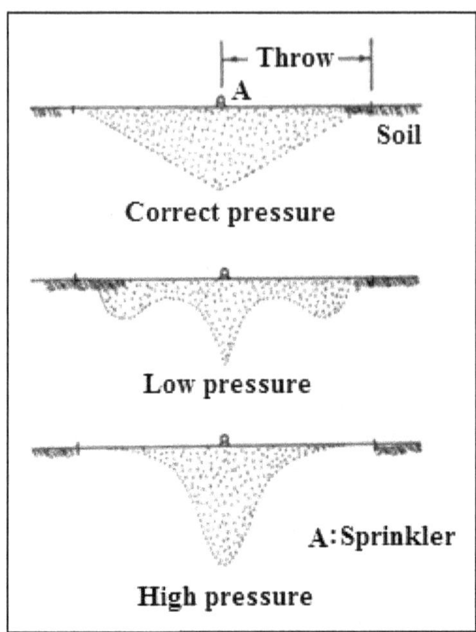

Figure 20.9: Water Distribution Profile around a Sprinkler at different Pressures.

To overcome the wind effect, several management options are applied to achieve uniform water application.

- Align sprinkler laterals across prevailing wind directions. The wind drift would be much less as compared to when the alignment is parallel to the wind.
- Select rotary sprinklers with a low trajectory angle
- Reduce the spacing between sprinklers
- Build extra capacity
- Provide straightening vanes under consistent windy conditions.

Irrigation System Performance

An irrigation system is designed to have an excellent performance so as to result in uniform application of water, in a right amount and at the right time. The performance of the irrigation system is significantly affected by the interactions between the application system characteristics and water supply characteristics. Four factors critical to achieve high levels of performance from an irrigation system are, irrigation timing, depth of application, uniformity, and water supply characteristics.

To assess the system's performance, a pressure gauge with pitot tube, catch cans to collect water, measuring cylinder and stop watch are required. Test procedure consists of the following steps:

- Select the appropriate site

- ✰ Operate the system for a few minutes to be sure that everything is working and operating correctly
- ✰ Measure the nozzle pressure by inserting the pitot tube directly in the nozzle as shown in Figure 20.8 and record the highest pressure reading
- ✰ Place catch cans in a grid pattern at regular spacing between sprinklers and laterals (Figure 20.10). The catch cans may also be placed in a circular grid, if desired or necessary
- ✰ About 20-24 cans are enough, but more the cans, the better it is
- ✰ Run the system for sufficient time, at least 30 min so that measurable volume of water is collected in the catch cans
- ✰ Record the water volumes in each can
- ✰ Because conditions (*e.g.* wind *etc.*) will be different each time a uniformity test is conducted, it is important to run the test under a range of conditions
- ✰ Reschedule the test if wind speed exceeds 12.5 kmph.

The data so collected are analyzed to determine application rate, application efficiency, coefficient of uniformity or distribution uniformity.

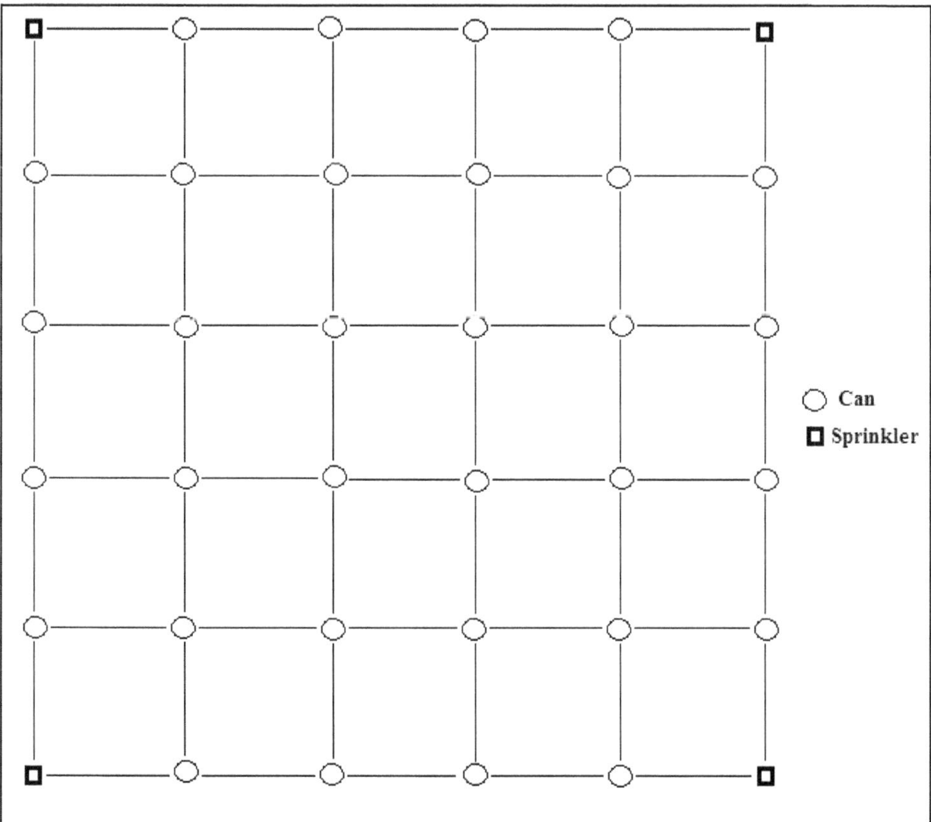

Figure 20.10: Set-up to Assess the Sprinkler System Performance.

Application Rate and Depth

Rate of water application is limited by infiltration capacity of the soil. If the application rate is more than the infiltration rate, it will cause runoff resulting in loss of water and poor water distribution. Although it is prudent to assess the rate for the fields for which systems are being designed, general guidelines are given in Table 20.4. One needs to collect information on nozzle pressure, soil type and root-zone depth to decide head and row spacing. While nozzle discharge is given by the manufacturers (Table 20.5), yet it may be prudent to measure the actual discharge in the field to assess the exact application rate.

Table 20.4: General Guidelines to Decide the Water Application Rates Based on Soil Type

Soil Texture	*Maximum Application Rates (mm/h)*
Coarse sand	20 to 40
Fine sand	12 to 25
Sandy loam	12
Silt loam	10
Clay loam/Clay	5 to 8

Table 20.5: The Manufacturers' Listed Discharge from Nozzles

Main Nozzle[x] *(Ø mm)*	*Pressure at Sprinkler (kPa)*	*Casting Radius (m)*	*Discharge*	
			m^3/h	*l/m*
3.5	200	11.50	0.67	10.28
3.5	300	11.90	0.75	12.60
4.0	200	11.50	0.81	13.44
4.0	300	12.10	0.98	16.46
4.5	200	11.60	1.02	17.01
4.5	300	12.25	1.25	20.83

[x]For nozzle having trajectory angle 25° and a connection of 1/2"

Depth of water application is calculated from the data collected by averaging the depth of water received in all the cans (See Example 20.3).

Irrigation Efficiencies

Irrigation efficiencies indicate as to how effectively the water extracted from a water supply source is used to produce the target crops for which water was withdrawn. Since, no irrigation system can be 100 per cent efficient, the efficiencies of irrigation systems vary from one project to another or within a project from one operational site to another. To make a relative comparison of various projects as well as to identify the spots where improvements can be made, irrigation efficiencies are calculated. Irrigation efficiencies have been defined in a number of ways and named accordingly. For on-farm tests, water application and distribution efficiencies are the most important numbers required to judge the system's performance (Calder, 2005).

Water Application Efficiency (AE)

The application efficiency is described by the relation (See Chapter 19):

$$AE = 100\ (Wb/Wf) \tag{20.3}$$

Here, Wb is the water used beneficially and Wf is the water delivered to field

Application efficiencies under different irrigation systems shown in Table 20.6 reveal that application efficiencies are more in drip followed by sprinkler and surface methods of irrigation.

Distribution Uniformity

The sprinkler uniformity is described mathematically by simple terms known as coefficient of uniformity (CU). For this purpose Christiansen coefficient of uniformity (CU) has been widely used. The second measure is the Distribution Uniformity (DU), proposed in one form or another by various workers. The DU is defined as the average water applied in 25 per cent of the area receiving the least amount of water, regardless of location, divided by the average water applied over the total area. For detailed discussions on these measurements, see chapter 19 of this book. However, the calculation procedure is again illustrated through the example 20.3.

Table 20.6: Irrigation Efficiencies under different Methods of Irrigation (per cent)

Methods of Irrigation	*Efficiencies of Irrigation*		
	Surface	*Sprinkler*	*Drip*
Conveyance efficiency	40-50 (canal) 60-70 (well)	100	100
Application efficiency	60-70	70-80	80-90
Surface water moisture evaporation	30-40	30-40	20-25
Overall efficiency	30-35	50-60	80-90

Source: Sivanappan (1998).

Example 20.3

Determine the uniformity coefficient from the following data obtained from a field test on a square plot bounded by four sprinklers:

Sprinkler : 4.365 × 2.381 mm nozzles at 2.8 kg/cm², Spacing : 24 m × 24 m, Wind : 3.5 km/h from south – west, Time of test: 1.0 h

Sprinkler	6.9 (0.42)	**5.6 (0.88)**	**4.6 (1.88)**	Sprinkler
6.1(0.38)ˣ	**5.6 (0.88)**	7.9 (1.42)	8.2 (1.72)	6.3(0.18)
6.9 (0.42)	7.1 (0.62)	7.1(0.62)	7.4 (0.92)	6.9 (0.42)
7.4 (0.92)	5.9 (0.58)	7.1 (0.62)	6.6 (0.12)	7.1(0.62)
Sprinkler	5.9 (0.58)	**4.6 (1.88)**	**4.8 (1.68)**	Sprinkler

( )ˣ Values of I(Xi-M)I in parenthesis, five spots where lowest amount of water is collected are in bold.

Mean = (6.1+ 6.90 +.........+ 7.1)/21 = 6.48

CU = 1- (17.76/(21 × 6.48)) = 1- 0.13 = 87 per cent

DU = (5.04/6.48) × 100 = 77.8 per cent (5 terms in bold are included to calculate the lowest depth of the quarter)

Design of Sprinkler System

The objectives of design of a sprinkler irrigation system are:

- ☆ Provide sufficient flow capacity to meet the irrigation demand
- ☆ Ensure that the least irrigated portions receives adequate water
- ☆ Ensure uniform distribution of water
- ☆ Minimize the cost.

Data Requirement

The values of following parameters shall be considered in the design of all sprinkler irrigation systems.

- ☆ Soil water holding capacity
- ☆ Water application rate
- ☆ Irrigation interval
- ☆ Water requirements of the area to be irrigated

The hydraulic design of the system should include:

- ☆ System capacity
- ☆ Pump and motor requirement (flow rate and total dynamic head)
- ☆ Main line sizing and pipe pressure requirements
- ☆ Lateral line sizing, spacing and pressure requirement
- ☆ Sprinkler nozzle sizing and spacing
- ☆ Appurtenance selection and sizing (pressure and air relief valves, thrust blocks, *etc*.).

Layout

System layout is governed by physical features of the site such as field shape and size, obstacles, topography and the type of equipment chosen. Where several possibilities of preparing the layout exist, a cost criterion is applied to these alternatives. Laterals should be as long as site dimensions, pressure variation and pipe diameter restrictions may allow. General rules for sprinkler system layout are:

- ☆ Water supply source should be nearest to the centre of the area
- ☆ Main should be laid up and down hill
- ☆ Lateral should be laid across the slope or nearly on the contour
- ☆ For multiple diameter laterals, laterals should not comprise of more than two diameters of pipes

- ☆ Layout should minimize the movement of laterals during the season
- ☆ Booster pump may be used where some portions of field require high pressure at the pump
- ☆ Layout should be modified if different rates and amounts of water need to be applied when soils of different textural group occur in the area
- ☆ The selected nozzle, operating pressure, discharge rate and sprinkler spacing must be shown on the plan. Irrigation interval, set time, application rate and net amount applied must be calculated and shown in the plan.

Water Supply

The water supply must be adequate and reliable as frequent irrigations are applied in sprinkler irrigation. Factors to consider are: quantity and reliability, water quality, the water level (if the source is a well), and the location of the source. If water supply is from a canal, one must have a reservoir of adequate capacity to meet the water requirement of the crops during the intervening periods of canal water supplies. Besides, the quality of water especially the amount of suspended material it contains must be considered in the design. Silt and debris may cause needless wear on the mechanical parts of the pump and sprinkler heads. Whenever water is pumped from a ditch or stream, a screen or strainer should be provided at the intake. It may be necessary to install a screening and desilting chamber or sediment trap under some situations. On the contrary, water from a well is usually free of debris, but in some cases it may contain sand. Debris or moss may clog or plug the nozzles of the sprinklers, requiring additional labor for cleaning. The water should be analyzed to make sure that it is suitable for irrigation. Water high in total salts, for example, may be injurious to some plants. Also important are the amounts of chlorides, the ratio of sodium to calcium and boron.

System Capacity of Sprinkler

The capacity of the sprinkler system is the flow rate needed to adequately irrigate an area and is expressed in liters per minute per hectare. The system capacity depends upon the peak crop water requirement during the growing season; effective crop rooting depth; texture and infiltration rate of the soil; the available water holding capacity of the soil; pumping capacity of the well. A sprinkler system must be designed to apply water uniformly without runoff or erosion. The capacity of the sprinkler system is based on the average peak demand. At each irrigation, the gross depth to be applied can be calculated as:

$$Da = f \times TAW/Ea \qquad (20.8)$$

Where Da is the gross average water application, mm, f is the management allowed deficit (MAD) or allowable soil moisture depletion expressed as a fraction, TAW is the total available soil moisture in the root zone, mm; and Ea is the application efficiency expressed as a fraction.

The frequency with which this depth must be applied is given as:

$$Ii = f \times TAW/Et \qquad (20.9)$$

Here Ii is the irrigation interval in days and Et is the design ET rate, mm/d.

For stationary sprinkler systems, the sprinkler application rate can be determined by:

$$d = \frac{Da \times N'}{Ii \times Td} = \frac{N' \times Et}{Ea \times Td} \qquad (20.10)$$

Such that d is the average application rate, mm/h (eq. 20.8), N′ is the number of sets per irrigation; and Td is the number of hours per day the system operates.

The total number of sprinkler in operation at one time, Ns, should be limited by:

$$Ns = Ql/ql \qquad (20.11)$$

Where Ns is the number of sprinklers, Ql is the discharge of manifold or lateral line, l/s and ql is the sprinkler discharge, l/s.

Pipeline Hydraulics and Design Equations

The flow of water in pipes is always accompanied by a loss of pressure head due to friction. The magnitude of the loss depends on the interior roughness of the pipe walls, the diameter of the pipe, the viscosity of the water, and the flow velocity. Based on experimental data, these factors have been lumped together into friction coefficient. Several equations have been developed to compute head loss in pipelines. Although this issue is beyond the scope of this course, readers may refer to Chapter 14 that introduces this issue in respect of underground pipeline irrigation.

Losses in Sprinkler Irrigation

Sprinkler irrigation is a highly efficient technology to save water and to improve crop productivity. In spite of that, water losses can be quite high if proper care is not exercised during design, operation and maintenance of the system. In areas where wind velocities and temperatures are high, evaporation losses from the sprinkler sprays can significantly reduce the expected savings of water (Davoren, 1995). The range and typical values of water losses in a sprinkler irrigation have been estimated (Table 20.7).

Table 20.7: Losses in Spray Irrigation Systems

Loss Component	*Range (Per cent)*	*Typical Values (Per cent)*
Leaking pipes	0-10	0-1
Evaporation in the air	0-10	<3
Wind drift	0-20	<5
Interception	0-10	<5
Surface runoff	0-10	<2
Excessive/uneven application depth and rates	5-80	5-30

Fertigation/Chemigation

Fertilizers are applied to the crops in one of the following ways:

Broadcasting: Uniform distribution over the whole cropped field often manually.

Placement: Application in bands or in pockets near the plants or plant rows using fertilizer-cum seed drills.

Foliar Application: In foliar application, fertilizers are dissolved in water and such diluted solutions are sprayed directly on the plants foliage. Hand operated sprayers are used for small holdings. On individual farms, tractor drawn low volume sprayers are used. On large areas, aircrafts can be used for foliar spraying.

The method of application is chosen on the basis of the characteristics of a particular nutrient, the crop, as well as methods of cultivation and irrigation. For pressurized irrigation systems, one of the options is fertigation.

Fertigation

Fertigation through sprinkler irrigation is a combination of foliar application and soil application as only a part of the material is retained on the foliage, remaining going to soil for absorption through the plant roots. Thus, some of the advantages of foliar application can be achieved through fertigation with sprinkler irrigation.

- To correct nutrient deficiencies of nitrogen, phosphorus, potassium, calcium, and magnesium
- To prevent nutritional disorders of micronutrients in fruit trees
- To improve the appearance of foliage and colour, size and quality of fruits

It may be noted that some fertilizers can be easily applied through the sprinkler system, certain ones present definite problems while others cannot be used. It depends on the following two conditions.

- The compound must be soluble in water or available in liquid form
- It should not be corrosive to the metals of the system, including the pump if the solution has to pass through the pump.

Fertigation is usually carried out using one of the following practices.

- Continuous application during the irrigation
- Fertigation only during the middle half of irrigation, also known as 25:50:25 rule

The latter option has the following advantages:

- The first quarter of irrigation ensures that the system is operating well at desired pressure and with desired uniformity of water application
- By withholding the injection during the last quarter, the chemical injected is flushed out of the system so that it does not cause corrosion or otherwise damage the fittings *etc.*

For more details on fertigation, readers may refer to Chapter 21 on drip Irrigation.

Limitations of Fertigation in Sprinkler Irrigation

Fertigation is a common practice under drip system. In the drip irrigation, it makes sense as the fertilizer is not sprayed into the air but supplied directly to the soil or plant roots. Therefore, it may be appropriate to understand the limitations of fertigation in sprinkler irrigation.

- ✰ Since most sprinklers produce mist, the chemical can easily drift to the adjacent areas. If appropriate care is not exercised, it can contaminate the environment and cause sickness to the people. Thus, fertigation needs to be conducted only under strict supervision.
- ✰ Because the fertilizers or chemical can be highly hazardous, a backflow preventer is needed to avoid contamination of water resource. It adds to the cost of the system.
- ✰ If runoff occurs especially in residential sprinkler systems, it can cause environmental problems as the runoff will move into the storm-water drainage system.
- ✰ Nitrogen, the primary nutrient can easily be volatilized (evaporated) into the air. It means some of the nitrogen would be lost in the air.
- ✰ If the fertilizer is not evenly mixed into the water, high concentration can cause burning of the foliage.
- ✰ The safe concentration of the fertilizer material in the solution is low such that repeated applications will be required to supply the needs of the plant. It may be especially true of nitrogen, phosphorus, and potassium.

Construction, Operation and Maintenance

Even a proper design of a sprinkler system is no guarantee to long-term good performance or trouble free operations. Improper maintenance of the systems may greatly impact the system performance. Besides, a careless approach might result in wastage of natural resources, cause water pollution, impact public health, jeopardize operator safety and inflate the costs. Therefore, appropriate precautions must be exercised during construction, operation and maintenance of the sprinkler systems.

Construction and Operation

To install pipes beneath the ground, trenches are made either manually or with the help of a JCB machine or a trencher. Trenching machines are easier and faster alternative. It may be prudent to verify all underground utilities before trenching. The trenches should be 20-30 cm deep. Begin with the main trench adding the branch lines later. Although it will require more work, a wider trench will make the job easier by providing more room to work with fittings. If necessary, soften the soil by irrigating the fields approximately two days before the digging operation.

It should be ensured that all the components beginning from prime mover and pump to the last sprinkler is laid in a scientific manner and maintained thereafter. The prime mover and the pump should be in alignment. The drive shaft and the pump shaft should lie at nearly the same height to prevent too great an angle on the universal shaft. While laying the main and lateral pipes, always begin at the pump. This ensures the correct connection of all quick coupling pipes. While joining couplings, ensure that both the couplings and the rubber seal rings are clean.

While operating the sprinkler system, the motor or engine should be started with the valves closed. The pump must attain the stated pressure. If not, then there may be some fault in the suction line. After the pump reaches the desired pressure, the delivery valve is slowly opened. Similarly, the delivery valve is closed after stopping the pumping unit. The pipes and sprinkler-lines are shifted as per requirement after stopping the pump. Dismantling the system should follow the reverse order (from that of assembling) *i.e.* begin dismantling from the far end from the pump.

Maintenance

In general, check all equipment at the end of the season and make any repairs and adjustments. Order the spare parts immediately, if required so that the equipment is in perfect condition for the next season. Follow general principles of maintenance of the pipes and fittings and sprinkler heads.

Pipes and Fittings

The pipes and fittings require virtually no maintenance, However, due attention must be given to occasionally clean any dirt or sand out of the groove in the coupler in which rubber sealing ring are fitted. Any accumulation of dirt or sand will affect the performance of the rubber sealing ring. Keep all nuts and bolts tight. Do not dump/store pipes on new damp concrete or on piles of fertilizer. Do not lay fertilizer sacks on the pipe.

Sprinkler Heads

The sprinkler heads being the heart of the system needs special care. Do not apply oil, grease or any lubricant to the sprinklers. Since these are water lubricated, any use of oil, grease or any other lubricants may affect their working. When moving sprinkler lines, make sure that the sprinklers are not damaged or pushed into the soil. Check the washers for wear once a season or every six months. It is especially important when water contains some sand as in the case of canal water. Replace the worn-out washers. After several seasons, the swing arm spring may need tightening. To increase the spring tension, pull out the spring end at the top and bend it.

Storage

The following points are to be observed while storing the sprinkler equipment during off season:

- Remove the sprinklers and store in a cool, dry place

- ☆ Remove the rubber sealing rings from the couplers and fittings and store them in a cool, dark place
- ☆ The pipes can be stored outdoors and should be placed in racks with one end higher than the other. Do not store pipes along with fertilizers
- ☆ Disconnect the suction and delivery pipe-work from the pump and pour in a small quantity of medium grade oil. Rotate the pump for a few minutes. Blank the suction and delivery branches. This will prevent the pump from rusting. Grease the shaft
- ☆ Protect the electric motor from dust, dampness and rodents.

Trouble Shooting

General guidelines to identify and remove some common troubles in the sprinkler systems are as follows:

Pump does not Prime or Develop Pressure

- ☆ Check that the suction lift is within the limits. If not, lower the pump so that it is closer to the water surface.
- ☆ Check the suction pipeline and all connections for air leaks. All connections and flanges should be air tight.
- ☆ Check the strainer on the foot valve for any blockage.
- ☆ Check that the flap on the foot valve is free and opens fully.
- ☆ Check the pump gland (s) for air leaks. If air leaks are suspected tighten the gland(s) gently. If necessary repack the gland(s) using a thick grease.
- ☆ Check that the gate valve on the delivery pipe is fully closed during priming and opens fully when the pump is running.

Sprinklers do not Turn

- ☆ Check pressure
- ☆ Check that the nozzle is not blocked. Preferably unscrew the nozzle or use a small soft piece of wood to clear the blockage. Do not use a piece of wire or metal as it could damage the nozzle.
- ☆ Check the condition of the washers at the bottom of the bearing and replace them if worn or damaged.
- ☆ Check that the swing arm moves freely and that the spoon which moves into the water stream is not bent. Compare it with a sprinkler which is operating correctly.
- ☆ Adjust the swing arm spring tension. Usually it should not be necessary to pull up the spring for more than 6 mm.

Leakage from Coupler or Fittings

The sealing rings in the couplers and fittings are usually designed to drain the water from the pipes when the pressure is turned off. This ensures that the pipes are automatically emptied and ready to be moved. With full pressure in the system the couplers and fittings will be effectively leak-free. If leakage is noticed, check for the following:

- ✰ See that there is no accumulation of dirt or sand in the grooves of the couplers in which the sealing rings fit. Clean out any dirt or sand and refit the sealing ring.
- ✰ The end of the pipe going inside the coupler is smooth, clean and not distorted.
- ✰ In the case of fittings such as bends, tees and reducers, ensure that the fitting are properly connected into the coupler.

Performance Studies in India

The performance of sprinkler irrigation has been widely tested in India over various agro-climatic and socio-economic endowments. It has been tested both under the canal as well as tube well irrigated lands. This section highlights only few experimental evidences to show that the system is technically feasible and socially and economically viable.

INCID (1998) made a comprehensive analysis of the data sets available on sprinkler studies in India and prepared a Table to illustrate the benefits of sprinkler irrigation over the surface methods of irrigation. In a recent compilation Planning Commission (2009) observed that water saving in various crops ranges from 16 per cent to 69 per cent and increase in crop yield from 3 per cent to 57 per cent over the traditional method (Table 20.8).

Table 20.8: Water Saving and Yield Increase with Sprinkler than Traditional Methods of Irrigation

Crops	*Water Saving (per cent)*	*Yield Increase (per cent)*
Pearl millet	56	19
Barley	56	16
Okra	28	23
Cabbage	40	3
Cauliflower	35	12
Chilies	33	24
Cotton	36	50
Cowpea	19	3
Fenugreek	29	35
Garlic	28	6
Gram	69	57
Ground nut	20	40
Sorghum	55	34
Lucerne	16	27

Crops	*Water Saving (per cent)*	*Yield Increase (per cent)*
Maize	41	36
Onion	33	23
Sunflower	33	20
Wheat	35	24

Source: Source: MicroIrrigation Division of Ministry of Agriculture, GOI as reported in Planning Commission, Government of India (2009)

At Tamil Nadu Agricultural University, water saving of 25 to 50 per cent has been worked out over the conventional surface irrigation methods (Table 20.9).

Table 20.9: Water Requirement of Field Crops with Sprinkler Irrigation System

Crop	*Actual Water Used (mm)*	*Water Saving over Surface Irrigation (per cent)*	*WUE (kg/ha-mm)*
Groundnut	390	24.7	5.13
Cotton	308	50.5	9.8
Soybean	380	50.0	4.77
Black gram	140	50.0	8.82
Lucerne	124	-	63.0

http:/www.tnau.ac.in/aecricbe/aetc/swc2.htm opened on 01.01.2014

Table 20.10: Effect of different Methods of Irrigation on Production and Water Use Efficiency

Methods of Irrigation	*Yield (t/ha)*	*Water Use Efficiency (kg/ha-cm)*
Groundnut		
Border irrigation	2.31	25.85
Check (basin)	2.38	26.45
Sprinkler	2.89	46.80
Chilly		
Furrow irrigation	1.89	45.03
Sprinkler	2.52	81.57

Comparisons have also been made between various methods of irrigation showing that water use efficiency is higher in sprinkler over border, check or furrow irrigation (Singh, 2002, Table 20.10). Similar comparison of surface, sprinkler and drip irrigation methods revealed that the efficiencies are in the order of drip> sprinkler> surface although the water use efficiency in one case was higher in sprinkler than drip irrigation (Table 20.11).

Table 20.11: Water Use Efficiency under different Irrigation Systems

Irrigation System	*Seed Cotton Yield (t/ha)*	*Water Applied (ha- cm)*	*Water Use Efficiency (kg/ha-cm)*
Drip	1.21	13.63	88.50
Sprinklerv[x]	1.15	11.81	96.90
Furrow	1.20	19.72	77.90

[x]At 0.6 IW/CPE (60 mm of water at 100 CPE); Source: Bhaskar *et al.* (2005)

Unit Cost of the System

Cost of system to an irrigator is calculated in a number of different ways depending upon the source of funding and the purpose for which economic analysis is being made. In this section, we directly list the cost of various variants of sprinkler irrigation (Table 20.12 and Table 20.13) as reported in the Government of India report (Ministry of Agriculture and Farmer´s Welfare, 2017). The indicative bill of quantities (BoQ) for portable and rain-gun systems are provided in Table 20.14 and Table 20.15 for illustration purposes. These can be used to calculate the market cost of the system at any time by multiplying with the cost of the items.

Table 20.12: Indicative Cost of Portable Sprinkler and Rain-Gun Irrigation System (Rs)

Area	*Portable, Pipe Diameter (mm)*			*Area (ha)*	*Rain Gun, Pipe Diameter (mm)*		
	63	*75*	*63*		*63*	*75*	*90*
Up to 1 ha	19542	21901	0	1	28681	34513	-
Up to 2 ha	28213	31372	0	2	-	43786	-
Up to 3 ha	-	-	42345	3	-	-	56818
Up to 4 ha	-	-	53404	4	-	-	65856
Up to 5 ha	-	-	60459	5	-	-	72322

Table 20.13: Indicative Cost of Micro-Sprinkler and Mini-Sprinkler Irrigation System (Rs)

Area (ha)	*Semi-Permanent System*	*Micro-Sprinkler, Spacing (m x m)*		*Mini-Sprinkler, Spacing (m x m)*	
		5 x 5	*3 x 3*	*10 x 10*	*8 x 8*
0.4	22557	29613	34637	41363	43023
1	36607	58932	67221	85212	94028
2	69804	103581	121138	160013	170118
3	94218	149305	172968	242982	263361
4	120392	201612	238845	312752	344013
5	146053	254872	290995	383123	425355

Table 20.14: Indicative Bill of Quantities (BOQ) for Portable Sprinkler Irrigation System

Sl.No.	Components/Area (ha)	Unit	Area (ha)					
			0.4	1	2	3	4	5
	With 63 mm coupler							
1	HDPE pipes with quick coupled (pipe of Class II; 3.2 kg/cm² IS:14151 Part II 63 mm diameter and 6 m long)	No	18	30	41	-	-	-
2	Quick coupled HDPE 63 mm foot batten assembly	No	3	5	9	-	-	-
3	GI riser pipe 3/4" diameter × 75 cm long	No	3	5	9	-	-	-
4	Sprinkler assembly	No	3	5	9	-	-	-
5	Quick coupled HDPE bend with coupler 900 (63/50 mm)	No	1	1	1	-	-	-
6	Quick coupled HDPE pump connecting nipple 63 mm	No	1	1	1	-	-	-
7	Quick coupled HDPE end plug (63 mm)	No	1	2	2	-	-	-
8	Quick coupled HDPE Tee with coupler (63mm)	No	1	1	1	-	-	-
	With 75 mm coupler							
1	HDPE pipes with quick coupled (pipe of Class I; 2.5 kg/cm² IS:14151 Part II, 75 mm diameter and 6 m long)	No	NA	30	41	NA	NA	NA
2	Quick coupled HDPE 75mm foot batten assembly	No	NA	5	9	NA	NA	NA
3	GI riser pipe 3/4" diameter × 75 cm long	No	NA	5	9	NA	NA	NA
4	Sprinkler nozzles (1.7 to 2.8 kg/cm²); IS 12232 Part I brass	No	NA	5	9	NA	NA	NA
5	Quick coupled HDPE bend with coupler 900 (75 mm)	No	NA	1	1	NA	NA	NA
6	Quick coupled HDPE pump connecting nipple, 75 mm	No	NA	1	1	NA	NA	NA
7	Quick coupled HDPE end plug (75 mm)	No	NA	2	2	NA	NA	NA
8	Quick coupled HDPE Tee with coupler (75 mm)	No	NA	1	1	NA	NA	NA
	With 90 mm coupler							
1	HDPE pipes with quick coupled (pipe of Class I; 2.5 kg/cm²; IS:14151 Part II, 90 mm diameter and 6m long)	No	NA	NA	NA	41	52	58
2	Quick coupled HDPE 90 mm foot batten assembly	No	NA	NA	NA	11	14	16
3	GI riser pipe 3/4" diameter × 75 cm long	No	NA	NA	NA	11	14	16
4	Sprinkler nozzles (1.7 to 2.8 kg/cm²) ;IS 12232 Part I brass	No	NA	NA	NA	11	14	16
5	Quick coupled HDPE bend with coupler 900 (90 mm)	No	NA	NA	NA	2	2	4
6	Quick coupled HDPE pump connecting nipple, 90 mm	No	NA	NA	NA	1	1	1

Sl.No.	Components/Area (ha)	Unit	Area (ha)					
			0.4	1	2	3	4	5
7	Quick coupled HDPE end plug (90 mm)	No	NA	NA	NA	2	2	2
8	Quick coupled HDPE Tee with coupler (90 mm)	No	NA	NA	NA	1	1	2

No: Number. NA: Not applicable.

Table 20.15: Indicative Bill of Quantities (BoQ) for Rain Gun Sprinkler Irrigation System

Sl.No.	Components/Area (ha)	Area (ha)				
		1	2	3	4	5
	With 63 mm coupler					
1	HDPE pipes with quick coupled (pipe of Class 3; 4 kg/cm^2 IS:14151 Part II 63 mm diameter and 6 m long)	30	NA	NA	NA	NA
2	Rain gun sprinkler 1.25" female threaded connection	1	N.A	N.A	N.A	N.A
3	Tripod stand with adapter to feeder line 1.25"x1.5 m	1	N.A	N.A	N.A	N.A
4	Quick coupled HDPE bend with coupler 900 (63/50 mm)	1	N.A	N.A	N.A	N.A
5	Quick coupled HDPE pump connecting nipple 63 mm	1	N.A	N.A	N.A	N.A
6	Quick coupled HDPE end plug (63 mm)	1	N.A	N.A	N.A	N.A
7	Quick coupled HDPE Tee with coupler (63mm)	1	N.A	N.A	N.A	N.A
8	Screen filter 20/25 m^3/hr	1	N.A	N.A	N.A	N.A
9	By-pass assembly - 2"x1.5"	1	N.A	N.A	N.A	N.A
	With 75 mm coupler					
1	HDPE Pipes with quick coupled (Pipe of Class 3; 4 kg/cm^2 IS:14151 Part II, 75 mm diameter and 6 m long)	30	42	NA	NA	NA
2	Rain gun sprinkler 1.25" female threaded connection	1	1	N.A	N.A	N.A
3	Tripod Stand with adaptor to feeder line 1.25"x1.5 m	1	1	N.A	N.A	N.A
4	Quick coupled HDPE bend with coupler 900 (75 mm)	1	1	N.A	N.A	N.A
5	Quick coupled HDPE pump connecting nipple, 75 mm	1	1	N.A	N.A	N.A
6	Quick coupled HDPE end plug (75 mm)	1	1	N.A	N.A	N.A
7	Quick coupled HDPE Tee with coupler (75 mm)	1	1	N.A	N.A	N.A
8	Screen filter 20/25 m^3/hr	1	1	N.A	N.A	N.A
9	By-pass assembly - 2"x1.5"	1	1	N.A	N.A	N.A
	With 90 mm coupler					
1	HDPE pipes with quick coupled (pipe of Class 3; 4 kg/cm^2; IS:14151 Part II, 90 mm diameter and 6m long)	NA	NA	45	52	60
2	Rain gun sprinkler 1.5" female threaded connection	N.A	N.A	1	1	1
3	Tripod stand with adapter to feeder line 1.5"x1.5 m	N.A	N.A	1	1	1
4	Quick coupled HDPE bend with coupler 900 (90 mm)	N.A	N.A	1	1	1
5	Quick coupled HDPE pump connecting nipple, 90 mm	N.A	N.A	1	1	1
6	Quick coupled HDPE end plug (90 mm)	N.A	N.A	1	1	1
7	Quick coupled HDPE Tee with coupler (90 mm)	N.A	N.A	1	1	1

Sl.No.	*Components/Area (ha)*	*Area (ha)*				
		1	*2*	*3*	*4*	*5*
8	Screen filter 30 m^3/hr	N.A	N.A	0	1	1
9	Screen filter 20/25 m^3/hr	N.A	N.A	1	0	0
10	By-pass assembly - 2"x1.5"	N.A	N.A	1	0	0
11	By-pass assembly - 2.5"x2"	N.A	N.A	0	1	1

Calculating Cost and Benefit for Investment Recovery

In general 3 methods described in the following sections are used for this purpose.

Equated Seasonal Installment

Borrowed capital to install sprinkler/drip irrigation along with interest on the borrowed sum needs to be paid back by the stakeholder in a specified period of time. The equated seasonal/annual installment (ESI) in such cases is calculated by the following formula.

$$ESI = (L \times I) \times [(1+I)^{M}/\{(1+I)^{M-1}\}] \quad (20.12)$$

Here, M is the loan period in seasons (Years × 2), I is the interest rate (Interest rate per annum/2)/100, L is the borrowed capital in Rs.

For a financially viable project, the seasonal/annual benefits from the system must exceed the seasonal/annual installment. For example; for a soft loan of Rs. 70,000 @ 6 per cent per annum, ESI for 10 years period will be Rs. 8206/-. Thus, the farmers must earn at least this much amount per season over and above the current income to repay the loan installment.

Seasonal Cost

The cost of a sprinkle/drip irrigation system can be calculated as:

Fixed costs, primarily dependent upon the life of the system and includes depreciation, interest, taxes, insurance and housing.

Variable costs, primarily comprising of the operation, maintenance and labour costs

Net additional profit of the farmer must be higher than the total cost per season/annum which is the sum of the fixed and variable cost.

Fixed Cost

Depreciation: Assuming that the value of the system reduces linearly over its economic life then:

$$\text{Average annual depreciation, } D = (P - S)/L \quad (20.13)$$

Here D is the depreciation for one year, L is the life of machine, P is the purchase price and S is the salvage value *i.e.* the value of the system after its life period.

Usually the life of a pump/motor *etc.* is taken as 10-15 years while pipes and nozzles are considered to last for about 8-10 years. Thus, the cost break-up must be known to arrive at the depreciation component. The salvage value of the pipes *etc.* may be taken as nil while that of the other components may be taken as 10-20 per cent of the initial purchase value. Thus eq. (20.13) can be written as:

$$\text{Average Annual Depreciation, } D = (P1 - S1)/L1 + (P2 - S2)/L2 \qquad (20.14)$$

Here P1 is the fraction of cost of components that have longer life L1 and P2 is the fraction of cost of components that have lesser life, L2. Salvage value S2 may be taken as zero.

Interest: The interest on investment is an expensive item of cost. It is charged for the use of funds to implement sprinkler/drip irrigation. A nominal interest rate is often used which is decided on the basis of interest rate on fixed deposits offered by the banks. Inflation rate is often neglected. When straight-line method of depreciation is used then

$$\text{Interest per annum, } I = \text{Nominal interest rate} \times 100/100\ AI$$

$$\text{Here } AI = \text{Average investment} = 0.5\,(P+S) \qquad (20.15)$$

Sometimes a value of 0.6 is used instead of 0.5 to calculate the AI. Taxes, Insurance, and shelter (H) may not be directly incurred in sprinkle/drip irrigation but should be charged. In the event of not carrying insurance, loss risk is assigned a value and those not providing shelter for pump *etc.* are likely to incur higher repair bills. Annual costs of insurance/taxes and housing are estimated to be 3 per cent of the initial investment of the sprinkler/drip irrigation.

Variable Cost

The major cost component in sprinkle/drip irrigation is the energy cost required to pump the water and includes both the fuel and lubrication (O). The next major component is due to maintenance that includes repairs and replacement of pipes and nozzles/emitters (M). Labour component sometimes is included in the operational cost while in some cases; it is accounted separately (L). The actual costs have to be assessed based on the pump HP and duration of pumping during the year. The maintenance cost is usually taken as 15-20 per cent of the initial purchase price. If all the costs are calculated on annual basis, then seasonal cost is calculated by:

$$\text{Total cost} = (D + I + H + O + M + L)/\text{Number of seasons} \qquad (20.16)$$

For sprinkle/drip irrigation to be a viable alternative, the net benefits over and above the normal cultivation practice must exceed the total cost calculated through eq. (20.16).

Or

One may also calculate the benefit cost ratio (B-C ratio) for each season as follows:

$$\text{B-C ratio} = \text{Gross income} / (\text{Total cost per season} + \text{Cultivation cost}) \quad (20.17)$$

Evaluation of Economic Parameters

Discounting

The costs and benefits of a sprinkle/drip irrigation project are incurred/received during different years of the project period. On the other hand, present worth of costs and benefits received say 10 years later is much less than received today. In other words, amount available today is worth more than the same sum of money received 10 years later. Such variation is taken into account by following the discounted cash flow method of analysis. The discount factors (DF) are calculated for a given discount rate as per the following relation:

$$DF = 1/(1 + Dr/100)^t \quad (20.18)$$

Where Dr is the discount rate (per cent) and t is the year say 1, 2, 3 ....15. For t=1 and Dr = 10 per cent, DF would be:

$$DF = 1/(1+0.1) = 0.909 \quad (20.19)$$

Discount rates are usually taken from a low of 6 per cent to about 12 per cent. Current bank rates on fixed deposits are good indicators of the discount rates to be adopted in economic evaluation. Since bank deposits rates fluctuate, it may be worthwhile to make the analysis using 3-4 discount rates (Less and more) to arrive at the most appropriate information. For example; for the current rate of interest of 8.5 per cent on fixed deposits, one could carry out the economic analysis for discount rates of 8.0 per cent, 10.5 per cent and 12.0 per cent. Discount factors for a discount rate of 10.5 per cent over a period of 15 years are given in Table 20.16.

Table 20.16: Discount Factors at a Discount Rate of 10.5 per cent

Year	*Discount Factor*	*Year*	*Discount Factor*
1	0.905	9	0.407
2	0.818	10	0.368
3	0.741	11	0.333
4	0.671	12	0.302
5	0.607	13	0.273
6	0.549	14	0.247
7	0.497	15	0.224
8	0.450		

Evaluation of Indices

For detailed economic analysis, economic parameters such as B-C ratio, internal rate of return, net present worth and payback period are calculated in the following manner.

Net Present Worth (NPW)

It is the difference between the present values of benefits over the present values of costs. A positive value indicates the worthiness of the project.

Benefit-Cost Ratio (BC ratio)

It is the ratio of the present worth of benefits divided by the present worth of costs. For an economically viable project, the ratio should be more than one.

Payback Period (PBP)

It is the time period for an investment to generate sufficient incremental cash to recover its initial capital outlay in full. Thus, the PBP measures number of years it will take for the net undiscounted benefits to repay the investment. If the cash flow is uniform over the years, the payback period, P is given by

$$P = (I/E) \tag{20.20}$$

Where I is the initial investments and E is the net annual benefits.

Internal Rate of Return (IRR)

IRR is used to find out rate of return which a project is likely to earn over its useful life. IRR is the discount rate which makes NPW zero and B-C ratio one. Thus, if the NPW of a project is positive at a given discount rate, its IRR would be more than the assumed discount rate. On the other hand, if the NPW is negative, the IRR would be less than the assumed discount rate.

IRR is compared with the minimum acceptable rate of return (say prevailing market rate). If the IRR is higher or equal to the minimum acceptable rate of return, the project is worthy of investment. In most irrigation/drainage projects, all the three indices IRR, NPW and B-C ratio are determined to assess the economic viability of the project.

The cost of several systems along with the cost-benefit ratio and the payback period for several crops is given in Table 20.17. While the cost-benefit ratio varied from 1.09 to a high of 5.16, the payback period range from1 to 3 years (Table 20.17)

New Innovations

Mini Sprinkler

A mini-sprinkler has smooth streamlined side ribs to secure the spinner and gives minimum obstruction to water jet. As such, it results in uniform and gentle precipitation, causing no damage to delicate items such as flowers/plants. These are easy to install having different mounting options, have wide flow range, 16-110 l/h at operating pressure of 1 kg/cm^2. These are most suited to under-foliage irrigation or for soils where adequate wetting is difficult to achieve. These are also used where water is available only for short periods. Mini-sprinklers are most appropriate for landscape, flower beds, shrubs, nurseries and/or large trunk trees like mango.

Table 20.17: Cost Benefits and Payback Periods of Micro-irrigation for Various Crops

Crops	*Spacing of Crops m*	*Cost of the System (Rs./ha)*	*Payback Period (Years)*	*B:C Ratio*
Banana	0.91 × 1.5 × 1.8 paired row	45,000	1-2	3.00
Grape	3.03 × 1.8	44,000	1	3.28
Pomegranate	4.3 × 4.3	30,000	2	5.16
Ber	4.5 × 4.5	30,000	1	4.56
Tomato	0.45 × 0.45 × 1.65 paired row	30,000	1	1.09
Papaya	1.81 × 1.81	40,000	2	4.09
Cotton	0.9 × 1.5 × 1.8 paired row	47,500	3	1.83
Sugarcane	0.83 × 1.66 paired row	47,500	1 –2	3.45

Jets

Jets like J-jets/Foggers/Misters operate at nominal operating pressure of about 1 kg/cm^2 and have a fan type spray jet with fine droplets resulting in uniform distribution. These can easily be installed with different mounting options like, stake, clip stake, rigid risers and 4 mm stake. These are useful for irrigation in orchards, nurseries, vineyards, green houses, delicate plants such as flowers, vanilla *etc.* and mature large trunk-trees or trees having wide spread root zone. These are also used to maintain optimum micro-environment in the canopy area.

Fogger

Foggers, having a very simple construction with no moving parts, are designed for misty spray of very fine droplets. Designed to operate at an operating pressure of about 3.0 kg/cm^2, these can be easily disassembled and cleaned. The foggers are recommended for orchards and green/glass houses requiring a fine mist spray for humidity control and/or crops that require optimum micro-climate in the canopy area. The maintenance cost of foggers is quite minimal.

Constraints in Spread of Sprinkler Technology

Sprinkler irrigation has been widely tested and adopted in India and around the globe. However, its application has not picked up the desired pace although currently nearly 4.5 M ha is estimated to be under drip irrigation. Following factors seems to have been inhibiting its wide application and adoptability by the farmers.

- High initial cost
- Lack of package of practices for irrigating various crops
- Lack of awareness
- Lack of social concern to save natural resources
- High water pressure required in sprinkler (>2.5kg/cm^2).

Chapter 21

Drip Irrigation and Fertigation

Introduction

Per capita land and water resources are decreasing at a very fast rate. Per capita land resource has decreased from 0.33 ha in 1951 to 0.13 ha in 2001. Similarly, per capita water availability assessed at more than 5300 m^3 in 1951 is likely to be less than 1500 m^3 by the year 2025. Besides, in the wake of resource exploiting green revolution, a degradation process has also set in as a result of over-exploitation of the resources. Therefore, to conserve water resource and to utilize degraded land and water resources in a most productive manner, improved irrigation techniques including drip irrigation will play a very prominent role.

Drip irrigation is a low volume, low pressure and low flow rate method of applying water to a limited cultivated area. It is characterized by water application at slower rate over a period of time as per water requirement, at frequent intervals and directly to the plant root zone. Drip irrigation is most suited to fruits trees, vegetables crops, and vine crops although currently its application has extended over a wide range of row crops. Because of the high capital cost of drip systems, high value crops are preferred when a switch over is made to drip irrigation. Drip irrigation in its present form has become compatible with plastics that are durable and easily moulded into a variety and complexity of shapes required for pipe and emitters. Drip irrigation is most suited to areas having the following problems or characteristics.

- ☆ Areas with limited water or where water is brought/pumped at high cost
- ☆ Where topography is uneven and levelling or land shaping for surface irrigation is difficult/costly especially in hilly terrain
- ☆ Areas used for commercial cultivation of cash or horticultural crops

- ☆ Where irrigation water is of poor quality such as saline/alkali or waste water
- ☆ Where soil physical properties hinder effective surface irrigation as in soils with low or high infiltration rates
- ☆ Where crops in sequence are amenable for planting in such a way so that drip layout need not be disturbed seasonally.

Scope and Extent

According to Sivanappan (1999), drip irrigation can be applied to an area of 28.5 M ha. Task Force on Micro-Irrigation constituted by Government of India also estimated that potential area for drip irrigation is around 27 M ha. To bridge the gap between the actual achievements at the end of 20th century and the potential, Government policies and programmes were tailored to encourage farmers to adopt drip irrigation.

Research experiments on drip irrigation in India were initiated in the early seventies in many State Agricultural Universities (SAUs) and research organizations. The area under drip irrigation increased from about Nil in 1970 to 355,000 ha in 2002 (Praveen Rao, 2002, Kumar and Singh, 2002). Current estimate (As on 31.03.2015) is that nearly 3,371,597 ha area is under drip irrigation (Table 21.1) and about 4,357,215 ha under sprinkler irrigation making a total of 7,728,812 ha under micro-irrigation (www.midh.gov.in/AtGlance/MI-AT-A-Glance.docx). Total area under micro-irrigation increased to 8,146,172 ha as on 8.03.2016 showing an increase of 0.42 M ha in about a year (http:/www.cicr.org.in/pdf/press/11-3-2016.pdf). India now leads the world with respect to area coverage under micro-irrigation. The largest fully automated community drip irrigation project in Asia or even in the world, the Ramthal project, in Karnataka covers about 24,000 ha across 55 villages with 14,660 beneficiary farmers. A total of 2,150 km length of pipeline was laid for the project costing Rs. 3815 million. No doubt, significant improvement in coverage has been made yet the area covered so far is much less than the potential area that can be brought under drip irrigation. With research results projected to show that application of drip irrigation to rice and wheat crops is possible, the potential area under drip will go for an upward revision. Therefore, we not only need to maintain the current pace but even upscale our efforts so as to cover all the potential areas in the shortest possible time.

Merits and Constraints

Drip irrigation offers many advantages; some of them are listed as under.

- ☆ Uniform application of water at low application rates with high application efficiency reaching as high as 90 per cent
- ☆ Enhanced plant growth, increased yield (40 to 50 per cent of the yield under conventional irrigation) and higher benefits
- ☆ Uniform and improved quality of produce
- ☆ Potential for use of saline/alkali waters including wastewaters

Table 21.1: State-wise Area Covered under Micro-irrigation (Drip and Sprinkler) as on 31-3-2015 (ha)

Sl.No.	State	Drip	Sprinkler	Total
1	Andhra Pradesh	834865	328441	1163306
2	Arunachal Pradesh	613	0	613
3	Assam	310	129	439
4	Bihar	4610	97440	102050
5	Chhattisgarh	15553	241420	256973
6	Goa	965	899	1864
7	Gujarat	411208	418165	829373
8	Haryana	22682	550458	573140
9	Himachal Pradesh	291	684	975
10	Jharkhand	6303	9919	16222
11	Karnataka	429903	417005	846907
12	Kerala	22516	6948	29464
13	Madhya Pradesh	166358	185759	352117
14	Maharashtra	896343	374783	1271125
15	Manipur	47	30	77
16	Mizoram	1727	425	2152
17	Nagaland	200	5005	5205
18	Odisha	18431	82147	100579
19	Punjab	30805	12161	42966
20	Rajasthan	170098	1514451	1684549
21	Sikkim	5544	2769	8312
22	Tamil Nadu	290009	30436	320445
23	Telangana	25299	5293	30592
24	Tripura	100	392	492
25	UP	15519	21164	36682
26	Uttarakhand	696	316	1012
27	West Bengal	604	50576	51180
28	Others	15500	31000	46500
	Grand Total	3371597	4357215	7728812

- Improved fertilizer and other chemicals use efficiency through fertigation/chemigation
- Reduced weed growth
- Reduced labour cost
- Suitable for difficult terrains, problem soils and undulating lands to minimize cost of land shaping
- Energy conservation

- ☆ Saving in land area
- ☆ Minimum damage to soil structure
- ☆ Frost protection
- ☆ Suited to complete mechanization of irrigation system
- ☆ Reduction in carbon emissions

The yield increase, water saving and water use efficiencies for various crops as an illustration are given in Table 21.2 (Saxena and Gupta, 2004).

Table 21.2: Average Crop Yield, Percentage Increase in Yield, Water Use Efficiency and Water Saving in Drip over the Conventional Irrigation System for Various Crops

Sl.No.	*Crop*	*Number of Experiments*	*Yield (t/ha)*	*Yield Increase (per cent)*	*WUE (t/ha- cm)*	*Water Saving (per cent)*
1	Acid lime	1	78.0	56.0	1.3	50.0
2	Baby corn	1	9.9	72.4	0.5	43.8
3	Banana	7	71.5	29.3	3.0	42.5
4	Bean	1	10.3	81.8	0.4	36.9
5	Beet root	1	48.9	7.0	2.8	79.0
6	Ber	3	71.0	27.7	0.7	34.3
7	Bitter gourd	4	2.7	44.4	1.4	69.5
8	Bottle gourd	1	55.8	46.8	1.0	35.7
9	Brinjal	7	16.0	44.6	1.5	42.6
10	Cabbage	5	50.5	37.5	3.2	37.4
11	Capsicum	1	22.5	66.6	0.8	43.1
12	Carrot	1	26.3	92.3	0.8	33.6
13	Castor	2	7.3	30.2	1.7	33.0
14	Cauliflower	3	19.5	39.7	0.7	37.1
15	Chickpea	1	3.8	66.6	1.6	42.6
16	Chilli	5	68.0	28.7	7.5	47.3
17	Coconut (Nos./plant)	2	181.0	7.1	6.9	50.5
18	Cotton	3	36.0	40.0	0.9	51.1
19	Cucumber	1	22.5	45.1	0.9	37.8
20	Grain Corn	1	6.5	52.9	2.2	45.0
21	Grape	5	29.9	20.9	1.0	43.0
22	Groundnut	2	3.5	62.5	1.0	32.4
23	Guava	2	25.5	63.0	3.5	9.0
24	Mango	3	19.5	80.7	2.4	28.9
25	Sweet lime, 1000 pcs	1	15.0	98.0	0.2	61.0
26	Oil Palm	1	-	-	-	21.0
27	Okra	12	20.1	20.7	1.9	44.7
28	Onion	3	17.0	42.6	1.2	36.7

Sl.No.	Crop	Number of Experi- ments	Yield (t/ha)	Yield Increase (per cent)	WUE (t/ha- cm)	Water Saving (per cent)
29	Papaya	5	56.6	72.0	0.9	68.0
30	Pomegranate, 100 pcs	3	44.7	55.7	0.5	57.3
31	Popcorn	1	5.5	75.4	2.1	42.0
32	Potato	5	28.7	50.0	2.8	24.6
33	Radish	2	17.0	27.5	5.0	64.0
34	Ridge gourd	3	17.4	14.5	4.4	43.4
35	Round gourd	1	36.6	24.0	0.5	0.0
36	Sapota	1	-	17.2	-	21.4
37	Sweet potato	1	50.0	39.0	2.0	68.0
38	Sugarcane	6	145.9	43.6	1.2	46.7
39	Sweet lime	1	15.0	50.0	2.3	61.4
40	Tapioca	2	54.6	12.6	0.6	23.4
41	Tomato	11	36.6	46.0	3.8	37.4
42	Turmeric	2	18.4	76.3	0.6	53.1
43	Watermelon	3	46.8	64.8	2.1	46.1

WUE: Water Use Efficiency; pcs: Pieces.

Constraints

Some of the constraints of drip irrigation are enumerated below.

- ✰ Technology is expensive requiring relatively more investments in the initial stages
- ✰ Need for external technical assistance to design, install and even to operate and maintain the system
- ✰ Either because of poor quality control or due to damage of lines and drippers by rodents, system life is short and that prolongs the payback period
- ✰ Risk of clogging of drippers, requires consistent efforts to control clogging
- ✰ Drip irrigation might pose problems when herbicides or top dressed fertilizers need other methods of irrigation for activation
- ✰ Since most drip systems are designed for high efficiency meaning little or no leaching fraction, salt accumulation at the wetting front is anticipated (Figure 21.1), leaching may be required to avoid salts build-up in the root zone
- ✰ Surface drip irrigation may interfere with intercultural operations like weeding, earthing-up and/or mechanical operations
- ✰ Restricted soil water distribution zone affects root development in deep rooted crops causing lodging/uprooting

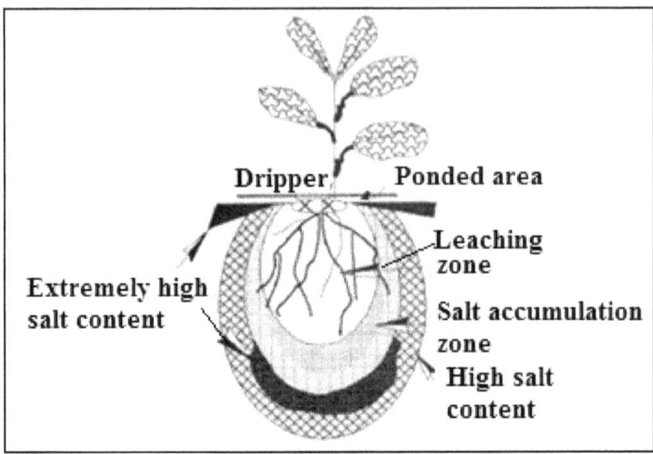

Figure 21.1: Salt Accumulation Zones at the Wetting Front.

- ☆ Rolling and re-laying cost in short duration crops and in crop sequences
- ☆ Might pose problems in canal irrigated areas due to clogging especially in run-off the river schemes.

Mechanisms of the Advantages of Drip Irrigation

Enhanced Plant Growth and Higher Yields

Vigorous plant growth and relatively higher yield in drip as compared to surface irrigation is due to fairly constant and optimal level of soil moisture in the root zone of the crops. The soil water potential remains fairly high to enable easy absorption of water by the plants. It helps to maintain high leaf water status. The soil water status in a surface, sprinkler and drip irrigation where frequency of water application differs are shown in Figure 21.2. It shows that moisture content in the drip remains fairly high as compared to other two application methods. Moreover, the water can be applied at rates and frequencies suited to a particular soil type. Since the matric potential remains high, it nullifies the adverse impact of high osmotic pressure in saline or saline water irrigated lands resulting in high yields.

Early maturity and lesser incidence of weeds, pest and diseases also enhance plant growth and results in improved quality of produce. A relatively better control of many disease causing pathogens is mainly due to modified micro-climate in drip irrigated fields.

Water Saving

Water saving in drip irrigation system is due to the following factors:

- ☆ High irrigation application efficiencies can be achieved up to 90 per cent as compared to 50-70 per cent in surface and about 80 per cent in sprinkler irrigation. It is attributed to reduced surface evaporation, reduced irrigation runoff, low or no deep percolation losses and fewer losses through transpiration by weeds.

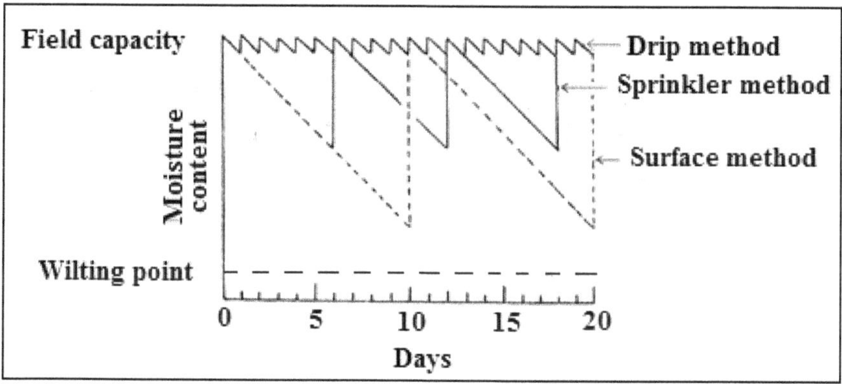

Figure 21.2: Moisture Content in the Root Zone as Affected by Irrigation Methods.

- Piped conveyance of water in the distribution network
- Irrigation to a fraction of the area, only 30-70 per cent of surface irrigated area is irrigated depending upon the crop geometry

High Nutrient Use Efficiency

In drip irrigation, it is possible to apply nutrients along with water commonly known as fertigation. Crops positively respond to nutrients applied through fertigation because their application directly into the root zone helps the crops to easily draw the nutrients for their use. Reduced leaching and other losses also help to improve nutrient use efficiency. This issue is also discussed in a forthcoming section of this chapter.

Reduced Salinity/Pathogenic Hazards

Adverse effects of salinity on the plants are reduced even when soils are saline or saline water is used for irrigation. Possible reasons are the dilution of salt concentration in the root zone by high frequency irrigation resulting in lesser variations in matric stress exerted by high frequency irrigations (Figure 21.2), lesser salt load due to reduced water application rates, movement of salts beyond active plant root zone by frequent irrigation and no contact of salt water with leaves (in drip irrigation) avoiding any scorching effect. Similar benefits from the use of waste water such as domestic sewage water and effluents from industrial units have been demonstrated. It is mainly attributed to lesser quantity of water use resulting in reduced pollutant loads, avoidance of salt injury and minimal contact of the irrigator with waste water.

Components of Drip Irrigation

The components of drip irrigation system are grouped under two heads namely head unit and field unit. A schematic diagram shows various components and their location within the system (Figure 21.3).

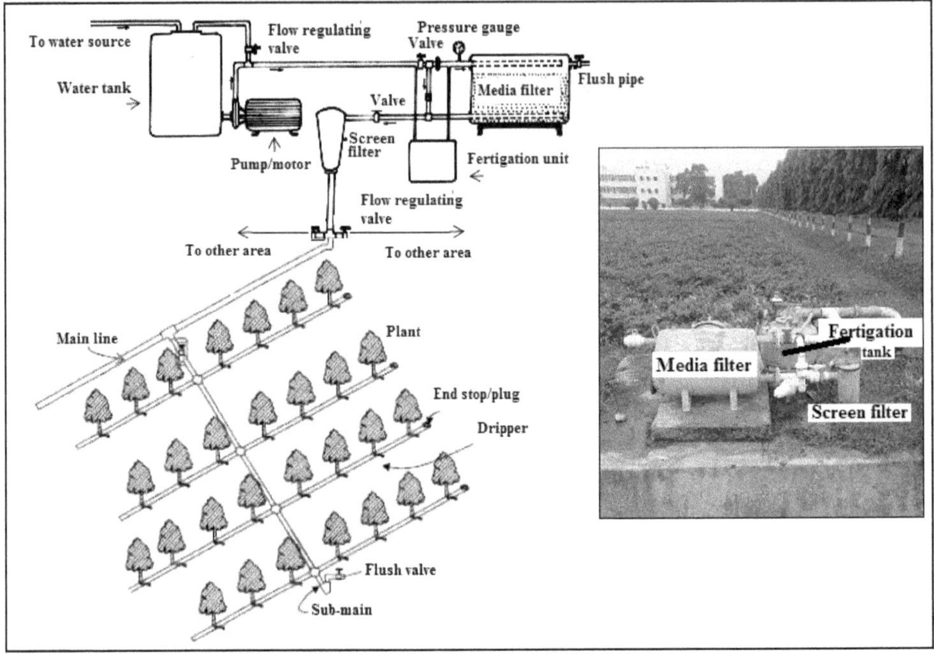

Figure 21.3: Components and their Location in a Drip Irrigation System (Line Diagram Left) and Pictorial View (Right).

Head Unit

- ☆ Water source
- ☆ Prime mover (Motor/Engine)
- ☆ Water pump
- ☆ Filters
- ☆ Fertigation/chemigation unit
- ☆ Water meter, if necessary
- ☆ Galvanized iron pipe fittings, pressure gauge
- ☆ Bypass valve

Field Unit

- ☆ Main line
- ☆ Sub-main line
- ☆ Manifolds
- ☆ Lateral line with emitters
- ☆ Air release valves
- ☆ Non-return valves

- ✰ Valves (Bypass, Flush and Air release)
- ✰ End plug

Criteria for components selection is based on topographic conditions (location, elevation, field boundaries, shape/slope, area, location of pumping unit *etc.*), soil condition (type, soil moisture holding capacity, depth, intake rate), water supply system (quantity, quality, temporal variation) and crop factors (crop, row-to-row and plant-to-plant spacing). Besides these factors cost criteria determines the kind of the system and layout plan *etc.* as both can result in cost savings.

Water Source

The water source can be a bore well, open well, subsurface tank or a reservoir. A reservoir is essentially required if the water source is a canal. The capacity of the tank is calculated from the water requirement of the crop, canal schedule, irrigation system dripper capacity, type of soil *etc.*

Pump/Overhead Tank

It is required to provide sufficient pressure in the system. Centrifugal pumps are generally used for low pressure trickle systems. Overhead tanks can be used for small areas or orchard crops with comparatively less water requirements.

Backflow Preventer

Check valves or non-return valves or backflow preventers allow water to go through it only in one direction so as to prevent unwanted flow reversal (Not shown in Figure 21.3). They are used to prevent return flow of chemicals and fertilizers from the system into the water source to avoid the contamination of the water resource. It is not necessary to have it in the system except under the following situations:

- ✰ When area is serviced by more than one water source and if it is not desirable to have mixing of the two water sources
- ✰ Required when fertilizer or chemical injectors are used
- ✰ Required when area being irrigated is at a higher level than the water source
- ✰ Commercial installations

Back-flow preventers have been categorized as follows:

1. Reduced pressure type (RP)
2. Double check type (DC)
3. Pressure vacuum breaker (PVB)
4. Anti-siphon valve (ASV)

Amongst these types, RP type backflow preventer assembly is commonly used. PVB type with standard globe valves is less expensive than anti-siphon valves.

Filters

The hazard of blocking or clogging necessitates the use of filters for efficient and trouble free operation of the micro-irrigation system. The filtration requirement depends on the size of the flow path in the emitter, quality of water in terms of total suspended solids (TSS) and flow in the mainline. Typically, all particles larger than 1/10th of the diameter of the orifice of the emitters should be removed/ filtered. For large systems, depending on water quality, different filters or their combinations are used. For large flows, filters are connected in parallel using manifolds so that pressure loss across the filters is minimal. Besides settling basin, which is quite necessary in canal water fed drip irrigation systems, four types of filters are commonly used.

1. Settling basins (If canal is a source of water)
2. Sand (Media) filter
3. Screen (Mesh) filter
4. Disc filter
5. Hydro-cyclone filter

Settling Basin

Settling basin is used to remove large volumes of sand and silt as may be expected in canal water. It is not required, if source of water supply is groundwater. Srivastava *et al.* (2010) observed that settling basin with other filters was essential to bring the turbidity of the canal water below 4 NTU. It has been found to be the above limit to minimize flow reduction due to clogging of emitters.

Gravel or Media Filter

This filter is provided as a primary filtration unit if irrigation water is drawn from an open water reservoir, canals or reservoirs likely to have heavy loads of very fine sands, suspended particles and organic impurities like algae. It is also called depth filter and is used in series with the screen filter. These filters are made of layered beds of graduated silica sand and gravel (usually 1.5 – 4.0 mm in diameter) free of calcium carbonate placed inside of one or more of mild steel/metal tanks (Figure 21.4). These are available in different size ranges as in screen filter. These filters require cleaning at regular intervals. These are easily cleaned by reversing the direction of water flow through the filter and bypassing the effluent. Pressure gauges are placed at the inlet and at the outlet ends of the filter to measure the head loss across the filter. Filter needs back washing, if the head loss exceeds 30 kPa.

Screen Filter

Screen filters, also called surface filters, are always installed after media filter as an additional safeguard against clogging. The minimum particle size retained by the screen is determined by its mesh size *i.e.* number of wires per inch (Table 21.3). These are made of plastic or metal and are used to remove fine sand or small amounts of algae. Screen filters made of cylinder screens of stainless steel or nylon are the most commonly used filters. These are available in different sizes and flow

Figure 21.4: A Pictorial View of Media Filter Tank.

rates ranging from 1 m^3/h to about 40 m^3/h. The aperture size of the screen opening should be between one seventh and one tenth of the orifice size of emission devices used. Many commercially available system require screening of 150-75 microns (100-200 mesh).

These filters are cleaned manually or by repeated washing. In large scale systems, even automatic cleaning systems are provided.

Table 21.3: Filtration Capability of different Screen Sizes

Mesh Number	*Filtration to Micron Size*	*Particle Designation*	*Equivalent Diameter (micron)*
16	1180	Coarse sand	>100
20	850	Medium sand	250-500
30	600	Very fine sand	50-250
40	425	Silt	2-50
80	175	Clay	<2
100	147	Bacteria	0.4-2.0
150	104	Virus	<0.4
170	90		
200	74		
270	53		
400	38		

Source: Thokal *et al.* (2012) Note: A 100 mesh screen will remove all particles above 147 microns.

Disk Filter

Disk filter comprises of stacks of grooved, ring shaped disks that capture the debris. These filters are quite effective in the filtration of organic materials including algae. During the filtration mode, the disks are pressed together with an angle in the alignment of two adjacent disks. It results in cavities of varying sizes. The flow through these filters is partly turbulent. The sizes of the groove determine the filtration grade (Figure 21.6). Disk filters are available in a wide range of sizes (25-400 microns). The disk filters are cleaned by back flushing, although unlike sand filters, it requires high back flushing pressure (2 to 3 kg/cm^2).

Figure 21.5: Screen Filters (Left) and Box Containing Screen Filters (Right).

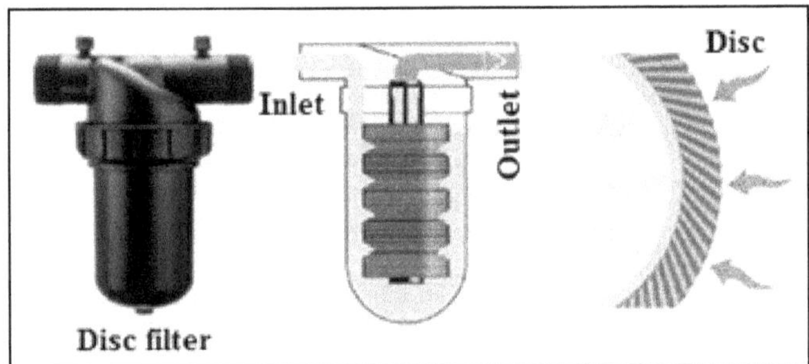

Figure 21.6: Disc Filter Partly Showing the Filtering Mechanism.

Hydro-Cyclone or Centrifugal Filter

Hydro-cyclone or centrifugal filters are made of mild steel having a conical shaped cylinder. Water is introduced tangentially at the top of a cone to create a circular motion resulting in a centrifugal force, which throws the heavy suspended particles against the walls. The separated particles are collected in a collecting vessel at the bottom. These filters effectively filter sand, fine gravel and other high density materials. These are used in conjunction with the screen filter and are placed upstream of the media, screen and disc filters.

Guidelines for Selection of Filter

The effectiveness of various filtering mechanisms discussed in the previous sections is reported in Table 21.4. Selection of filters on the basis of water quality is included in Table 21.5.

Table 21.4: Effectiveness of Various Filtering Mechanisms

Filter Type	*Size Range (micron)*
Sediment basins	>40
Slotted cartridge	>152
Sand media	5-100
Sand media	>20
Screen (100-200 mesh)	75-150
Screen (150 mesh)	>100
Screen (200 mesh)	>75
Separator	>74
Separator (2 stages)	>44

Separators remove 98 per cent of the particles larger than the size indicated.

Table 21.5: Selection of Filters as per Irrigation Water Quality

Water Quality	*Type of Filter*	*Remarks*
Good without any physical and biological impurities	Screen	Screen filters shall be used only if the physical impurity do not call for cleaning of filter element more than once a day
Water sources with heavy physical and biological impurities	Only screen filter will not be sufficient	Additional filter is required depending upon the type of water impurity
Water sources with sand and other heavier particles	Hydro cyclone separator or Hydro cyclone filter of matching flow capacity	Disc/screen filter shall be provided after hydro cyclone filter
Water sources with heavy biological impurities such as algae, trash and other debris	Media/sand filter	Disc/screen filter should be provided after media filter
Water sources with heavy sand and other biological impurities such as algae and trash	Combination of hydro cyclone followed by a sand filter	Screen/disc filter should be provided after sand filter

Distribution Network

It mainly constitutes main line, sub-main lines and laterals with drippers and other accessories. The main line made of rigid Polyvinyl Chloride (PVC) and High Density Polyethylene (HDPE) transports water within the field and distributes it to the sub-mains. Pipes of 65 mm diameter and above with a pressure rating 4 to 6 kg/cm^2 are used as main lines. Sub-mains distribute water evenly to a number of lateral lines. For sub-mains, rigid PVC, HDPE or LDPE (Low Density Polyethylene) of diameter ranging from 32 mm to 75 mm having pressure rating of 2.5 kg/cm^2 are used. Laterals distribute the water uniformly along their length by means of drippers or emitters. These are normally made of LDPE or LLDPE with wall thickness varying from 1 to 3 mm for pipes of 10, 12 and 16 mm internal diameter.

Emitters/Drippers

Emitters function as energy dissipaters, reducing the inlet pressure head (0.5 to 1.5 atm) to zero atm at the outlet. The commonly used drippers are online pressure compensating, online non-pressure compensating, in-line dripper, adjustable discharge type drippers, vortex type drippers and micro tubes of 1 to 4 mm diameter. These are manufactured from Poly- propylene or LLDPE.

Long-Path Emitters

These emitters route the water through a very long, narrow passage that circles around and around a barrel shaped core or tube. The small diameter and long length of this path reduces the water pressure and creates a more uniform flow. Long path emitters tend to be fairly large in size.

Short-Path Emitters

These are similar to the long path emitters but have a shorter and smaller water path. These are relatively cheaper and work even on very low-pressure systems such as low pressure gravity flow drip systems. Two major problems encountered are the clogging and non-uniform distribution of water. Therefore, these are applied only on small systems, where cost is a critical issue.

Tortuous-Path or Turbulent Flow Emitters

Tortuous-path and/or turbulent-flow emitters have a long path having turns and obstacles causing turbulence in the flow. It reduces both the flow and the pressure. By using the tortuous path, the emitter water passages can be of a shorter length and of larger diameter. It helps to considerably reduce the clogging problem.

Vortex Emitters

Vortex emitters use the principle of swirling the water around the outlet hole so as to cause pressure drop and reduced flow through the hole. Most vortex emitters have very small inlet and outlet holes. Although inexpensive and smaller in size, these emitters are prone to clogging.

Diaphragm Emitters or Pressure Compensating Emitters

Pressure compensating drippers, also known as PC or flow regulated drippers can self-regulate the flow-rate regardless of system pressure. These emitters are manufactured with high quality flexible rubber diaphragm or disc inside the emitter that changes shape according to operating pressure to deliver uniform discharge (Figure 21.7). In this way, pressure-compensating devices are able to deliver the designed flow rate over a fairly wide range of inlet pressures and within that range their flow rates are relatively constant. These are most suitable on slopes and difficult topographic terrains. Pressure compensating emitters are costly and the elastomeric material used in their construction has a tendency to change its properties and wear with time. Depending on the material, the elastomer may absorb water, lose elasticity or creep under prolonged stress altering the performance of the device over time.

Thus, it is extremely important that highest quality elastomeric material is used in the emitter. These emitters should be used only if absolutely necessary being costly. It may be useful to use pressure compensating emitters in undulating fields, if an elevation difference of over 1.5 m is encountered in the area being irrigated. The pressure compensating emitters often do not work at all with very low water pressure. Therefore these should not be used with very low water pressure such as gravity flow systems.

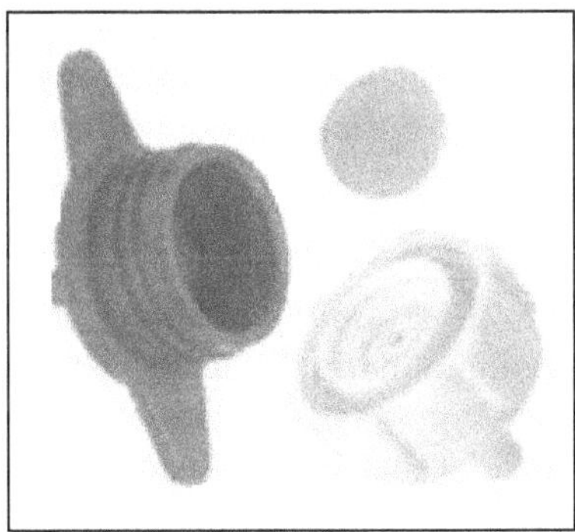

Figure 21.7: A View of a Pressure Compensating Emitter.

Adjustable Flow Emitters

Adjustable flow emitters have some control for adjusting the flow rate. Most of these are very similar in design to the short path type of emitters.

Flushing Emitters

These emitters are designed to have a flushing flow of water to clear the discharge opening every time the system is turned on. Continuous flushing emitters allow continuous passage of large solid particles during operation. Some on-line emitters and drip tape are manufactured with flushing capabilities.

Multi-Exit Emitters

Some on-line emitters are capable of supplying water to two or more points through small diameter auxiliary tubing. These are used in orchards where each tree having large canopy area may require several emission points. These are usually more expensive than single exit emitters.

Porous Pipe, Laser Tubing

Porous pipe and/or laser tubing are various adaptations of the "extremely small holes in a pipe" type of drip system. They just have very small holes drilled (usually using a laser) into a tube or are made from materials that create porous tubing walls. Water slowly leak out of the porous wall. The obvious advantage is their low cost. The disadvantage is that the tiny holes are easily clogged and/or water uniformity may not be to the extent desired. These types of systems are often used in landscaping and in portable irrigation systems. These have fairly limited lifespan and work well with water having low mineral levels.

Drip Tape

It has large cross-sectional, hydraulically designed high vortex, turbulent flow-path that makes it a clog resistant drip system. Off center slit outlets restricts the water movement along the tape. It also ensures that not only each drop falls on the soil but also prevents entry of the soil through the outlet. To use the drip tape, a minimum 100 micron filtration of water is recommended. Low tension during winding avoids kinking. Excellent coefficient of variation of flow rate and emitter pressure-discharge exponent makes it possible to have long lengths of run. It can be used in surface as well as subsurface drip installations. It is recommended for crops like cotton, vegetables, sugarcane, floriculture, strawberries, peppers and melons *etc.* It is quite economical.

On-line Drippers

On-line drippers are mounted on top of lateral pipes (Figure 21.8 left). These are installed in the pipes using a manual punch. The on-line drippers may be on-line pressure compensating drippers (diaphragm type) or any other non-pressure compensating drippers (NOPC) such as simple thread type, labyrinth type, zigzag path, vortex type flow path or have float type arrangement to dissipate energy (all discussed in previous sections). The on-line NPC emitters are usually designed so that working pressure differences between the various drippers does not exceed 7.5 per cent.

Figure 21.8: Fixing of an Online Emitter with an Emitter Shown as Inset (Left) and In-line Dripper (Right).

In-line Drippers

In line drippers, drip line, dripper line and other variations of these names are used to describe a drip tube with factory pre-installed emitters (Figure 21.8 right). In-line emitters are an integral part of the dripping laterals, that creates a smooth and bump less duct, allowing quick and easy deployment and collection of the drip lines. These are fixed along with the line, *i.e.* the pipe is cut and dripper is fixed in between the cut ends, such that it makes a continuous row. These include in-line tube with cylindrical dripper, in-line tubes with patch drippers, or porous tapes or bi-wall tubes. Often the emitters are moulded inside the tubing with an outside hole from which water is released. The emitters are typically the simple thread type or tortuous-path or diaphragm type but may be other types as well. The emitters are uniformly spaced along the tube; often several different spacing options are available. They are provided with independent pressure compensating water discharge mechanism and extremely wide water passage to prevent clogging. It is easier to install the system beacuse the emitters are pre-installed,.

Selection of Emitter

The emitter is the most important part of a drip irrigation system because it delivers water at the desired rate to the plant and maintains water application uniformity over the irrigated area. Thus, selection of the emitter is quite crucial in assembling an efficient system. Some of the criteria that are applied in the selection of dripper are as follows.

- ☆ Emission uniformity
- ☆ Reliability against clogging and malfunctioning
- ☆ Simplicity in installing and maintenance
- ☆ Pressure compensation in case of undulated terrain
- ☆ Percentage wetted area
- ☆ Flow rate
- ☆ Operating pressure
- ☆ Cost

The pressure variation within the system and the flow characteristics of the emission device, both influence the uniformity of water distribution termed as emission uniformity (EU) of a drip irrigation system. While the pressure variation can be controlled by proper system design, the selection of the emitter device itself is vital to achieve high EU. Moreover, emitter selection is critical to specify the water treatment and hence the filtration equipment for the drip system. Other important items to be considered in emitter selection are the percentage area wetted, which is related to delivering the required amount of water to the plant at the design pressure, and the reliability of the emitter against clogging and malfunctioning. Besides what is stated, user preference, personal and local experience and cost of the emitter may influence the choice. The information given in Table 21.6 can be used to select the emitters.

Table 21.6: Selection Criteria for the Emitters

Type of Emitter	*Flow Rate (l/h)*	*Operating Pressure (m)*	*Application to Type of Crop and Terrain*
Micro-tube, on-line dripper, in-line drippers	1-10	1-10	Vegetable and fruit crops on flat terrain
Self or pressure compensating dripper	1-10	10-30	Vegetable and fruit crops on uneven land
Line source tube/thin walled tape	1-5	1-15	Long row crops

Example 21.1

Emitters of 1.2 l/hr are selected to irrigate a crop that requires 4.3 l/day of water. If the irrigation interval is once in 2 days and 4 emitters are used per plant, how many hours the irrigation system must run to meet the crop water requirement?

Water amount needed per irrigation : 4.3 l/day/plant × 2 days = 9.6 liter/plant

The total discharge of the emitters per plant = 4 emitters × 1.2 l/hr = 4.8 l/hr-plant

Hours the irrigation system should be run = 9.6/4.8 = 2 hr.

Flush Caps and Flush Valves

Flush caps or valves are provided in the system to flush debris from the pipe to minimize emitter clogging. While flush valves are provided at the main and sub-mains, flush caps are placed at the ends of all irrigation laterals. These are installed in such a way that their location is easily identified for easy access and for maintenance. Installation of the flush device in a valve box is preferred. Folding and clamping the pipe is commonly used but it can result in leaks as well as does not look good to the viewers.

Other Items

Besides the major components discussed in the preceding section, some of the components used and their functions are:

Information on accurate flow rate is indispensable for the analysis of crop response to water and nutrients and for monitoring the performance of the irrigation system over short and long-term basis. System flow meters are used to collect the information on flow rates and total water applied during a given time period.

Since visual monitoring is impractical or impossible, pressure gauges are used to monitor the health of the system. In experimental systems, pressure gauges placed at the pump station, before and after the filter station, and upstream and downstream of each control valve provide an immediate indication of system performance. Too low pressure may indicate a leak, a filter plugged or a valve jammed. If pressures is too high, the system might be plugged or a valve may be set incorrectly. For field systems, one pressure gauge may suffice (Figure 21.9).

Zone control valves are used to control the flow of water to the various blocks or network zones. They can be simple manually operated valves, or fully automated solenoid activated on/off or pressure reducing valves.

Pressure relief valves, regulators or bypass valves are installed at any point where possibility of excessively high pressures either static or surge pressures exists. A bypass arrangement is simplest and cost effective means to avoid problem of high pressure than costly pressure relief valves. Air vacuum relief valve(s) are provided at the zones of highest point(s).

Wetted Diameter

The wetted diameter and the vertical movement of water below the dripper

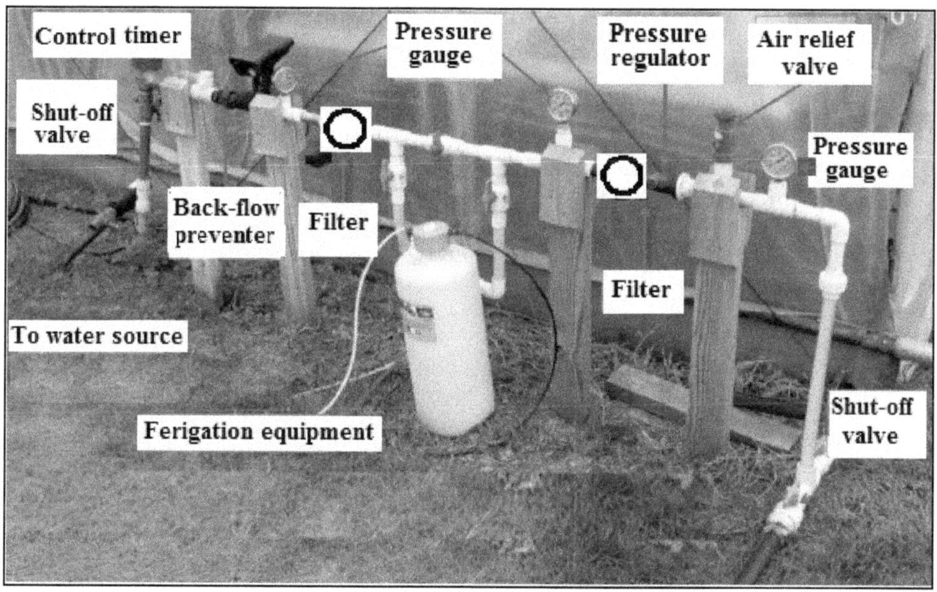

Figure 21.9: Some of the Accessories Commonly Used in Drip Irrigation System.

largely depend upon the soil texture. The wetted diameter is much smaller in sand than a clay soil (Figure 21.10). The wetted volume is oblong in sand but may be ellipsoid in a clay (Figure 21.10).

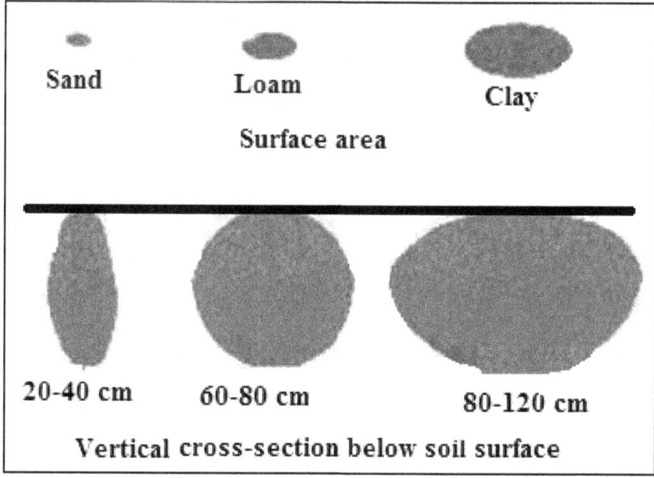

Figure 21.10: Relative Horizontal and Vertical Movement of Water in Various Soil Texture Groups.

Design of Drip Irrigation

Focus of any system design is to ensure that it is capable to meet the following conditions.

- ☆ Apply water to meet peak crop water requirement
- ☆ Maintain application and uniformity efficiencies at a desired level
- ☆ Energy and water efficient to keep initial capital and operation cost as low as possible
- ☆ Simple in operation and maintenance so that farmers can use the system without extensive training

The inputs required to make an effective customized drip irrigation system are as follows:

- ☆ Layout of the area
- ☆ Details of the water source and soil type
- ☆ Agronomic details (plant spacing, crop period, season, canopy, *etc.*)
- ☆ Climatic data (rainfall, temperature and evapotranspiration, *etc.*)

With any irrigation system, the design process starts at the lower end (plants) and processed upwards. In practice, the design of a drip Irrigation system involves estimation of the following parameters.

- ☆ Detailed layout of the system in the field
- ☆ Peak water requirement of a plant per day. To estimate total water requirement of an area, the number of emitters required per plant, amount of water discharged per hour through each emitter and the total number of hours water is available should be known/estimated (See example 21.1)
- ☆ Size and lengths of distribution system such as lateral drip lines, sub-mains and mains. This depends upon the flow rates and friction head loss in pipes
- ☆ Operational schedule for irrigation
- ☆ Water requirement depends upon hydro-geological conditions in the area and water requirement of plants/crop
- ☆ Horse Power of the pump set. This depends upon discharge and total head including friction losses over which water is to be lifted/pumped
- ☆ Cost and economic analysis (See last section of chapter 20 and this chapter)

Layout of Drip Irrigation System

Water source and pumping plant should be located as close to the center of the irrigated area as possible. The main line/manifold in a drip system should follow land contour as closely as possible. The slope should be used to compensate pressure differences due to change in elevation. When water flows down slope, it allows longer laterals for a given pipe size or smaller pipe for a given length of lateral. A fall of 1 m in elevation is equivalent to an increase in pressure of about 0.1 bar (1 bar = 10.2 m of water = 14.05 psi = 0.987 atm = 100 kPa = 1.02 kg/cm^2). Wherever possible, avoid running laterals uphill.

Rule of Thumb: Drip emitters should be selected and installed in such a manner so as to minimize damage and deterioration from vandalism, insects, animals, ultraviolet light, weather and field operations. For large plants, install emitters 600 mm apart at 80 per cent of the leaf canopy away from the plant stem. That's where most of the roots are located. In highly permeable soils emitters are installed 300 mm to 450 mm apart. A loop of 3 to 4 emitters equally spaced is also placed around each plant to ensure the required wetted area. For example; 2 emitters are placed at 180°, 3 emitters at 120° and 4 emitters at 90° apart. If a single emitter is provided, it must be placed 15-30 cm from the base of the plant.

Installation of Drip Irrigation

In general, the following activities are involved in the installation of drip irrigation system.

- ☆ Draw and/or study installation sketch
- ☆ Give layout for water tank/filter platform and trenches for pipes if required
- ☆ Check components at site as per the list of materials in the users' manual
- ☆ Install water storage tank and filter on the platform
- ☆ Connect filter to the water source/pump and the main line
- ☆ Lay out the main line, sub-main and lateral pipes
- ☆ Place/fix the emitters (if micro-tubes require inflated lateral pipes, then fill the pipes with water then punch holes and fix micro-tubes)
- ☆ Start the pump/open the valves and fill the pipes with water
- ☆ Release all end caps/flush valves to clean the system of dirt
- ☆ Check pressure and discharge and ensure all emitters are working
- ☆ Cover the pipe trenches, if required
- ☆ Operate according to schedule

Installation of Head Equipment

The following points should be considered while fixing the head equipment.

- ☆ Use minimum of fittings such as elbows and bends
- ☆ Connect sand/screen filter to the mainline
- ☆ Make arrangement of water bypass
- ☆ Provide sufficient space for the easy operation of filter valves
- ☆ Hard surface or cement concrete foundation needs to be made for sand filter so that it does not collapse due to vibration and load. For screen filter, strong support needs to be provided by using GI fittings to avoid vibrations
- ☆ Apply proper mixture of M-seal uniformly over the joints to avoid leakage
- ☆ Fix the pressure gauges at the inlet and outlet of the filter

Connecting Mains and Sub-mains

- Mains and sub-mains should be laid at a depth of more than 30 - 45 cm so as to avoid damage during inter-cultivation
- The pipes should be fitted using solvent cement with the help of brush
- A gun metal gate/PP Ball valve should be provided at the start of sub-main
- Provide flush valve at the end of main and sub-main such that it faces towards slope
- Apply uniform pressure vertically over the drill while drilling in the sub-main. It will give a smooth and round hole.
- Fix the rubber grommets in the holes made in the sub-main in such a way that the groove in it goes inside the pipe
- Fix the take-off position such that its arrow or the chamber faces towards the gate valve of the sub main for the easy flow of water. See that the take-off is fixed tightly in the grommet. The loose fitting of take-off indicates the breakage of grommet
- Flush the sub-main so that the PVC piece/mud fallen in the sub-main while making holes is flushed. Otherwise this scrap will block the drippers.

Laying of Laterals and Drippers

- Pass water through the poly tube and get it flushed so that it is bulged. It is easy to punch the holes in the bulged pipe
- Punching should begin from the sub-main
- Punch the lateral sideway from the yellow strip
- The dripper position should be fixed according to design, soil and water report and water required during the peak summer
- Do not fix drippers unless a complete lateral line is punched
- While fixing the dripper, push it inside the lateral and pull it slightly
- Close the end of lateral by fitting end cap.

Standard Procedure to Assess Drip Performance

- Check installation according to approved design layout
- Start the pump
- Flush the filters
- Allow the drip system to be loaded with water for 10 min
- Note the pressure from the pressure gauge at the inlet and outlet of sand and screen filters
- Record the dripper discharge as per the format
- The discharge and pressure readings have to be taken in the below mentioned locations

- ❖ First, middle and last dripper of a lateral
- ❖ For laterals at beginning, ¼, ½, ¾ and end of sub-main

- ☆ Laterals on anyone side of the sub-main can be selected in case of plain land or alternative laterals on either side in case of slight slope in the direction along the lateral
- ☆ Measure the volume of water collected in a can for 36 seconds
- ☆ Measure the pressure at start and end of laterals
- ☆ If the emission uniformity is less than 85 per cent then the issue needs to be discussed with the designer for modifications.

Chemigation/Fertigation

Chemigation is an inclusive term referring to the application of a chemical into or through an irrigation system. It includes the application of fertilizers, acids, chlorine and pesticides. Fertigation is a specific term used for the application of fertilizer (plant nutrients) through an irrigation system. The term connotes, Fertilizer + Irrigation = Fertigation. Fertigation is a sophisticated and efficient method of applying fertilizers in which irrigation system is used as the carrier and distributor of crop nutrients. Since the root system development is extensive in a restricted volume of soil with drip irrigation, application of fertilizer or any other chemical through fertigation can efficiently place plant nutrients in the zone of highest root concentration. The synergy of the combination of water and nutrients also leads to efficient use of nutrients by the plants. Although, liquid fertilizers are commonly used being easy to handle and convenient to use, solids fully soluble plant nutrients can also be used in fertigation. The use of these fertilizers has proven to be economical. The solubility of various fertilizers is reported in Table 21.7. A filtering unit should be provided after the fertigation unit.

Table 21.7: Fertilizers Suitable for Fertigation

Name	*Chemical Form*	*N-P_2O_5-K_2O Content (Per cent)*	*Solubility (g/l at 20°C)*	*Remarks*
Ammonium nitrate	NH_4NO_3	34-0-0	1920	Incompatible with acids
Ammonium sulphate	$(NH_4)_2SO_4$	21-0-0	750	Clogging with hard water
Urea	$CO(NH_2)_2$	46-0-0	1100	-
Diammonium phosphate	$(NH_4)_2HP_2O_5$	18-46-0	575	Contains phosphorous at high solubility
Potassium chloride	KCl	0-0-60	347	Cheapest K source but chloride can be toxic to some crops
Potassium nitrate	KNO_3	13-0-44	316	Expensive, high nitrate
Potassium sulphate	K_2SO_4	0-0-50	110	Excellent source of sulphur, clogging with hard water
Phosphoric acid	H_3PO_4	0-52-0	457	Incompatible with calcium

Characteristics of the Fertilizers

A good quality solid fertilizer should have the following characteristics:

- High nutrient content in a form readily available to plans
- Fully and quickly soluble in water at field temperature conditions
- Low insoluble contents (0.02 per cent) so as to minimize chances of clogging of filters and emitters
- Minimum content of conditioning agents
- Compatible with other fertilizers
- Minimum interaction with chemicals present in irrigation water
- No heat generation during dissolution as it may soften the plastic pipe tubing
- No drastic change in EC and pH of water
- Non-corrosive to various parts of the system.

Besides what is stated, the non-nutritional elements in the fertilizers such as Na, Cl, F, Al that may cause toxicity should be given due importance. Fertigation is quite problematic if the water pH is high.

- The soil is negatively charged at high pH and PO_4^- will be precipitated with Ca^{++} and absorbed with clay. Thus, availability of P is quite low.
- Availability of micro-nutrients (Fe, Mn, *etc.*) also reduces due to their precipitation as well as low uptake. Hence iron and zinc chelates are applied to prevent their precipitation.

Advantages

- Quick and convenient, eliminates manual application
- Uniform application to all the plants
- High fertilizer use efficiency resulting in saving of fertilizers, can be as much as 50 per cent higher compared to soil application
- As water and fertilizer are supplied evenly to all the crops through fertigation there is possibility of getting higher yield by about 25-50 per cent
- Reduced leaching of fertilizers resulting in no or minimum pollution of water resource
- Better penetration of P and K in the soil layers
- Efficient co-ordination of nutrition requirement with crop stage or development
- Dosage can be easily controlled
- People and non-target crops are not exposed to inadvertent chemical drifts
- Others chemicals such as herbicides, pesticides, acid, *etc.* can also be applied. Chemigation minimizes manual handling, mixing and dispensing of potentially hazardous materials.

Limitations

- ☆ May cause toxicity to field workers
- ☆ Chance of backflow into water source, requiring back flow prevention equipment such as non-return valve or vacuum valve
- ☆ Insoluble fertilizers are not suited to fertigation (super phosphate)
- ☆ Corrosive effect of fertilizer on system components
- ☆ Phosphate may get precipitated in the pipeline and clog dripper at high pH
- ☆ High cost of liquid fertilizers.

Kinds of Fertilizers

Liquid Fertilizers

Preparation of nutrient stock solutions from dry fertilizers may require considerable time and effort and can generate sediments and scums as waste products. Therefore, commercially available liquid fertilizer solutions (true solutions, not suspensions) that are completely water soluble are often used. Liquid fertilizers are available in a variety of grades (4-0-8, 7-0-7, *etc.*) and can be purchased with or without micro-nutrients. A liquid formulation of calcium nitrate (9 per cent N, 11 per cent Calog) is also available. Liquid formulations such as this are very convenient, because they can be injected directly (without mixing in water) with a variable rate injection pump. Although transportation costs make liquid formulations a little more expensive, they save time and labor and help prevent problems associated with poorly made "home mixes." Also, they eliminate the problems caused by insoluble materials found in some dry fertilizers. Special fertilizers like mono ammonium phosphate (Nitrogen and Phosphorus), poly feed (Nitrogen, Phosphorus and Potassium), Multi K (Nitrogen and Potassium), Potassium sulphate (Potassium and Sulphur) are highly suitable for fertigation as they are highly soluble in water. Fe, Mn, Zn, Cu, B, Mo are also supplied along with special fertilizers.

Even with liquid formulations, one has to be careful when injecting fertilizers containing phosphorus or sulphur into drip systems. Phosphorus and sulphur may react with calcium and/or magnesium in the irrigation water to form mineral precipitates that can clog emitters. Micro-nutrients can also cause problems due to precipitation. If micro-nutrients must be injected, use soluble chelated forms that are less likely to precipitate.

Solid Fertilizers

Completely soluble granular fertilizers are readily dissolved in stock solutions. These fertilizers should be completely dissolved in the nutrient stock solution before injection is undertaken. Granular formulations that are not completely water soluble, can cause serious clogging problems in drip systems and result in variable (non-uniform) application of water and nutrients. Some fertilizers suitable for drip irrigation and their solubility are reported in Table 21.7.

Nitrogen

Nitrogen (N) is the major plant nutrients. Its availability is usually limited in the soil compared with other plant nutrients because its various forms can be leached, volatilized, denitrified or fixed in the organic fraction of the soil. One of the favoured forms of N for use in drip system is urea, because it is a highly soluble and does not react with water to form ions unless the enzyme urease is present. Although water quality must be considered when N is applied through a trickle irrigation system, it is less of a problem in this case than other nutrients such as phosphorus. The injections of anhydrous ammonia or aqua ammonia into irrigation water will bring about an increase in pH that may be conducive to the precipitation of calcium, magnesium and phosphorus, or the formation of complex magnesium ammonium phosphates, which are insoluble. This can be especially serious if bicarbonate is also present in the irrigation water. Nitrogen injected in the form of ammonium phosphate can cause serious clogging of the irrigation system especially when calcium and magnesium are present in the irrigation water.

Phosphorus

Generally, injection of phosphorus (P) fertilizer through a trickle irrigation system is not recommended. Most P fertilizers are prone to chemical or physical precipitation problems and subsequent clogging of the trickle irrigation system. Further, the fixation rate of P by soils is high and subsequent movement from its point of placement is limited. The possibility of precipitation of insoluble phosphate is extremely high in calcium and magnesium rich environment, be it through irrigation water, or soil or through application of other chemicals. However, phosphoric acid has proved to be a good source of phosphorus besides addressing the clogging problems.

Potassium

No adverse chemical reactions are anticipated with the potassium (K) fertilizers if added alone to water. However, reduced solubility and/or fertilizer incompatibility may be encountered if more than one kind of fertilizers are mixed. For example, when calcium nitrate and potassium sulphate are mixed, it may form insoluble calcium sulphate.

Compatibility

At certain times, mixing of two or more than two fertilizers may be needed. In that event their compatibility needs to be examined. Besides the compatibility of fertilizers to salts present in irrigation water also needs to be ensured. It is because in certain cases, it results in the formation of precipitates that may clog the emitters. Mixing should only be done for fertilizers, which are compatible except in emergency where even low compatible fertilizers can be mixed with care. Non-compatible fertilizers should never be mixed. To test the compatibility, one can perform a jar test. To do this, fill a jar with the irrigation water and the two fertilizers in the same proportion as is proposed to be used during fertigation. Wait and see for any precipitation. If precipitation is formed, the two fertilizers are incompatible.

Irrigate-Fertigate-Irrigate

Fertigation is normally done during a part of an irrigation cycle. Initially, irrigation is applied for a part of the cycle followed by fertigation and again by irrigation, thus ending the cycle with water alone (Hagin *et al.*, 2003). It is called irrigate-fertigate - irrigate. The usual proportion of the 3 cycles is 25 per cent, 50 per cent and 25 per cent resulting in its designation as 25-50-25 rule. This management strategy ensures the build-up of the appropriate pressure when irrigation commences, and the flushing out of the nutrients from the irrigation system, towards the end of the irrigation period. It ensures that no fertilizer solution remains in the system as it can be corrosive to certain parts of the system

Equipment and Methods for Fertilizer Injection

Injection of fertilizer and other agro-chemicals into the drip irrigation system can be accomplished by using equipment such as bypass pressure tank, venturi system and direct injection pump system.

Bypass Pressure Tank

It employs a tank into which the dry or liquid fertilizers are kept. The tank is connected to the main irrigation line by means of a bypass so that some of the irrigation water flows through the tank and dilutes the fertilizer solution (Figure 21.11). This bypass flow is brought about by a pressure gradient between the entrance and exit of the tank, created by a permanent constriction in the line or by a control

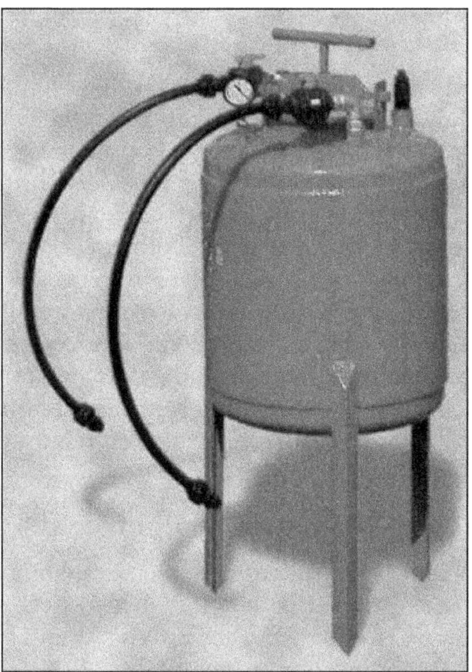

Figure 21.11: Commonly Used Device for Fertigation: A Fertilizer Tank.

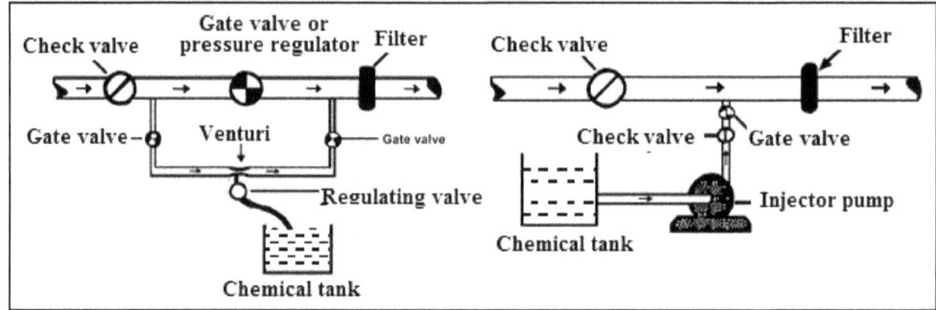

Figure 21.12: Fertilizer Application Equipment, Venturi (left) and Injection Pump (Right).

valve. The concentration of fertilizer in the tank thus, becomes gradually reduced resulting in non-uniform application of water.

Venturi Injector

It is a simple and relatively inexpensive method of fertilizer application. As water flows through the tapered venturi orifice, a rapid change in velocity occurs. This velocity change creates a reduced pressure (vacuum), which is sufficient to suck fertilizer solution from an open reservoir into the water stream (Figure 21.12 left). Each venturi must be calibrated for local conditions, since the injection rate vary with the pressure differential across the venturi. A precise regulating valve and a flow meter are used to calibrate the system. It usually requires a large stock tank.

Direct Injection, Pump System

With this method a pump is used to inject fertilizer solution into the irrigation line (Figure 21.12 right). The capacity of the injection system depends on the concentration, rate and frequency of application of fertilizer solution. Positive displacement metering pumps powered by small electric motors or by hydraulic drive systems are used. An electric pump with an automatically controlled arrangement is the most convenient to use. However, its use is limited by the availability of electrical power. Hydraulic drive systems use the water pressure in the system to power the pump. Unlike past models, injection rates of some of the more recent models can be adjusted with a variable frequency drive. This drive varies the speed of the injection pump with the flow rate of the irrigation system. Another advantage of using hydraulic pump for fertigation is that if the flow of water stops in the irrigation system, fertilizer injection is automatically stopped. It is the most perfect equipment for fertigation.

Maintenance of Drip Irrigation System

The biggest problem of any micro-irrigation system is clogging of emitters. Therefore, periodic and preventive maintenance is necessary for the smooth functioning of the system. The following general checks are carried out periodically depending on the local condition and water quality.

- ☆ Clogging of emitters and resulting variations in the wetting pattern
- ☆ Placement of emitters/micro-tubes
- ☆ Leakages in pipes, valves, filter, fittings, *etc.*
- ☆ Flushing and cleaning of filter by opening and cleaning the screen
- ☆ Flushing of sub-mains and laterals by releasing the end caps/flush valves.

Regular Maintenance

The following activities are performed as routine maintenance activities at least once a month.

- ☆ With the system on, visually inspect for leaks or damage to emitters, connectors, or tubing. Visually inspect each drip emitter for clogging or high flow. Emitters with abnormally high or low flow rates should be replaced immediately
- ☆ Check plants and soil for signs of under watering or over watering
- ☆ Inspect the irrigation controller(s) to make sure it is working and that the watering schedule is properly set
- ☆ Ensure that all emission points are visible and above the finished grade
- ☆ Plug emission points where plants have died, or replant
- ☆ Bury exposed lines, as exposure to sun may damage the pipes.

Semi-annual Maintenance

The following maintenance activities are performed at least twice a year:

- ☆ Replace batteries in controllers (electronic models). Controllers can lose the programme or revert to default programme, if power is lost
- ☆ Ensure all flush valve/cap locations are visible
- ☆ Ensure valve boxes are visible and can be opened
- ☆ Inspect valves, filters, and pressure regulators for damage or leaks. Remove excess dirt and debris in valve box
- ☆ Move drip emitters/emission points away from the bases of the plants. In most cases, emission points should never be closer than 30 cm from the base of a mature plant. Most of the emission points should be located near the drip line (canopy edge) of the plant. Watering too close to the base of the plant can cause root rot. Moving emitters away from the base of the plant promotes greater root support.

Annual Maintenance

The following maintenance activities are performed at least once a year. Flushing and cleaning may need to be performed more often depending on water characteristics.

- ☆ Clean valve boxes of dirt and debris

- Flush filters. A hose can be attached to the flush cap to keep water out of the valve box. Debris can be drawn into the system if they are not flushed prior to opening the filter assembly
- Flush laterals
- Since the root zone expands with age, make sure plants have adequate numbers of drip emitters
- Test PVB and RP backflow preventers.

Curative Maintenance Activities

The maintenance of drip irrigation system is very essential for its trouble-free functioning.

Daily Maintenance

- Clean the sand and screen filters for 5 min before starting the system
- Ensure all drippers are working properly without any leakage
- Before the irrigation is stopped, backwash the sand filter for about 5 min.

Sand Filter

The contaminants in water accumulate and clog pore space in the sand bed of a media filter that may reduce its efficiency. Regularly, observe pressure difference across the sand filter and see that it never exceeds 0.30 kg/cm^2. As soon as it happens one should perform the backwashing. Backwashing is the processes in which water flow is reversed so that collected dirt flows out of the filter through the backwashing valve. Normally, daily backwashing of sand filter is recommended (Figure 21.13 left). It is important to regulate the backwash flow because excessive flow may even remove sand itself out of the filter. Normally to avoid loss of sand, back flush the system at a pressure of 0.5 kg/cm^2. Insufficient backwash flow on the other hand, will not properly clean the filter. Sequence of backwash operation comprises of

- Open the backwash valve
- Close the outlet valve
- Open the bypass valve
- Close the inlet valve

Allow all the water along with dirt to flow out of the sand filter.

Once in few days, backwash water may be allowed to come out through the lid instead of backwash valve (Figure 21.13 right). Stir the sand in the filter bed up to filter candle without damaging the candles. Whatever, dirt is accumulated deep inside the sand bed will be released and it will get out with the water through the lid. Close the lid for normal operation.

Screen Filter

Flushing scheduled at daily interval is necessary to maintain a screen filter. Practically, flush the screen filter, whenever pressure drops more than 0.5 kg/cm^2.

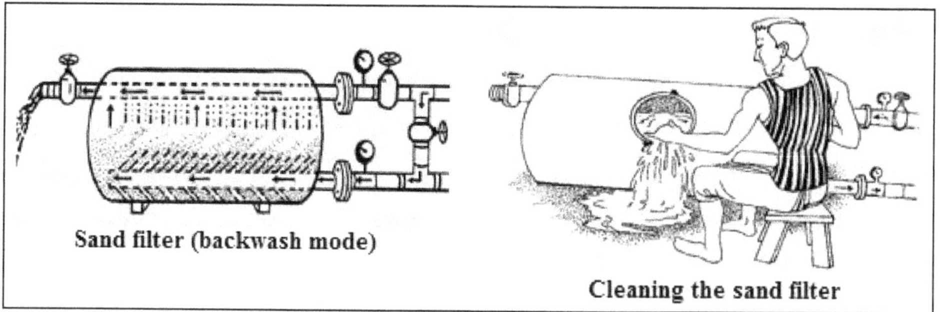

Figure 21.13: Cleaning of Sand Filter.

Flushing is done by simple opening of the drain valve so that the force of water flushes out the dirt through drain valve. For regular cleaning of the screen filter, open the screen filter lid and take out the filter elements. Clean it in flowing water by rubbing with cloth or soft nylon brush. Take out the rubber seals and clean them from both sides. Ample care should be exercised while replacing the rubber seals so that these remain in position. Check for any mechanical damage. Never use stones (hard substance) to rub the screen surface or acid to clean the screens.

Weekly Maintenance

- ☆ Clean the sand filter by hand (Figure 21.13 right)
- ☆ Flush the sub-main by opening the flush valve for 5 minutes
- ☆ Flush laterals 5 numbers at a time for 5 minutes

Monthly Maintenance

Treat the system with chlorine/acid. The frequency of chemical treatment depends on the degree of problem at a particular site.

Chemical Treatment

Clogging or plugging of drippers may be due to precipitation and accumulation of certain dissolved salts like carbonates, bicarbonates, iron, calcium and manganese *etc.* The clogging may also result from the presence of micro-organisms and the related iron and sulphur slimes or due to algae and bacteria. Although efforts should be made to prevent the clogging of drippers, yet if it happens, then the drippers need to be cleaned by chemical treatment. Chemical treatments commonly used in drip irrigation systems include acid and/or chlorine treatment.

Acid Treatment

Hydrochloric acid (HCl) or sulphuric acid is used for this purpose. The acid is injected into drip systems at a pre-decide rate. The acid treatment is continued till a pH of 4 is observed. At this stage, the system is shut down for 24 hr. The system is flushed after 24 hr by opening the flush valve and lateral ends. It should be ensured that no human or animal comes in contact with this water or it is not drained into some surface water source.

Chlorine Treatment

For chlorine treatment, bleaching powder is used as a means to inhibit the growth of organisms like algae or bacteria. The bleaching powder is dissolved in water and this solution is injected into the system for about 30 minutes. The recommended level of free chlorine is 1 to 2 ppm at the end of the irrigation system. Then the system is shut off for 24 hours. After 24 hours, the lateral ends and flush valves are opened to flush out the water containing impurities. Bleaching powder can directly be added into the water source at a rate of 2 mg/l or through the venturi assembly.

Repair of the System

Repairs should be undertaken immediately after the detection of any damage. The following procedures are used for some of the common types of repairs.

Clogged Emitters

- ✰ Drip emitters that cannot be unplugged or are damaged should be replaced. Replaced and repaired drip emitters should be checked to verify that their flow rate is as per design requirement
- ✰ Flush tubing before emitters are reattached
- ✰ Emitter/tubing connections should be checked for leaks.

Broken, Torn, or Crushed Lines

- ✰ Dirt and debris should be cleared away from the damaged area before making repairs
- ✰ Lines should be flushed immediately after the repair is made
- ✰ Repairs should be checked for leaks before backfilling.

Leaky Emitter or Connector Barb Connections

- ✰ Damaged portion of tubing should be removed and replaced
- ✰ Use sealing plugs to seal holes in polyethylene pipe where connectors have been removed. Tap new holes for the connectors
- ✰ Flush tubing before emitters are reattached
- ✰ Connections should be checked to ensure they are secure (hold together)
- ✰ Repaired connections should be checked for leaks
- ✰ Sections where large numbers of holes need to be tapped should be replaced with new lateral tubing.

Drip Irrigation System Troubleshooting

Some of common problems encountered during the operation of the system, their causes and troubleshooting procedures are presented in Table 21.8.

Table 21.8: Causes and Troubleshooting for Commonly Encountered Problems

Sl.No.	Problems	Causes	Troubleshooting
1.	Leakage of water at the joint between sub-main and lateral	Damaged joints	Repair damages
2.	Water is unevenly distributed down the slope	Pressure increases down the hill causing more water at the bottom than at the top	Refit and run the line at an angle down the slope. Use pressure regulators at intervals along the line
3.	Water not coming out of emitter	Emitter may be closed	Clean or change the emitter
4.	Leakage in the poly tube	Damage of poly tube by farming activities/rat	Block the holes by goof plug Use poly joiners at cuts
5.	Water not flowing up to lateral end	Holes/cuts in laterals Bends in laterals	Close the holes and cuts Remove the bends
6.	White mixture comes on removing the end plug	More salinity in water Lateral are not properly cleaned	Remove the end plug Clean the laterals fortnightly
7.	Under flow or over flow from laterals	Clogging of drippers Unclosed end plug	Clean the sand and screen filters Close the end cap
8.	Oily gum type material comes out on opening the lateral end	Algae or ferrous material in water	Clean the laterals with water or give chemical treatment
9.	More pressure drop in filters	Accumulation of dirt in filters	Clean filters every week Backwash the filters for 5 min daily
10.	Pressure gauge not working	Rain water might have entered inside. Corrosion in gauge Pointer damage	Provide plastic cover and fix pointer properly
11.	Drop in pressure	Leakage in main Opened outlet Low water level in well	Arrest the leakage and close the outlet Lower the pump with reference to well water level
12.	More pressure at the entry of sand filter	No bypass in the pipeline/ bypass not opened Displacement of filter element Less quantity of sand in filters	Provide bypass before filter and regulate pressure Place filter element properly Fill required quantity of sand
13.	Venturi not working during chemical treatment and fertigation	Excess pressure on filters Improper fitting of venturi assembly	Bypass extra water to reduce pressure Repair the venturi assembly
14.	Leakage of water from air release valve	Damaged air release valve ring	Replace the damaged ring

Sl.No.	Problems	Causes	Troubleshooting
15.	Water is leaking from the body of the filter	-	Remove the O-ring, clean it, see that it fits in the groove, re-assemble the filter If the leak persists, replace the O-ring or see whether there is another source of leak

Cost of Drip Irrigation System

The cost of drip system depends on a number of factors. Some of them are: the type of crop, plant to plant and row to row spacing, water requirement, proximity to water source *etc.* Cost of drip irrigation system for various crops as per plant to plant and row to row spacing has been prepared which gives an insight into the cost of the system (Ministry of Agriculture and Farmers Welfare, 2017; Table 21.9). The exact cost, however, has to be calculated on case to case basis. The indicative bill of quantities (Table 21.10) may be multiplied by the unit cost to calculate the total cost of the system (Ministry of Agriculture and Farmers Welfare, 2017). The life of the system varies from 6 to 10 years.

Table 21.9: Indicative Cost of Drip Irrigation System for Calculation of Subsidy (Rs.)

Spacing (m × m)	*0.4 ha*	*1 ha*	*2 ha*	*3 ha*	*4 ha*	*5 ha*
12 × 12	15853	21643	34417	53437	66480	84653
10 × 10	16419	23047	37171	57647	72205	91806
9 × 9	16826	24035	39145	60610	76238	96852
8 × 8	17351	25332	41650	64500	81527	103459
6 × 6	19096	30534	51045	82472	100016	125498
5 × 5	20674	34664	59154	85484	108635	145964
4 × 4	21414	36562	64084	99965	130884	155778
3 × 3	23055	42034	72759	112065	140936	176457
2.5 × 2.5	31156	60065	109345	167011	234396	286297
2 × 2	36358	73138	141957	206232	286504	351667
1.5 × 1.5	41369	85603	163137	243633	336484	414002
2.5 × 0.6	30810	63145	116042	177345	246276	302318
1.8 × 0.6	37845	80599	152551	229637	312784	389511
1.2 × 0.6 (or lower spacing)	50388	112237	213400	323019	435788	545181

Cost of Drip Irrigation to Stakeholder

To calculate the seasonal or annual cost or to evaluate the various economic/ financial parameters, procedures are described in chapter 20. The same procedures will be applicable for drip irrigation system as well.

Table 21.10: Drip Irrigation System–Indicative Bill of Quantities (BOQ) per ha

Sl. No.	*Component/Lateral to Lateral × Dripper Spacing (m × m)*	*Unit*	*12 × 12*	*10 × 10*	*9 × 9*	*8 × 8*	*6 × 6*	*5 × 5*	*4 × 4*	*3 × 3*	*2.5 × 2.5*	*2 × 2*	*1.5 × 1.5*	*2.5 × 0.6*	*1.8 × 0.6*	*1.2 × 0.6*
	Control Unit															
1	Screen filter 20/25 m³/hr	No.	1	1	1	1	1	1	1	1	1	1	1	1	1	1
2	Venturi and manifold (2")	No.	1	1	1	1	1	1	1	1	1	1	1	1	1	1
3	Air release valve 1"	No.	1	1	1	1	1	1	1	1	1	1	1	1	1	1
4	Non return valve - 1.5"	No.	1	1	1	1	1	1	1	1	1	1	1	1	1	1
5	Bypass assembly - 1.5"x1.5"	No.	1	1	1	1	1	1	1	1	1	1	1	1	1	1
	Field Unit															
6	PVC pipe 75 mm, class-II; 4 kg/cm²	m	0	0	0	0	0	0	0	54	54	54	54	54	54	54
7	PVC pipe 63 mm, class-II; 4 kg/cm²	m	0	0	0	0	54	156	156	102	102	102	102	102	102	102
8	PVC Pipe 50 mm, class-III; 6 kg/cm²	m	156	156	156	156	102	0	0	0	0	0	0	0	0	0
9	Lateral 16 mm, Class II; 2.5 kg/cm²	m	833	1000	1111	1250	1667	2000	2500	3333	4000	5000	6667	60	83	125
10	Emitting pipe 16 mm; Class II: (0.6 m × 1 to 4 lph)	m	0	0	0	0	0	0	0	0	0	0	0	4040	5611	8417
11	Emitter/dripper 4/8 lph	No.	278	400	494	625	1111	1600	1275	2267	3232	5050	4489	0	0	0
12	Control valve 75 mm	No.	0	0	0	0	0	0	0	1	1	1	1	0	0	0
13	Control valve 63 mm	No.	0	0	0	0	1	1	1	0	0	0	0	1	1	1
14	Control valve 50 mm	No.	1	1	1	1	0	0	0	0	1	1	1	1	1	2

Sl. No.	*Component/Lateral to Lateral × Dripper Spacing (m × m)*	*Unit*	*12 × 12*	*10 × 10*	*9 × 9*	*8 × 8*	*6 × 6*	*5 × 5*	*4 × 4*	*3 × 3*	*2.5 × 2.5*	*2 × 2*	*1.5 × 1.5*	*2.5 × 0.6*	*1.8 × 0.6*	*1.2 × 0.6*
15	Flush valve 63 mm	No.	0	0	0	0	1	1	1	1	1	1	1	1	1	2
16	Flush Valve 50 mm	No.	1	1	1	1	0	0	0	0	0	0	0	0	0	0
17	Throttle valve - 2"	No.	1	1	1	1	1	1	1	1	1	1	1	1	1	1
18	Fittings and accessories @5 per cent	set	5 per cent	5 per cent	5 per cent	5 per cent	5 per cent	5 per cent	5 per cent	5 per cent	5 per cent	5 per cent	5 per cent	5 per cent	5 per cent	5 per cent

Section IV

Soil and Water Conservation Engineering

Chapter 22

Soil Erosion: An Introduction

Soil Erosion Defined

Soil conservation in a wider connotation is the prevention of soil loss from erosion (water or wind) or reduced fertility caused by over usage, acidification, salinization or other soil chemical contaminats. Action taken to manage soil compaction, low organic matter, loss of soil structure, poor internal drainage, salinization and soil acidity problems also falls in realm of soil conservation. It is now known that these problems also create serious soil degradation and can accelerate the soil erosion processes. In a restricted sense, soil conservation is to prevent or reduce soil erosion and soil depletion by protective measures against water and wind erosion in order to maintain agricultural productivity. Soil erosion is often the effect of many natural causes, such as water and wind but several anthropogenic factors such as construction activities, unscientific cultivation practices and other human interventions enhance the soil erosion rate.

Soil Erosion: Historical Background

The term erosion is derived from the Latin word *'erodere'* meaning eat away or to excavate. Currently, erosion is defined as any process by which soil is transported from one place to another. Soil erosion might have begun even before the advent of agriculture and therefore, it is easy to surmise that soil erosion of agricultural fields might be as old as agriculture. The adverse impact might not have been prominent because land was in abundance and cheap. In the Indian context, Aryans settled mainly in the plains along the river front, and didn't fiddle much with the forest areas. It was only after during the British rule, large-scale denudation of forests occurred resulting in intense erosion. Human intervention in the form of increased construction activities such as road building, canal networks and laying of rail tracks blocked drainage ways in the flatter areas resulting in water logging and loss of cultivated lands. Loss of agricultural lands forced the people to encroach on to steeper slopes and marginal lands for agriculture. Besides, increasing natural disasters greatly increased the rate of soil erosion in most regions. Eroded hill slopes, increasing areas under ravines and gullies, deposition of sediments in irrigation reservoirs and lakes, foot slopes, valleys, flood plains, and river deltas are all evidences of heightened erosion through unscientific land use and climate change impacts during the last few decades. It is believed that further expansion of cultivation into ecologically fragile areas unless carefully managed will significant increase the soil erosion. Although few works of soil conservation appeared here and there, the professional soil conservation came quite late in the eighteenth century. The first essay on soil conservation appeared in Germany in 1815. The real impetus to soil conservation activities was noticed in the late nineteenth and early twentieth century.

Realizing that land degradation especially the soil erosion is indeed a potentially serious threat to food production and rural livelihoods, Government of India started several programs after India's independence. Soil and Water Conservation programs were first launched during the First Five Year Plan. The emphasis was on development of technology for problem identification, enactment of appropriate legislation and constitution of policy coordination bodies. The Centrally-sponsored Scheme of Soil Conservation in the catchments of River Valley Projects (RVP) was started in third Five Year Plan. Drought Prone Areas Program (DPAP) was launched by MoRD in 1972-73 which mainly relied on conservation works. In 1977-78, the Ministry of Rural Development (MoRD) started a special program named Desert Development Program (DDP) to address the issues of hot desert areas of Rajasthan, Gujarat and Haryana and cold desert areas of Jammu and Kashmir and Himachal Pradesh. Another scheme of Flood- Prone Rivers (FPR) was started in the Sixth Five Year Plan keeping in view the magnitude of floods in the year 1978. Later, during Ninth Five Year Plan both these schemes were clubbed together and further subsumed under Macro Management mode in 2000. A scheme for propagation of water harvesting/conservation technology was launched by MoA for rainfed areas in 1982-83 covering 19 locations. Adopting this approach, MoRD extended the program in 22 other locations in rainfed areas in 1984. The scheme of Watershed Development Project for Shifting Cultivation Areas (WDPSCA) was launched in seven north-eastern States during the Eighth five year plan. The scheme aimed at overall development of *jhum* areas on watershed basis through treatment packages.

Since soil and water conservation is integral part of any watershed management program, a look at these programs will also be useful. The concept of integrated watershed development was first launched as the National Watershed Development Program of Rainfed Areas (NWDPRA) in 1990, covering 99 districts in 16 states. A Technical Committee under the Chairmanship of Prof. C.H. Hanumantha Rao was appointed in 1994. Its tasks were to appraise the impact of the work done under DPAP/DDP, to identify the weaknesses of the program and to suggest improvements. Besides several recommendations, the Committee called for substantial augmentation of resources for watershed development by pooling resources from other programs being implemented by the MoRD. The Committee formulated a set of "Common Guidelines", for five different programs under the

MoRD, namely, DPAP, DDP and Integrated Wastelands Development Program (IWDP), as also the Innovative- Jawahar *Rozgar Yojana* (I-JRY) and Employment Assurance Scheme (EAS), 50 per cent of the funds of both of which were to be allocated for watershed works. In 2000, the Ministry of Agriculture revised its guidelines for NWDPRA, making them "more participatory, sustainable and equitable. These were called WARASA – JAN SAHABHAGITA Guidelines. The Common Guidelines of 1994 were revised by MoRD in 2001 and then again modified and reissued as "Guidelines for *Hariyali*" in April 2003. Based on several evaluation studies conducted by various organizations, it emerged that successes so far have been sporadic and intermittent, the overall impact at the state and national levels being generally inadequate. Besides, additional challenges were emerging. National Rainfed Area Authority (NRAA), under the overall guidance of the Planning Commission, developed Common Guidelines for Watershed Development Projects. It will have a unified perspective by all Ministries (NRAA, 2011).

Why Soil and Water Conservation?

Soil and water conservation is a combined term used to describe the two intricately inter-connected terms namely soil conservation and water conservation. For a better perspective, let us look at the issue under the two separate heads.

Soil Conservation

Soil erosion is the most serious environmental problem being faced by the mankind. If we do not protect our soils from this enemy, it may be nye impossible to feed the ever increasing population as is evident from the following projections.

Assessment made by various investigators reveals that in India, out of a total geographical area of 329 M ha around 137 M ha is subjected to water and wind erosion. It is expected that by 2024, India's population will be about 1.44 billion, which will further rise to 1.51 billion by 2030. The per capita cropped area therefore, will shrink to 0.093 ha against 0.15 ha in 2000. It may shrink even less than this value as a result of rapid urbanization and industrialization. Thus, to feed about 18 per cent of the global population with a limited land resource of about 2.3 per cent will be an uphill task. Clearly, the limited land resource has to be managed carefully so that it is neither degraded by any degrading agent nor its productivity is reduced on a long-term basis.

Even on a global scale the problem is quite serious. According to the UN Food and Agriculture Organization, the world is losing an equivalent of 5-7 M ha of farmland through erosion every year (See section on extent and distribution in this chapter).

The soil erosion process more or less is always in operation and cannot be eliminated completely. At naturally occurring rates, land typically loses about 2.5 cm of topsoil in more than 100 years (0.025 cm per annum). Soil erosion rate occurring at 55-91 kg/ha per year is considered tolerable. If the rate exceeds this limit, it becomes a serious issue. Soil erosion, besides causing loss of precious soil, also causes the loss of organic matter and nutrients that go with the soil adversely impact the land productivity. Several off-site effects appear in the form of sedimentation in dams and reservoirs. The fertilizers and pesticides from the eroded soils pollute rivers and lakes. To overcome all these adverse impacts, especially when the soil loss exceeds the tolerable limits, one has to go for soil conservation. In essence, it means to use the land according to its needs and capabilities.

Water Conservation

Water is finite, inelastic, inadequate and unreliable (having wide variations in time and space) source globally, India being no exception. India, on an average receives 4000 km^3 of water, which is even less than 4 per cent of the global replenishable water resource (See chapter 16). Amount of water available per capita in India is likely to reduce to1401 m^3 and 1191 m^3 by the years 2025 and 2050. These values are much less than the 1700 m^3 per capita requirement for comfortable living. The degree and extent of the problem can be gauged from the fact that 21 cities in India may run out of groundwater by 2020. Our water requirement is likely to increase by about 2 times the current requirement by 2030. If we look at another angle, average annual rainfall in India is more than the world average. But the rainwater use efficiency in India is only to the extent of 20-30 per cent. Thus, it has emerged that our water woes are mainly related to mismanagement of water resource rather than its inadequacy. It calls for a management approach that takes into account the integrated use of rain, canal, ground and waste waters.

According to the dictionary meaning, water conservation is the preservation, control and management of water resources. From soil conservation point of view, control and management being crucial. Water management begins as soon as precipitation begins, and continues through its transportation until it is safely stored for future use. All this time, it should be ensured that water preserves the sanctity of soil on and through which it flows so that soil erosion does not exceed the tolerance level as discussed in the previous section. The role of soil and water conservationist in this respect is crucial.

Objectives of Soil and Water Conservation

Soil is the most fundamental and basic natural resource for all life to survive. Erosion, on the other hand, is almost universally recognized as a serious threat to human well being. Erosion reduces the productivity of crop land, is an important social and economic problem and an essential factor in assessing ecosystem health and function. It is a slow process and initially goes unnoticed. But can also occur at alarming rates, resulting in serious loss of the topsoil. For sustained productivity at a reasonably high level, soil erosion must be prevented, and water conserved for its judicious use without impairing its quality. Sustainable production, which is hallmark of any agricultural programme, implies that agricultural practices should ensure highest economic gains without impairing the quality and usefulness of the soil for future use. The objectives of a soil and water conservation programme are briefly stated as:

- Promotion of proper land use
- Prevention of soil erosion
- Restoration of the productivity of eroded land

- ☆ Maintenance of soil productivity
- ☆ Control of runoff
- ☆ Regulation of water resource for and through irrigation and drainage.

In the pursuit of achieving these objectives, a professional soil conservationist applies the principles of several specialized disciplines such as agricultural engineering, soil science, agronomy, forestry, or agriculture. He plans and conducts surveys and carries out investigations to decide and implement measures required to restore or maintain proper soil management. He coordinates the implementation of soil erosion control and moisture conservation practices for sound land use. He assesses the relative merits of soil management practices, such as crop rotation, reforestation, permanent vegetation, contour ploughing, or terracing *etc.* He undertakes watershed management activities including engineering interventions. He prepares soil conservation plans in cooperation with local government, farmers, foresters, planners at the state and central level.

What is Soil Erosion?

The uppermost weathered and disintegrated layer of the earth's crust is referred to as soil. The soil layer is composed of mineral and organic matter with a mix of air and water in its pores necessary to sustain plant life. The soil depth may vary from practically nil to several meters. The top soil layer is continuously exposed to the actions of atmosphere. Wind and water in motion are two main agents, which act on the soil layer and dislodge the soil particles and transport them to deposit at some point downstream. The process of loosening of the soil from its place and its transportation from one place to another is known as soil erosion. Accordingly, erosion has been described as a two phase process, one the detachment of individual soil particle from soil mass, and two its transportation from one place to another (by the action of one or a combination of erosion the agents namely water, wind, ice or gravity). Some consider soil erosion as a three phase process. According to them, as soon as sufficient energy is not available to transport a particle, a third phase known as deposition of soil particles (sediments) begins. Hence soil erosion is defined as a process of detachment, transportation and deposition of soil particles (sediment) as described below. It is evident that sediment is the end product of soil erosion.

Detachment is the dislodging of the soil particle from the soil mass by erosive agents. Water, a major erosive agent, impacts the top soil through the raindrops (detachment) and runoff water flowing over the soil surface entrains and moves the detached soil particles (transportation) from their original location. Sediments move from the upland sources through the stream system and may eventually reach the ocean. Not all the sediment reaches the ocean; some are deposited at the base of the slopes, in reservoirs and flood plains along the way.

Agents of Soil Erosion

Soil erosion, as already defined, is the detachment of soil from its original location and transportation to a new location. Although water is the main erosion agent, at many locations wind and gravity assume significance as eroding agents (Figure 22.1). Glacier erosion is sometimes included as a sub-group of water erosion (Figure 22.1), but is included as a separate group in few textbooks.

Types of Soil Erosion

On the Basis of Origin

Soil erosion can broadly be categorized into two types namely geologic erosion and accelerated erosion.

Geological Erosion

Under natural undisturbed conditions, equilibrium is normally established between the climate of a place and the vegetation that protects the soil layer. Vegetative cover (trees and forests) retard the transportation of soil material and act as a check against excessive erosion. A certain amount of erosion, however, does take place even under the natural cover. This erosion, called geologic erosion, is a slow process and is compensated by the formation of soil under the natural weathering process. The effects of geological erosion have little consequence so far as agricultural lands are concerned.

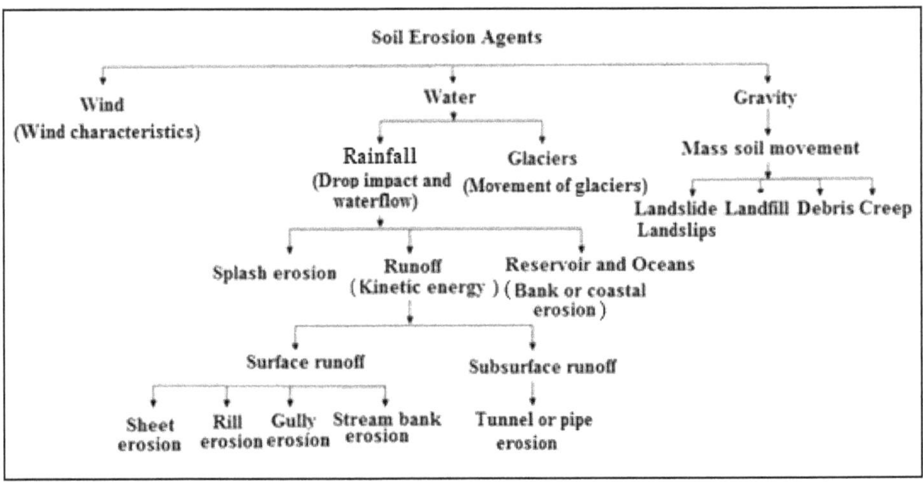

Figure 22.1: Description of Various Soil Erosion Agents and Kinds of Soil Erosion.

Accelerated Erosion

When land is put under cultivation, the natural balance existing between the soil, its vegetation cover and climate is disturbed. Under such condition, the removal of surface soil due to natural agents takes places at a faster rate than it can be built by the soil formation processes. Erosion occurring under these conditions is referred to as accelerated erosion. Its rate is higher than geological erosion. Accelerated erosion causes depletion of soil fertility of the agricultural lands.

On the Basis of Eroding Agents

Soil erosion is broadly categorized into different types depending on the agent which triggers the erosion activity. Mentioned below are the four main types of soil erosion categories on the basis of eroding agents. Two natural eroding agents namely water and wind, are constantly at work, cover large areas and are visible to the mankind. Therefore, these eroding agents are categorized as major eroding agents. The other two are categorized as minor eroding agents

Water Erosion

Rainfall and running water is the most common agent of soil erosion. This includes rainwater which erodes various landforms, rivers which erode the river basin, and the sea waves which erode the coastal areas. The two principal erosive agents that become active on land during rainstorms are the falling raindrops and flowing water. The energy of falling raindrops is applied from above, and their principal function is to detach soil particles from the soil mass. The energy of flowing water is usually applied parallel to the surface; its principal function is to transport the soil material. Both raindrops and flowing water are complete erosive agents in themselves, but in the erosion process they work together. Based on different actions of water responsible for erosion, water erosion is further classified as: (i) raindrop erosion, (ii) sheet erosion, (iii) rill erosion, (iv) gully erosion, (v) stream bank erosion, and (vi) slip erosion.

Wind Erosion

In arid and semi-arid regions having little rainfall, the wind acts as a powerful agent of soil erosion causing heavy loss to production and productivity of agricultural lands. Winds blowing at considerable speed, remove the fertile, arable, loose soils leaving behind a depression devoid of top soil. Wind erosion is accentuated when strong winds are blowing, the soil is dry, weakly aggregated, and devoid of vegetation cover. Even modest wind velocities can keep individual particles of humus, clay and silt in suspension. The best examples of wind erosion are sand dunes typically found in deserts, which are formed and transported from one place to another.

Glacial Erosion

Glacial erosion, also referred to as ice erosion, is common in cold regions at high altitudes. When soil comes in contact with large moving glaciers, it sticks to the base of moving glaciers. The soil is eventually transported along with the glaciers. As the glaciers melt, the soil is deposited in the course of moving chunks of ice.

Gravitational Erosion

Although gravitational erosion is not as common a phenomenon as water erosion, it can cause huge damage to natural, as well as man-made structures. It is basically the mass movement of soil assisted by gravity. Some of the forms of mass movement include soil creep, earth flow, slumps, landslips, landslides and rock avalanches.

Earth flow: Slabs of soil become liquid and flow downhill

Slumps and landslips: Usually caused by rotational failure of sub-soil layer

Rock avalanches: Comprises of collapse of rock structures

Soil creep: Slow downhill creeping and rolling of soil particles

Terracetting: Minor form of mass movement caused by livestock traversing hillsides

While landslides and slumps happen within seconds, phenomena such as soil creep take a longer period for occurrence. Landslips, landslides and earth flows occur on steep slopes with high clay content. These are triggered by over-saturation of soils from excessive rainfall. As the soil weight increase following a rain event and layers within the soil get lubricated, a land slip occurs. Slips usually occur along a weak plane of highly lubricated soil layer. Landslips, slides and earth flows may also be triggered by erosion at the base of the slope. Although relatively infrequent, large mass movement events are dramatic, resulting in permanent loss of houses, roads and agricultural land.

Landslip prevention and control usually requires a combination of vegetative and engineering measures. Where landslips occurs as a result of clearing of vegetation, re-vegetation of cleared areas with densely planted tree species above the slip zone is necessary to de-water the area (evapotranspiration) and to stabilize soils. Surface and subsurface drains need to be provided within and above the affected areas. In some cases engineering structures such as soil pins, retaining walls, rock filling and slope grooming may also be needed to prevent or control landslips.

Factors Affecting Soil Erosion

No single unique factor is responsible for soil erosion. There are several factors that cause or aid the process of soil erosion. Some of these factors are natural (caused by nature) while others are due to human intervention (Anthropogenic). From a general point of view, the factors that affect soil erosion can be clubbed under the following groups.

Energy

Some kind of energy is required to detach and transport the soil often referred as erosivity of the eroding agent. Depending upon the kind of erosion, it is derived from water or wind or other processes such as waves in case of shore erosion. The greater the energy to erode the soil, the higher will be the amount of eroded soil. High rainfall erosivity, high winds and high runoff volumes increase soil erosion. The groups of factors that enhance the eroding power also enhance the soil erosion. Some of these include relief, slope length, length of wind fetch *etc.* Topography, especially the slope, is a major contributor in affecting the erosivity of the eroding agents. Erosion is accelerated with increasing slope since the flow velocity increases with increasing slope. All factors that reduce the power of the eroding agent help to control the soil erosion.

Resistance

Under the same conditions, some soils erode more readily than others. The main factor that influences the resistive forces is the soil type. Fine textured and alkali soils are more erodible. Sandy soils are relatively less erodible by water because water is absorbed readily due to its high permeability. On the other hand, these are more erodible if wind is the eroding agent. All physical and chemical properties of the soils can be counted under the group of factors that impact the soil erodibility. These include soil texture, structure, organic matter content, nature of clay and the amount and kind of salts present in the soil. Low infiltration rate and low organic matter content increase the soil erodibility. On the other hand, high organic matter content improves granular structure of the soil and water holding capacity resulting in lesser erosion. Besides, all cultivation practices (soil management) that enhance or reduce soil erodibility are also included in this group.

Protection

The presence of vegetation (ground cover) is known to retard soil erosion. Lack of vegetation, on the other hand, creates erosion permitting condition. Soils protected by good cover have higher surface roughness, absorb rain splash, intercept energy, provide soil reinforcement (bind the soil), promote good soil structure, reduce runoff volume, create more storage in the soil and trap sediments. Forests and grasses are more effective in providing soil cover than cultivated crops.

Anthropogenic Interventions

Anthropogenic factors that aid in the erosion impact one of the three factors *i.e.* increases the energy, reduces the resistance or changes the protective cover. Some of such activities are listed as under:

- Cultivation along the land slope instead of across the slope
- Unscientific planning and alignment of development activities such as roads, rail and other networks and buildings
- Faulty methods of irrigation and drainage
- Destruction of natural protective cover through
  - Indiscriminate cutting of forests
  - Over grazing
  - Shifting (*Jhoom*) cultivation
- Forest fires that remove the protective cover
- Keeping the land barren subjecting it to the action of rain and wind
- Growing crops that accelerate soil erosion
- Removal of organic matter and plant nutrients by injudicious cropping patterns and residue burning
- Climate change, which is also anthropogenic in nature, is likely to affect soil erosion especially by water through its effect on rainfall intensity, soil erodibility, vegetative cover, and patterns of land use. Besides, areas where climate change is predicted to lead to more droughty soils under

increasing temperatures will become increasingly vulnerable to wind erosion.

A detailed discussion on some of these issues is included in the sections on factor affecting water and wind erosion in respective chapters.

Ill-Effects of Soil Erosion

No soil phenomena are more destructive than soil erosion. Erosion damages the site on which it occurs. Additionally, it has undesirable off-site affects and has economical and environmental implications.

On-Site Effects

- Soil erosion removes the top soil, the most productive layer of the soil profile. Soil quality is significantly reduced as a result of loss of the nutrient-rich upper layer and reduced organic matter adversely impacting the ability of the land to regenerate new flora or support crops. Under severe conditions, crops yield are low or even a complete loss of the standing crop(s) is not ruled out
- With the top soil gone, eroded soils may have rills and gullies that may pose additional problems during crop cultivation
- Soil compaction reduces the ability of soil to absorb water resulting in high runoff, increasing the risk and rate of soil erosion. Too much compaction in sandy soils can cause crusting and sealing of the surface layer
- Sub-soil layers often having poor physical and chemical properties are exposed
- In extreme cases of water and wind erosion, there may be change in the soil texture impacting the water-holding capacity of the soils. Reduced water-holding capacity makes the soils more susceptible to extreme conditions such as drought
- Wind erosion is very selective, may reduce the capacity of the soil to store nutrients and water, thus making the environment drier
- Breakdown of aggregates, removal of smaller particles or entire layers of the soil and loss of organic matter may weaken the soil structure
- The chances of other problems such as acidity, water logging (surface stagnation) and soil salinity to appear increase manifold
- Seeds and plants can be disturbed or completely removed by the erosion that may impact crop emergence and growth
- May spread plant diseases and non-friendly organisms
- Adversely impacts the pasture lands especially where overgrazing is practiced
- The disappearance of vegetation may trigger avalanches, landslides and floods

- May undercut pavements and foundations resulting in increased expenditure on their upkeep
- New seeds and seedlings are buried or destroyed. This may impact future crop production
- Damage to the sea coast.

Off-Site Effects

The soil that is detached by water or wind erosion is transported to far and near places and most of time ends up at undesirable places. This gives rise to 'off-site problems' i.e away from the sites where soil erosion is taking place.

- The movement of sediments into water sources can lead to silting-up of dams, disruption of the ecosystems of lakes, and contamination of drinking water
- Siltation of rivers (eroded sediments get deposited on river beds) may result in increased downstream flooding
- Sediments can also accumulate down-slope and damage the infrastructure such as roads, buildings *etc.*
- Sediments that reach streams or water courses can accelerate bank erosion, obstruct stream and drainage channels, fill in reservoirs, damage fish habitat and degrade downstream water quality
- Increasing bank erosion leads to loss of valuable land, and threatens the infrastructure on or along theses channels
- Increased turbidity of water prevents penetration of sunlight that may impact the aquatic life
- Runoff and sediments from agricultural lands may be laden with chemical contaminants such as from fertilizers or pesticides that may pollute downstream water sources, wetlands and lakes. The nutrient ions may result in eutrophication of these resources
- Soil quality, structure, stability and texture of the soils can be altered due to deposition of the soil on the top layer
- Increase in disease and other public health hazards
- The most spectacular off-sit effect of wind erosion is movement of dunes - mounds of more or less sterile sand. It settles at far off distances affecting the productivity of the lands, and may even bury the oases and important social and cultural sites
- Sheets of sand travelling close to the ground (30 to 50 m) can degrade standing crops on its way. Fruit and foliage is most susceptible to wind blast
- Soil particles in the air contribute to air pollution having close links to human respiratory and skin diseases.

Economic Effects

Both on-site and off-site effects have enormous economic consequences. Some of these are:

- Soil erosion may lead to crop failures resulting in lost farm income in extreme cases even the livelihood resource
- Recurring floods and droughts may adversely impact the income of the farmers besides resulting in huge losses in the short and long-terms
- The losses may lead to migration (temporary or permanent) causing immense cultural and economic shock to migrants
- Loss of soil and vegetation may blow up the adverse impacts of climate change, disturbing the ecologic balance and loss of natural resources.

Extent and Distribution of Erosion Problem

The total land area subjected to human-induced soil degradation is estimated at about 2 billion ha (UNEP, 1997). Water erosion is by far the most important type of soil degradation affecting about 1100 M ha (Oldeman *et al.*, 1991) worldwide (56 per cent of the total area affected by human-induced soil degradation). The most widespread form of water erosion is the loss of the topsoil (on 920 M ha), while terrain deformation (rills and gullies) occurs on 175 M ha. Wind erosion is a widespread phenomenon in arid and semi-arid zones, usually on coarse textured soils with a limited vegetative cover. The global extent of soils affected by wind erosion is around 550 M ha (28 per cent of the total terrain affected by soil degradation). Similar to water erosion, the loss of topsoil by wind forces is by far the most important type of wind erosion (on 455 M ha), while terrain deformation (82 M ha) and over blowing (12 M ha) are relatively speaking less important in extent, although these types of wind erosion may have very serious effects on human activities.

The extent of the water erosion in India has been estimated by different agencies, the extent varying from 87 to 111 M ha. As per NBSS and LUP soil degradation classes, derived from 1:250,000 soil map (1985–1995), it emerged that water erosion is affecting an area of 83.31 M ha resulting in loss of top-soil while terrain deformation is reported from 10.37 M ha. According to this statistics, wind erosion covers an area of 9.48 M ha. As per NRSA wasteland classes (1986–2000), gullied/ravine land covers an area of 2.06 M ha. All these figures have now been reconciled by ICAR-NAAS (2010) and it is estimated that total area affected by water erosion (> 10 t/ha-year) is 126 M ha (Table 22.1) out of which 12.96 M ha has soil loss exceeding 80 t/ha-year. An area of 11 M ha suffers from wind erosion of various intensities.

From erosion point of view, India can be divided into the following regions:

Himalayan Region: This region is most vulnerable to erosion. High degree of seismicity of the area, very steep slopes, weak geological formation and improper land use practices accelerate erosion losses. Gully formation, landslides and slips are most common. Floods, stream bank cutting and sand deposition have degraded lands of the northeast region that experiences intense storms during the monsoon season.

Table 22.1: State-wise Area Affected by Water Erosion (> 10 t/ha-yr)

State	*Water Erosion Affected Area (km^2)*	*Percent of Total Geographical area*
Andhra Pradesh	109,275	40
Arunachal Pradesh	60,839	73
Assam	51,604	66
Bihar	12,189	23
Chhattisgarh	88,028	65
Delhi	330	22
Gujarat	31,364	16
Haryana	2,918	7
Himachal Pradesh	18,038	32
Jammu and Kashmir	35,024	16
Jharkhand	51,750	65
Karnataka	93,978	49
Kerala	5,927	15
Madhya Pradesh	182,870	59
Maharashtra	105,269	34
Manipur	11,898	53
Meghalaya	17,472	76
Nagaland	14,468	87
Odisha	49,468	32
Punjab	3,369	7
Rajasthan	89,769	26
Sikkim	2,616	37
Tamil Nadu	25,660	20
Tripura	4,027	38
Uttar Pradesh	142,772	59
Uttarakhand	31,069	58
West Bengal	17,919	20
Total	**1,259,910** **~ 126 M ha**	**39**

Note: Andaman and Nicobar Islands, Goa and Mizoram have not been evaluated.

Gangetic Plains: Major problems in the region are riverine erosion, drainage, saline and alkali soil conditions.

Peninsular Region: Main problems of this region are rill and gully erosion.

Arid Regions: These areas are severely eroded as a result of wind erosion.

Semi-arid Regions: These are subjected to sheet and gully erosion. Ravines are serious problem in Yamuna and Chambal river catchments. South and south east are characterized by undulating terrain with severe erosion in black and red lateritic soils.

Chapter 23

Water Erosion

The soil erosion caused by water is known as water erosion. In this case, water acts as an agent to dislodge and transport the eroded soil particle from one location to another. Water erosion is classified as splash erosion, sheet erosion, rill erosion, gully erosion, stream bank erosion and sea-shore erosion. Sometimes land slide erosion is also included in this group because water is the main agent in this kind of erosion.

Types of Water Erosion

Splash Erosion

Splash erosion is the first stage of water erosion (Figure 23.1). It is also known as raindrop erosion because it is caused by the impact of raindrops on the exposed soil surface. In this case, rain drops behave as little bombs when falling on exposed or bare soil. When raindrop strikes, a crater is formed. The explosive impact breaks up soil aggregates so that individual soil particles are 'splashed' onto the soil surface (Figure 23.2 right). It is accomplished by forming a blast which bounces the water and soil up and returns back around the crater. The soil may be splashed into the air up to a height of 50 to 75 cm depending upon the size of rain drops. At the same time the soil particles also move horizontally as much as 1.50 m on level land surface. The soil particles so detached and transported also block the spaces between soil aggregates, forming a crust that seals the soil surface. It causes water to accumulate and result in high runoff because of reduced infiltration. The compacting action by the falling raindrops also causes soil to suddenly lose its capacity to absorb water. All these factors result in high runoff during heavy rains that imparts enough energy to transport the soil particles. Besides the soil particles, the plant nutrients are also removed and transported to adversely impact the land productivity. It may be noted that splash erosion alone may not be able to move enough quantities of soil downhill to any appreciable extent. But because of a thin pre-channel surface flow

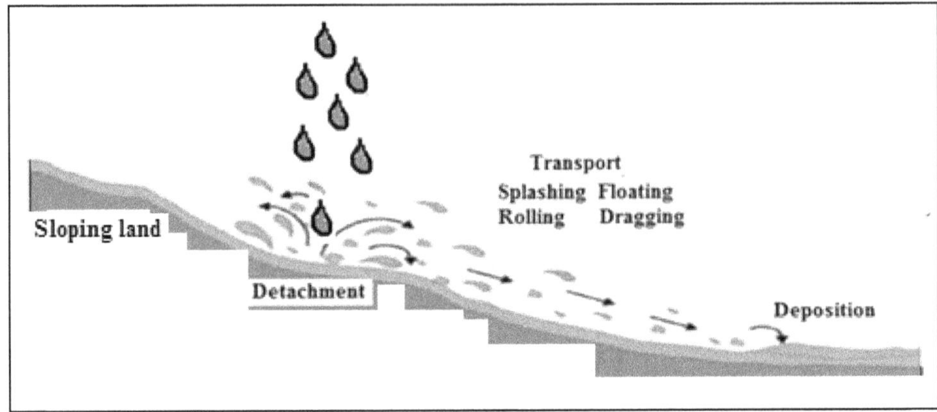

Figure 23.1: Soil Detachment and Transport of Detached Particles on Sloping Lands.

of water and some of the soil splashed remaining in suspension in the surface flow, moves the soil particles into rills and then to larger channels. The combined action of raindrop splash and the flow of water accounts for nearly all of the sediment removal from individual fields.

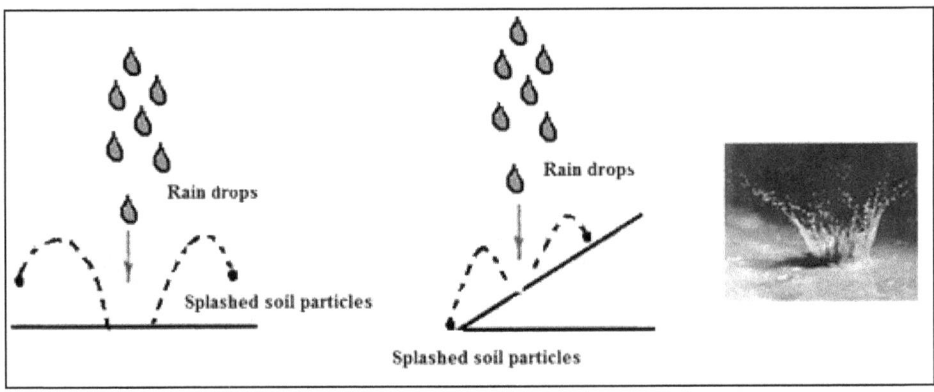

Figure 23.2: Splash Erosion: Flat Lands (Left), Sloping Lands (Middle) and a Pictorial View (Right).

Factors Affecting Splash Erosion

The potential ability of rainfall to cause soil erosion is called erosivity. In evaluating rainfall erosivity, rainfall should first be perceived as an aggregate of different drops of water. It can further be perceived as the summation of the individual drops. While individual drops play a major role in detaching the soil, the cumulative drops affect runoff, transport and deposition of detached soil. Therefore rainfall characteristics which influence splash erosion or rainfall erosivity are:

- ☆ Amount, duration and intensity
- ☆ Drop size and drop size distribution

- ☆ Terminal velocity
- ☆ Kinetic energy

Amount and Intensity

Rainfall amount, duration and intensity vary from one storm to another. These values are obtained from the measurements using non-recording/recording rain gauges.

Drop Size

Rainfall is made up of all possible drop sizes; the proportion of the different drop sizes varying from one rain event to another. Raindrop size hardly exceeds 5 mm, but when they coalesce they may momentarily exceed this value. In general, rains in the tropics have relatively bigger drops than those in temperate regions. Drop size distribution of rainfall is described by a parameter called the median drop size or median volume drop diameter (D50). D50 refers to the drop diameter at which 50 per cent of the volume of the rain occurs in the shape of drops with a smaller diameter and 50 per cent as bigger diameter drops. The median drop size is affected by several factors such as type of rain, amount of rain, duration of rain, rainfall intensity and wind velocity accompanying a rain event. High rainfall intensity comprises of high median drop size or form more drops per unit area per unit time. Median drop size generally increases with increase in rainfall intensity up to a certain limit according to the relationship: $D50 = 2.23\ I^{0.182}$, the relationship is valid up to rainfall intensities of 75 mm/hr. The general relationship is written as $D50 = a\ I^{b}$. Here D50 is in mm and I is in mm/hr. The constants vary with rain types.

Terminal Velocity

A body falling freely under the force of gravity will accelerate until the frictional resistance of the air is equal to the gravitational force. The raindrop velocity depends upon the height of fall up to a height of about 10.5 m after which it attains the terminal velocity. Under natural condition, terminal velocity increases with an increase in drop size and may vary from 4.5 to 9 m/s. Soil loss increases with the increase of the terminal velocity. The raindrops sometimes fall at a high speed of 50 kmph. In that event soil particles may be splashed to a height of 60 cm and move laterally to a distance of 150 cm. The terminal velocity of the rainfall is related to the diameter of the rainfall by the following relation (Slastikhin, 1964).

$$Vt = 13\sqrt{d} \qquad (23.1)$$

Here Vt is the terminal velocity of the falling raindrop, m/s and d is the drop diameter, cm. Clearly, the bigger the size of rain drops, the greater the kinetic energy and greater is the rain drop erosion.

Kinetic Energy (KE) and Splash Erosion

The process of raindrop erosion is related to the kinetic energy of the rain drops. The kinetic energy of falling raindrop (a function of size, mass, shape, velocity and direction of the raindrops) dislodges the soil particle and the resultant runoff transports soil particles. KE is the product of the mass of rain falling per unit time multiplied by the square of its terminal velocity. The KE of a raindrop is easily expressed in terms of its properties as:

$$KE = 0.5\ m\ V^2 = \pi\rho\ D^3\ V^2/12 \quad (23.2)$$

where ρ is the water density, *D* is the diameter of the equivalent spherical drop and *V* is its terminal velocity. So, the *KE* of a rainfall event per unit surface can be directly computed knowing the *D* and the *V* of each raindrop falling on a unit area. These features are not normally measured in field studies on erosion. Therefore, relations between *KE* and rainfall intensity are normally used to assess KE.

The kinetic energy of rainfall can be measured directly using equipment such as acoustic sensors, piezoelectric sensors and pressure transducers. Since these instruments are not often used routinely, empirical relations are used to estimate KE. Amongst several relations used for this purpose, the most commonly used equation is that of Wischmeier and Smith (1958).

$$KE = (11.9 + 8.73 \log I)\ I\ \Delta t \quad (23.3)$$

Here KE is the kinetic energy, J/m^2 of the rainfall, I is the intensity, mm/h falling over a unit surface area during a time step Δt, h. The simplified version of this equation is

$$KE = 210.3 + 89 \log I \quad (23.4)$$

Here KE is in metric tons per ha per cm of rain and I is the rainfall intensity in cm/h.

These equations rely on average relationships between I and the drop size distributions. This equation also forms a part of Universal Soil Loss Equation. The kinetic energy of a raindrop amounts to 10^4 ergs for drops of 2 mm radius. The kinetic energy contained in a drop of 2.5 mm radius is sufficient to raise by 1 cm a body weighing 46 grams.

Besides the rainfall characteristics, other factors that influence soil erosion are as follows.

Soil Type

Sandy soils are easier to detach than clay soils. On the other hand once detached, clay particles may be easier to transport to farther distance. Besides the size of soil particles, other soil characteristics such as organic matter content, presence of cementing materials such as Fe and Al oxides, cations present on the exchange complex (Na is more harmful from this point) and structural stability all influence the detachability.

Wind Velocity

The wind velocity and air resistance also affects the raindrop velocity. Direction and strength of wind velocity deflect the rain drops and may reduce their kinetic energy and thereby the rain drop erosion. On the other hand, high wind velocity in the direction of slope causes higher splash.

Topography

In a level plane one may not anticipate much problem since the soil is just shifted from one place to another. But in sloping lands the detached soil particles are easily transported down the slope. In between these particles may join a rill or a gully to move further downward as the movement is made easier. Compared to upslope, a greater amount of soil material is splashed towards slope because of the force of gravity (Figure 23.2 middle). It means the soil is transported down slope even if there is no runoff. The amount of soil carried down slope mainly depends on the drop size, wind velocity, slope steepness and surface conditions. The same soil particles may be splashed more than once. As a result, they are detached from the main soil body and are easily carried away by the runoff water. During a heavy rainfall event, as much as 200 t/ha of soil is splashed into the air. On a sloping land, more than half of the splashed particles move down with the runoff (Figure 23.1).

Vegetation

Erosion due to rain drop splash reduces if soil surface is rough and causes obstructions to the falling drops or runoff. Raindrops falling on plants, crop residues or other mulches lose their energy before striking the soil. Canopy height and extent of the vegetative cover influence the splash erosion. Ground cover maintained at a level exceeding 70 per cent prevents the detachment of soil particles and reduces initiation of sheet and rill erosion. For erodible soils or having high gradients, ground cover at levels between 80-100 per cent needs to be maintained to reduce soil erosion.

Sheet Erosion

When the rainfall intensity exceeds the infiltration rate of the soil, water begins to flow over the sloping surface. It carries the detached particles in the form of a thin sheet. The removal of more or less uniform thin layer or sheet of soil from a given area of land surface by the action of water is called sheet erosion. Soil deposits on the high side of obstructions such as fences are indicative of active sheet erosion. Two basic erosion processes involved in the sheet erosion are 1) soil particles are detached from the soil surface by falling raindrops and 2) detached soil particles are transported away by surface runoff from the original place. The detaching process is already discussed in the splash erosion. The transportation of detached particles by the flowing water is known as wash erosion. It may be noted that even the flowing water is capable of detaching soil particles from the land surface. The eroding and transporting power of sheet flow depends on the depth and velocity of flowing water for the given size, shape and density of soil particles. Sheet erosion causes the loss of the finest soil particles that contain most of the available nutrients and organic matter. Since the soil loss in this kind of erosion is so gradual, most of the time it goes unnoticed. Moreover, the surface flow that causes sheet erosion rarely flow for more than a few meters before concentrating into rills (Figure 23.3). It is why it is sometimes called an inconspicuous type of soil erosion. But the cumulative impact may result in huge soil losses. The term inter-rill erosion is many times used for the combined erosion resulting from the splash and sheet erosion.

Rill Erosion

Sheet erosion mainly occurs on a smooth and uniformly sloped land. When the land surface is irregular, the water tends to accumulate in the surface depressions and begins to flow over the least resistant path. In this kind of water movement, large amounts of soil particles are eroded from the sides and bottom of the flow path. The soil particles so eroded gets mixed in the flowing water. As the surface flow containing soil particle in suspension form moves ahead, it forms micro-channels or rills. The rills are shallow drainage lines less than 30 cm deep and 50 cm wide. If erosion continues unchecked for a sufficient time, and silt laden run off moves over the land surface, numerous finger shaped grooves may develop all over the area (Figure 23.3). The whole pattern resembles that of the twigs, branches and trunk of a tree. The whole process is called rill erosion. Rill erosion is common in bare agricultural land, particularly overgrazed land, and in freshly tilled soil where the soil structure is quite loose. The rills can be easily removed with ordinary farm machines. Detachment and transportation of soil particles in rill erosion always exceed than that in the sheet erosion. Rill formation is an intermittent process as rills can easily get transformed to gullies (gully erosion) if no precautionary measures are taken. In fact the advance form of the rill is the initial stage of gully formation.

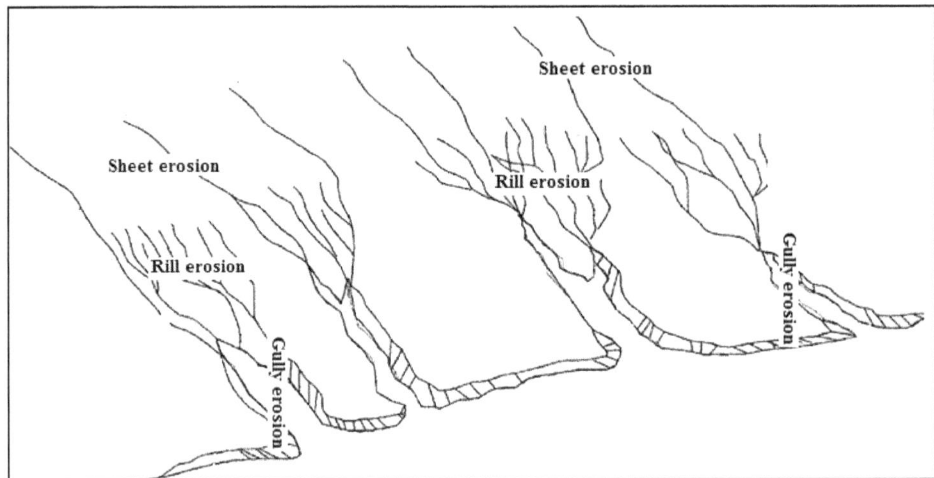

Figure 23.3: Sheet, Rill and Gully Erosion.

In rill erosion, the detachment of soil particles is from the energy of flowing water and not by the raindrop impact as in the case of sheet erosion. Detachment of soil particles by flowing water in rill erosion (D) varies with square of velocity of flowing water (V^2) and ability of flowing water to transport soil particles (T) varies with fifth power to the velocity of flowing water (V^5).

$$D \propto V^2 \text{ and } T \propto V^5 \tag{23.5}$$

Gully Erosion

If the rills formed in the field are overlooked by the farmers, rills are enlarged in size and shape as a result of prolonged flow through them. Rills can even extend

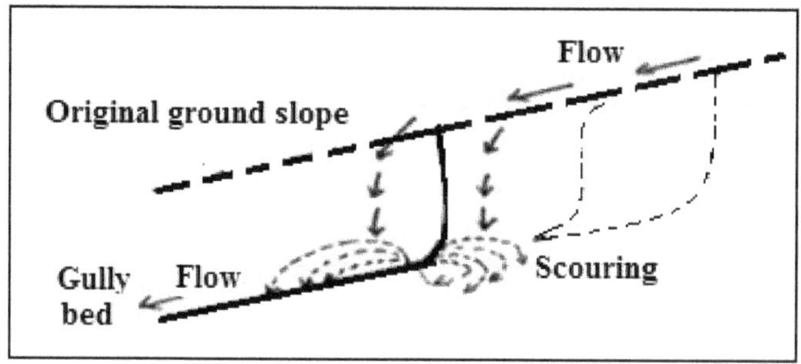

Figure 23.4: Formation of Gully with Flow Lines at the Gully Head.

to sub-soil. If the rills reach a stage such that these cannot be removed by normal tillage operation, then these are called gullies (Figure 23.3). The process of gully formation is initiated when the depth and width of the rills is more than 50 cm. A ravine is formed, whenever a gully bed cuts into the soil with an immediate drop of 3 to 4 m that gradually flattens out. The network of such gullies is known as ravine area. It is the last and advanced stage of water erosion. Some of the major causes of gully erosion are: steepness of land slope, soil texture, rainfall intensity, land mismanagement, biotic interference with natural vegetation, and unscientific cultivation practices, *etc.* Gully erosion gets initiated when the longitudinal profile of an alluvial land becomes too steep as a result of sediment deposits. Gullies further advance as a result of soil removal by the flowing water at the base of a steep slope, or a cliff (Figure 23.4). In the absence of proper control measures, slowly the gullies extend to nearby areas and subsequently engulf the entire region with a network of gullies of various sizes and shapes. The sizable soil loss occurs during gully formation which subsequently causes sedimentation downstream in rivers or other water resources.

Stream Bank Erosion

Stream bank erosion is defined as the removal of stream bank soil by water either flowing over the sides of the stream or caused by wave action or in the form of scouring and undercutting of the soil below the water surface. Stream bank erosion is a serious problem where development activities have limited the meandering nature of streams or where streams have been channelized, or where stream bank structures (like bridges, culverts, *etc.*) are improperly located at places where they can actually cause damage to downstream areas. Stream bank erosion is also caused by the occurrence of flood in the stream. The stream bank erosion is a continuous process in perennial streams. It is aggravated due to removal of vegetation, over grazing or cultivation on the area close to stream banks. The three main processes in this kind of erosion are termed as scour, mass failure and slumping. Bank scour is the direct removal of bank materials by the physical action of flowing water and is often dominant in smaller streams and the upper reaches of larger streams and rivers. Mass failure, which includes bank collapse and slumping, is where large chunks of bank material become unstable and is toppled into the stream or river in single events. Mass failure is often dominant in the lower reaches of large streams and often occurs in association with scouring of the lower banks. Since the management required to slow or prevent the stream bank erosion differ from one process to another, it is necessary to determine which processes or the combination of processes are operating at a particular site.

Tunnel Erosion

Tunneling is an insidious form of subsurface erosion, insidious because erosion is happening much before its surface manifestations are evident. It is known by different names such as tunnel-gully erosion, soil piping, piping erosion or simply the piping. Tunnel erosion is likely to occur in areas having rolling to steep slopes where there is a variation in the permeability within the soil profile such as easily drained soil or sub-soil overlying an impermeable layer (Figure 23.5). It is also common in semi-arid regions where soil and geological substrata are highly fissured. It is also caused by the movement of excess water directly into the subsoil via surface cracks, rabbit burrows, or old root holes. The dispersive nature of the sub-soils further aids in increasing the rate of tunnel erosion. Compacted bare areas generate runoff which flows in the subsoil, causes the sodic clays to disperse and form a suspension or slurry. The slurry flows beneath the soil surface. Once a tunnel is formed, it continues to enlarge with each wet period in future. Eventually tunnels reach a point where the roof collapses resulting in potholes and formation of erosion gullies. Tunnel erosion appears as a series of tunnels formed beneath the soil surface. Tunnel erosion disrupts soil continuity, increases soil drainage, reduces water holding capacity and considerably reduces productivity. Clay transported from the tunnels is highly mobile and can readily reach streams and reservoirs to cause sedimentation.

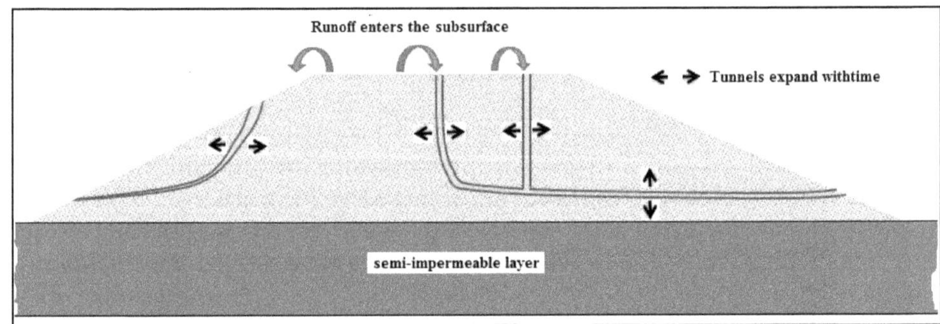

Figure 23.5: Process of Tunnel Erosion under a given Setting.

Land Slide Erosion or Mass movement

The erosion agent in this case is gravity, the soil being moved as a result of gravity force. Soil movement caused by gravity is known as landslide, mudflow, slip, slump, soil creep, surface creep and rock avalanche. It is the movement of the soil (earth, rock and soil material) through gravity down slope both slowly (few millimeters per year) and suddenly (*e.g.* rock falls). Soil covered by natural forest even on steep slopes usually remains in equilibrium with its environment, and this makes soil movement to be extremely slow. When soil cover is destroyed or greatly reduced by overgrazing, logging, cultivation, surface mining or construction *etc.*, during periods of prolonged and heavy rainfall, water entering the permeable soils is stopped by a barrier such as bedrock or a clay-rich soil horizon. The heavy weight of this saturated soil can slide down slope if it is sitting on a rock surface loosened by the build-up of water in the soil. The water seeped through the surface acts as a lubricant and causes the outer layers to detach. Then gravitational forces take over and cause the mass to slide rapidly to the foot of the hill or the bottom of the gully.

Sea Shore Erosion

Coastal erosion is another form of erosion caused by constant pounding of tidal waves and currents on the sea coast. Coastal erosion wears away the land and the removal of beach or sand dunes sediments by wave action, tidal currents, wave currents, and drainage or high winds. Waves generated by storms, wind, or fast moving motor craft, can cause coastal erosion, which may take the form of long-term losses of sediment and rocks, or merely the temporary redistribution of coastal sediments. Erosion in one location may result in accretion nearby or the sand may ultimately enter the sea. Coastal erosion is quite pronounced in the monsoon season and during storms and cyclones. Coastal areas in Gujarat, Maharashtra, Karnataka, Kerala, Tamil Nadu, Andhra Pradesh and Odisha suffer heavily at the hands of sea erosion. Of the 560 km long coast of Kerala, about 32 km stretch consisting of sandy beaches is subjected to severe sea erosion. It is now known that sea level rise will lead to increase in coastal erosion.

Other Kinds of Erosion

Some other types of erosion encountered are: rock erosion, waterfall erosion, pothole erosion and snow slide.

Mechanics of Water Erosion

Erosion, as already discussed in the previous sections comprises of the following three processes:

- ✰ Detachment
- ✰ Transportation
- ✰ Deposition

Detachment

The impact of rainfall results in splash erosion that causes the detachment of soil particles Detachment or loosening of soil particles is also caused by flowing water and freezing and thawing of the top soil. The degree and extent of these processes is governed by the type of soil, organic matter content, moisture and nature of detaching agents (energy). The following processes are involved.

Hydraulic Action: The hydraulic action takes place when water runs over the soil surface that compresses the soil. As a result the air present in the voids exerts a pressure on the soil particles, also called hydraulic pressure. It leads to the detachment of soil particles. The soil particles so detached are scoured by the running water. The hydraulic action assumes greater significance when the soil is in a loose condition.

Abrasion: Soil particles mixed with the running water create an abrasive power in the water which increases the capacity of flowing water to scour more soil particles. Due to this effect, larger soil particles are broken and eroded by the flowing water.

Attrition: This form includes mechanical breakdown of clods due to collision of particles with each other while running along the moving water. The broken particles are moved along with the flow velocity, which generate abrasion effect on the bottom and banks of the water course. This effect pronounces the water erosion.

Solution: This form is associated with the chemical action between running water and the soil or the rocks. This type of condition is observed in areas where existing rocks or soils are easily dissolved in the running water.

Transportation

Transportation of soil particles occurs due to floating, rolling, dragging, and/ or splashing. These processes are mainly governed by the size, density and shape of detached material and velocity of the transporting agent. Even large sized particles may be easily transported at high velocities. Besides the velocity, carrying capacity of the running water, load present in the water (lesser the load greater the transportation) and impediments or obstacles encountered on the way affect the soil transportation. Following processes are involved.

Solution: The water soluble contents such as chemical constituents present in the water are transported by the water in the solution form.

Suspension: Finer soil particles that are present in suspension form in the flowing water are transported through the process of suspension.

Saltation: It involves transportation of medium sized soil particles that are not able to stand in suspension form, but are mixed in water and flow over the stream bed in the form of mud.

Surface Creep: Also known as traction, is the transportation of coarser soil particle, which takes place through jumping, collision and creeping.

Deposition

Soil that is eroded from a particular location is always deposited somewhere else downstream. It happens because of some obstruction encountered along the path or due to the reduced velocity of water. It may be near to its place of origin or far off; the longest distance being down to the sea. It may also settle anywhere between the place of origin and to the sea. The exact location is dictated by the size of the soil particles and the velocity of the water. Coarser particles move the shortest distance and are first to settle. Fine sand and silt settle next as run-off water slows down. Some very fine silt settle out only when the water has come to a standstill state. Very fine clay and colloidal humus will not settle even in standing water but stay suspended in the water for a long time to come.The curves of meandering channels or winding course of the streams may result in deposits of soil load because flow velocity is significantly reduced at these points.

Factors Influencing Soil Erosion by Water

The factors that influence soil erosion by water are described in the following sections.

- ☆ Climatic factors
- ☆ Soil characteristics
- ☆ Vegetation: presence of crops and forests including vegetation management
- ☆ Topography: slope length, slope steepness and slope shape
- ☆ Anthropogenic/human interventions

Climatic Factors

Climatic factors that influence the soil erosion are: rainfall, temperature and wind velocity. Besides, the general climate of an area also influences the erosion as it determines the kind of crops grown and the periods when the land remains fallow. If the land remains fallow during critical periods, erosion will be far more than if the land is fallow during non-critical periods. Moreover, in highlands, it may be difficult to maintain the desired ground cover because of low temperature and short growing season of plants. In such areas, the intense rain is likely to cause severe erosion.

The characteristics of the rainfall that influence the soil erosion are the rainfall amount, intensity, frequency of occurrence and duration. All these characteristics, as already discussed, affect the rainfall erosivity, runoff and soil loss. If the rainfall intensity is less, it may not cause much runoff even if the total amount of rainfall may be more. Moreover, it may not be able to detach the soil particles much because of the smaller drop size. On the other hand, higher the intensity, bigger is the size of rain drops causing higher detachment of the soil particles. The bigger drops not only cause more detachment, but are also able to break the aggregates into finer particles that are susceptible to move with the runoff water. An area having frequent rains will have higher antecedent moisture content resulting in higher runoff and soil loss.

A strong wind adds a horizontal velocity component to raindrops. Wind partially reduces the air resistance by moving air horizontally away from the drop. This reduced resistance and the actual drop of the wind combine to accelerate the drop. Wind also tends to break large drops (> 5 mm) into smaller drops. Wind-driven rains are usually more erosive than non-wind driven rains. The magnitude of the difference depends on wind velocity, rainfall intensity, and slope characteristics.

Temperature is an important parameter that indirectly influences the soil erosion. It often determines the level of ground cover that helps to protect soil from erosion. The temperature also has a profound effect on the organic matter content in the soils. The soils of the arid regions are low in organic matter because warm temperature results in rapid decomposition of organic matter. The low organic matter content in soils makes the soils more susceptible to erosion especially during intense storms. Since the slopes facing south gets more sun energy than the slopes of other aspects, soils of southern aspects are low in organic matter than those facing northern slopes. Because of the low organic matter content in soil and sparse vegetation cover in southern aspects, the southern aspects are more susceptible to erosion.

Soil Characteristics

Soil texture and particle size distribution are important in sediment detachment, dispersion and transportation. A light textured soil will have high infiltration rate resulting in smaller runoff for a given storm. Similarly, a compacted soil will have lesser infiltration and higher runoff, but the detachment of the particles may be difficult. Soils with large stable particles such as sand grains or iron cemented soil particles are difficult to be detached and transported. The bigger the soil particles, the more the force required for their transportation. For example, the clay particles are difficult to detach than sand but easier to transport. The detaching capability of flowing water or runoff is also less for bigger size of soil particles.

Vegetation

Vegetation plays an important role in increasing or decreasing the rate of water erosion. It is attributed to following reasons.

- A part of the rainfall that could have reached the soil surface is intercepted by the vegetative foliage and is unable to reach to the soil surface
- Besides interception, the kinetic energy of the rain drops reaching the ground surface is greatly reduced and thereby their impact on land surface is much less than on a bare soil surface
- Vegetation improves the soil structures by adding organic matter into the soil. Soils with high organic matter content will be more porous and have high infiltration rate and water holding capacity. Further, the humus layer acts as a sponge that absorbs moisture. Soil having low organic matter and humus layer are susceptible to erosion because of high runoff from such soils
- In the vegetative areas, the root network, the channels created by root decay, animal burrows *etc.* help to dissipate runoff water
- The root system below the soil surface binds and aggregates the soil through mechanical actions. It helps to prevent soil erosion and landslides
- High surface friction due to leaf litter or increased roughness of the ground due to vegetation tends to spread out the runoff laterally dissipating its energy
- A part of the water is transpired by the vegetation resulting in reduced surface runoff and soil erosion.

Physiography

As the land slope increases from mild to steep, the soil erosion can increase manifold. It is because of higher velocity of water that causes more detachment and transportation of soil particles. The velocity of flowing water increases about twice and kinetic energy about four times as the land slope increases by about four times. The transportation capacity of the water increases about 32 times. Moreover, soil stability and slope stability decrease with increasing slope. Thus, the possibility of soil displacement in the downward direction increases with increasing slope. Steep slopes are also susceptible to higher erosion because surface detention of water is low and breaking of rain drop energy is difficult to form in steep slopes. The slope length is also an important factor affecting soil erosion, longer the slope length more is the erosion. On longer slopes, there will be large accumulation of water at the base.

The relationship between soil loss (erosion) and slope length states that erosion (E) is approximately equal to the square root of the slope length ($L^{1/2}$) *i.e.* soil loss varies with square root of slope length. Shape of the slope also has significant influence on erosion, since the base of a slope is more susceptible to erosion tan the top. It is due to the fact that runoff gains momentum as it approaches the base. On convex slopes (this phenomenon is magnified than on concave slopes, because in the former the steepness increases towards bottom compared to a concave slope. For complex shapes, where slope varies from convex to concave or vice-versa at different places, the soil loss is in the order of convex > complex > concave.

Anthropogenic/Human Interventions

These factors are briefly discussed in Chapter 22 and in detailed manner in Chapter 24.

Principles of Water Erosion Control

The water erosion control measures are devised as per the rain drop impact and kind of soil erosion *i.e.* splash erosion, sheet erosion, rill erosion and gully erosion. While most of the control measures are discussed in the following chapters, three basic principles for prevention and control of water erosion are as under.

- ✰ Use land according to its capability
- ✰ Protect the soil surface with some form of cover
- ✰ Control runoff and soil transport.

Land Capability

The land capability is determined mainly by the land's position, soil type and the land slope. These together also determine the land vulnerability to erosion. For example, if a land is not suited to agriculture then some other use that might limit the soil erosion must be considered. Clearly, soil erosion can be prevented or minimized by using the soil as per its capability.

Soil Cover for Energy Dissipation

Lands without vegetation are vulnerable to raindrop, sheet and gully erosions. Therefore, planting a cover crop during off season helps to minimize rain drop erosion. The cover crop should be selected so that besides providing the cover, it helps to increase soil organic matter, improve water infiltration and reduce runoff. If it is infeasible to grow vegetation, spreading of crop residues or stubble on the soil surface can serve the same purpose. The stubble should be retained over the ground and should not be plowed under the soil.

To establish vegetation on sloping lands is a difficult proposition. Some interventions are needed to vegetate the hilly terrain. For example while seeds are broadcasted over the terrain, woven or non-woven geotextile mats may be spread so that seeds are retained until they germinate and vegetate the surface. Bio-degradable mats such as that of coir can be used to take care of environmental issues under the fragile eco-systems.

Control Runoff and Soil Transport

Since the soil erosion is a function of the runoff water velocity, methods that slow down the speed of eroding water from hill slopes should be adopted. Most agronomical measures (discussed in Chapter 25) such as conservation tillage, cultivation along the contour or right angle to natural slope and strip cropping and construction of barriers such as contour bunds and terraces help to reduce the flow velocity. It has emerged that about 80 per cent of soil loss resulting from poor crop cover can be trapped in the fields using contour bunds. The graded bunds are capable of channeling the runoff at low speed into grassed waterways. Since complete control of soil erosion is not possible, methods that trap the sediments within the area should be adopted. For this purpose, plan and construct sediment traps and basins at appropriate locations.

Classification of Soil Erosion on the Basis of Intensity

Smith and Stamey (1964) defined the permissible erosion as the erosion where rate of soil destruction is equal to rate of soil formation. It is also known as compensating erosion. Depending upon the degree of soil loss exceeding the compensating erosion (harmful erosion), soil erosion has been classified into 5 classes (Table 23.1).

Table 23.1: Classification Criteria on the Basis of Soil Loss per Year

Classification as per Smith and Stamey (1964)		*Classification as per ICAR-NAAS (2010)*	
Type of Erosion	*Soil Loss per Year (mm)*[x]	*Type of Erosion*	*Soil Loss per Year (t/ha)*
Weak	0.05 -0.5 (0.5-5.0)	Moderate	10-15
Medium	0.5 – 1.5 (5-15)	Moderately severe	15-20
Serious	1.5-5.0 (15-50)	Severe	20-40
Severe	5.0-20.0 (50-200)	Very severe	40-80
Catastrophic	>20 (>200)	Extremely severe	>80

[x]Values in parenthesis are in m^3/ha

Erosivity and Erodibility

Amount of soil erosion is dependent upon i) the ability of rain to detach the soil particles (erosivity of rainfall), and ii) on the susceptibility of soil to get eroded due to raindrop (erodibility of soil). These two processes are known as rainfall erosivity and soil erodibility respectively. In the case of wind erosion it will be the ability of the wind to detach the soil particles (wind erosivity) while soil erodibility will be the susceptibility of soil to get eroded due to wind.

Erosivity

Erosivity is defined as the potential ability of the rain to cause erosion. It is a function of the physical characteristics of the rainfall. For given soil conditions, one storm can be quantitatively compared with another on a numerical scale and the values of erosivity within a given time period can be developed.

Since the rainfall erosivity is related to the kinetic energy of rainfall, two methods are widely used to compute the erosivity of rainfall. These are the EI30 Index method and KE > 25 Index method.

EI30 Index Method

EI30 index method is based on the assumption that the product of kinetic energy of a given storm and the 30-minute maximum rainfall intensity gives the best estimate of soil loss (Wischmeier and Smith, 1958). The greatest average intensity experienced in any 30 minute period during the storm is computed from recording rain gauge charts by locating the maximum amount of rain which falls in 30 minute period. Same values are then converted to intensity in mm/hour. EI30 index is calculated by the following step wise approach.

- ☆ From the recording rain gauge total amount of rainfall and rainfall intensity for various times say for each five minute period is deciphered
- ☆ Kinetic energy for each time is calculated using eq. (23.4) and summed for the whole storm
- ☆ The recording rain gauge chart or the data tabulated in step 1 is used to find out the maximum rainfall during any 30 minute period. It is used to calculate the rainfall intensity (cm/hr). It will be just double the value of the depth of rainfall during this period.

Rainfall erosivity index, R is the value of kinetic energy of the storm calculated in step 2 times the maximum rainfall intensity during the 30 minutes period calculated in step 3 as:

$$R = KE \times I30 \qquad (23.6)$$

EI30 index can be computed for individual storms, and the storm values can be added over periods of time to give weekly, monthly or yearly values of erosivity.

The EI30 index developed under American condition is not suited to tropical and sub-tropical zones for estimating the erosivity. Therefore, KE > 25 Index method is used for these conditions.

KE > 25 Index Method

KE > 25 index method, introduced by Hudson (1971), conceptualize that erosion can occur only if rainfall intensity exceeds a minimum threshold value. From experiments, it has emerged that the rainfall intensities less than 25 mm/h are not able to cause soil erosion in significant amounts. Thus, this method accounts only for those rainfall intensities, which are greater than 25 mm/h. It is why the method is named as KE > 25 Index method. The procedure to calculate R using this procedure is similar to the EI30 method except that in step 2, period during which intensity of rainfall is less than 25 mm/hr are left out while summing the KE of the storm.

Erodibility

Some soils are more resistant than others to erosion. Erodibility is defined as the vulnerability or susceptibility of the soil to erosion. Erodibility of the soil can be subdivided into three parts: i) the fundamental or inherent characteristics of the soil- the mechanical, chemical and physical composition, ii) the topographic features- slope, and iii) the management of the soil. Amongst the various soil properties, soil erodibility varies with soil texture, aggregate stability, shear strength, infiltration capacity and organic matter content of the soil. While the effect of the physical properties can be evaluated more precisely, effect of the management practices is difficult to assess. Therefore, attempts have been made to evaluate the soil erodibility on the basis of texture-based indices. Some of the indices often used are as follows:

Bouyoucos (1935) Index

(Per cent sand + per cent silt)/per cent clay (23.6)

Dispersion Ratio

100 × (Silt + Clay in undispersed sample)/(Silt + Clay in dispersed sample) (23.7)

Erosion Ratio

(Dispersion ratio)/(Colloid content/Moisture equivalent) (23.8)

Here colloid content refers to the < 2 μm particle content and moisture equivalent is represented by the water retention at pF 3. An erodible soil has a higher dispersion ratio and erosion ratio.

Clay Ratio or Mechanical Ratio

Sand/(Silt+Clay) (23.9)

Amongst these indices, Bouyoucos (1935) index is most commonly used to assess soil erodibility. However, none of these has been found to be universally applicable.

Chapter 24

Gully Erosion and Control

According to the Soil Conservation Society of America a gully as defined as "a channel or miniature valley cut by concentrated runoff but through which water commonly flows only during and immediately after heavy rains". It may be dendritic or branching or it may be linear, rather long, narrow and of uniform width. Gully erosion is a highly visible form of soil erosion that affects soil productivity, restricts land use and can threaten roads, fences and buildings. Gully erosion is an advanced stage of rill erosion in which the size of the rill is so enlarged that it cannot be rectified by ordinary tillage implements. Gullies are formed where many rills join and gain more than 50 cm depth. The distinction between ravine, gully and rills is that of size. A gully is too large to be filled by normal tillage practices. A ravine is a deep narrow gorge. It is larger than a gully and is usually worn down by running water. Large gullies and their networks are known as ravines. In India about 2.06 M ha land has been severely affected by gully erosion.

The process of gully formation usually follows sheet and rill erosion but it also occurs when runoff volume from a sloppy land increases sufficiently to cut deep inversions or when concentration of runoff water is long enough in the same channel. It can be formed on natural depressions where runoff water accumulates. It can also be formed on the track formed by movement of vehicles.

Formation of Gullies

A gully develops in three distinct stages; waterfall erosion; channel erosion along the gully bed; and landslide erosion on gully banks. Correct gully control measures must be adopted according to these development stages.

Waterfall Erosion

Waterfall erosion begins with sheet erosion that develops into rills, which gain depth and reach the B-horizon of the soil. The gully reaches the C-horizon and a gully head often develops where flowing water plunges from the upstream segment to the bottom of the gully. The falling water from the gully head scarves a hollow at the bottom of the gully through its direct action and splash (See Figure 23.4). When the excavation becomes too deep, the steep gully-head wall collapses. This process repeated over time makes the gully head to progress backwards to the upper end of the watershed. This process is also called gully-head advancement. As the gully head advances backwards and crosses lateral drainage ways caused by waterfall erosion, new gully branches develop. Branching of the gully may continue until a gully network cover the entire watershed.

Channel Erosion along Gully Beds

In channel erosion, soil is scoured away from the bottom and sides of the gully by flowing water. The length of the gully channel increases as waterfall erosion causes the gully head to advance backwards. At the same time, the gully becomes deeper and wider because of channel erosion.

Land-Slide Erosion on Gully Banks

Channel erosion along gully beds is the main cause of landslides on gully banks. During the rainy season, when the soil becomes saturated, and the gully banks are undermined and scoured by channel erosion, big soil blocks start sliding down the banks and are washed away through the gully channel. After landslides have occurred on all gully banks, a considerable number of new branch gullies may begin along the disturbed banks. During the third stage of gully development, gullies become deeper and longer as well as wider.

Land-slide erosion on gully banks also occurs in regions with temperatures that alternate between freezing and thawing. When the temperature drops below zero (Celsius), wet gully banks freeze. After the temperature rises above zero, the banks thaw, the soil loosens, and the loose gully banks easily slide during the first rainy season.

Stages of Gully Development

In the course of its formation and development, a gully passes through four stages. These stages continue unless the gully is stabilized by structural control measures and re-vegetation.

Stage 1: Initiation Stage

During this stage the channel erosion and deepening of gully bed takes place. This stage normally proceeds slowly where the top soil is resistant to erosion.

Stage 2: Development Stage

In this stage, both depth and width of the gully are increased due to waterfall action of the flowing velocity. The loose soil materials are washed away causing upstream movement of the gully head and enlargement of the gully in width and depth. It may also result in under cutting by the flowing water that may cause caving in of the soil mass. Lot of soil erosion takes place during this stage.

Stage 3: Healing Stage

After the development stage, further degradation slowly comes to halt and the healing stage begins. At this stage, the vegetation is established slowly both on the bed as well as on the side slopes of the gully.

Stage 4: Stabilization Stage

In this stage, gully stabilization takes place. While the healing process is in operation, the channel approaches a stable bed gradient and gully walls attain a stable slope. Vegetation grows abundantly to cover the soil surface. New top soil begins to form.

Classification of Gully

Gullies are classified on the basis of shape, size and state of the gully.

Based on Shape

'U' Shaped Gully

These gullies are formed in alluvial plains where surface and sub-soils are equally and easily erodible. Such gullies are flatter in shape and are formed on leveled land. The runoff flow undermines, and the gully banks collapse. It results in the formation of vertical side walls and formation of 'U' shaped gullies (Figure 24.1 left).

'V' Shaped Gully

These gullies develop where the sub soils are tough and resist the rapid cutting of soil by runoff flow. As resistance to erosion increases with depth, the width of cut is decreased resulting in the formation of 'V' shaped gullies (Figure 24.1 middle). In hilly terrain 'V' shaped gullies are quite common because high flow velocity with flow volume results in development of V-shaped gullies.

Trapezoidal

These gullies are formed where the gully bottom is made of more resistant material than the topsoil (Figure 24.1 right). In this case, sub-soil layer at the bottom of gully being much more resistant to erosion restricts further development of depth of gully. Thus, a trapezoidal gully is formed.

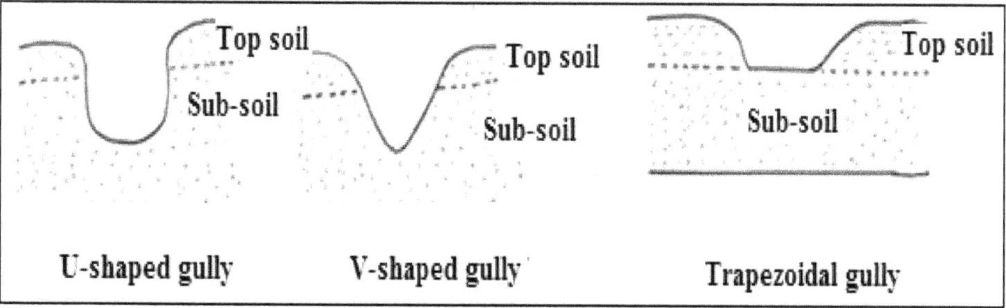

Figure 24.1: Classification of Gullies on the Basis of Shape.

Based on State of the Gully

Active Gully

In these gullies, the dimensions of the gully continue to enlarge with time. The size enlargement is dictated by the soil characteristics, land use and volume of runoff passing through the gully. The gullies found in plain areas are active in nature.

Inactive Gully

The active gullies that have attained constant dimensions and whose dimensions do not change with time are inactive gullies. The gullies found in rocky areas are inactive, because rocks are very tough to erode by the runoff flow.

Based on the Dimension of the Gully

This gully classification system is based on size - depth and drainage area. Several classification criteria to classify gullies on the basis of size have appeared in the literature. The criterion used herein classifying the gullies as small, medium and large appears to be the simplest (Table 24.1).

Small Gully

These gullies can be easily crossed by farm implements. These can also be removed by ploughing and smoothing operations and stabilized through vegetation. The depth is usually 1 m or less.

Medium Gully

Medium gullies are those that cannot be easily crossed by farm implements. They can be controlled by terracing or ploughing operations. The sides can be stabilized by growing vegetation on them. The depth of the gullies is in the range of 1-5 m.

Large Gully

These gullies are more than 5 m deep and are considered to have gone beyond their reclaimable stage. Tree planting is an effective method to stabilize such gullies.

Table 24.1: Size Criteria for Gully Classification

Classification	*Depth (m)*	*Drainage Area (ha)*
Small	< 1	< 2
Medium	1 to 5	2 to 20
Large	> 5	> 20

Gully Classes Based on Continuation

Continuous Gullies

The continuous gullies comprise of many branched gullies. A continuous gully has a main gully channel and many mature or immature branch gullies. A gully network (gully system) is made up of many continuous gullies. A multiple-gully system may be composed of several gully networks.

Discontinuous Gullies

These gullies may develop on hillsides following landslides. They are also called independent gullies. At the beginning of its development, a discontinuous gully does not have a distinct junction with the main gully or stream channel. Flowing water in a discontinuous gully spreads over a nearly flat area. After some time, it reaches the main gully channel or stream. Independent gullies may be scattered between the branches of a continuous gully, or they may occupy the whole area without having any continuous gullies.

Factors Affecting Gully Formation

The factors affecting gully formation are categorized into two groups, man-made factors, and physical factors.

Man-made Factors

Shifting Cultivation (Improper Land Use)

In developing countries, rapidly-increasing populations migrate to upland to occupy forests or rangelands. These migrants cut trees, burn litter and grasses, and cultivate hillsides without practicing any worthwhile conservation measure(s). After few years, as the productivity of the soil is lost because of sheet, rill and gully erosion, the land is abandoned. This kind of cultivation, (slash and burn or shifting cultivation) is repeated by farmers on other hillsides until the land loses its productivity at the new site as well. Thus, the whole area may be completely destroyed by gullying as the gully heads advance to the upper ends of the watershed. Shifting agriculture has caused maximum soil erosion in tribal areas of Assam, Meghalaya, Tripura, Nagaland, Mizoram, Kerala, Andhra Pradesh, Odisha, Madhya Pradesh, Chhattisgarh *etc.* It has been estimated that more than 1.5 M ha of forest land is cleared for shifting agriculture every year, and the total area affected has been estimated at 4.5 M ha. It is reported that about 0.33 M ha in Odisha, 0.21 M ha in Assam, 0.042 M ha in Tripura and 0.022 M ha in Manipur are subjected to shifting cultivation.

Forest Fires

Forest fires are caused by the uncontrolled burning used in shifting cultivation. These fires can easily spread into the forest and destroy the undergrowth and litter. Grass fires are usually ignited by farmers near the end of the dry season in order to obtain young shoots for their livestock or new land for cultivation. On slopes, the soil that is exposed after forest and grass fires is usually, gullied during the first rainy season.

Overgrazing

Overgrazing removes too much of the soil's protective vegetal cover while trampling by the animals compacts the soil. These together reduce the infiltration capacity of the land. The increased run-off, caused by the insufficient water holding capacity of the soil and compactness produces new gullies or enlarges the old ones.

Mining

Underground mining is another cause of gully formation. Initially, cracks in the ground and soil creep (a kind of gravity erosion) are observed in the mining areas. During rainy seasons, these enlarge to take the form of gullies. Gully formation is also a big problem in many open-pit mining areas.

Road Construction

If cuts and fill slopes during road construction are not re-vegetated during or immediately following road construction, gullies may be formed on both sides of the road. Inadequate drainage capacity along the roads (small number of culverts, insufficient capacity of road ditches, *etc.*) can be a major cause of gully formation.

Livestock and Vehicle Trails

Gullies are also formed on livestock and vehicle trails that run along hillsides. It is because the traffic compact the soil and reduces its water holding capacity.

Destructive Logging

In forest regions, logging with tractors down slopes can lead to gully erosion, because the runoff becomes concentrated along the skid trails. Highland logging with slack cables also causes gully formation on forest land.

Physical Factors

The main physical factors effecting the rate and amount of surface runoff are precipitation, topography, soil properties and vegetative cover.

Precipitation

Monthly Distribution of Rainfall: The duration of wet and dry seasons can influence the chances of gully formation. Monthly distribution of rainfall is more significant than the total annual rainfall because of its effects on the growth of vegetation. In humid regions with uniform rainfall distribution, surface erosion, including gully formation, may not be a serious problem because vegetation grows throughout the year. However, in areas that do not have uniform rainfall, the vegetation (especially grass) dries up during the prolonged dry season (3 to 5 months or more). If the land is not properly used, or if forest or grass fires occur during the dry period, it is unable to hold rainwater. Increased surface runoff in the rainy season produces large scale landslides and gullies.

Rainfall Intensity and Run-off: There is a relationship between rainfall intensity, rate of runoff, density of vegetative cover, and the size of a catchment area. This relationship is generally expressed in the form of mathematical equations. The Rational formula which is used in engineering designs for gully and torrent control is a good way to demonstrate this relationship. If the amount of rainfall is more than the holding capacity of the soil, there will be an increase in surface runoff, followed by surface erosion and gullying. In some tropical and sub-tropical regions, after the soil is completely saturated, almost all of the rainfall turns into runoff during the wet months, which include the monsoon season, tropical cyclones and especially typhoons. If intensive rains occur for two or three days (typhoon period), the resulting high runoff may cause landslides, huge gullies and devastating floods.

In continental and temperate climate, prolonged rains of moderate intensity (duration several days) or short, intensive rain storms lasting from 15 to 90 minutes (maximum rainfall intensity about 3 mm/minute), cause landslides, gullies and floods because of the increased runoff in the watersheds. Torrential floods, which generally occur after the short, intensive rain storms, destroy agricultural lands, residential areas, roads, irrigation ditches and canals at the base of the valley below a deteriorated watershed.

Rapid Snowmelts: Rapid snowmelts that turn into high runoff, acts as a cutting agent and produces gullies. Like prolonged rains of moderate intensity and short intensive rain storms, rapid snowmelts cause destructive floods.

Freezing and Thawing: Clay soils experiencing distinct seasonal change that result in freezing and thawing (F/T) of the soil, are prone to gully erosion. In the absence of residue on the soil surface, increasing frequency of F/T impacts the aggregate stability, which in turn impacts the potential for soil erosion during rainfall events.

Topography

The size and shape of a drainage area, as well as the length and gradient of its slopes have an effect on the runoff rate and amount of surface runoff. Therefore, all topographic characteristics should be studied in detail before gully control works begin.

Shape of Catchment: If two catchments have the same area and have symmetrical drainage patterns, the distance to the outlet in a long catchment will be greater than in the short one, due to differences in their shapes. The time of concentration will be longer in the long catchment. The peak runoff rate in that case will be lower. This explains why, if all other factors are equal, long narrow catchments have fewer flash floods than square or round catchments.

Size of Catchment: The larger the catchment, the greater the amount of runoff. To understand its impact one needs to determine the area of the catchment. The catchment area of a gully can be easily and accurately measured by using a 1/10 000 scaled map. Even a new map can be prepared for the catchment area of the gullies and torrents using a theodolite or transit. Mapping continuous gullies or gully networks can also be undertaken by surveying the closed traverses with a clinometer (0-90°), hand compass (0-360°) and 50 m measuring tape.

Length and Gradient of the Slope: On long slopes, there is tendency of water to accumulate towards the base. To prevent the gully formation, the water (runoff) should be safely conducted downhill to stable, natural water courses or vegetated outlets. The other option is to store the water within the catchment so that it gets infiltrated into the ground. Land treatment measures such as contour ditches (infiltration trenches), level terraces (gradoni), wattling, staking, *etc.* can be adopted. Moreover, steeper the slope, the higher the velocity of running water and higher the erosive power of the runoff. Watershed treatment measures should not only reduce the amount of surface water, but they should also help in decreasing its velocity so as to reduce its erosive power.

Soil Properties

Soil properties especially the soil texture has a great bearing on soil erosion. Let us consider seven soil classes based on their soil texture *i.e.* sand, loamy sand, sandy loam, loam, silt, loam, clay loam and clay. The infiltration rate decreases as one move from sand to clay. Contrarily, the resistance against erosion increases from sand to clay.

Vegetative Cover

The role of vegetative cover is to intercept rainfall, to keep the soil covered with litter, to maintain soil structure and pore space, and to create openings and cavities by root penetration. This is best achieved by an undisturbed multistory forest cover. Under special conditions, however, a well-protected, dense grass cover may also provide the necessary protection. Protection provided against erosion rather than the type of the vegetative cover determines its effectiveness in gully control. Therefore, vegetation well-adapted to local conditions having vigorous growth may be used. For example, while broadleaf species may prove good under some conditions, in others conifers, tall grasses, *etc.* may prove better. In critical areas, even one may have to protect the vegetation. However, whenever possible it is desirable to establish a vegetative cover to serve the dual purpose. Vegetation besides providing the cover, be able to meet the requirement of fodder, fuel wood and/or fruit, *etc.*

Principles of Gully Control

Gullies are mainly formed as a result of practices that cause an increase in surface runoff. Therefore, minimizing surface runoff is an essential element of any programme on gully control. One must remember that it is always better to prevent gully erosion than to control it at a later stage. Proper land management practices can help to control gully erosion.

- ✰ Regular inspection of land helps to detect early signs of gully erosion
- ✰ Control of illegal wood cutting in plantations and natural forests
- ✰ Prevention of forest fires including grass fires
- ✰ Control of grazing, and re-vegetation of open and grass lands
- ✰ Maintenance of soil fertility and stability on land which is under agro-forestry or agriculture
- ✰ Regulating road construction and mining activities
- ✰ The early action to stabilize moderate sheet and rill erosion, and incipient gullies in forest, rangeland and cultivated areas.

For gully control, following three methods are applied in the order given as follows.

- ✰ Improvement of gully catchments to reduce and regulate the run-off rates (peak flows)
- ✰ Diversion of surface water above the gully area
- ✰ Stabilization of gullies by structural measures and accompanying re-vegetation.

When the first and/or second methods are applied especially in temperate climates, small or incipient gullies may be stabilized without having to use the third method. On the other hand, in tropical and sub-tropical regions characterized by heavy rains (monsoons, typhoons, tropical cyclones, *etc.*), all three methods must be implemented to successfully control the gully erosion.

Improvement of Gully Catchments

The most practical way is to sub-divide the catchment into appropriate land classes. Thereafter, grazing and cropping practices are applied that most suits to each class. General guidelines for catchment area treatment depending upon the slope are:

1 - 3 per cent slope - Contour bunds, vegetative bunds

4-10 per cent slope - Graded bund, graded trench

11-15 per cent slope - Bench terraces, inward sloped bench terraces, outward sloped bench terraces

16-33 per cent slope - Contour stone wall and staggered stone walls

> 33 per cent slope - Afforestation

Diversions

Most effective control of gully erosion requires complete elimination of runoff into the gully or the gullied area. This may often be accomplished by diverting runoff from above the gully and causing it to flow in a controlled manner to some suitably protected outlet. The basic aim of diversions is to reduce the surface water entering into the gully through gully heads and along gully edges, and to protect critical planted areas from being washed away. Small diversion ditches constructed either alone or with other structures such as earth plugs and check dams, are commonly used in gully control. A diversion channel is constructed around the slope and given a slight gradient for water to flow to the desired outlet. The diversion should be located far enough above the gully over fall so that sloughing of the gully head does not threaten the diversion. The capacity of the diversion channels should be based upon the estimated peak runoff for the 10-year recurrence interval. Bottom widths and side slopes may vary with soil, land slope, and individual preference, but side slopes of 4:1 and the bottom widths sufficient to permit mowing of grasses are desired. Water may be discharged into pastures, woodlands, or vegetated outlets. Construction and maintenance of diversions is similar to that described for grassed waterways (Chapter 29).

Temporary and Permanent (Engineering) Measures

Biological or Vegetative Measures

Biological measures are the most economical and easily adoptable measures to control gully erosion.

Vegetating gully bed helps to reduce the flow velocity. If that happens, some of the sediments will be deposited on the gully head. As a result, thick vegetation would begin to come up on the haed as well as side beds. If biological measures prove ineffective, then one may go for mechanical measures.

Anti-Erosion Crops

Crops differ in their capability to minimize erosion. While some crops such as cover crops and soil binding crops may prove to be anti-erosion crops, other may not help much to reduce soil erosion. Thus, cultivation of anti-erosion crops is the simplest way to stabilize gullies. Besides, gully stabilization, crops provide supplementary income to the farmers.

Conversion of Gully into Grassed Waterway

Small and medium size gullies can be easily converted into grassed waterways. In practice, gully is appropriately shaped and suitable species of grasses are grown. Channel cross-section is kept broad and flat, so that water spread is uniform over a wide area. Sloping of gully banks is done only to the extent required for establishment of vegetation or for facilitating tillage operations. Where trees and shrubs are to be established, rough sloping of the banks to about 1:1 will be sufficient. Where gullies are to be reclaimed as grassed waterways, sloping of banks to 4:1 or flatter is desirable. While sloping of banks of small gullies may be accomplished with plows, disks, and other farm tools, large gullies may require heavy earth-moving equipment.

Sodding

Sod is a grass along with the soil beneath it. The sod is held together by its roots or another piece of thin material. In British English, such material is usually known as *turf*, but for soil conservation and agriculture, term "sod" is commonly used. Sod in most cases is specially grown and usually harvested 10 to 18 months after planting, depending on the climate. It is harvested using specialized equipment and precisely cut to standardized sizes. Sod is typically harvested in small square or rectangular slabs, or large 1.2 m rolls. These are then used to control soil erosion as sod helps in vegetating the gullies quickly and profusely.

Sod Flumes

This technique has proved to be successful where gully over fall head are less than 3 m and watershed areas are less than 10 ha. The design of sod flume is shown in Figure 24.2. It prevents further waterfall erosion by providing a protected surface over which the runoff flows into the gully. Slope varies with the soil type, size of watershed, height of over fall and type of sod used. steepest recommended slope

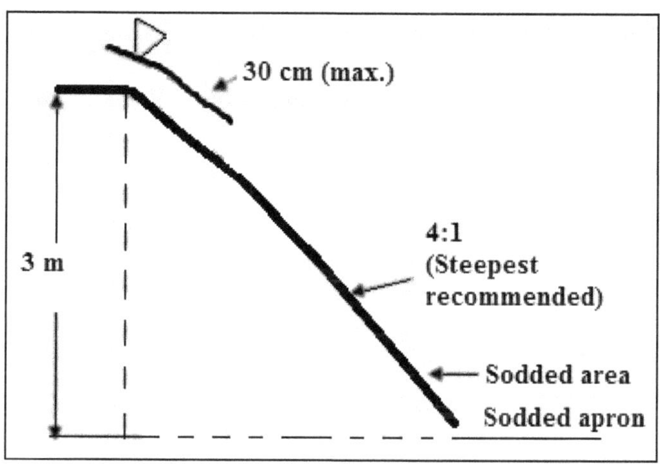

Figure 24.2: Description of a Sod Flume.

is 4:1. Flume is made wide enough to ensure a non-erosive velocity. The maximum depth of flow over the flume should not exceed 30 cm.

Sod Strip Checks

Sod checks are commonly used to stabilize gully channel until the intervening areas gets vegetated. These checks are best adapted to small gullies with small to medium sized watersheds. The checks are not used in gullies with very steep grades. Strips are laid across gully channel (Figure 24.3) having a minimum width of 30 cm. The strips should extend to gully sides by at least 15 cm. Strip spacing usually varies from 1.5 to 2.0 m.

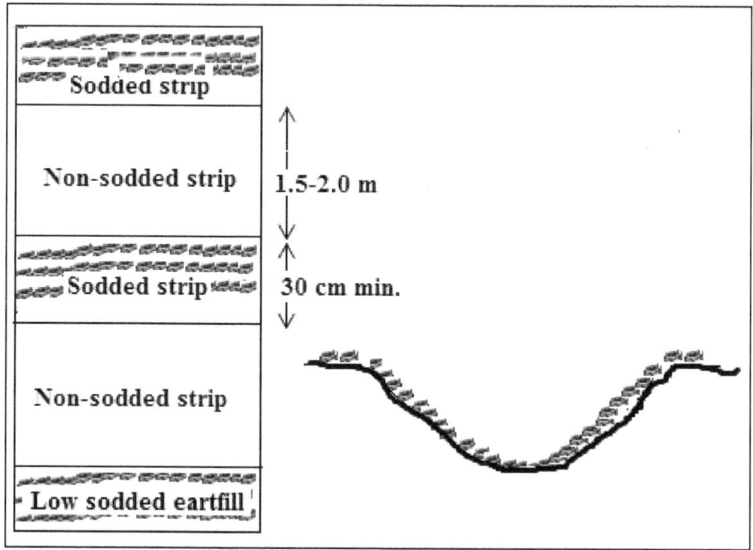

Figure 24.3: Sod Strip Checks, Plan (Left) and Cross-Section (Right).

Low Sodded Earth-Fills

These are used as substitutes for temporary gully control structures in small and medium sized gullies. Already growing sods are cut along with soil mass and combined together to form earth fill dams (Figure 24.4). These have been usefully employed for small and medium gullies with drainage areas of about 10 ha. They are constructed with a maximum height of 45 cm, upstream (u/s) side slope of 3:1 and downstream (d/S) side slope of 4:1. The top of the fill should be low in the middle gradually curve upwards to meet the gully banks. The spacing should be such that top of the last earth-fill should be as high as base of the next earth-fill.

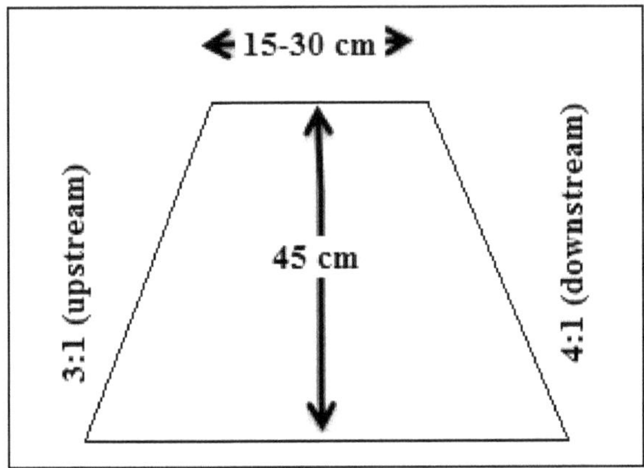

Figure 24.4: Low Sodded Earth-Fills.

Vegetation

Large gullies may be controlled by creating an environment conducive to reestablishment of natural vegetation. or by plantings of trees or vines. The opportunity to provide protective cover by natural re-vegetation is often overlooked although such an attempt can save huge expenditure required for plantings and/ or construction of structures.

Natural Re-vegetation

If the runoff resulting in the formation of gully is diverted, and the area is suitably fenced to avoid entry of the livestock, plants begin to come naturally. A gradual succession of plant species eventually will protect the gullied area with grasses, vines, shrubs, or trees native to the area. In some cases additional measures such as fertilization and mulch to conserve moisture and protect young volunteer plants may help. Vertical gully banks may be roughly sloped to prevent caving and provide improved conditions for natural seeding.

Artificial Re-vegetation

A gully at the lower end of a grassed waterway can be stabilized more readily and at less cost with trees than by building a dam. In severe cases of gully formation in small watersheds, trees may be used to stabilize the greater part of the drainage way or drainage area. Selection of vegetation to be established artificially in a reclaimed gully is governed by the intended use for the planted area. Grasses and legumes may be planted if the vegetation is to be used for hay or pasture crop. Where gullies are reclaimed as drainage ways in cultivated fields, sod-forming vegetation should be selected to permit crossing of the drainage way with farm machines. In some areas trees and shrubs are easier to establish in gullies than grasses, particularly if the gully is not to be shaped to permit operation of farm implements. Trees and shrubs should be planted in accordance with local recommendations for control of erosion and establishment of wildlife refuges. A plant spacing of 1 × 1 m, 1.2 × 1.2 m or a maximum of 2 × 2 m should be maintained. Locally adapted trees planted on gullied areas may be utilized for fence posts or rough timber.

Mechanical Measures

Normally, these structures are constructed from masonry and related materials across the gully as well as along the side slopes of the gully.

Check Dams

Check dams are constructed across the gully. A check dam consists of body wall, side walls and an apron on d/s side to prevent scouring.

Retaining Walls

Retaining walls are constructed in vulnerable places along the side walls of the gullies. Retaining walls may or may not be plastered. If un-plastered, it is called 'dry stone pitching'

Temporary Gully Control Structures (TGCS)

TGCS are pretty effective where the amount of runoff is not too large. These have a life span of 3 to 8 years. These are made of locally available materials. Basic purposes of these structures are to retain more water, to retain soil for proper plant growth and prevent channel erosion until sufficient vegetation is established on the upstream side of the gully. Some of the TGCS given below are discussed in the following sections.

- Woven wire check dams
- Brushwood dams
- Loose rock dams
- Plan or slab dams
- Log check dams
- Boulder check dams

Permanent Gully Control Structures (PGCS)

If the erosion control cannot be achieved through TGCS, then one should go for bigger structure *i.e.* PGCS. Some of these are:

- Drop spillway

- ✰ Drop-inlet spillway
- ✰ Chute spillway
- ✰ Permanent earthen check dams

Construction Criteria of TGCS

General guidelines common to all the TGCS structures are:

- ✰ The effective height of about 30 cm is considered sufficient but under ordinary conditions in no case the height of temporary checks should be more than 75 cm. Sufficient freeboard should be included.
- ✰ The spillway capacity should be capable of handling peak runoff that may be expected once in 5 to 10 year return period.
- ✰ Since the purpose of check dams in gully control is to eliminate grade in the channel, theoretically check dams should be spaced in such a way that the crest elevation of one will be same as the bottom elevation of the adjacent dam up-stream.
- ✰ An apron or platform of sufficient length and width must be provided at the down-stream end to catch the water falling over the top so as to conduct it safely without scouring.

Woven Wire Check Dams

Woven-wire check dams are small barriers that hold fine material flowing in the water passing through the gully. As such, the runoff should not carry rocks and boulders. These are commonly used in gullies of moderate slopes (not more than 10 per cent) with small drainage areas. It helps in the establishment of vegetation that in turn serves to control further erosion.

These kind of dams are built in half-moon shape with the open end up-stream. The amount of curvature is arbitrary fixed but an off-set equal to $1/6^{th}$ of the width of gully at the dam site is optimum. To construct a woven-wire dam, a row of posts is set along the curve of the proposed dam at about 1.2 m intervals and 60-90 cm deep (Figure 24.5). Heavy gauge woven wire is placed against the post with the lower part set in a trench (15-20 cm deep), and 25-30 cm projected above the ground surface along the spillway width. Rocks, brushwood or sod is placed approximately up to a length of 1.2 m to form the apron. To seal the structure, straw, fine brush or similar material is placed against the wire on the upstream side up to the height of spillway.

Brushwood Dams

The brushwood dams (BWDs) are quite cheap, easy to build, but are the least stable of all types of check dams. These are also used to control gullies with small drainage area. Center of the dam is kept lower than the ends to allow water to flow over the dam rather than around it (Figure 24.6). The objective of BWDs is to retain sediments and slowdown runoff, and enhance the re-vegetation of gully areas. They are constructed either in single or double rows. The sides of the gully are

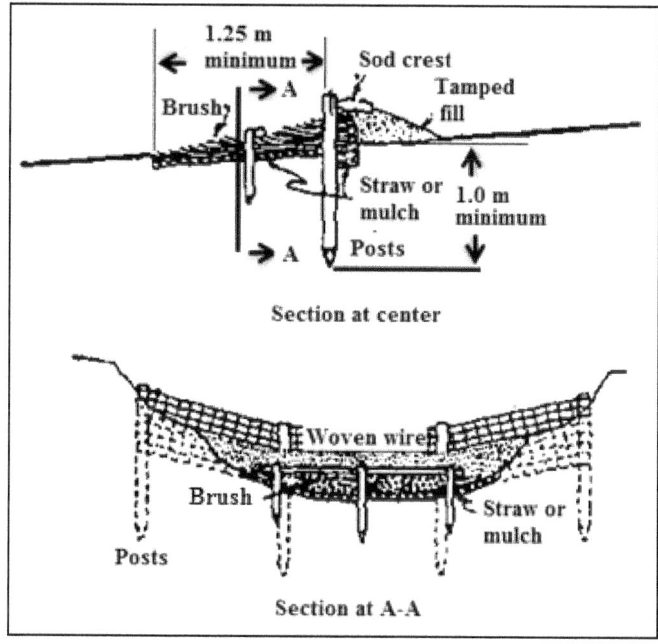

Figure 24.5: Woven Wire Check Dam.

sloped to 1:1 and the gully bed is deepened by 15 cm depth along the gully width to be covered with BWD. For a distance of 3-4.5 m along the site of the structure, sides and bottom of the gully are covered with thin layer of straw or similar fine

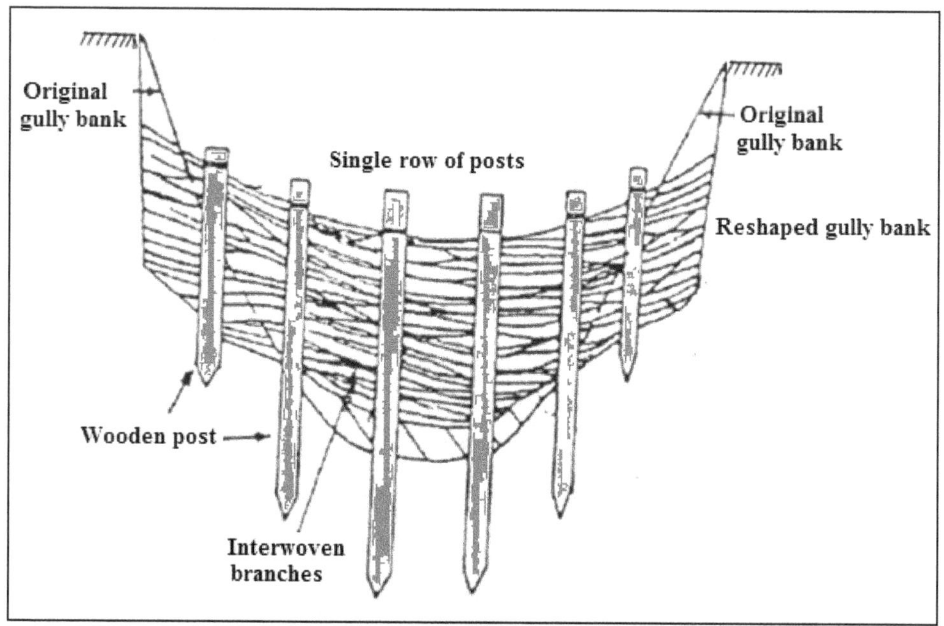

Figure 24.6: Single Row Brushwood Dam.

mulch. Brushes are then packed closely together over the mulch to about one half of the proposed height of dam. Several rows of stakes are then driven crosswise in the gully, with rows 60 cm apart, and stakes 30-60 cm apart in the rows. Heavy galvanized wire is used to fasten the stakes in a row, as well as to firmly compress the brushes in places. Sometimes large stones are also placed on top of brush to keep it compressed and in close contact with the bottom of the gully. It is difficulty to prevent the leaks and constant attention is required to plug openings with straw as they develop.

A double row brushwood dam is used for gullies that are 2-2.5 m deep and about 6 m wide. The watershed contribution may extend up to 40 ha. To construct the check dam a trench of about 90 cm wide and 30 cm deep is dug across the gully. The sides are also shaped to a slope of 1:1. The wooden posts of about 15 cm diameter are embedded to a depth of 75 cm at 180 cm apart in two rows along the trench. Another line of stakes is also driven at about 150-180 cm on the downstream. Two layers of brushwood material are laid along the flow of water and tied to the post and stakes with the help of galvanized wire (Figure 24.7). At the required height a notch is made so as to conduct the flow downstream.

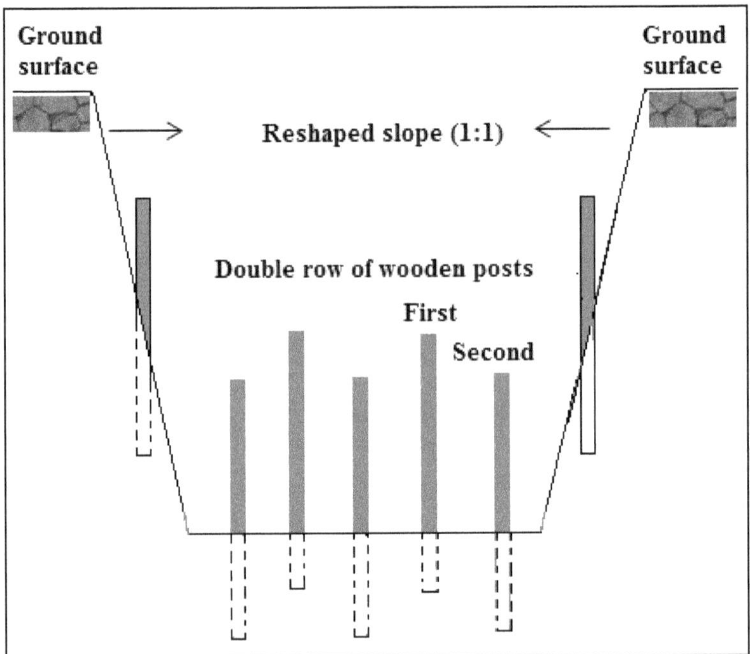

Figure 24.7: A Double Row Brushwood Check Dam.

Loose Rock Dam

Loose rock dams made of relatively small rocks are placed across the gully (Figure 24.8). The main objectives of these dams are to control channel erosion along the gully bed, and to stop waterfall erosion by stabilizing gully heads. Loose stone check dams are used to stabilize the incipient and small gullies and the branch

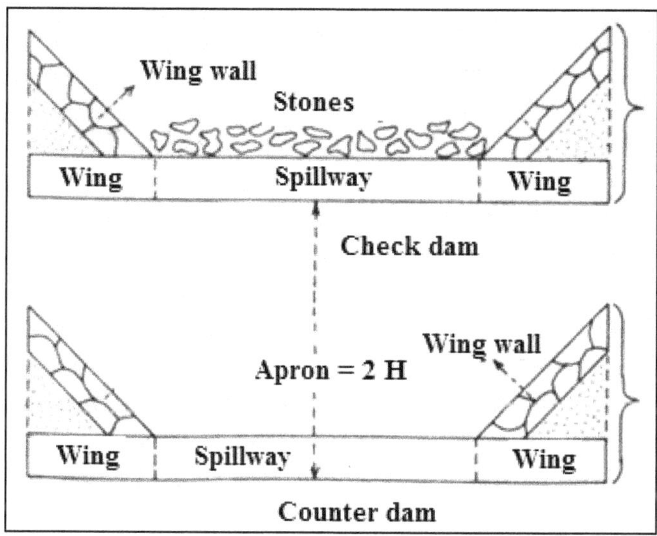

Figure 24.8: Cross-Section of Loose Boulder Check Dam.

gullies of a continuous gully or gully network. The length of the gully channel is not more than 100 m and the gully catchment area is 2 ha or less. These dams can be used in all regions especially in areas where stones or rocks of appreciable size and suitable quality are available. Flat stones are the best for making the dams of this kind.

A trench is made across the gully to a depth of about 30 cm. This forms the base of the dam on which the stones are laid in rows and are brought to the required height. The center of the dam is kept lower than the sides to form spillway. Stones are laid in such a way that the entire structure is keyed together. If round or irregular shaped stones are used, structure is generally encased in woven-wire so as to prevent stones from being washed away. If the rocks are small, they should be enclosed in a cage of woven-wire. Several large flat rocks may be countersunk below the spillway, extending about 1 m down-stream from the base of the dam to serve as an apron.

A series of loose boulder checks are constructed on each drainage line. The minimum vertical interval between two successive checks on a drainage line should equal the height of the structure, so that the water temporarily stored in one check will reach the toe of the check upstream.

Through years of experience in watershed development, the maximum height generally accepted for loose boulder checks is 1m (Figure 24.8 and Figure 24.9). The design height of 1m means that the top of the check in the middle of the stream is 1m above ground level. The top width of the boulder check is usually 0.4-0.5 m (GSWMA, 2011). As the material used in the check has a high angle of repose, the upstream slope of the check should be fixed at 1:1in general, to be varied only in exceptional cases where the structure has to handle very high volume of runoff flowing at a high velocity. The downstream slope of the boulder check can vary from

1:2 to 1:4 depending on the volume and velocity of runoff. The higher the volume and velocity of runoff, flatter the slope. Since the boulder check is composed of highly porous material it is not expected to hold water for a long period.

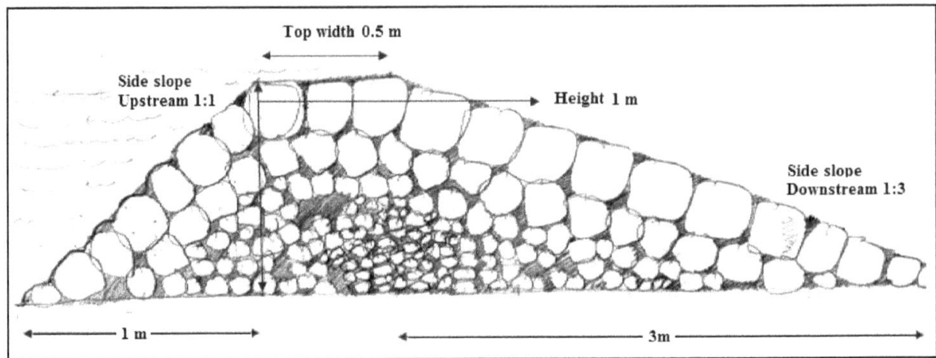

Figure 24.9: Top View of a Loose Boulder Dam with Counter Dam.

Plank or Slab Dam

These dams are used to control gullies with large drainage area. These dams are most suited for areas where timber is plentiful. The dams can be constructed with much less cost as compared to other types of temporary structures. To construct the dam, planks are placed across the gully. If the planks are not close fitting, straw or grass is used for initial sealing purposes. A spillway notch is provided over the headwall. On the up-stream face, a well-tempered earth fill is made. On the down-stream, an apron is provided that is made of loose rock, brush, sod or planks.

Log Check Dam

The main objectives of log check dams are to hold fine and coarse material carried by flowing water in the gully, and to stabilize gully heads. They are used to stabilize incipient, small and branch gullies generally not longer than 100 m and with catchment areas of less than two hectares. These are similar to plank or slab dams but made of somewhat stronger and durable material such as planks, heavy boards, slabs, poles or old railroad ties. Logs and posts like plank dam are placed across the gully. The maximum height of the dam is 1.5 m from the ground level. Both, its downstream and upstream face inclination are 25 per cent backwards. A rectangular spillway is provided with 1-2 m in length and 0.5 to 0.6 m deep (Figure 24.10).

Boulder Check Dams

In a gully or a multiple-gully system all the main gully channels of continuous gullies can be stabilized by boulder check dams. These dams can be built in almost all the conditions. The total height of the dam should not exceed 2 m. The thickness of the dam at spillway level is 0.7 to 1.0 m (average 0.85 m), and the inclination of its downstream face is 30 per cent (1:0.3 ratio). Accordingly, the thickness of the base is calculated. Foundation depth must be at least half of the effective height.

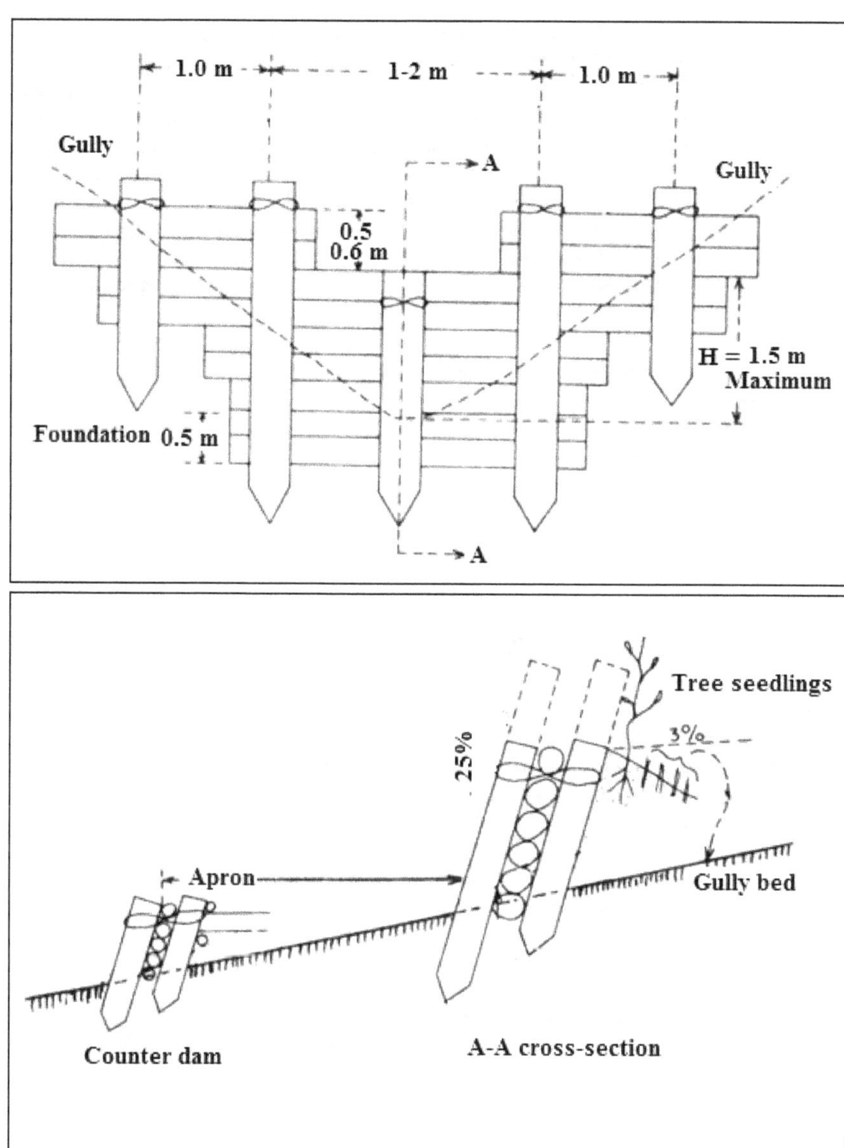

Figure 24.10: Front View of the Log Check Dam (Top) and A-A Cross-Section of the Check Dam and Counter Dam (Bottom).

The upstream face of the dam is usually vertical. A trapezoidal spillway is generally used and its dimensions (Figure 24.11) are computed according to the maximum discharge of the gully catchment area.

Rock Walls

These are provided in the upper gully segments where structures of low height are required. These are built from locally available materials. These are also used

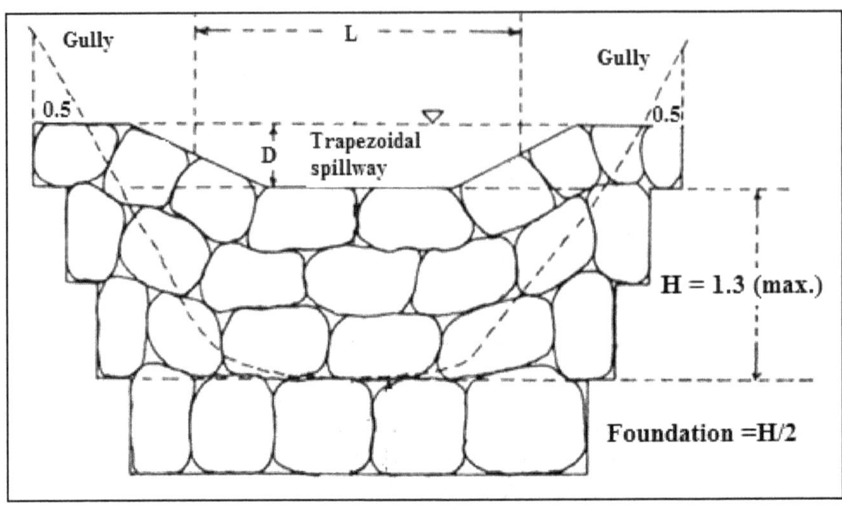

Figure 24.11: Front View of the Boulder Check Dam.

on eroded channels on the gully side slopes. These are loose rocks packaged to give a trapezoidal shape. The walls may be provided with or without vegetation.

Permanent Structures: An Introduction

Drop Spillways

Drop spillways also called drop structures are installed in gully bed so as to establish permanent control elevations below which an eroding stream cannot lower the channel floor. These are categorized as:

- Straight type
- Curved type and
- Box-inlet type

The first one is the most commonly used structure and is similar to weir type check dam.

These structures control the stream grade not only at the spillway crest itself but also through the ponded reach upstream. Drop structures placed at intervals along a channel can stabilize it by changing its profile from a continuous steep gradient to a series of more gently sloping reaches. Where relatively large volumes of water must flow through a narrow structure at low head, the box-type inlet is preferred. The curved inlet serves a similar purpose and also gives the advantage of arch strength where masonry construction is used. Drop spillways are usually limited to drops of 3-4 m, flumes or drop-inlet pipe spillways being used for greater drops.

In a drop spillway, the flow enters through a weir and falls over an apron or a stilling basin (Figure 24.12). Here, the energy of the flowing water is dissipated through the creation of hydraulic jump. The flow thereafter passes off to the outlet channel.

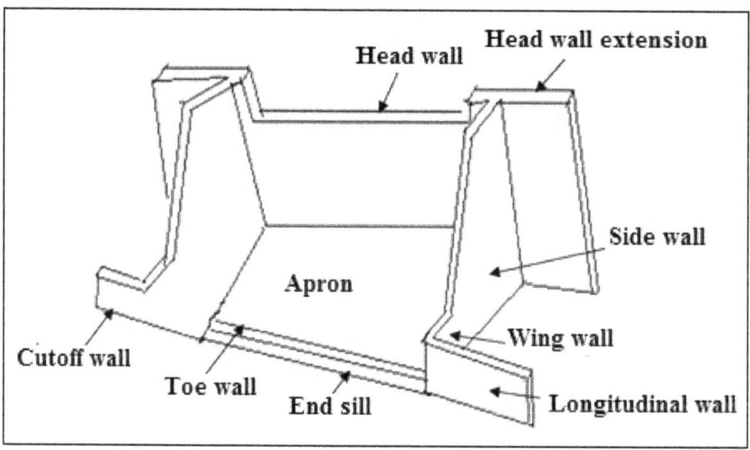

Figure 24.12: A Straight Type Drop Spillway.

Drop Inlet Spillway

A drop inlet spillway is a closed conduit generally designed to carry water under pressure from above an embankment to a lower elevation. These spillways are commonly used to remove water from reservoirs or other water storage structures including those created in gullies (Figure 24.13). A drop inlet spillway, also known as pipe spillway like other spillways consists of an inlet, a conduit and an outlet. Depending upon the type of inlet used the drop inlet spillways have been categorized as:

- ✰ Straight inlet
- ✰ Upstream flared inlet
- ✰ Flared inlet

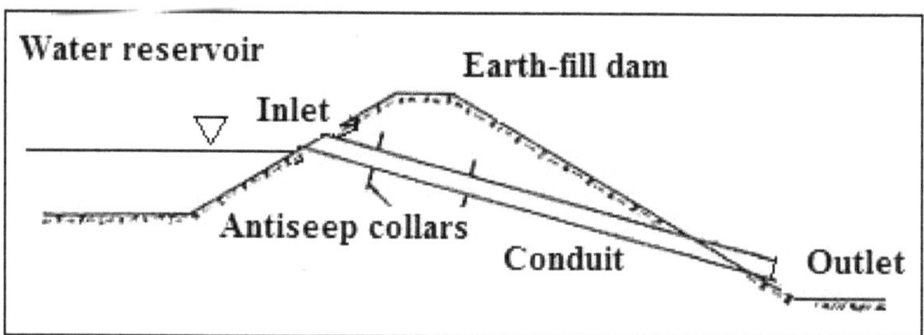

Figure 24.13: Schematic Diagram of a Drop Inlet Spillway

Chute Spillway

In situations where construction of a drop spillway or drop inlet spillway is going to be costly, chute spillway structures, which are relatively cheaper, are adopted. A chute spillway is an open channel with a steep slope, in which flow is carried at supercritical velocities. It usually consists of an inlet, vertical curve section (Usually when length of the conduit is large), steep-sloped channel and outlet (Figure 24.14). Flow passes through the inlet and down the paved channel to the floor of the outlet. Chute spillways are generally used to drop water at reaches which are much farther. In gully control, they can be used for the control of gully drops up to 6 m. The main design consideration would be to fix the longitudinal bed profile of the channel and its sectional dimensions. The energy of the flow is dissipated at the outlet, before the flow enters the downstream channel. However, there is the danger of the structure being undermined by rodents and in locations with poorly drained soil; its foundations may be threatened by seepage. Necessary precautions need to be taken during design.

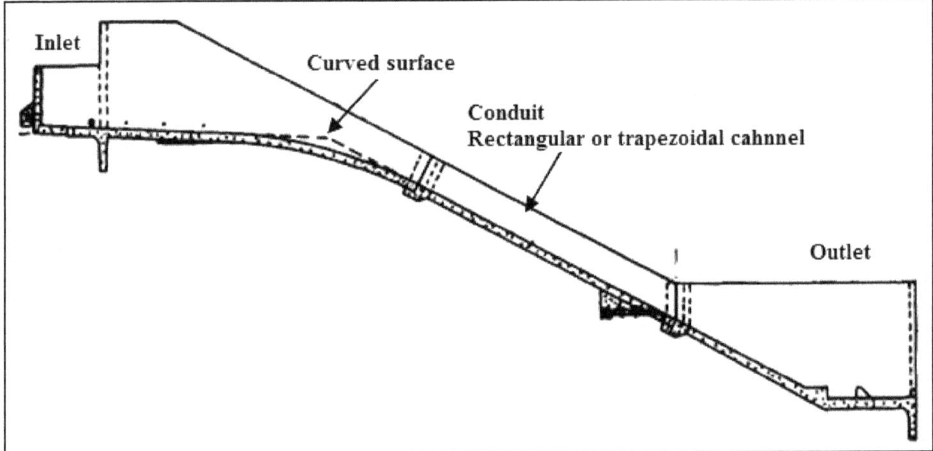

Figure 24.14: A Sketch of a Chute Spillway.

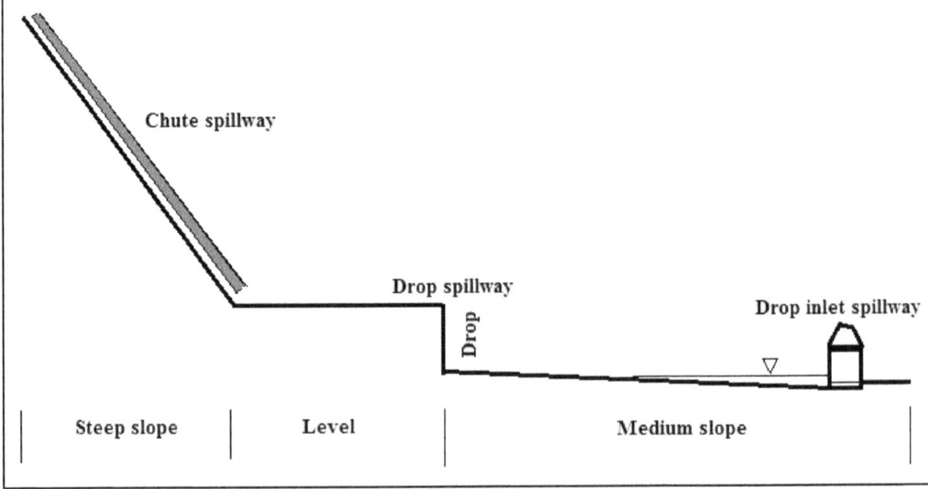

Figure 24.15: Gully Control Structures with their Location in the Gully.

Other Structures

Besides the permanent control structures discussed in the previous section, several other permanent structures are used. These are: rubble masonry dam, concrete dam, gabions, silt trap dam *etc.*

Location of Gully Control Structures

Gully control structures are laid in the gully section along gully length as per the profile of the gully bed. The general layout of the permanent gully control structures is shown in Figure 24.15. It shows that chute spillways are located at the gully head where steep slope is encountered. The drop structures are provided in series to act as control points so that the gully bed is not eroded below the crest level of the structure. Drop inlet spillways are provided at places where there are depressions and there is a scope to store water.

Chapter 25

Agronomical Measures of Erosion Control

Soil conservation in general comprises of a bundle of techniques, which are applied either alone or in combination to eliminate or minimize soil deterioration or its loss within their capabilities. The conservation techniques while protecting the soil also helps to improve its quality. In any soil and water conservation program several kinds of conservation measures are taken. These are categorized as agronomical or biological measures and engineering measures, which are further categorized as temporary and permanent.

Agronomical or Biological Measures

Agronomical or biological or vegetative measures are preferred in watershed development programs as they are eco-friendly, sustainable and cost effective. These measures are normally adopted on lands having mild slope, less runoff and less sediments. These can be adopted singly or in combination with mechanical measures depending upon the intensity of the erosion problem. If managed properly, these measures can have long lasting effect. That means one should plant such species, which are capable of holding the soil strongly and can survive even under very adverse conditions.

The functions of any biological treatment are:

- ☆ To reduce splash erosion by moderating the impact of rain drops
- ☆ To reduce runoff velocity and increase infiltration opportunity time
- ☆ To slow down wind speed
- ☆ To trap fine soil and nutrients, and
- ☆ To improve physico-chemical properties of the soil and help in conserving moisture in low rainfall area.

Commonly used biological measures consist of vegetative barriers, contour farming, land configurations, alley cropping, strip cropping, tillage, mulching and residue management.

Vegetative Barriers

Vegetative barriers also known as contour vegetative hedges or live bunds are useful in preventing soil erosion and conserving soil moisture. These can be implemented with low investments. Once the surface runoff from the fields reaches the vegetative hedge, its velocity is reduced, water spreads out evenly in the hedge row and dumps most of its silt load within hedge row. Velocity and amount of water and sediments in water passing out of the hedge are much less since a large portion of runoff is soaked into the soil. Vegetative barriers are generally planted on contours. The functions of vegetative barriers are:

1. To break the slope length, reduce runoff velocity and increase infiltration
2. To reduce the eroding and transport capacity of surface runoff
3. To cause sedimentation, trap absorbed nutrients and induce terracing
4. To help cultivation and planting operations on the contour.

The advantages of vegetative barriers are:

1. Once properly established, such live bunds need almost no maintenance and continue to protect the land from erosion for years, as they build up natural terraces
2. Land occupied by vegetative barriers is much less as compared to the mechanical measures
3. Land under the hedge is used for biomass production as well as growing of suitable plants for fodder and commercial value.

While constructing vegetative barriers, first of all small cross-section bunds (0.06-0.15 m^2 size) are constructed as per alignment before planting the vegetative hedges. It helps in the establishment of vegetation as well as in checking runoff and soil loss during the initial first one or two years. On steeply sloping lands hedges are provided at a vertical interval of 1 - 2.5 m. On moderately sloping arable lands these may be spaced at 30-40 m horizontal interval. The choice of plants that will form an effective barrier in the shortest possible time is crucial for the success of contour vegetative hedges.

Contour Cropping

Contour cropping is a conservation farming technique commonly used on slopes to control soil losses due to water erosion. Contour cultivation refers to applying all tillage practices, such as, ploughing, planting, cultivation and harvesting on the contour instead of the conventional up and down the slope (Figure 25.1). The furrows between the ridges developed by contour tillage catch and hold the water, thereby, checking the high water velocity, helping to minimize soil erosion through sheet, rill or gully erosion. It is most effective on slopes ranging from 2-7 per cent, just effective on slopes less than 1 per cent or between 7 per cent and 15

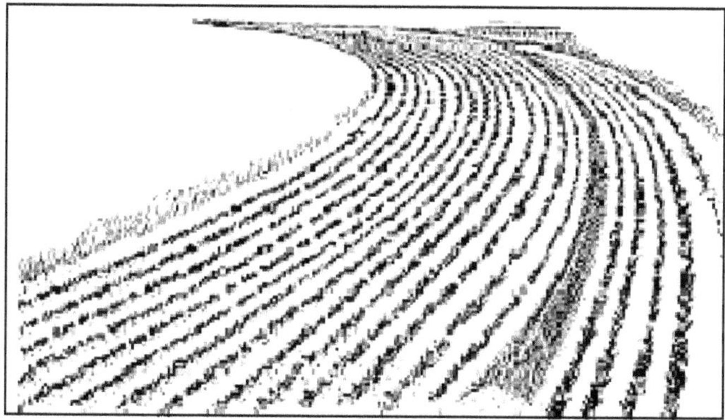

Figure 25.1: A View of Contour Cropping on a Sloping Land.

per cent but is not much effective at slopes exceeding 18 per cent. Contour cropping reduces the velocity of runoff water, induces more infiltration and thereby protects the valuable top soil from erosion. Contour cropping is quite effective on long and smooth slopes, as it shortens the slope length to reduce the flow velocity, which is usually high under such situations. Results have shown that intensity of soil erosion is reduced by about 30 per cent with contour cultivation as compared to ploughing up-and-down the slope. Contour cropping, if supplemented with furrow cultivation, is even more effective. Contour farming is also an efficient method of *in-situ* moisture conservation on lands where other engineering/mechanical measures such as contour *bunding*, terracing, *etc.*, have been adopted for soil and water conservation.

Mixed Cropping

In this system two or more than two crops are raised in the same land at the same time. Mixed cropping is also known as inter cropping or co-cultivation. Mixed cropping system offers several benefits which include better utilization of soil nutrients, low weeds and insect pests, resistance to climate extremes, increase in overall productivity and use of limited resources to the fullest extent.

Cover Cropping

Cover crops are grown as a conservation measure either during off-season or for ground protection under trees. These crops besides providing good cover against erosion also provide hay or fodder for animals. These crops add organic matter to the soil and serve as soil building crops. The cover crops are also grown under trees to protect the soil from the impact of water drops falling from the tree canopy. It is quite effective under tall trees where height of fall is more.

Crop Rotation

It may be defined as a more or less regular succession of different crops being grown on the same piece of land. For example, leguminous crops like pulses are grown alternately with wheat, barley or mustard. Figure 25.2 shows a typical layout of crop rotation system. Rotation of crops has to be decided on the basis of its objectives. It can be used to reduce erosion and/or increase the fertility of soil.

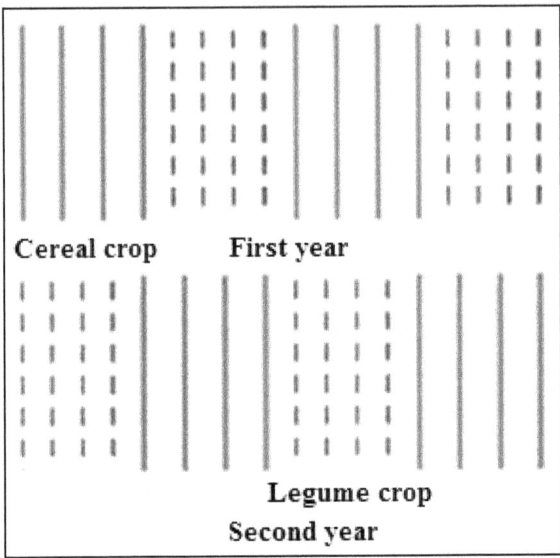

Figure 25.2: Crop Rotation on different Parcel of Land over Years (In third year the rotation of first year can be used).

In the former erosion resistant and cover crops need to be included in the rotation. A crop rotation with a combination of inter-tilled and close growing crops, farmed on contours, provides food, fodder and conserves soil moisture. As such, it ensures higher returns from the field. In the latter legume crops and crops of different rooting depths should be included. The latter strategy enables the plants to draw their plant nutrient requirement from different layers of soil. Crop rotations help to increase crop yield and net profits. A well planned crop rotation may reduce the use of chemicals that in turn may reduce the pollution of water resources.

Strip Cropping

Strip cropping is the practice of growing strip of crops having poor potential for erosion control, such as root crop, cereals, *etc.*, alternated with strips of close growing crops having high potential for erosion control, such as fodder crops, grasses, *etc.* in the same field (Figure 4.2). The crops are grown in strips at right angles to the slope of land. Close growing crops act as barriers to flow and reduce the runoff velocity generated from the strips of inter-tilled crops to eventually reduce soil erosion. Strip cropping is a more intensive farming practice than contour farming. It is desirable to rotate the strips from one year to another by shifting the erosion resisting strips in a regular sequence. On very erodible soils, it may be necessary to retain same strips as permanent vegetation. Generally, a strip width of 15, 20, 25 and 30 m is recommended on lands having fairly high infiltration rates and slopes of 15-20, 10-14, 6-9 and 2- 5 per cent, respectively (MPUA & T, 2016). The farming practices that are included in this type of farming are contour strip farming, cover cropping, farming with conservation tillage and suitable crop rotation. Strip cropping is laid out by using any of the following methods:

Contour Strip Cropping

In contour strip cropping, alternate strips of crop are sown more or less following the contours, similar to contouring (Figure 25.3). Suitable rotation of crops and tillage operations are followed during the farming operations.

Field Strip Cropping

In strip cropping, strip of uniform width are laid out across the prevailing slope, not necessarily on contour. Such a practice is generally followed in areas where the topography is very irregular, and the contour lines are too curvy for strict contour farming.

Buffer Strip Cropping

Buffer strip cropping is practiced where uniform strip of crops are required to be laid out for smooth operations of the farm machinery, while farming on a contour strip cropping layout (Figure 25.3). Buffer strip of legumes, grasses and similar other crops are laid out between the contour strips as correction strips. Buffer strips provide good protection and effective control of soil erosion.

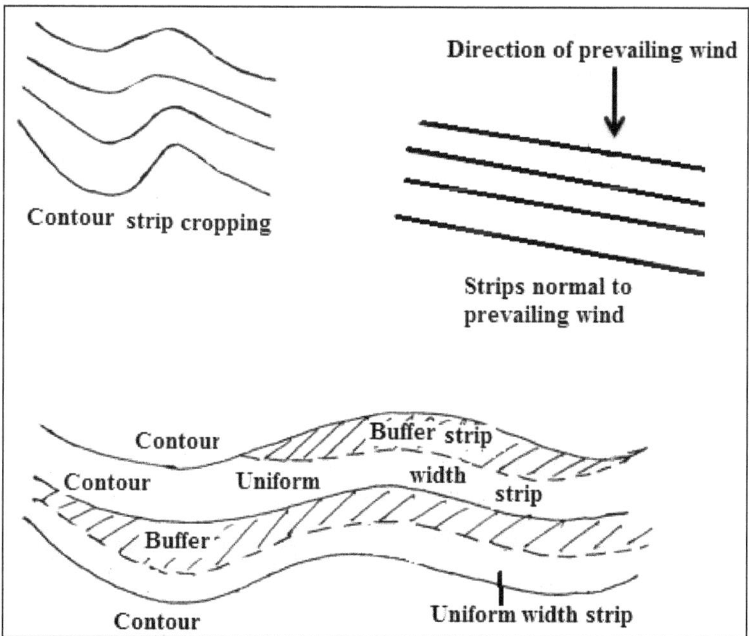

Figure 25.3: Strip Cropping: Contour Strip Cropping (Top Left), Wind Strip Cropping (Top Right) and Buffer Strip Cropping (Bottom).

Wind Strip Cropping

To protect the soil from wind erosion, strips are laid out across the prevailing direction of wind without regard of the contour (Figure 25.3). These are used only on flat or nearly level topography where wind is the major cause of soil erosion.

Mulching

Mulch is any material that is spread or laid over the surface of the soil as a covering. Mulching on the other hand is a process of covering the soil that makes more favorable conditions for plant growth and development. Mulches are used to minimize rain splash, conserve moisture, reduce evaporation, control weeds, and moderate the temperature to a level conducive to microbial activity. Mulches help in breaking the energy of raindrops, prevent splash and dissipation of soil structure, obstruct the flow of runoff to reduce their velocity and prevent sheet and rill erosion. Organic mulches also help to improve the infiltration capacity by maintaining a conductive soil structure at the top surface of land. It is also useful as an alternative to cover crop in dry areas where a cover crop would compete for moisture with the main crop. The mulches serve the following functions.

- Protects the soil from erosion
- Reduces compaction from the impact of heavy rains
- Conserves moisture, reducing the need for frequent watering
- Maintains a even soil temperature
- Prevents weed growth
- Keeps fruits and vegetables clean
- Modifies the micro-environment of the soil
- Mulches can also provide a barrier to soil pathogens and repel certain whitefly and thrips
- Provides an aesthetic look to the garden

A wide variety of mulches are used such as: cut grasses or foliage, straw materials, farm residue, wood chips, saw dust, paper, stones, plastics and even glass wool, metal foils and cellophanes. Only the most important amongst these are discussed in the following sections.

Conventional or Organic Mulch

Tree branches, twigs, leaves, leaf litter, grasses, weeds, hay or straw and farm residue *etc.* can be used as organic mulch. These are more effective than the other kind of mulches in respect of moisture conservation, and in reducing evaporation and runoff. These mulches also reduce the fluctuations in soil temperature, protect the soil from rain drop impacts, and hold the excess surface water in contact with the soil. Increase in infiltration rate due to application of these mulches helps to reduce runoff and soil erosion. During day hours these mulches absorb as much insolation as a bare soil does, but little energy is conducted downward. This causes the surface of the mulch to become hot while the underlying soil to remain cool. On the other hand, during night hours, the mulch cools down while the soil remains warm. A study revealed that application of organic mulch @4 t/ha effectively reduced runoff from 37 to 15 per cent and soil loss from 18 to 5.4 t/ha in maize crop on 18 per cent slope and increased the yield of subsequent crop from 1.9 to 2.4 t/ha. Application of mulch @2 t/ha also reduced the soil erosion loss by about 50 per cent.

Paper Mulch

The paper mulch, a kind of conventional mulch, is reported to have produced remarkable results. Paper mulch increased the soil temperature, especially of the surface soil layers. Evidences have been generated to show that paper mulch performed better in improving the soil condition, besides promoting the earthworm activity. Necessary precautions however, should be exercised since some toxic elements may be produced because of the chemicals leaching out of the paper mulch.

Plastic Mulch

Plastic films made of Polythene or polyvinyl are more widely used as mulch. They help in maintaining higher water content in soil resulting from reduced evaporation, induced infiltration, reduced transpiration from weeds or combination of all these factors. They are relatively expensive and difficult to manage under large scale field conditions for low value crops. Although plastic mulch is used in a similar fashion as other kind of mulches, yet it is easier to mechanize the laying operation.

Synthetic Mulch

It includes organic and inorganic liquids that are sprayed on the soil surface to form a thin film to control various atmospheric agents acting on the soil surface. The different synthetic mulching materials are: resins, asphalt emulsions, latex and cut back asphalt, canvas *etc.*

Petroleum Mulch

The petroleum mulches are easier to apply, are less expensive and are long lasting. These mulches are available in the form of emulsions of asphalt in water, which can be sprayed on the soil surface at ambient temperature to form a thin continuous film that clings to soil, but does not penetrate deep inside the soil. The mulch film promotes uniform and rapid seed germination and also plays a significant role in vigorous growth of seedlings. An ideal surface film is also stable against erosion, sufficiently porous to allow water into the soil, yet insoluble in water and resistant enough to the forces of weather.

Stone Mulch

It is a very old practice, and has been commonly adopted in arid zones to minimize wind erosion. It comprises of spreading stone pieces on the ground surface to conserve the moisture besides reducing wind erosion. Soil under the stones tends to retain more moisture although the temperature of the soil may be slightly higher. The soil below the stones harbors micro-organisms. The stone mulching also helps to trap dew, particularly in locations where significant dew fall occurs. Central arid zone research institute Jodhpur, has reported the use of rubble mulch, which is simply combination of small fragments of stones and bricks. This mulch provides better results on moisture conservation compared to the stone mulch, synthetic mulch and mulch made of straw materials.

Inorganic Mulch

Also known as soil mulch, is an important mulch for arid and semi-arid regions.

To prepare the soil mulch, the soil is first ploughed to the desired depth, left over for some time to dry followed by planking that planes the land. Thus, a loose and dried soil layer of 5 -8 cm thick is established on the soil surface. The soil mulch helps to conserve moisture as it breaks the contact with moist soil layers. It also increases the non-capillary pore space.

Vertical Mulch

Quite high runoff is expected in areas where rainfall occurs as high intensity storms and soils have low to moderately low infiltration rate. This runoff can be stored in the soil profile itself. For this purpose, a new technique named as vertical mulch has been designed. In this case, suitable trenches across the slope are made so that sufficient surface area is available for water absorption. Such trenches at suitable intervals (5 m for black soils) provide low density areas with high intake rates to absorb water at relatively higher rates. The trenches 15-20 cm wide and 30 cm deep are ideally suited for this purpose. To avoid silting of these trenches, these are filled with organic farm waste material like straw stubbles or stalks. The filter material used should not easily decompose and should provide service at least for 3 - 4 years. Water percolating into the trench gets distributed in the soil profile. The width of trench is so adjusted that least area goes out of cultivation. For example, if trenches are accommodated between crop rows, practically no area is wasted in trenches.

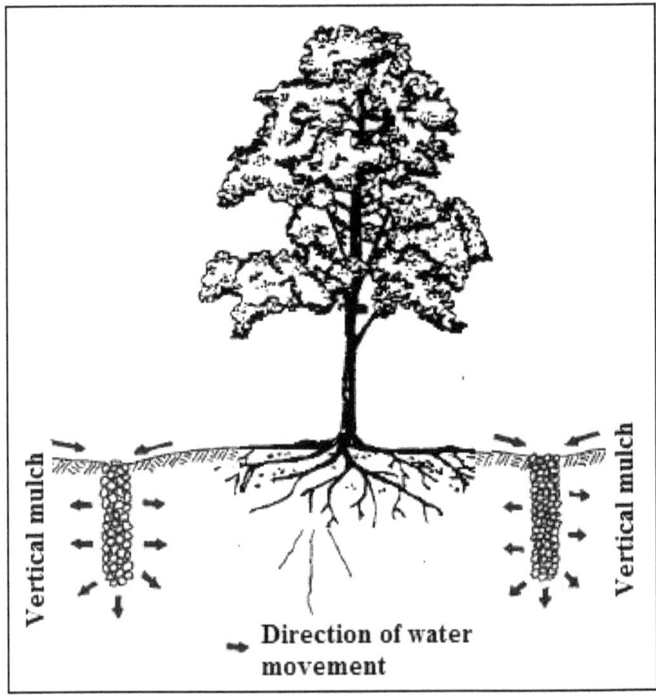

Figure 25.4: Vertical Mulch for Water Storage in the Profile.

Tillage

Tillage refers to mechanical manipulation of the soil to create necessary soil conditions congenial for plant growth. It facilitates conservation of moisture, breaks hard pan, helps root proliferation, controls weeds, improves aeration and provides a good seed-bed. Tillage operations like ploughing and chiseling open up the hard layers in the profile, increase infiltration rate and thereby increase the depth of moisture penetration into the profile. Tillage operations carried out for soil conservation are a bit different than commonly carried out field operations. These comprises of:

- ✰ Till no more than necessary
- ✰ Till only when the moisture is at the optimum level for the kind of soil
- ✰ Vary the depth of ploughing.

Conservation Tillage

Conservation tillage is any method of soil cultivation that leaves previous crops residue on field before and after planting the next crop. It decreases soil erosion, runoff, water pollution, and CO_2 emissions. Conservation tillage methods include **no-till, strip-till, ridge-till** and **mulch-till. No-till** involves planting crops directly into residue that has been not been removed at all, whereas in strip-till narrow strips are tilled and the rest of the field left untilled (strip-till). In Ridge-till row crops are planted directly on permanent ridges of 10-15 cm by clearing previous year's crop residues from ridge tops only. **Mulch-till** is any other reduced tillage system that leaves at least one third of the soil surface covered with crop residue.

Agro-forestry

Forestation/plantation plays major role in erosion control on gullied areas and areas prone to landslides. It may comprise of fresh plantation or restocking the existing depleted forests and woodlands either naturally or intentionally. Trees can also be incorporated within a farming system by planting them on terraces, contour bunds or on contour lines (Figure 25.5) and as ornamental around the homestead. It

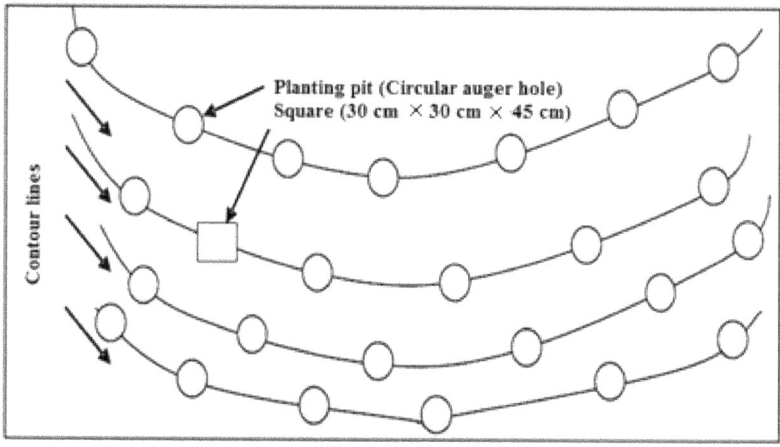

Figure 25.5: Contour Planting of Agro-forestry Plants.

not only reduces soil erosion but also meets some of additional needs of the farmers. Afforestation on a bigger scale helps to improve air quality, mitigate global warming and rebuilds ecosystem by controlling soil erosion.

Alley Cropping

Alley cropping is an agro-forestry system in which food crops are grown in an alley formed by hedge rows of trees or shrubs spaced at regular intervals. The essential feature of this system is that hedge rows are cut back and kept pruned during cropping to prevent shading and to reduce competition with field crops. The main criteria for the selection of species to be planted are their amenability to lopping besides being multipurpose and fast growing. Some of the hedge row species are: *Leucaena leucocephala, Gliricidia sepium, Sesbania grandiflora, Robinia pseudocacia, Cassia siamea* and *Pongamia pinnata*. The alley width varies with the choice of the species and watershed location in a given climatic region. By and large, alley width has been found to vary from 2-10 m; wider the alley, the lesser is the adverse effect on crop performance.

Chapter 26

Contour Bunding and Trenching

Mechanical measures play a vital role in controlling and preventing soil erosion on agricultural lands. These measures are mainly adopted to supplement the agricultural practices (Agronomical methods) *i.e.* when these alone fail to control soil erosion. Three of the mechanical measures includes contour bunds, graded bunds and contour trenches. Contour and graded *bunding* in India has been practiced for a long time. From the experience gained so far, it has emerged that these bunds can be useful in shallow, medium and medium deep soils. On the other hand, cracks may develop in deep black soil upon drying, and bunds may fail as water may continue to flow through these cracks, which may turn into big breaches. This results in severe damage to the fields' downside. This chapter describes the practice and design of contour bunds, graded bunds and contour trenches.

Types of Bunds

Bund is simply an embankment like structure, constructed across the land slope. Different kinds of bunds are used for erosion control and moisture conservation. The choice of the bund depends upon the land slope, rainfall, soil type and the purpose of the bund for which it is constructed. Commonly used types of bunds are:

- Contour bunds
- Graded bunds

Both kinds of bunds are further sub-categorized as narrow based and broad based bunds. The narrow based bunds cannot be crossed by the farm machinery since for the same cross-section a narrow based bund has to have greater height. The area under the bund is not available for cultivation. Thus, considerable area is lost besides creating difficulties in farm mechanization. The cross-section is also impacted by rain drops, thus requiring frequent maintenance.

Farm implements can easily cross the broad based bunds. The area under these bunds can be put under cultivation. However, with repeated passes of the farm machinery, the bunds gets loosened and size of the bunds may get reduced in a short period.

Besides the main contour and graded bunds, other kinds of bunds are also constructed and are designated as follows.

Peripheral Bunds: These are constructed to encircle the boundaries of areas.

Marginal Bunds: These are constructed in the lowest part of the catchment without any consideration of contour

Side Bunds: These are constructed along the slope on the two sides of contour bunds to check the outflow of ponded water from the sides of the strips. The height of these bunds is usually 30 cm.

Lateral Bunds: These bunds are constructed along the slopes in between two side bunds to reduce the length of contour if it exceeds the recommended length of 300 m. This helps to reduce accumulation of runoff along one side. These are also provided if the contour bunds cross the boundaries of the property of the owner so that water from the other fields does not enter into the *bunded* fields. No special planning is made for these kinds of bunds. If these bunds are constructed, a lump sum allowance of about 30 per cent of the calculated earthwork for the main bunds is used to calculate the cost *etc.* No farming is done on these bunds expects at few places, where some types of stabilization grasses are planted to protect the bund.

Contour Bunds

The bunds constructed along the contours with some minor deviation to adapt to practical situation, are known as contour bunds. Contour bunds are laid out in those areas which are characterized by low annual rainfall (< 600 mm) and agricultural fields have permeable soils. These are used where land slope is less than 6 per cent. The major objectives of contour *bunding* are:

- ☆ To reduce the velocity of runoff
- ☆ To intercept the flowing water or runoff
- ☆ To hold water and maintain moisture in the catchment for a longer period
- ☆ To allow more water to infiltrate
- ☆ Orderly disposal of runoff
- ☆ To reduce soil erosion

Design of Bunds

The following parameters are considered during design of a bund.

Type of Bund

The type of bund (contour or graded bund) to be constructed depends upon the rainfall and soil condition as discussed in previous sections.

Spacing

The basic principles to decide the spacing of bunds are:

- ✰ The seepage zone below the upper bund should meet the saturation zone of the lower bund
- ✰ The bunds should check the water at a point where the water attains erosive velocity
- ✰ The bund should not cause inconvenience in the agricultural operations.

To determine the spacing, following formula is used:

$$VI = [(S/a) + b] \times 0.3 \quad (26.1)$$

Here, VI is the vertical interval between consecutive bunds, m, S is the land slope, per cent and a and b are constants, depending upon the soil and rainfall characteristics of the area. The values of a = 3 and b =2 are taken for medium and heavy rainfall zones and a = 2 and b =2 for low rainfall zones. For graded bunds the values of a are slightly higher varying from 3 to 4, the value of b remaining the same.

It has been recommended that spacing can be increased by 25 per cent, if conservation measures are applied and reduced by 15 per cent if conservation measures are not adopted or conditions are unfavourable.

USDA with appropriate values of a and b arrived at the following equation such that VI is in cm and S in per cent.

$$VI = 7.5\,S + 60 \quad (26.2)$$

The above equation is area specific. It has been further modified for areas with different rainfall amounts.

For areas having high rainfall:

$$VI = 10\,S + 60 \quad (26.3)$$

For areas having low rainfall:

$$VI = 15\,S + 60 \quad (26.4)$$

Cox suggested that the VI also depends upon the rainfall, infiltration rate and crop cover. Accordingly he suggested that

$$VI = (X \times S + Y) \times 0.3 \quad (26.5)$$

Here X is the rainfall factor, S is the land slope, per cent and Y is the infiltration and crop cover factor. The value of X and Y are taken from Table 26.1. Here VI is in m.

Table 26.1: Values of X and Y for Use in Cox Formula

Values of X		
Rainfall Condition	*Average Rainfall (cm)*	*X*
Scanty	64	0.80
Moderate	64-90	0.60
Heavy	>90	0.40

Values of Y		
Intake rate	*Crop Cover during Erosive Period of Rainfall*	*Y*
Below average	Low coverage	1.0
Average	Good coverage	2.0
One of the above factor favourable and another unfavourable		1.5

Since it may be difficult to locate the bund spacing on the ground on the basis of vertical interval, horizontal interval (HI) or spacing is usually determined as it can be easily measured on the land surface. For this purpose, the relationship between horizontal and vertical spacing is given as follows:

$$HI = VI/S \tag{26.6}$$

Here, HI indicates the horizontal distance of the bund and VI is the vertical interval (Figure 26.1).Bund spacing should not be so wide as to cause excessive soil erosion between adjacent bunds. General recommendations on spacing of contour bunds are given in Table 26.2.

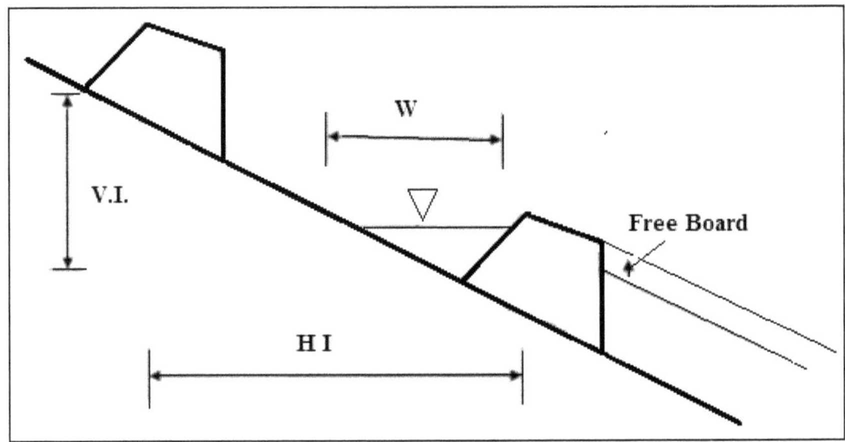

Figure 26.1: Schematic View of Two Bunds Showing VI and HI.

Table 26.2: Spacing of Contour Bunds

Land Slope (per cent)	*Vertical Interval (m)*	*Approximate Horizontal Distance (m)*
0 to 1	1.05	105
1 to 1.5	1.20	98
1.5 to 2	1.35	75
2 to 3	1.50	60
3 to 4	1.65	52

Size of the Bund

The size of bund includes its height, top width, side slopes and bottom width.

The height of bunds mainly depends upon the slope of the land, spacing of the bunds and the maximum intensity of rainfall expected in the area. Once the height of the bund is determined, other dimensions of the bund *viz.*, base width, top width and side slopes can be determined on the basis of the nature of the soil. Depending on the amount of water to be intercepted, the height of the bund can be calculated.

Let X be height of the bund, L the distance between bunds, V the vertical interval between bunds, and W the width of water spread, then from Figure 26.2, we have:

$$W/L = X/V \quad (26.7)$$

$$W = LX/V \quad (26.8)$$

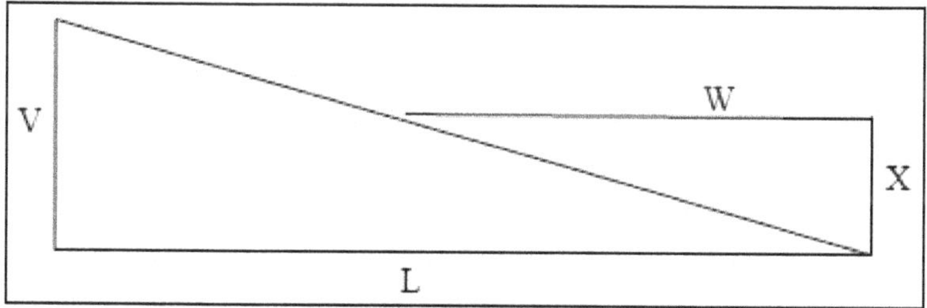

Figure 26.2: Basic Diagram for Deriving the Height of Bund.

Considering 1m length of the bund, amount of water stored = ½ WX

Substituting W from eq. (26.8)

$$\text{Amount of water stored} = 0.5\,(LX^2/V) \quad (26.9)$$

Assuming that the maximum rainfall, which the bunds have to withstand at any time is 15 cm, then the water retained by 1 m length of the bund is

$$\text{Amount of water stored} = 15\,(L \times 1)/100 = 3L/20 \quad (26.10)$$

Equating the two values of amount of water stored, we get

$$0.5\,(LX^2/V) = 3L/20$$

$$\text{Height of bund} = X = (3V/10)^{1/2} \quad (26.11)$$

When, the land slope is expressed in per cent.

$$V = LS/100 \quad (26.12)$$

$$X = (3LS/1000)^{1/2} \quad (26.13)$$

The height of the bund to store 24 hr rainfall excess of a given return period is calculated by the following formula.

$$\text{Height of bund} = X = (Re \times V/50)^{1/2} \quad (26.14)$$

Here Re is the 24 hr rainfall excess in cm and V is the vertical interval between consecutive bunds in m.

Here L and X are in m. It is the theoretical height and a suitable free board may be added to arrive at the real height of the bund. Usually, 20 per cent of the calculated X is added as freeboard.

Base width of the bund depends upon the hydraulic gradient of water in the soil. It is desirable to know the approximate line of seepage or hydraulic gradient line (HGL). The following guidelines may be used to determine the HGL.

Clayey soil: 3H:1 to 4H:1V

Clayey sand or sandy loam: 5H:1V

Sandy soil: 6H:1V

Base width is calculated through the following procedure.

Depending upon the soil, hydraulic gradient is selected. Let us say value of the hydraulic gradient is assumed as n_1H:1V. Then to pass the HGL through the toe of the bund, the part of the bund width will be n_1.d (Figure 26.3). Similarly, the part of the bund will be decided on the basis of side slopes of the bund. Side slopes are dependent upon the angle of repose of the soil. Side slopes of the bund recommended for different soils are given in Table 26.3 (Murthy, 1994). If the side slope is n:1, then the total base width of the bund is given as:

$$\text{Base width} = n.d + n1.d \tag{26.15}$$

Table 26.3: Side Slopes of the Bunds Recommended for different Soil Types

Side slopes	1.5 to 1	2 to 1	2.5 to 1
Soil types	Red Gravel	Light Sandy loam	Sand
	Light red loam	Clay	
	Black loam	Black cotton soil	
	White gravel	Soft decomposed rock	

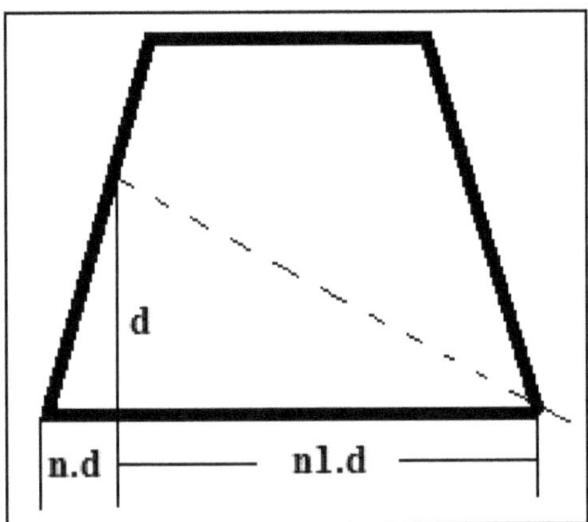

Figure 26.3: Calculation of Base Width on the Basis of Hydraulic Grade Line and Side Slopes.

The other way to calculate the base width is to take the top width as minimum 50 cm so that base width is calculated using side slopes as per the soil type. A minimum cross-sectional area is taken as 1 m^2.

The third way is use the general recommendation on cross-section areas (Table 26.4). Usually a higher size of the bunds than required by the hydraulic considerations is adopted to allow for the settlement and poor maintenance by the cultivators.

Table 26.4: Recommended Dimensions of the Contour Bunds

Depth of Soil (cm)	*Base Width 'm'*	*Top Width 'm'*	*Height 'm'*	*Side Slope*
Shallow (7.5 to 22.5)	2.67	0.38	0.75	1.5 : 1
Medium (22.5 to 45)	3.12	0.6	0.85	1.5 :1
Medium deep (45 to 90)	4.25	0.6	0.9	2 : 1

Length of Bund

The length of bund is determined by the horizontal interval of the bund. The length of bund per hectare area of land is given as:

$$L = 10000/HI = (10000 \times S)/(VI \times 100) = 100(S/VI) \qquad (26.16)$$

Earth Work

The earthwork required to construct a bund is obtained by multiplying the cross-sectional area, A to its total length. The earth work for complete *bunding* system includes the sum of earthwork made in main bunds, side bunds and lateral bunds. The total earthwork is given by the following equation.

$$E_t = E_m + E_s + E_l \qquad (26.17)$$

Here, E_t is the total earthwork, E_m is the earthwork in main bunds, E_s is the earthwork in side bunds and E_l is the earthwork in lateral bunds. The value of $E_s + E_l$ is taken as 30 per cent of earth work of main contour bund (E_m) by assuming that the length of side and lateral bund is 30 per cent of the length of main bund and their cross-sectional area is also equal to main bund.

$$E_m = A \times \text{total length of bund} = (100S/VI) \times A$$

$$\text{Therefore, } E_s + E_l = ((100S/VI) \times 30/100) \times A$$

$$\text{Therefore, total } E_t = (100S/VI + 30S/VI) \times A = 130S/VI \times A$$

$$E_t = 1.3 \times (100S/VI) \times A \qquad (26.18)$$

Area Lost due to Bunding

It is calculated by multiplying the length of contour bund per hectare with its base width. *i.e.*

$$A_L = (10000/HI) \times b = (100S/VI) \times b \qquad (26.19)$$

Where, b is the base width of bund.

This equation computes only the area lost due to main contour bund and not the area lost due to side and lateral bunds. Usually, the area lost due to side and lateral bunds is taken as 30 per cent of the area lost due to main contour bund. Thus, the total area lost due to contour *bunding* is:

$130 \times b \times S/VI$

The area lost as percent of the total area is given as:

$$A_L \text{ (per cent)} = 1.3 \times S \times b/VI \quad (26.20)$$

Here A_L is the area lost in per cent of the total area.

Surplus Water Outlets

In contour bunds water is impounded between the contour and side bunds. If the water head at any time reaches 30 cm, the earthen bund is likely to breach. Thus, an arrangement is made to discharge the excess water such that water head behind the bund does not exceed 30 cm. These arrangements can be any one of the following types:

- Weirs
- Pipe outlets
- Grassed outlets.

Limitations of Contour Bunds

While contour bunds save soils from erosion, maintain soil fertility and considerably increase water infiltration into the soil, few limitations of these bunds are identified.

- Contour *bunding* is not technically feasible on the land slopes exceeding 6 per cent
- Contour *bunding* is not suited to areas where rainfall exceeds 600 mm
- No cultivation is allowed on the earthen embankments of contour bunds. Therefore, an area of about 5 per cent is lost under the bunds and is not available for cultivation
- Contour bunds in deep black soils (heavy clay soils) have been a failure because of the nature of soil, which cracks during hot weather and cakes during the monsoon. The stability of contour bunds in these soils is doubtful. Moreover, due to poor drainage properties of deep black soils, water stagnation against contour bunds is quite prolonged that may also make the bunds unstable.
- Contour bunds are also not successful in very shallow soils, when soil depth is less than 7.5 cm.

Graded Bunds

The bunds constructed with some slope are known as graded bunds. Graded bunds are constructed in areas where land is susceptible to water erosion, the soil is less permeable and the area is prone to water logging. Graded bunds are used to safely dispose off the excess runoff. Excess water flows through the graded channels constructed on upstream side of bunds to an outlet or a grassed water way. The gradient can be either uniform or variable.

The uniformly graded bunds are used where bunds are of shorter lengths and expected runoff is likely to be small. A grade of 0.1 per cent to a maximum of 0.4 per cent is given. The variable graded bunds are used where bunds are longer, and cumulative runoff increases towards the outlets. In these bunds, grade varies at different sections of the channel along length of the bund. While doing so, runoff velocity is controlled within the desired limits so that soil erosion is avoided.

The design procedure of a graded bund is similar to the contour bund. Since the graded bund is required to drain off the runoff, additional calculations as listed below are performed.

- For the hydrological design, peak rate of runoff is determined by the rational formula discussed in chapter 29 or some other technique for which readers' may refer any textbook on hydrology.
- Manning´s formula is used for the hydraulic design and to calculate the velocity of flow.
- The discharge capacity of the channel should match the peak runoff rate.

Construction and Maintenance of Bunds

As far as possible, bunds should be aligned on contour with minor deviations, if necessary. Construction should begin from the ridge and proceed down the valley. It helps to protect the bunds from being washed away due to rains occurring during construction. The area comprising of the base width of the bund should be cleared of all vegetation. The soil in this area should also be scraped so as to establish a good bond with the soil used to form the bund. The borrow pits normally located on the upstream side of the bund should be continuous with no breaks. These should have a uniform depth of 30 cm while the width may be varied as per earth requirement. The borrow pits should not be located in a gully or depression. During construction, the earth should be put in layers 15 cm thick and suitably consolidated by trampling. The templates of the specified dimensions are used to check the bund section. The bund section should be finally shaped, trimmed and slightly rammed on the top and the sides. After the bund is completed, the field and the borrow pit should be ploughed.

To construct a graded bund, location of outlet should be decided first so as to drain the surplus water. The channel capacity of the graded bund should be sufficient to take care of the peak runoff. If graded bund is constructed with a variable grade, then the low grade is given at the upper while higher grade is given at the downstream end.

The maintenance requirement of bunds depends on the type of construction. Earthen ridges need to be rebuilt to their original height after each rainy season. To minimize the maintenance requirement, appropriate ramming of the fill material should be carefully done at the initial stage of the construction. Vegetating the bunds with appropriate grass can enhance their stability as the roots help to fix the soil. Construction made of stones, on the other hand, is quite resistant to erosion and will require only limited repairs.

The effectiveness of contour bunds can be enhanced by placing planting pits between the bunds and/or constructing earthen ties. The ties are meant to create micro-catchments along the bund. These can be built manually or with the help of special machines. If excess rainfall is anticipated, cut-off drains may be provided to let off the surplus runoff.

Example 26.1

Calculate the VI of contour bunds to be constructed on a 4.5 per cent land slope. The rainfall is moderate and infiltration rate of the soils is medium. A good vegetative cover exists on the land surface.

1. Using Ramsar formula

   $V I = 0.3 (S/3 + 2)$ where S = Land slope (per cent) $= 0.3 (4.5/3 + 2) = 1.05$ m

   Add 25 per cent extra spacing because the ground cover is good

   Therefore, $VI = 1.05 + 0.26 = 1.31$ m

2. Using Cox's formula $VI = (X \times S + Y)\ 0.3 = (0.6 \times 4.5+2)0.3 = 1.41$ m

Example 26.2

Calculate the total length and earthwork of contour bunds per hectare if the contour bunds are constructed at a horizontal interval of 30 m on a land slope of 6 per cent. Top and bottom width are 50 and 120 cm while the height of bunds is 90 cm. What would be the change in earthwork if lateral and side bunds are also formed? Also calculate the area lost per hectare due to *bunding*.

Length of bunds = 10000/30 = 333.3 m

Cross sectional area of the bund = 90 (50+120)/2 = 7650 cm^2 or 0.765 m^2

Earthwork for the main bund = $0.765 \times 333.3 = 255.13$ m^3

If side and lateral bunds are formed then

Earthwork for the side and lateral bunds = $0.3 \times 255.13 = 76.54$ m^3 (30 per cent of the main bund)

Total earth work = $255.13+76.54 = 331.67 \approx 332$ m^3.

Area lost = Total length of contour bund × base width = $333.3 \times 1.2 = 399.96$ m^2 (main bund)

Total area lost = $400 + 0.3 \times 400 = 520$ m^2 (additional 30 per cent due to side and lateral bunds)

Contour Trenching

The areas having steep slope are prone to soil erosion. These areas exhibit the characteristics of "high rainfall, quick drainage" resulting in very little retention time for the runoff to infiltrate into the soil profile. The overland flow velocity often surpasses the safe limit resulting in excessive soil erosion. In such regions, trenches are constructed to address the problem of soil conservation. These trenches act as flow barrier (restricting the flow velocity within the safe limit from soil erosion point of view) and facilitate *in-situ* water conservation for the establishment of vegetation.

Contour trench is an excavated trench constructed along the contour across the slope of the land in the upper and middle reaches of watershed. It is constructed both on hill slopes as well as on degraded and sloping wastelands as a measure of soil and water conservation and to establish vegetative cover. Trenches are constructed below the contour line in such a way that its upper edge coincides with the contour line. The soil excavated from the ditch is used to form a small bund, about 20 cm away downhill side of the trench. The gap between the trench and bund is known as berm. The rainwater retained in the trenches travels down and benefits land in the lower reaches of the watershed. The bunds are planted with some permanent vegetation (native grasses, legumes) to stabilize the soil. Besides the roots and foliage are able to trap sediment that might overflow from the trench in the event of a heavy rain storm.

Objectives

Contour trenches are used to achieve the following objectives.

- Break up the soil slope
- Slow down velocity of runoff
- Allow more time for water infiltration
- Trap sediments
- To intercept the rills
- Seeding of the trenches and berms (often used in conjunction) to improve soil moisture profile.

Types of Trenches

The trenches are constructed in different geometrical configurations namely continuous trenching, in line trenching and staggered trenching. The selection of the type of trench depends on the site characteristics and rainfall intensity.

Continuous Contour Trenches

These trenches are also known as graded trenches are identical to graded terrace. Continuous contour trenches (CCT) can be of any length across the slope without any break in between. These trenches are suited to high rainfall areas. The cross section of the trench generally varies from 30 × 30 cm to 45 × 45 cm (GSWMA; 2011). The life of trench is usually considered as 5 years. De-siltation is required every alternate year under normal rainfall and every year under excess rainfall conditions.

Staggered Contour Trenches

In medium rainfall areas with highly dissected topography, staggered contour trenches (SCT) are used. The length of these trenches is short; around 2-3 m and the spacing between the rows may vary from 3-5 m. Experience of watershed programs has shown that it is better to construct stagger contour trenches, because aligning the trench along the contour over long distances is invariably prone to errors. Besides, construction of CCTs demands high skill and SCTs are less prone to breaches than CCT. The staggered trenches can be laid in series (In-line trenches) or staggered format (staggered trenches), the later is normally preferred over the former.

In-line trenches: A major limitation of CCTs is due to the inconsistent deposition of soil along the length. The discontinuous in-line trenches overcome this limitation since these are maximum 3-5 meter long. The cross section of these trenches is similar to continuous trenches. The gap between two in-line trenches should not be more than 2 m (Figure 26.4). A major limitation of these trenches is that these fail to collect runoff flowing between the gaps of two trenches. They are constructed mainly for moisture conservation in low rainfall areas receiving storm of mild intensities. No cross disposal drain is required since the trenches are discontinuous.

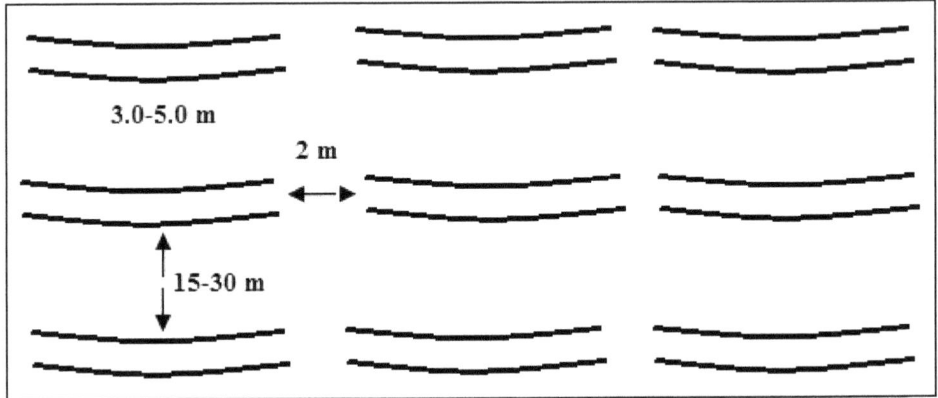

Figure 26.4: In-line Trench Layout.

Staggered Trenches or Interrupted Trenches: The staggered trenching involves the excavation of trenches of shorter length (2-3 m) in a row along the contour with interspaces between them as in the case of in-line trenches. But these trenches are arranged in straight line in a staggered manner (Figure 26.5). In the alternate row, the trenches are located directly below one another. The trenches in successive rows are thus staggered, with the trenches in the upper row and the interspaces in the lower row being directly below each other. Suitable vertical intervals between the rows are kept so as to impound the runoff without overflow. The length of the trench and the interspaces between the trenches in the same row should be suitably designed such that no long unprotected or uninterrupted slope remains untreated. The cross sectional area of these trenches is designed to collect the runoff expected from intense storms having recurrence interval of 5-10 years.

Design of Contour Trenches

To design a contour trench, one needs to determine the spacing and cross sectional area so as to collect the desired amount of runoff generated from the catchment area. Steps involved in design of contour trenches are as follows:

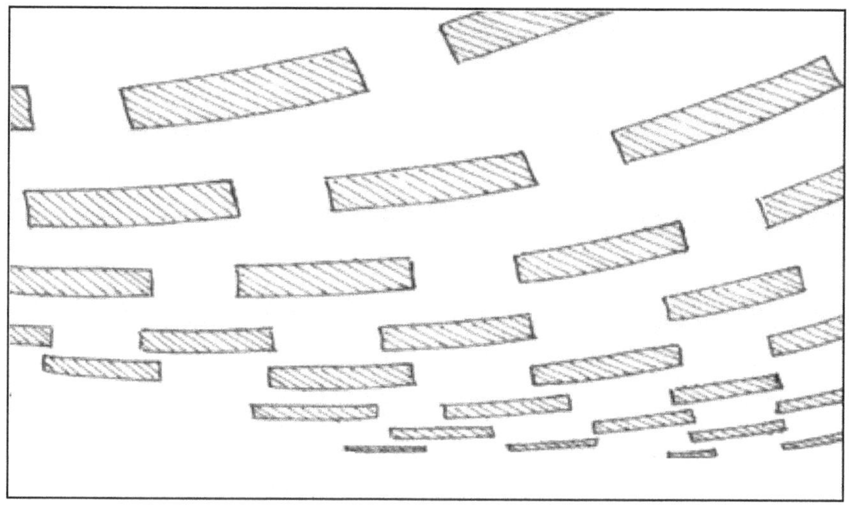

Figure 26.5: A Schematic Diagram of the Staggered Trenches.

Determination of Spacing

Spacing is expressed in terms of horizontal and vertical interval. Vertical interval is defined as the elevation difference between the upper or lower edge of successive contour trenches. A definition sketch of contour trenching is given in Figure 26.6. The vertical interval between two adjacent contour trenches is calculated by the following formula:

$$VI = S/C \tag{26.21}$$

VI is the vertical interval, m and C is taken as 12 for the medium rainfall. S is in per cent.

VI can also be calculated by the formula used for contour bunds with values of 'a' and 'b' respectively taken as 3 and 2 for soils with good infiltration rates. For soils of low infiltration rates the corresponding value are 4 and 2, respectively.

Relation between the horizontal spacing of contour trenches, runoff depth and dimension of trenches is given below:

$$HI = \text{Cross sectional area/Runoff depth} = A/Q \tag{26.22}$$

Assuming trench to be rectangular

$$HI = W \times D/(100 \times Q) \tag{26.23}$$

Where, HI is the horizontal spacing, m, W is the width of trench, cm, D is the depth of trench, cm and Q is the runoff depth, cm.

Size and Layout

The cross section area of the trenches can be square, rectangle, trapezoidal or triangular V-shape. The size of trenches is dictated by the soil depth at the site. In

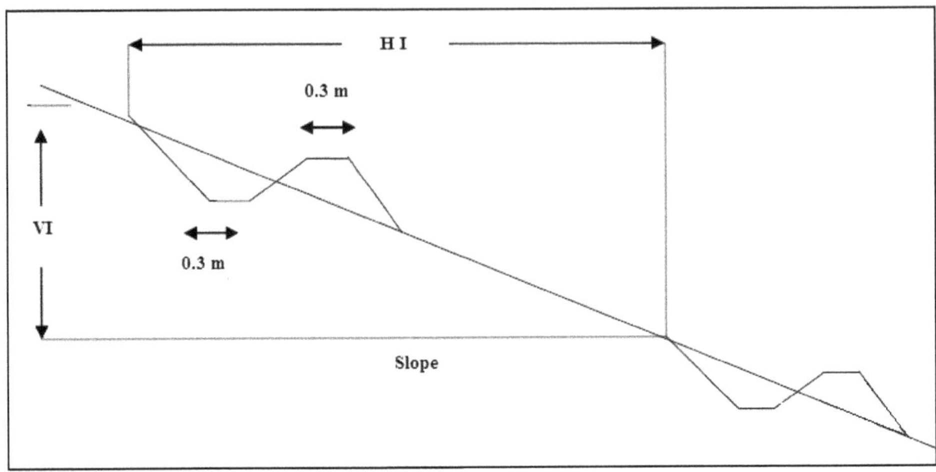

Figure 26.6: Definition Sketch of Contour Trench for Design.

relatively deeper soils, trench depth is generally fixed at 40-50 cm while for shallow soils; trench depth may be reduced to about 15-20 cm. As far as length of trench is considered, shorter lengths 3 to 7 m are adopted for convenience of layout and construction. Normally 1,000 to 2,500 cm^2 cross sections are used. The trenches may be of 30 cm base and 30 cm top width and square in cross section. For trapezoidal cross sections, side slopes of ½: 1 or 1:1 can be used (Figure 26.6). A general layout of a trench with bund and berm is shown in Figure 26.7. Based on the quantum of rainfall to be retained, it is possible to calculate the size and number of trenches.

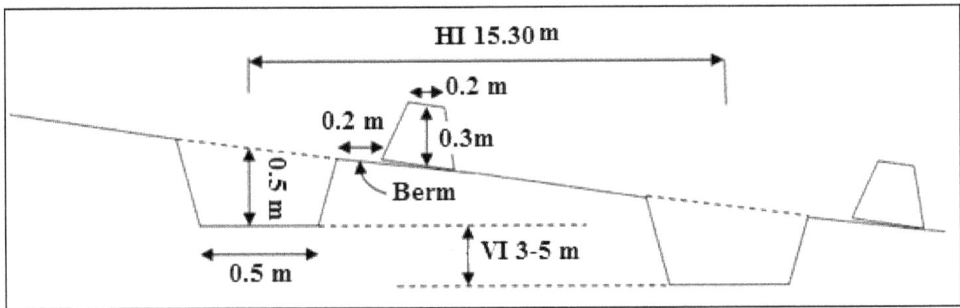

Figure 26.7: A General Layout of Contour Trench with Bund and Berm.

Bund Height

The height of the bund constructed on the downstream side of the trench is determined by the following formula:

$$H = 0.3(1+1.81 \sqrt{D}) \qquad (26.24)$$

Here H is the bund height, m and D is the vertical drop of the contour trench, m. Appropriate free board should be added.

In order to decide the number of rows of contour trenches (N), we have to divide the longest section of ridge area (L_1) by the distance between successive rows of contour trenches (d).

$$N = L_1/d \qquad (26.25)$$

Direct Runoff Volume

Trenches are designed to hold a part normally 60-70 per cent of the runoff expected from a storm of 5-10 years recurrence interval. The volume of runoff from the design storm is then computed by the formula. $Q = C \times R \times A$ such that C is the runoff coefficient, R is the daily rainfall and A is catchment area for that particular structure. It is further multiplied by a factor of 0.6- 0.7 depending upon the percentage of runoff to be stored.

Suggestions for Effective Design

- ☆ Trenches are recommended for slopes up to 25 per cent although in Tamil Nadu trenches have been constructed on slopes exceeding 33 per cent. In such cases, planting of grasses, shrubs and trees on berms may be required
- ☆ For slopes less than 10 per cent, construct contour or graded bunds rather than contour trenches
- ☆ Do not excavate trenches across large streams or drainage lines
- ☆ Begin the layout from the longest section within the largest area of uniform slope rather than from the shorter section
- ☆ Trenching should be avoided in shallow soils less than 20 cm deep as well as in black soils.

Example 26.3

Design a continuous contour trench in an area of 30 ha having daily rainfall of 120 mm. The runoff coefficient of this area is 0.45. It is proposed to store 65 per cent of the runoff in the contour trenches. Assume the trench gets 2 refills a day. The longest section of the ridge area is 2500 m.

$A = 30$ ha or $300000\ m^2$

$R = 120\ mm = 0.12\ m$

Therefore, $Q = 0.45 \times 0.12 \times 300000 = 16200\ m^3$

Only 65 per cent of runoff is proposed to be stored. Therefore,

$Q_1 = 16200 \times 65/100 = 10530\ m^3$

Assuming the cross-section of contour trenches as 0.5 m × 0.5 m, the cross sectional area is

$0.25\ m^2$ and given that 2 refills per day are possible, then

$L = 10530/(0.25 \times 2) = 21060\ m$

Distance between successive rows of contour trenches (d),

d = A/L = 300000/21060 = 14.24 m

Number of rows of contour trenches (N)

N= L_t/d = 2500/14.24 = 175.6

Chapter 27

Terraces for Water Erosion Control

Mechanical measures are engineering practices used to control erosion from slopping land surfaces through land surface modifications for retention and safe disposal of runoff. In the design of such practices, the basic approach is i) to increase the retention time of water to increase its infiltration into the soil, ii) to decrease the effect of land slope on runoff velocity by intercepting the slope at several points so that the velocity is less than the critical velocity, and iii) to protect the soil from erosion due to runoff. The mechanical measures adopted for soil and water conservation are: *bunding*, trenching and terracing, *etc.* While *bunding* and trenching have been described in the previous chapter, this chapter describes the practices used in terrace design and construction.

Terraces: Functions and Types

A terrace is an earth-embankment, constructed across the slope, to control runoff and minimize soil erosion. A terrace acts as an intercept to land slope, and divides the sloping land surface into short strips. As a result, the slope length is reduced. Since soil loss is related to the square root of the slope length, soil erosion is reduced through the shortening of the length of run. Besides, the scour of soil surface gets initiated once the runoff attains a velocity above the critical value. Since terraces are designed to control the runoff velocity to less than the critical value, these structures prevent soil erosion owing to scour.

Several types of terraces are used depending upon the objectives, functions, climate and crop conditions (Figure 27.1). Commonly used terraces are classified into two major types: broad-base terraces and bench terraces. Broad-base terraces are adapted where the main purpose is either to remove or retain water on sloping land suitable for cultivation whereas, the purpose of bench terraces is mainly to reduce the land slope.

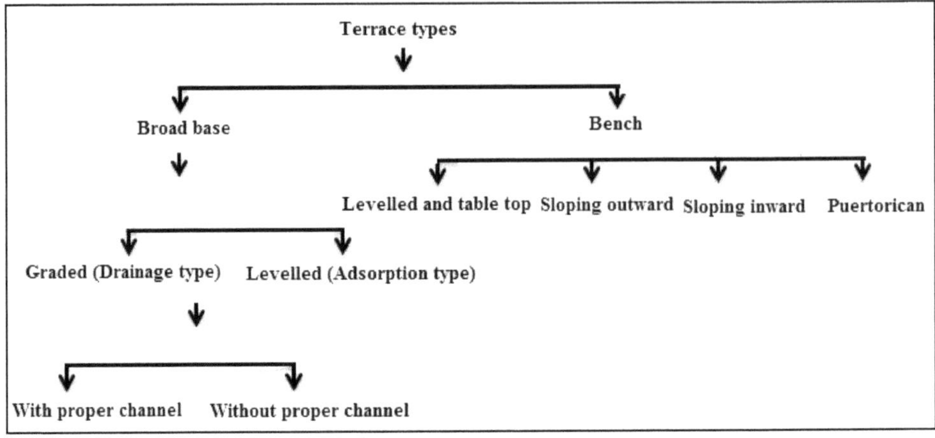

Figure 27.1: Classification Diagram of Terraces.

The functions or benefits of terraces can be listed as under:

- Protects the soil from erosion by shortening the length of slope and conducting the runoff water on a non-erosive grade
- Minimize sediments and water pollution
- Reduce run-off water, improve drainage to control flood damage and provide better sites for cultivation
- Intensify land use
- Stimulate improved farming practices and increased production
- Facilitate mechanization of steep slopes
- Conserve moisture and maximize irrigation benefits
- Enable a wider choice of crops and cropping pattern
- Encourage established farming system and reduce shifting cultivation and forest fires
- Create labor intensive programs and create local job opportunities
- Beautify landscape adding to its aesthetic value.

Some of the limitations of terraces are as follows.

- Terraces cannot be constructed on soils that are too stony, sandy or shallow as these limitations may not permit practical and economical construction and may require regular and costly maintenance
- It is not advisable to terrace lands where the slope is either too slight or excessive, or the topography is extremely irregular. The cost of construction and maintenance of terraces is difficult and farming them may pose insurmountable problems
- With increasing slopes, the cultivable strips become narrower and only a limited level area is available for cultivation
- High risers are needed on steep slopes increasing the risk of their failure.

Broad Based Terraces

These terraces, earlier known as ridge type terraces, remove from or retain water on the sloping lands. These are categorized as broad based (BB) terraces or narrow based terraces depending on the width of the base. Broad based terraces are sometimes referred to as magnum terraces after the inventor, Priestly Magnum, who introduced BB terrace by widening narrow ridge. A broad base terrace has a ridge 25 to 50 cm high with gently sloping sides and a dish shaped channel along the upper side to control erosion by diverting runoff at a non–erosive velocity. It may be level or have a grade towards one or both ends. Based on grads; it is classified as:

- ✰ Graded or drainage terrace
- ✰ Level terrace

Graded terrace is constructed in high rainfall (>600 mm) areas, where excess rainfall needs to be removed quickly. A graded terrace has a constant or variable grade along its length and is used to convey excess runoff at safe velocity into a vegetated waterway or channel. In some cases, whole strip is used as a channel for conveying the water. It is used in areas where expected runoff is less and water can be allowed to stay without affecting the crop yield. While former is known as BB terrace with proper channel, later is known as without proper channel.

A level terrace follows the contour line and is recommended for areas having permeable soils. If soils are quite permeable, level terraces can be used on slopes up to 20 per cent. In such cases, their main function is largely to prevent soil erosion. In semi-arid regions, the level terraces are also used for moisture conservation on low slopes of 2 per cent or less. On these flatter and uniform slopes, level terraces may be laid out parallel to each other. Narrow based terraces having widths in the range of 1.2 to 2.5 m are also known as contour bunds in India. These have been discussed in the previous chapter.

Broad based terraces are not recommended on lands with slopes exceeding 10-12 per cent. However, under special conditions these may be used on slopes up to 20 per cent but in no case if the slopes exceed 20 per cent.

Terrace Spacing

Many empirical relations are used to determine vertical interval. The most common is the following.

$$VI = 0.3\,(a.S+b) \qquad (27.1)$$

Here VI is the vertical interval in m, a is a constant depending upon the geographical location and its value is less than 1.0. S is the average land slope above the terrace, per cent and b is a constant that depends upon the soil erodibility and crop cover condition during the critical periods. The values vary from 1 for low and 4 for high intake rates.

More common way to determine the terrace spacing is the guidelines given by US SCS (1953) as reported in Table 27.1. These values are recommended for land slopes varying from 10-15 per cent.

Table 27.1: Recommended Spacing for Graded Terraces

Land Slope (Per cent)	*Vertical Interval (m)*	*Horizontal Interval (m)*
10	2.10	21.0
11	2.25	20.4
12	2.40	20.1
13	2.55	19.8
14	2.70	19.2
15	2.85	18.9

Terrace Grades

The channel grade should be sufficient to provide good drainage and should be able to remove runoff at non-erosive velocities. While level terraces have zero grade, a minimum grade is desirable for graded terraces from the standpoint of soil loss. Terraces may have uniform or variable grades. For the uniform graded terrace, the slope remains constant throughout its entire length. A grade of 0.3 per cent is common in many regions; however, grades may range from 0.1 to 0.6 per cent, depending upon soil and climatic factors. Generally, steeper grades are used for impervious soils and shorter terraces. Uniform graded terraces are best in areas where drainage is a problem and where the terraces are short. Sediments are unlikely to settle in the upper portion of the channel because of higher velocities in this portion of a uniform graded terrace.

The variable graded terraces are more effective as their capacity increases with a corresponding increase in runoff toward the outlet. The grade usually varies from a minimum at the upper portion to a maximum at the outlet end (Table 27.2). The resulting reduced velocity in the upper reaches helps greater absorption of runoff in the terraced fields. Variable gradient adds greater flexibility in design of a terrace. For instance, either constant velocity or constant capacity can be ensured by varying the channel grade. Such design manipulations are sometimes required in large diversion terraces.

Table 27.2: Recommended Variable Grades for Varying Terrace Lengths

Terrace Length (m)	*Channel Grade (per cent) Starting from Outlet*			
	First 25 per cent	*Second 25 per cent*	*Third 25 per cent*	*Fourth 25 per cent*
30-120	0.3	0.3	0.2	0.2
121-149	0.35	0.3	0.2	0.2
150-240	0.4	0.3	0.2	0.2
241-359	0.45	0.35	0.25	0.2
≥ 360	0.5	0.4	0.3	0.2

Terrace Length

Size and shape of the field, outlet possibilities, runoff rate as affected by rainfall, soil infiltration rate, and channel capacity all influence the terrace length. Extremely long terraces should be avoided. If necessary, length may be reduced by dividing the flow midway in the terrace length by constructing outlets at both the ends. Depending on local conditions, maximum length for graded terraces ranges from about 300 to 500 m. The maximum applies only to that portion of the terrace that drains toward one of the outlets. With a good layout and design, the number of outlets can be minimized.

Terrace Cross Section

Cross section of a terrace is described by side slopes of the channel and ridge, channel width (decreases with increasing slope), ridge width, and ridge height. The terrace cross section should be enough to provide adequate capacity for the water flow, should have broad farmable side slopes, and be economical to construct with available equipment. Recommended cross sections may vary from one region to another. A typical design cross section is shown in Figure 27.2 (top), which after few years of farming may reduce to a section shown in Figure 27.2 (bottom).

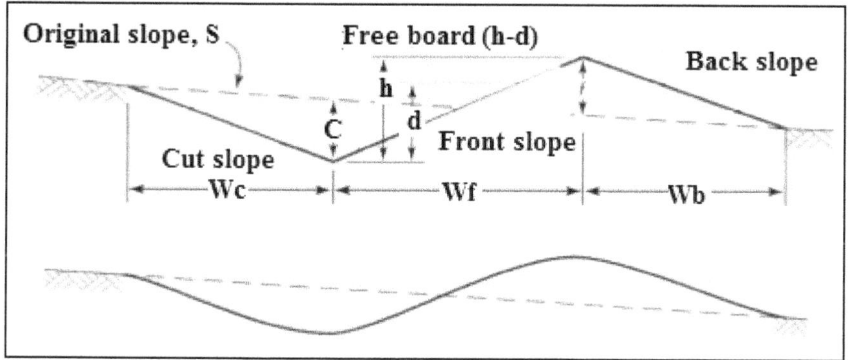

Figure 27.2: Designed Cross Section and Cross Section after Few Years of Cultivation.

Terrace Channel Capacity

Recommended dimensions for terrace cross sections as a function of slope are shown in Table 27.3 (USDA, 1953). Depending upon channel capacity and runoff, the cross-sectional area of the channel should be at least 0.6 to 1.0 m^2. Besides depth of channel, d, and side slopes should be properly decided since farming may be difficult with increase in these dimensions. For practical reasons, a terrace of uniform cross section from the outlet to the upper end is constructed, although it results in overdesign of the upper portion of the channel.

Peak Rate of Runoff

The peak rate of runoff is estimated using rational formula with time of concentration derived using the Kirpich formula. The area is taken as length multiplied by width of the terrace.

Table 27.3: Recommended Dimensions of Graded Terraces

Land Slope (per cent)	*Ridge Height at Terrace Length of*				*Recommended Slope Ratios**		
	60 m	*120 m*	*180 m*	*320 m*	*Wc*	*Wf*	*Wb*
2	24	27	30	36	10:1	10:1	10:1
4	21	27	30	33	6:1	8:1	8:1
6	21	24	27	30	6:1	8:1	8:1
8	21	24	27	30	4:1	6:1	6:1
10	18	24	27	30	4:1	6:1	6:1
12	18	24	27	30	4:1	4:1	4:1
14	18	21	27	30	4:1	4:1	4:1

* Refer Figure 27.2.

Construction Equipment

A variety of equipment is available for terrace construction. Terracing machines are classified on the basis of size as light equipment adapted to power available on the farm or heavy equipment designed for a number of earth-moving jobs. Depending upon the earth movement, terracing machines are also classified as lift and roll, throw, scrape and push, and carry. An ideal terracing machine should have the following characteristics.

- ☆ Displace soil laterally to the desired place in the ridge
- ☆ Rate of terrace construction should be fast
- ☆ Build terraces effectively on all slopes up to 20 per cent
- ☆ Place topsoil on or near surface of the ridge
- ☆ Have low initial and operating cost.

Outlets

All terraces must have adequate outlets that convey runoff water to a point where it does not cause damage. Grassed waterways or naturally vegetated drainage ways may be used as outlets. The capacity of the vegetated outlet must be large enough so that the water surface in the outlet is below the water surface in the terrace at the design flow. It is always desirable to construct and stabilize the grassed waterway prior to the construction of the terrace. Thus, the terrace will have a stable outlet immediately after it is constructed. Tile drainage outlets may be provided if grassed waterways require abnormally large areas.

Bench Terracing

A bench terrace system consists of a series of flat shelf-like areas that convert a steep slope into a series of level, or nearly level benches in a step like formation (Figure 27.3). These platforms are separated at regular intervals by vertical drop or by sloping risers protected by vegetation and sometimes packed by stone retaining walls. In fact, bench terrace converts the long un-interrupted slope into several small strips so that protected platform is available for farming.

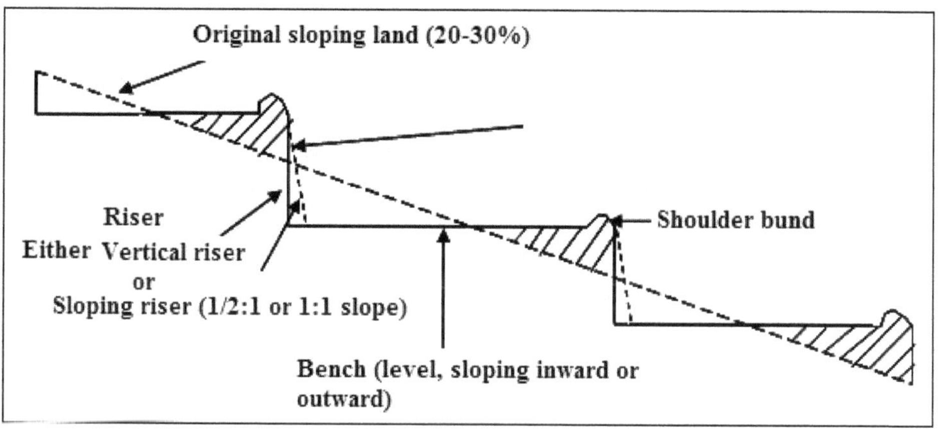

Figure 27.3: Components of the Bench Terrace.

Types of Bench Terraces

Depending on the purpose for which they are used and on the basis of slope of the benches, bench terraces are classified into several groups (Table 27.4) and are briefly described in the following sections.

Table 27.4: Classification of Bench Terraces

Based on Use/Application	*Based on Slope*
1. Hill type	1. Level and table top
2. Irrigated type	2. Sloping inwards
3. Orchard type	3. Sloping outwards
	4. Puertorican or California type

Based on Use

Hill-type Bench Terraces: These are used in hilly areas where reverse grades towards the hill are encountered.

Irrigated Bench Terraces: The terraces having level benches are often adopted for irrigated lands.

Orchard Bench Terraces: These are special kind of terraces of narrow width (About 1 m) for individual trees. These are also referred to as intermittent terraces or step terraces when a series of step like formations are made.

Based on Slope

Level Bench Terraces: These are table top level terraces (Figure 27.4 top). The bench is surrounded by a dyke (shoulder bund) of about 20 cm to retain water and to give stability to the outer edge of the terrace. Since the benches are level, most of the rainfall falling over the area is absorbed by the soil with no or very little water appearing as runoff. To facilitate some water flow, a notch is cut in the dyke so that excess water flow down to the next terrace or bench. These terraces

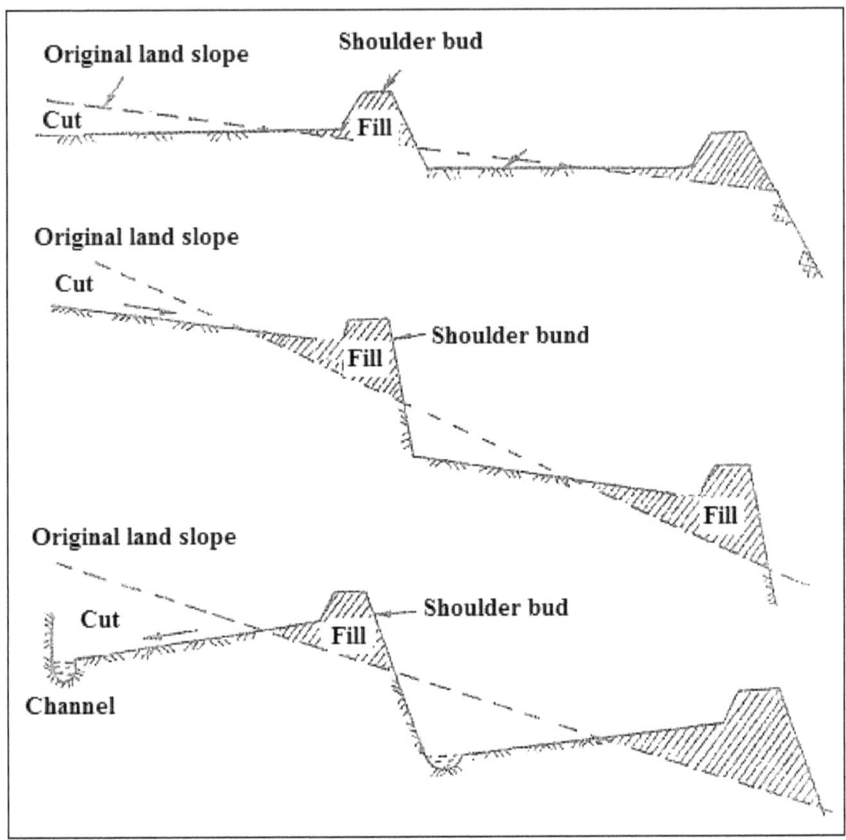

Figure 27.4: Kinds of Bench Terraces: Level (Top), Outward Sloping (Middle) and Inward Sloping (Bottom).

are usually constructed in areas having medium rainfall and highly permeable soils mainly to grow paddy. These terraces are also known as irrigated bench terraces (since adopted in irrigated lands) and conservation bench terraces (since water is conserved in the system).

Outward Sloping Terraces: These terraces slope outwards (Figure 27.4 middle). A shoulder bund is provided on the outer end of the terrace. Although a shoulder bund is provided in other types of terraces as well, it is an integral part of this kind of terrace. A drain is provided either along the top line or at the toe of the riser to collect runoff. These terraces are used in semi-arid areas where rainfall intensity is not high. The amount of earthwork in this type is less than for the level terrace. Since runoff flows down the fill portion of the riser, these terraces are easily eroded.

Inward Sloping Terraces: Bench terraces with slopes inwardly towards the hill are adopted in areas experiencing heavy rainfall or on soils with low permeability, such that a major portion of the rainfall needs to be drained (Figure 27.4 bottom). A suitable drain at the inward end of each terrace is provided to take the runoff to a suitable outlet. These terraces are usually preferred for crops, which are extremely susceptible to water logging. These are also known as hill-type terraces. Since there is no overflow over the risers, grass cover is enough to protect them from rain drop impact.

Puertorican Terraces: This terrace is gradually developed by pushing the soil downhill against mechanical or vegetative barriers, which are established across the land at suitable intervals. The soil is excavated each time during the ploughing operation in a way that a bench develops slowly over time. It is also called California type of bench terrace.

Design of Bench Terraces

The design exercise comprises of determining the following:

- Type of the bench terrace
- Terrace spacing or the depth of the cut
- Terrace length and width
- Terrace cross section

Basic design parameters includes in any such exercise are: width, vertical interval (VI), length and grade, riser and reverse height, net area for cultivation, cross-section and volume of earth work. Since the type of bench terrace to be used depends upon the rainfall, soil conditions, land slope and crops to be grown, information on these parameters should be collected before any design exercise in initiated. Guidelines to select type of terrace are given as follows.

- Irrigation or level bench terraces are used for crops, such as rice, which need flood irrigation and impounding of water.
- Upland bench terraces are used for rainfed crops or crops which only require irrigation during the dry season. They are generally sloped to ensure proper drainage.
- In humid regions, use inward sloping terraces
- In arid or semi-arid regions, use outward sloping terraces.

Terrace Spacing

Terrace spacing is generally expressed as the vertical interval between two terraces. The vertical interval, VI also denoted as D is dependent upon the depth of the cut. Since the cut and fill are balanced, VI is equal to double the depth of cut (Figure 27.5). The depth of cut, limited by the soil depth and the land slope, should not be too high so as to expose the bed rock which may make the bench terraces unsuitable for cultivation. Moreover, on highly sloping lands, high depth of cuts may increase the height of embankments that may make them unstable.

The width of the bench terraces (W) is decided on the basis of the purpose for which these terraces are constructed. Once the width of the terrace is decided, the depth of cut required can be calculated using the following formulae.

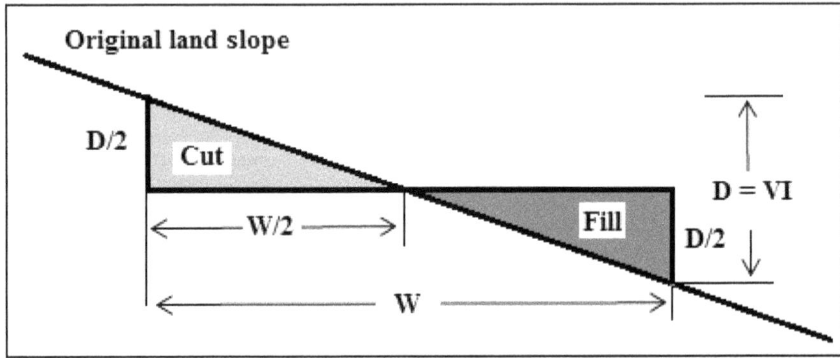

Figure 27.5: Cut-Fill and Vertical Interval Relation.

Case 1: When the terrace cut is vertical

$$D = W \times S/100 \quad (27.2)$$

S is the land slope, per cent and W is the width of terrace. The depth of cut is D/2.

Case 2: When the batter slope is 1:1

$$S/100 = (D/2)/(W/2 + D/2) \quad (27.3)$$

$$D = WS/(100\text{-}S) \quad (27.4)$$

Case 3: When the batter slope is ½: 1

$$S/100 = (D/2)/(W/2 + D/4) \quad (27.5)$$

$$D = 2WS/(200 - S) \quad (27.6)$$

In general, $D = WS/(100 - S \times U)$ such that U is the batter slope.

Terrace Cross-section

The terrace cross section requires decisions on: 1) the batter slope, 2) dimensions of the shoulder bund, 3) slope of the terrace and the dimensions of the drainage channel in terraces sloping inward, and 4) slope in case of terraces sloping outward.

Vertical cuts are used in very stable soils or when the depth of the cut is small (up to 1 m). Under other situations, batter slope is provided to ensure stability of the fill and the embankment. Batter slopes of ½: 1 can be used in loose and unstable soils. The flatter the batter slope, the larger the area lost due to bench terracing. The size of the shoulder bunds in terraces sloping inward is nominal. In terraces with flat top and sloping outwards, shoulder bunds of larger sections are required as water will stand against these bunds. Although, bund cross section may depend upon the terrace width and soil conditions, commonly used dimensions of bunds in all kinds of terraces are: height (30 cm), bottom width (75 cm), side slopes (1:1) and freeboard (15 cm). The inward slope of the terrace may be from 1 in 50 to 1 in 10 depending upon the soil conditions.

Alignment of Bench Terraces

Alignment of bench terraces should begin from the ridge and progress towards the valley. The average land slope of the area to be terraced should be determined from engineering survey and accordingly specifications of the terrace should be worked out. Contour lines may also be marked with the help of a leveling instrument. Taking a contour line as the center line, the terrace width may be marked on the ground. The alignment may now be examined and suitable adjustments may be made wherever necessary taking into considerations the local conditions like depressions, sharp turns, field boundaries *etc.*

Construction of the bench terraces should start from the ridge and proceeded downwards. By this method, the top soil and the subsoil get mixed up and the top soil may not be available for spreading over the terrace surface. If the subsoil condition is not good as it may happen under several situations, it is necessary to keep the top soil apart and again spread it on the terrace. Since it is a difficult proposition, it can be accomplished by beginning the construction from the lower most terrace. After the construction of the first terrace, the top soil from the second terrace may be spread on the first terrace. The process is continued for subsequent terraces. In bench terraced areas, suitable outlets should be provided to safely dispose of the runoff. One of the sides of the hill slope where vegetation is well established can be used as an outlet. Where such outlets are not available or feasible, waterways need to be constructed for this purpose.

Area Lost for Cultivation due to Bench Terracing

The area lost for cultivation due to bench terracing of a slope is calculated in the following manner.

Let D be the vertical interval of the benches to be laid out on a land with a slope of S per cent along AB (Figure 27.6). Let us also assume that the batter of the risers is 1:1. If L is the horizontal interval between the benches *i.e.*, projected length of AB on horizontal plane, then actual length of AB is given by:

$$AB = (L^2 + D^2)^{1/2} \qquad (27.7)$$

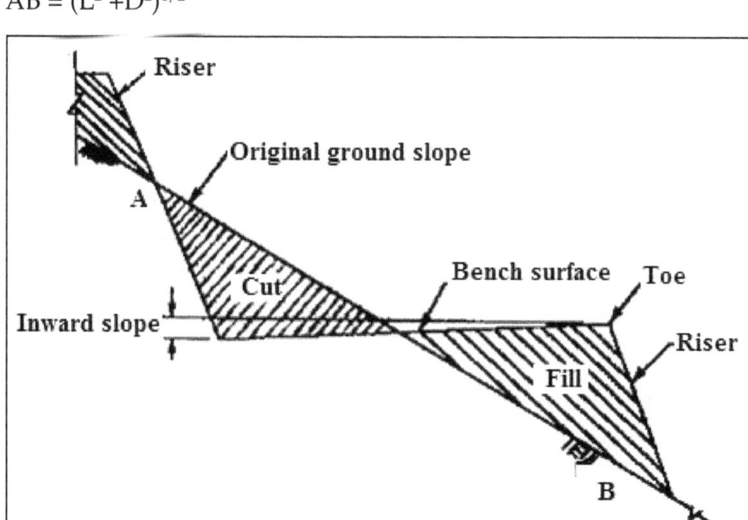

Figure 27.6: Schematic Diagram of a Bench Terrace.

Taking out L from the square root and L being larger than D, one can expand the term in bracket to write:

$$AB = L\ (1+D^2/2L^2+ D^4/8L^4+...) \quad (27.8)$$

Retaining only the first term, we have

$$AB = L + D^2/2L \quad (27.9)$$

Since L = 100D/S

$$AB = 100\ D/S + (D^2/2)\ (S/100D) \quad (27.10)$$

$$AB = 100\ D/S + DS/200 \quad (27.11)$$

If W is the width available for cultivation after terracing:

$$L = W + D = 100D/S \quad (27.12)$$

$$W = 100D/S - D = (100\ D- DS)/S \quad (27.13)$$

Width not available for cultivation after terracing (from eqs. (27.11) and (27.13)) = AB-W

$$AB-W = 100\ D/S + DS/200 - 100D/S + DS/S$$

$$= DS/200 + DS/S \quad (27.14)$$

Width loss in percentage of original inclined width AB

$$(DS/200 + DS/S) \times 100/(100\ D/S + DS/200)$$

$$(S/200 +1) \times 100/(100/S +S/200) = (S+200) \times S \times 100/(20000+S^2)$$

Simplifying, we get

$$\text{Area lost in per cent} = (S+200)/(200/S + S/100) \quad (27.15)$$

Similarly, it can be shown that for a batter slope of ½:1, the area lost in per cent will be:

$$\text{Area lost in per cent} = (S+100)/(200/S + S/100) \quad (27.16)$$

The percentage width lost can be taken as the percentage area lost. When the batter is vertical, the length of bench terrace per hectare in meters will be 10000/W where W is in m. When the batter slope is 1/2:1 or 1:1 the length per hectare will be 10000/(W + Wr); Wr is given as D×U such that U is the batter slope. Both Wr and W are in meters.

Earthwork

The earthwork for a leveled bench terrace is determined using Figure 27.5. Since cut is equal to fill, we consider volume of cut per unit length of the bench terrace. It is given as:

$$[½ \times (D/2) \times (W/2)] \times 1.0 = DW/8 \quad (27.18)$$

Earthwork per hectare is given as:

$EW = D \times 10^4/8$ (Since W.L is the area of 1 bench which is replaced by 1 ha area *i.e.* 10^4 m^2)

$$EW = 1250\ VI = 12.50\ W.S\ (\text{Since } VI = WS/100) \quad (27.19)$$

For the complete length of the bench, it is given as LDW/8 such that L is the length of the bench, m.

For 1:1 or ½ :1 batter slope, the value of W is incorporated in eq. (27.19) such that the earthwork respectively can be given as:

For 1:1 batter slope, $EW = D^2 (100-S)/8S$ (27.20)

For 1/2:1 batter slope, $EW = D^2 (200-S)/16S$ (27.21)

For inward sloping lands, the earthwork is given as:

$EW = 1250\ VI = 12.50\ W.(S+s)$ (Since $VI = W(S+s)/100$) (27.22)

For outward sloping lands, the earthwork is given as:

$EW = 1250\ VI = 12.50\ W.(S-s)$ (Since $VI = W(S-s)/100$) (27.23)

Here s is the inward or outward slope of the bench in per cent.

Cost/ha = Earth work × cost per unit of earth work

Other Related Terms

Area available for cultivation, $Ac = 100(100-n.S)$ (27.24)

Batter area to be planted with grass, $Ag = 100\ (1+n^2)^{1/2}$ (27.25)

Here n is the batter slope.

Maintenance of Bench Terraces

Following actions may help in the stability and protection of terraces for their long-term productivity.

- New terraces should be protected at its riser, outlets and waterways for at least 2 years.
- Depending upon the drop height, risers can be made of earth or dry stone riprap. Former, with a 1:1 slope, is used for riser having drop height of $<$ 1.3 m whereas the other kind is used when the riser height is more than 1.3 m.
- After construction, the riser should be shaped and planted with grasses. Sod-forming or rhizome-type grasses are better than those of the tall or bunch-type. If stones are available, these can also be used to protect and support the risers. Hydro-seeding is also used in protecting the risers.
- If water flows from the higher terraces onto those below in case of level bench terraces, it is useful to make a small platform using rock/stone on the lower terrace to dissipate the energy of the falling water.
- Terrace outlets (point where the runoff leaves the terrace and goes into the waterways) needs to be protected by sodding, bricks, stones or cement blocks.
- Keep the toe-drain open, and do not permit any accumulation of water in any part of the terrace. Any kind of obstruction such as continuous mounds or beds must be removed at regular intervals to allow water to pass to the toe drain.

- ☆ Ploughing should be done carefully not to destroy the toe drains or creating reverse grades.
- ☆ Cattle should not be allowed to trample on the risers or graze the grasses.
- ☆ Do not allow runoff flow over the risers. If necessary, dykes along the edge of bench should be built.
- ☆ Any small damage or break of the riser should be repaired immediately.
- ☆ Make sure that the water flows through the outlets instead of going around them.

Maintaining the Soil Productivity

Deep ploughing, ripping or sub-soiling may be required to improve the structure of the soils on the cut part of the bench terraces. Green manuring, compost or sludge should be applied in the initial years so as to improve soil fertility. Soil productivity should be maintained by means of proper crop rotation and the use of fertilizers.

Example 27.1

On an 18 per cent hill slope, it is proposed to construct bench terraces. Calculate (i) length of terrace per hectare, (ii) earth work required per hectare, and (iii) area lost per hectare both for vertical cut and batter slope of 1:1, if the vertical interval of terraces is 2 m. The cut is equal to fill.

Using the equation for vertical cut, and estimating the width of bench terrace (W)

$W = 100D/S = 100 \times 2/18 = 11.1$ m

Length of the terrace per ha = 10000/10 = 1000 m

Earthwork = $LDW/8 = 1000 \times 2 \times 11.1/8 = 2775$ m^3/ha

Since it is a vertical cut, no area is practically lost except for the shoulder bunds.

If batter slope is equal to 1:1 then

$W = D\,(100\text{-}S)/S = 2(82)/18 = 9.1$ m

Length per ha = 10000/(9.1+2) = 900.9 m

Earthwork = $LDW/8 = 900.9 \times 2 \times 9.1/8 = 2049.5$ m^3/ha

Area lost in per cent = $(S+200)/(200/S + S/100) = (18+200)/(200/18 + 18/100)$ = 19.31 per cent

Example 27.2

Compute the width of the terrace, if the vertical interval of bench terrace is 1.5 m, land slope is 25 per cent and batter slope is 0.50:1.0. If width is fixed at 5.0 m, what will be the vertical interval for a batter slope of 1:1?

Since $D = 2WS/(200\text{-}S)$, we have

$1.5 = (2W \times 25)/(200\text{-}25)$

$W = 175 \times 1.5/50 = 5.25$ m

For a batter slope of 1:1, D = WS/(100-S), such that vertical interval is

$D = 5.0 \times 25/(100\text{-}25) = 1.666$ m

Example 27.3

A 15 per cent hilly land is proposed for construction of bench terraces. Calculate the following parameter of bench terrace if vertical interval is 2.5 m and batter slope is 1:1.

a) Total width of the bench (Wt)
b) Length/ha
c) Earth work
d) Area lost

In a bench terrace, VI = Height of the riser, Hr = 2.5

$VI = (S \times Wb)/(100 - S \times U)$

Here U is the slope of riser.

$(15 \times Wb)/(100 - 15 \times 1) = 2.5$

Wb (Width of bench) = 14.16

Wr (Width of riser) = $D \times U = 2.5 \times 1 = 2.5$ m

Total width of bench Wt = Wr + Wb = 2.5 + 14.16 = 16.66 m

Terrace length/ha, L is

L = 10000/Wt = 10000/16.66 = 600 m/ha

Earthwork (volume) = $(L \times Wb \times Hr)/8$

$= (600 \times 14.16 \times 2.5)/8 = 2655\ m^3$

Area lost due to slope = (S + 200)/(200/S + S/100)

= (15 + 200)/(200/15 + 15/100) = 15.95 per cent

Example 27.4

On a 16 per cent hill slope, inwardly and outwardly sloping level terraces are to be constructed with a vertical interval of 2.5 m. Determine for each case the length of the bench terrace, earthwork per ha, cost of the earthwork. Assume that the cost of earthwork is Rs. $3/m^3$ and inward and outward slopes are 0.2 per cent.

Outwardly Sloping Case

$W = 100\ D/(S\text{-}s) = 100 \times 2.5/(16\text{-}0.2) = 250/15.8 = 15.82$ m

Length of bench terrace = 10000/15.8 = 632.9 m

Earthwork = $12.50\ W \times (S\text{-}s) = 12.5 \times 15.82 \times (16\text{-}0.2) = 3124.5\ m^3$

Cost of earthwork = 3124.5×3 = Rs. 9373.5.

Inwardly Sloping Case

W = 100 D/(S+s) = 100 × 2.5/(16-0.2) = 250/15.8 = 15.43 m

Length of bench terrace = 10000/15.43 = 6648.1 m

Earthwork = 12.50 W × (S+s) = 12.5 × 15.82 × (16+0.2) = 3203.3 m^3

Cost of earthwork = 3203.3 × 3 = Rs. 9609.9.

Chapter 28

Wind Erosion and Control

Definition

Wind is simply air in motion. Air has mass and when mass is in motion, it has energy. It is this energy that moves the soil during wind erosion. Thus, wind erosion is defined as the detachment of soil particles from a place, their transportation and deposition at any other place with wind as the eroding agent. One comes across the evidence of wind erosion whenever one encounters an atmospheric haze of dust comprising fine mineral and soil organic particles. The most spectacular forms are dunes - mounds of more or less sterile sand - which move as the wind displaces them from one place to another. It may bury the most productive fields. In the past, instances are quoted where even the oases and ancient cities were buried due to wind erosion. Now-a-days, Wind erosion is a serious environmental problem.

The risk of wind erosion increases with increasing wind speed. To initiate the wind erosion, air movement must attain a certain velocity (with enough speed to generate visible movement of particles at the soil level) before it can generate deflation and transport of particles. Winds, with velocities less than 12-19 km/hr, can seldom impart sufficient energy at the soil surface to dislodge and put into motion sand-sized particles. Drifting of highly erosive soil usually starts when the wind attains a forward velocity of 25-30 km/hr. Besides, a loose, dry and smooth soil surface, large fields having sparse or no vegetative cover, all increase the risk of wind erosion. Wind erosion tends to occur mostly in low rainfall areas when soil moisture content is at wilting point or below. In fact all drought-stricken soils are at the risk of wind erosion.

Adverse Impacts of Wind Erosion

A major adverse impact of wind erosion is loss of soil productivity. Both the areas of erosion and deposition suffer because of the variations encountered in the soil characteristics. Such changes may occur slowly and go unnoticed for many years especially if mixing by tillage masks the effects of wind erosion. Following factors adversely impact the soil productivity.

- ☆ Wind erosion is very selective process that carries the finest particles, particularly organic matter, clay and loam. This effect is known as the winnowing of light particles. The continued loss of fine particles impacts soil quality. It results in reduced water-holding capacity of the soils
- ☆ Since organic matter contains most of the soil nutrients, its loss decreases the fertility of the soil
- ☆ In shallow soils and soils with a hardpan layer, wind erosion results in decreased root zone depth
- ☆ While soil can be blown away at virtually any height, the majority (over 93 per cent) of soil movement/transportation takes place at or within one meter height from the land surface. Thus, sheets of sand travelling close to the ground can smother crops and impact the productivity during the season (particularly millet or cotton seedlings in semi-arid zones)
- ☆ Wind erosion reduces the capacity of the soil to store nutrients and water, thus making the environment drier
- ☆ Wind erosion may disseminate weeds
- ☆ Cover the plants and seed with soils during the deposition process
- ☆ Adverse effects on field machinery.

Besides what is stated, wind erosion adversely impacts the general quality of life in the following manner.

- ☆ Minute soil particles (< 60 μm) blown by wind in the desert are one of the major sources of dust pollution and cause serious health hazard. Nearly 30 per cent of the eroded aeolian mass comprises of particulate matter, PM10 (< 10μm). It is one of the parameters to describe air quality
- ☆ Deposition of sand on roads, railway lines, canal networks and other infrastructure impacts the inhabitants and largely increase the maintenance cost of these structures.

Areas Vulnerable to Wind Erosion

Wind erosion is a common phenomenon occurring mostly in flat, bare areas; dry, sandy soils; or anywhere the soil is loose, dry and finely granulated and where high velocity wind blows. In the Indian context such areas are characterized as follows.

Desert Areas

The Thar Desert (also known as the Great Indian Desert) is a hot desert covering an area of 208,110 km^2. About 85 per cent of the Thar Desert is located within India covering about 170,000 km^2. It is spread over four states namely Punjab, Haryana, Rajasthan, and Gujarat. Most of the Thar Desert is situated in Rajasthan, covering 61 per cent of its geographic area. The Thar Desert extends between the Aravalli Hills in the north-east, the Great Rann of Kutch along the coast and the alluvial plains of the Indus River in the west and north-west. Most of the desert is covered by huge shifting sand dunes that receive sediments from the alluvial plains and the coast. Wind erosion is a serious problem in this arid zone receiving scanty rainfall, devoid of vegetation cover and having sandy soil. It is estimated that 3.4 million tons of fertile soil is removed by wind every year in the districts of Jodhpur, Bikaner, Kota, Jaipur, Bharatpur, Kishangarh *etc.* in Rajasthan.

Semi-arid Regions

The areas at the boundary of the arid zone also suffer from wind erosion because of the advancement of the desert. Besides Punjab, Haryana, Rajasthan, and Gujarat, areas in Madhya Pradesh and Andhra Pradesh suffer from wind erosion. The severity of the problem is relatively less because of relatively more rainfall, relatively less temperatures and wind velocities and a slightly better land cover.

Coastal Sandy Areas

Coastal sandy areas covering about 1.47 M ha area lie along the sea coast. High wind velocities originating from the sea shifts this sand from one place to another resulting in formation of sand dunes near the coast. A major problem of wind erosion is that winds blowing are salty in nature and adversely affects the vegetation, houses and other structures.

Other Areas

Besides what has been stated, the areas found along the plain of major rivers in India, and the cold deserts of Ladakh, Lahul and Spiti also experience the problems of wind erosion.

Factors Affecting Wind Erosion

The factors that affect wind erosion can be grouped under the four groups namely the climate, soil, vegetation and biotic. Within these groups one can identify several factors as listed herein.

1. Climatic factors: Nature of wind (velocity, direction *etc.*), precipitation, humidity and temperature
2. Soil physical characteristics such as texture and structure, density of particles, moisture condition, organic matter (OM), and surface roughness or cloddiness *etc.* and chemical characteristics such as salinity, alkalinity, and acidity *etc.*
3. Surface features such as
   - Types and density of vegetation and ground cover
   - Topographic features such as terrain characteristics (flat, undulating, rolling and continuous)
4. Anthropogenic factors such as land use, over use, over grazing, and plant residue *etc.*

Some of the important factors that affect wind erosion under the four groups are discussed in the following sections.

Climate

Climate plays a major role in soil loss either through the wind or the water. In wind erosion, wind velocity is the major controlling factor.

Wind Velocity

The ability of the wind to erode the surface is determined by two factors, namely the air density and wind velocity. The erosive power of the wind (E) is given as:

$$E = V^3\rho \tag{28.1}$$

Here V is the wind velocity, and ρ is the air density. It is clear that the erosive power of the wind varies with the third power of velocity. It means doubling the wind velocity increases the erosive power by 8 fold while a tripling of velocity produces a 27 fold increase. Velocity of wind increases dramatically above the ground surface.

Besides the velocity, wind turbulence also affects the wind erosion. Wind, strong enough to cause erosion, is always turbulent with eddies moving in all directions at a variety of velocities. Turbulence increases with increase in friction velocity, with increasing surface roughness, and with pronounced changes in surface temperature. It is more pronounced near the soil surface than higher in the wind stream. Turbulence keeps the soil grains suspended in air.

The wind also indirectly affects the wind erosion as it increases water vaporization and tends to exhaust the soil water content more rapidly to make it dry.

Rainfall, Temperature and Humidity

Wind erosion is dominant factor of soil loss in dry climates. In fact, scanty rainfall, high temperature and dry winds are normally sure prescriptions for high wind erosion because soil moisture will be usually low under these conditions. Along with low humidity, all these factors result in drying of soil that makes it vulnerable to soil erosion. Moreover, these factors also decide the kind of vegetation and its coverage, the vegetation coverage being much less under these conditions.

Soil Characteristics

Soil Texture

The proportion of soil particles *i.e.* sand, silt and clay decides the soil erodibility. Sands erode easily because they contain a great proportion of particles of saltation size, but little binding material. Soils made up of very fine particles especially with diameters less than 0.01 mm are very resistant to movement, because cohesive forces are much greater for these particles. Moreover, clay particles are too small to protrude above a laminar and viscous layer of air close to the surface of the ground. Fine dust is lifted from the ground mainly by impact of larger grains, which are more erodible because they protrude farther into the fast moving, turbulent currents of higher air. But, clay particles once detached can travel large distances in suspension as compared to sand that moves mainly through the saltation process.

The cohesiveness of surface materials affects the resistance of the surface to erosion. Clay-rich soils tend to resist erosion by wind more than less cohesive materials. Thus, clays require a much higher threshold velocity for detachment. Sometimes, cohesiveness may be provided by some cementing agent such as calcium carbonate and other salts commonly present in desert soils. Chemical properties of the soil may also influence wind erosion. An alkali soil having dispersed clay may be more prone to wind erosion.

High sand percentage is also not conducive to clod formation. Soils, high in clay and silt, are rather stable because they are coherent (clay acting as binding agent) and easily form a protective crust at the surface. Clods formed by clay and sand are harder and less subject to abrasion by wind borne sand than those formed by silt and sand. Although silt has the capacity to form clods, but the clods are softer and more readily abraded than those formed from clay and sand. Soil containing 20 to 30 per cent clay, 40 to 50 per cent silt and 20 to 40 per cent sand produced a greater proportion of non-erodible and mechanically stable clods.

Medium textured soils containing a high proportion of silt are prone to crusting. The crusts may be less than 1/6 to more than 1/2 cm in thickness. Soil crusts formed by dispersion of surface soil due to raindrop impact and subsequent drying are more compact and mechanically stable against wind action than the soil below. Sandy soils containing low silt and clay are less subject to crusting and more erodible by wind. It may be noted that on crusted soils wind erosion may be slow in the beginning but may accelerate as the surface crust is broken, exposing a weakly consolidated soil below the crust.

Water Stable Aggregates

The amount of soil loss caused due to wind erosion depends on the size of soil particles. For fine textured soils, the particles may form water stable aggregates among the structural units resulting in greater mechanical stability. Mechanical stability reduces wind erosion by resisting the breakdown of soil particles.

Soil Moisture

In general, wind erosion can only happen when the soil surface is dry or only slightly moist, because surface tension holds the soil particles together when wet. Moisture film between individual particles increases the cohesive force (force between same particles). Adhesive forces because of the adhesion between water and soil particles are also high in a wet soil that restricts the blowing of the wet soil. Wind velocities must create a force in excess of these film forces in order to cause soil movement.

Therefore, soils that have a tendency to retain moisture and to conduct it to the surface are fairly resistant to drifting. It is why, the severity of wind erosion increases during periods of drought. The wind erosion rate decreases exponentially with increasing soil moisture content. In a study, 6 per cent soil moisture content was a turning point where wind erosion turned from severe to light. Since the moisture content is controlled by the vegetation, the topsoil moisture content under different land use patterns was in order of natural grassland > rainfed farmland >shrub land. However, these results may vary from one soil to another and as well as from one region to another.

Organic Matter Content

The organic matter content in the soil reduces soil erosion as a result of its binding properties. It has been observed that soil microbes produce some cementing agents such as humic substance and their derivatives in the soil. These agent acts as sticking gum and bind the soil particles together as a clod to reduce the process of soil erosion. Organic matter and other cementing agents enhance aggregate stability

Surface Roughness or Soil Cloddiness

Soil clods being large in size increase the soil roughness. Since rough surfaces reduce the wind velocity up to a certain point, it results in less soil loss. Besides soil clods act as a barrier and trap the erodible material especially during the saltation process. In general, greater the soil roughness, lower the wind velocity against the ground and lower is the rate of erosion. Surface roughness can be enhanced through tillage, ridging and/or mulching

Surface Features

Topography

Soil surface characteristics depend upon the soil topography. An undulating land will have much less soil loss than a plain area. Although, slope may not directly influence the wind erosion, but if the direction of wind is along the slope then the wind erosion will be more then against the slope.

Field Length

The intensity of the soil movement increases progressively with distance across eroding areas. The intensity of the movement at the windward edge of an eroding area, whether it is the edge of a field or the beginning of a highly erodible spot in the field, is approximately zero. From this point the amount of soil being moved may increase progressively up to certain limits if no obstruction or trap of any kind is encountered. Field with their broad sides at right angles to, and their narrow sides parallel with the prevailing wind direction will have overall lesser rate of erosion. Field orientation is of little consequence where erosive winds blow equally from all directions. It also indicates the importance of trapping the soil where it first starts to move.

Vegetation

The roughness length of the surface, a parameter based on the size and distance between objects in a group, has a critical control over the velocity of the wind. Grass, shrubs, and trees, all impart a drag on the wind to reduce its erosive force. Wind velocity approaches zero near the soil surface in a vegetated area. A surface without a cover of vegetation exposes soil to the direct force of the wind making erosion more effective. In addition, plant roots bind the soil. Thus, dry regions lacking a protective cover of vegetation display the effects of wind erosion more than humid climates. The factors that results in less erosion in vegetated than a barren land can be listed as:

- ☆ Reduced wind velocity because of higher roughness
- ☆ Vegetation acts as a barrier to wind
- ☆ Vegetation collects the moving sediments
- ☆ Vegetation also acts to bind soil particles to the surface
- ☆ Higher organic matter content in the soil.

Anthropogenic Factors

Crops differ in their resistant to erosion. Wind erosion is more if farming techniques that favour erosion are applied. Similarly, cultivation of crops sensitive to wind erosion will also result in more wind erosion. For example; repeated working of the soil and excessive soil fragmentation during the dry season tend to increase the erosion. On the other hand, cultivation techniques which maintain or increase soil surface roughness (ploughing, dyking) have a protective effect.

Overgrazing

Grazing is yet another destructive biological factor for the soil erosion. Cattle and sheep during the summer graze the rangelands, pastures and forest vegetation and make the soil bare resulting in higher losses.

Forms of Wind Erosion

Wind erosion is the process by which loose soil material is picked up and transported by wind. The movement of soil by wind is a complex process that is influenced by many factors discussed in the previous section. The soil loss by wind movement involves two processes:

1. Detachment
2. Transportation

Depending upon the direction of the wind with respect to the stratum on which wind is acting, several other forms of wind erosion also come into play. There are five ways in which soil particles are loosened and transported although in a way these are closely interdependent.

Deflation

The process of removing, lifting and blowing away dry and loose particles of sand and dusts by wind is called deflation, which is derived from Latin word deflatus, meaning blowing away. This process is further splitted in the following forms.

The detrusion refers to the disintegration of soil clods and wearing away of rock and land projection by wind pressure or by the bombarding action of highly erosive grains coming from the windward side.

The term effluxion refers to the removal of soil grains ranging from 0.05 to 0.5 mm in diameter, and is initiated and maintained by the direct pressure of the wind. The removal is almost entirely by saltation, but a minor portion of the soil may be removed in surface creep, and some fine particles may be picked up directly by the wind and carried away in true suspension. Both detrusion and effluxion occur particularly when the direction of the wind is either tangent or diagonal to the surface.

Extrusion is the forward thrust of soil particles which are too coarse to be removed by direct wind pressure. If a field composed of these particles has on its windward side an area containing particles removable by effluxion, many of these coarse fractions may be removed as a result of bombardment by the smaller grains. Extrusion is carried out chiefly by surface creep.

Another form of deflation is efflation, which is the process of very fine particles being carried off by the wind. Attrition involves mechanical tear and wear of particles suffered by them while they are being transported by wind through the processes of saltation and surface creep.

Some or all of these forms of erosion may be operating at the same time. However, none of them can exist without effluxion. In other words, effluxion is a prerequisite to and cause of the other forms of wind erosion. Therefore, a program of wind erosion prevention and control should be based on either reducing the amount of particles ranging from 0.05 to 0.5 mm in diameter in a soil to an allowable minimum, or protecting the soil from the erosive force of the wind.

Removal of soil/sand particles through deflation results in a loss of the surface soil layer and appearance of a stony surface and exposure of plant roots. The left out rocks, gravel-sized particles that are not easily moved by wind, are called desert pavement. Deflation basins, called blowouts, are hollows formed by the removal of particles by wind. Blowouts are generally small, but may be up to several kilometers in diameter.

Abrasion or Corrasion or Sandblasting

Deflation sometimes is also accompanied by corrasion. Corrasion is the wearing away of rock by abrasion. When the wind is laden with soil particles, its abrasive action is greatly increased. Wind armed with entrained sand grains as tools of erosion attacks the rocks and erodes them through the mechanisms of abrasion, fluting, grooving, omitting and polishing. The combined effects of these mechanisms are collectively called abrasion or sandblasting. The impact of these rapidly moving grains dislodges other particles from soil clods and aggregates. It may be noted that deflation can take place without corrasion but corrasion cannot take place without deflation.

Mechanism of Wind Erosion

The occurrence of wind erosion can be categorized under the following 3 heads.

- ✰ Initiation of the soil movement

- ✰ Transport of the soil particles
- ✰ Deposition of the soil particles

Initiation of Soil Movement

Initiation of soil movement refers to the movement of soil particles by wind forces. The movement of soil particles is initiated by wind forces exerted against or parallel to the ground surface. The surface wind with a velocity more than 1 m/s is turbulent in nature at all heights except in a very thin layer along the surface where the flow is laminar. As wind moves over the surface, a gradient in velocity exists with lowest velocity near ground surface that increases in proportion to logarithm of the height above the surface. This gradient in velocity determines the magnitude of forces exerted. For each specific soil type and surface condition there is a minimum velocity required to move soil particles, called the threshold velocity. It is the minimum mean velocity required to initiate the soil movement by direct action of the wind. Soil particles will move only when the wind speed is more than the threshold velocity. A wind speed of 5 m/s or less at 30 cm height is usually non-erosive for mineral soils. For the most erodible soils of particle size about 0.1 mm; the required threshold velocity is 16 km/h at a height of 30 cm above the ground.

Transportation

Soil material, in order to be transported by wind, must first be loosened from its position on the surface of the land. It may then be lifted, rolled, slid or bounced along the surface of the ground. These processes are largely the result of wind turbulence, mainly eddies and irregularities of wind movement. Once the threshold velocity is reached, the quantity of soil moved depends upon the particle size, the roughness of the soil surfaces (cloddiness) of particles, and wind velocity itself. The quantity of soil transported can be calculated using the following formula:

$$Q = (V - Vt)^3 d^{1/2} \quad (28.2)$$

where Q is the quantity of soil moved, V is the velocity, Vt is threshold velocity, and d is diameter of soil particles.

Soil material in the air is carried away by three ways. All three types of movements usually take place simultaneously.

- ✰ Suspension
- ✰ Saltation
- ✰ Surface creep

Suspension

Suspension represents the floating of finer particles dislodged by saltating particles. Suspension refers to particles less than 1/10 of a millimeter (usual range 0.02 to 0.1 mm in diameter) - smaller than the diameter of a human hair - that are lifted far above the surface and carried great distances. These particles include very fine grains of sand, clay particles and organic matter. Once the suspended particles enter the turbulent layer when the soil is bombarded by the saltation movement, they ate lifted up high due to action of upward eddy current. These particles remain suspended in air for a long period of time and are often carried away many kilometers before being deposited. Larger dust particles (0.05 to 0.1 mm) may be dropped within few km of the erosion site. Particles of the order of 0.01 mm may travel hundreds of km and 0.001 mm sized particles may travel thousands of km. Some of these particles form the dust clouds that have been traced across continents, oceans, and around the world. Dusty storm is an example of this kind of movement. Suspension often mars the visibility. A small fraction of the suspended particles when inhaled may cause health problems. The particulate matter equal to less than 10 microns in size is known as PM_{10} often used as a parameter to describe air quality. The other parameter is $PM_{2.5}$ *i.e.* particulate matter equal to or less than 2.5 microns in size. Soil transported through the process of suspension constitutes about 3 to 38 per cent of the total weight of the soil eroded (Figure 28.1).

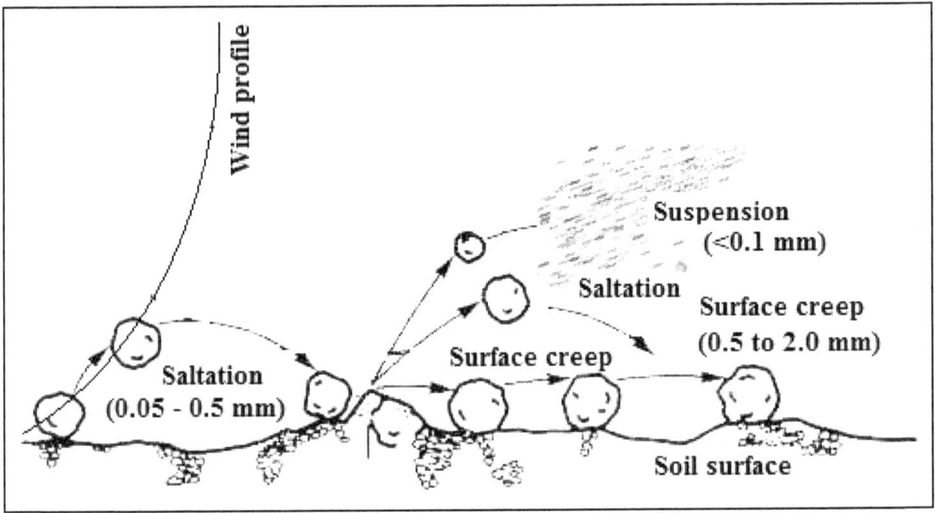

Figure 28.1: Soil Movement through the Processes of Suspension, Saltation and Surface Creep.

Saltation

Saltation is the process, where particles of intermediate size (0.05 to 0.5 mm in diameter) are lifted from the surface and follow distinct trajectories under the influence of air resistance and gravitation force acting on it (Figure 28.1). The movement continues for some time and then descends almost in a straight line with an angle of descent in the range of 6° to 12° with the horizontal. After they strike the ground, they may rebound and continue their movement by the saltation process. Thus, soil particles move with the wind in a series of jumping and falling at short distance. During this process a chain reaction of particle movement takes place as when a lifted particle returns to the surface, it may rebound or become embedded to the surface, and in either case, it initiate movement of other particles. The major portion of soil is carried away by this kind of movement. The movement of soil by wind is not only dependent on the force of the wind acting on the particles but also on the velocity distribution of the wind to the height of saltation. The height of movement is limited and therefore the wind velocity above certain height has no influence on the soil movement. Most saltation occurs within 0.3 m of the surface. This height of jump varies with size and density of soil particles, surface roughness and velocity of the wind. Thus, it may be noted that saltation process is dependent not so much on the wind force as on the velocity distribution of the wind in relation to height of saltation. These particles also cause abrasion on the soil surface and attrition (the breaking of particles into smaller particles). Soil transported through the process of saltation constitutes about 55 to 72 per cent of the total weight of the soil eroded.

Saltation is often responsible for damaging young plants, threatening their survival and can damage their fruit, reducing their marketability. Like particles under surface creep, saltating particles continue to move until the wind slows or they're trapped in sheltered areas.

Surface Creep

Coarser soil particles having a diameter range of 0.5 to 2 mm are too heavy and cannot be lifted up by wind action. Therefore, they can move neither by saltation nor by suspension. When the particles moving due to saltation strike them, they are pushed along the ground surface. This type of rolling or sliding of heavy particles along the ground surface is known as surface creep. It may be noted that particles in saltation receive their impact energy from the direct action of the wind pressure, whereas, in surface creep the particles are set in motion by the kinetic energy derived from the impact of saltating particles. Surface creep results in these larger particles moving only a few meters. Surface creep accounts for 5 to 25 per cent of soil movement due to wind erosion.

The particle size involved and the relative contribution to transportation by the 3 processes is summarized in Table 28.1.

Table 28.1: Particle Sizes Involved and Relative Contribution to Soil Movement

Process	*Particle Size (mm)*	*Relative Contribution (Per cent of total weight of soil eroded)*
Suspension	0.02 -0.10	3-38
Saltation	0.05-0.50	55-72
Surface creep	0.50-2.00	5-25

Deposition

The particles set in motion by any of above mechanism are transported and finally deposited at some different location. Deposition of eroded soil materials occurs whenever the wind speed is reduced either due to surface roughness or some other natural cause. It also depends upon the weight of the soil particles which is directly related to gravitational force. Deposition occurs when gravitational force is greater than the winds force holding the soil particles. Coarse and dense materials are settled in a short distance whereas fine material will remain suspended in the air and can be carried for long distance before being deposited.

Besides, these particles also get deposited once an obstruction is encountered on the way. Because of their retentive ability, plants are particularly efficient in collecting drifting sand and other material moving along the surface. A clump of plants in an area of moderate sand drift will thus collect blown material around it, forming a small mound and continue to accumulate soil until a heap several feet high may be formed.

Wind Erosion: Control Measures

There is no single recipe for erosion control as many factors affect the outcome. However, with an understanding of how soil is eroded, strategies can be devised to minimize erosion. The basic methods used to control wind erosion can be categorized as:

- ☆ Vegetative measures to reduce wind velocity that can be achieved by providing suitable vegetative cover on the surface such as shelterbelts, windbreaks, growing grasses *etc.*
- ☆ Tillage measures for soil aggregation (addition of organic manures, tillage at optimum moisture condition, crop rotation, *etc.*)
- ☆ Improving soil characteristics so as to conserve moisture
- ☆ Mechanical or structural measures

Vegetative Measures

Application of vegetative measures is one of the most effective and economical means to reduce wind erosion. Vegetative measures roughen the soil surface to prevent soil movement by reducing the wind speed at the surface or to quickly trap bouncing soil particles. Vegetation not only retards the velocity near the ground surface but also holds the soil against tractive force of wind thereby helping in reducing the soil erosion. The vegetative measures are often classified into two groups namely temporary and permanent vegetative measures. Temporary measures employ different crop management practices that help to maintain a good vegetative cover for most of the time. The permanent vegetative measures include planting the trees, shrubs and grasses including planting of wind breaks and shelterbelts. Many times, windbreaks and shelterbelts are also categorized as mechanical measures. Irrespective of this classification, some of the activities that help to control wind erosion are:

- ☆ Follow a cropping pattern that keeps the vegetation on the land surface most of the time especially during critical periods (summer season in India)
- ☆ Avoid removal of vegetation
- ☆ Maintain plant residue/stubble on land
- ☆ Plant shrubs/grasses
- ☆ Access the spacing of shrubs and trees to reduce wind velocity
- ☆ Maintain good balance between grass, shrubs and trees
- ☆ Maintain shelterbelts/windbreaks

The degree of control depends upon the stage of growth, density of cover, row direction, kind of crops and climatic conditions. The vegetation used in wind erosion control should be able to grow fast and should also be able to withstand for a long time. The crop cultivation on arid lands without causing wind erosion is possible provided one or more appropriate basic land use principles are observed. For example, a cover crop with quick and luxurious growth can provide protection against soil erosion during the cropping season. Besides, a scientifically designed crop rotation, apart from improving soil fertility, helps to avoid the undue exposure of the soil to dryness and wind erosion. Grain followed by a legume, then by a row crop or fallow, and then back to grain adds residues to the soil which bind soil particles. Strip cropping reduces the width of the field area where the soil is exposed to wind erosion. Alternating strips of crops and fallow or strips of clean cultivated crops with strips under grass goes a long way toward reducing wind erosion. Intercropping is widely used in many arid regions, and several systems have been developed. Other systems employ the planting of rows of fast-growing protection crops (maize) with crops that are sensitive to wind damage (tomatoes, beans). This system is employed in areas where water stress on the protected crop is not a problem. If irrigation water is available, the fields should be cultivated before the critical periods.

It is not necessary that vegetative cover is provided only by a growing crop. It could be through, standing stubble or other crop residues on the soil surface. Crop residues can be retained after harvest, which generally coincides with the dry season, to provide soil cover and reduce wind erosion. Most soils require about 30 per cent ground cover to prevent wind erosion. For cereal crop residues, it is equivalent to about 900 to 1100 kg/ha of residue. Highly erodible soils may require double this amount.

Multi-story cropping systems make optimum use of solar energy, which is abundant in arid regions, by utilizing side light rays for photosynthesis. The plants in the upper layers provide shelter from excessive heat and ameliorate the micro-climate for crops at the ground level. As a result, overall demand on limited soil moisture is reduced and crop production per unit area of land is increased or one is able to pursue diversified cropping.

Permanent vegetation is the ultimate means of protecting soils from wind erosion. Properly managed grasses are the most reliable method of reducing erosion on deep sandy soils. But, if the grass is overgrazed, erosion can be very severe, particularly during prolonged droughts. One advantage of grasses is that the plant crowns and root systems provide some cover as well as binding force to blunt the severe erosion.

Among the different land use units in the rainfed agro-eco-system, the permanent pastures are the most degraded and neglected. In many instances these are devoid of even the basal vegetation cover. Since the land does not belong to any individual, there is no concern for it. In the pasture and rangelands, overgrazing reduces long-term pasture productivity and leaves soil prone to erosion. Grazing must be managed to leave adequate plant cover. In general, for perennial pastures under continuous grazing, leave about 50 per cent of the current growth for native range and about 25 per cent for tame pastures. Controlled grazing by regulating the number of animals and periods during which rangelands are grazed helps to protect and perpetuate vegetation cover. Reseeding of rangelands and the planting of scattered trees and shrubs in them has not only improved the carrying capacity of such lands but also contributed to the reduction of the wind erosion hazard. In the arid areas, reseeding through pelletted seeds of grasses such as *Lasiurus sindicus*, *Cenchrus ciliaris* and *Cenchrus setigerus* improved the carrying capacity of an average permanent pasture from 2.5 sheep/ha to as much as 4.5-6.9 sheep/ha in sandy soils and 9.0 to 13.8 sheep/ha in loamy sand soils. Besides, the fodder obtained from trees and shrubs can provide a critical element in the diets of animals, especially during years of drought when other annual forage species are not available.

Alley cropping can also be adopted to control wind erosion. It is defined as the planting of two or more sets of single or multiple rows of trees or shrubs at wide spacing, creating alleys within which agricultural, horticultural, or forage crops are cultivated. Alley cropping protects fragile soils through a network of roots produced by the trees and supplemental ground cover resulting from fallen leaves and the companion crop. Rows of trees, shrubs, and/or grasses planted on the contour of a slope also serve to reduce soil movement down the slope.

To improve the permanent vegetation cover, multi-purpose trees and shrubs such as *Prosopis cineraria, Ziziphus nummularia, Capparis decidua*, and *Acacia nilotica* are suitable for silvipasture. The areas so planted need to be protected from grazing for at least two years. The controlled rotational grazing can then be introduced.

Tillage Practices for Increasing Surface Roughness

Bare fields are especially at risk of wind erosion during the fallow period, and immediately after seeding of the next crop. It is because; smooth soil surfaces offer little resistance to the wind force. Keeping the surface rough by tilling when the soil is moist enough (after the rain) to form large clods reduces wind erosion. Although clods help to control wind erosion, but large clods are not desirable for a good seedbed. Ripping clay soil using spikes can bring up non-erodible clods to create a rough surface. Ploughing, especially with a disc plough is helpful in developing a rough soil surface which in turn reduces the impact of erosive wind velocity. The lister and mould board plow produce the highest number and most stable clods. These are the dominant tillage implements used to reduce wind on coarse textured soils with no residue. The increased cloddiness from listing or mouldboard plough can even last for 12 months as these are quite stable in the absence of rainfall or freezing and thawing. Intense rainstorms will break down clods, particularly on coarse-textured soils. On the other side, excessive and frequent tillage can gradually reduce roughness of soil by breaking clods and aggregates that resist erosion. Tilling dry soils may form dry dust, which can aggravate the erosion problem. Tilling summer fallow fields dries out the surface soil and leaves it prone to erosion.

A tillage operation that creates ridges and depressions alters wind speed by absorbing and deflecting part of the wind energy away from erodible soil. For ridges to be effective, these must be nearly perpendicular to the direction of prevailing winds. Rough surfaces also trap moving particles. This reduces abrasion and the normal build-up of eroding materials downwind. Research shows that creating ridges 50-100 mm high, at correct angles relative to the wind, reduces erosion. But erosion increases if height of ridges exceeds 100 mm.

The best way to conserve soil moisture and prevent wind erosion is to leave as much crop residue cover as possible during the fallow period. The stubble-mulch tillage seems to be a good option for conserving water and to control erosion. The crop residue cover so left on the surface reduces evaporation, and standing stubble traps eroded soil or snow. At the same time, crop residues reduce wind speed at the soil surface, and standing stubble anchors the soil. Spreading straw at 2 to 4 t/ha also protects eroding soil. The straw should be shredded, with a disk implement that may also anchor it against the wind. In cold climates, apply the straw before freeze-up because anchoring straw into the ground is very difficult once the soil is frozen. The effectiveness of tillage in reducing wind erosion is improved when residue crops or crops with extensive root systems are used as these help to increase the organic matter content of the soil. In sandy soils, this tillage practice may not prove to be effective, because of light texture and lack of cohesiveness.

Mechanical Measures

Moisture Conservation

High moisture content in the soil is important to cultivate crop in arid regions having scanty rainfall, limited surface and groundwater and high temperature. High moisture content is also an antidote to wind erosion. Thus, any activity that helps to conserve rainfall and reduce evapotranspiration will keep the soil moist. Following activities are helpful.

- ☆ Activities that increase infiltration rate notably tilling before the onset of monsoon
- ☆ Mulching
- ☆ Cultural practices based on watershed management approaches

Application of Manures, Chemicals and Stabilizers

Covering soil with manure or straw is an important cultural practice to reduce wind erosion. Manure is preferred as a soil cover because it also enhances soil fertility and tilth. Depending on the soil, a rate of 30 to 60 t/ha is required to protect the soil. Spread the manure evenly, and do not work it into the ground.

Application of chemicals for stabilizing soil surfaces against wind erosion can be useful. Several polymers and resin-in-water emulsion reduced erosion but the effect lasted only for two months. Application did not adversely affect plant emergence or growth, and were easily applied without special equipment. It was observed that the application of chemicals was more expensive than equally effective anchoring residues with a rolling disk packer. There is a need to develop chemicals that may form films on the soil surface that are resistant to raindrop impact yet allow water and plant penetration and are environmentally safe. Methods must also be developed to apply large volumes of these chemicals rapidly and safely.

Mechanical Obstacles

These methods consist of some mechanical obstacles, constructed across the prevailing wind, to reduce the impact of blowing wind on the soil surface. These obstacles may be fences, walls, stone packing *etc.*, either in the nature of semi-permeable or permeable barriers. The semi-permeable barriers are most effective, because they create diffusion and eddying effects on their downstream face. Terraces and bunds also obstruct the wind velocity and control the wind erosion to some extent. More often mechanical measures adopted to control the wind erosion are the wind breaks and the shelterbelts.

Windbreaks and Shelterbelts

Windbreaks and shelterbelts are vegetative barriers also referred as environmental buffers designed to reduce wind speed and provide sheltered areas on the leeward (the side away from the wind) and windward (the side toward the wind) sides of these structures. Another terminology is snow fencing, a *snow* control barrier. Although, locally there may be some differences between the windbreaks and shelterbelts but in terminological sense there is not much difference. In the literature, these terms have often been used interchangeably. Some apparent differences are mainly on the basis of scale.

- ☆ Windbreaks are constructed of one or maximum two rows of shrubs and trees across the prevailing wind. On the other hand a shelterbelt is a longer barrier comprising of more than two rows of shrubs and trees.
- ☆ The windbreak acts as a fencing wall around the affected area. Windbreaks are structures which break the wind flow and reduce wind speed. Shelterbelts are rows of trees or shrubs planted to protect crops against wind.

Advantages

- ☆ To deflect air currents to reduce the velocity of wind
- ☆ To protect the leeward areas (having cultivated field crops/livestock) from cold/hot wind
- ☆ To prevent or reduce wind erosion
- ☆ Reduce the wind speed for reducing evaporation losses so as to makes more water available for crop use
- ☆ Reduced transpiration from crops
- ☆ Protection to farm buildings
- ☆ To improve the micro-climate including moderation of temperature
- ☆ Improve soil fertility
- ☆ An habitat for birds and insects
- ☆ Provide some products such as fuel, fodder, biomass and/or wood
- ☆ Serves as fencing, boundary demarcation and has ornamental value

Wind Breaks

Wind-breaks are strips of trees and or shrubs planted to protect fields, homes canals or other areas from wind and blowing soil or sand. It is almost a permanent and most effective vegetative or mechanical measure. A wind break is defined as any type of barrier either mechanical or vegetative to protect the area from blowing winds. The wind break acts as fencing wall around the affected areas. Since wind erosion is proportional to the cube of wind speed, the reduction in wind speed by 1/2 will reduce erosion by over 80 per cent. A wind break comprises of one row or maximum two rows of trees across the prevailing wind direction. Some of the main features of wind break are described as follows.

Height

The higher the height of the windbreak, the longer is the horizontal distance protected on the leeward side. It is believed that a windbreak offers protection over a distance of about 10 to 20 times the height of the windbreak. The recommended spacing between windbreaks is 60 m. For a single row windbreak, trees should be spaced about 2.5 m apart. In a double row windbreak, the recommended spacing is 2 m between rows and 3 m between trees within a row.

Permeability

The primary purpose of windbreak is to filter and break up the force of the wind. Permeable windbreaks, which allow some wind to pass through, are the most suitable. Although dense windbreak may greatly reduce wind speed, its impact lasts only for a short distance from the base. On the other hand, the impact of a porous windbreak extends over a larger distance from the base. Studies have shown that a porosity of 50 to 80 per cent offers the best combination of wind speed reduction and snow accumulation. The desired permeability is obtained by carefully selecting tree and shrub species. Regular pruning is carried out to maintain the required porosity.

Orientation

Windbreaks should be raised at right angles to the direction of wind. N - S direction is good compromise. They should give better shading to adjacent crops and pastures. Adequate protection is provided with an orientation up to 45 degrees to one side or the other.

Maintenance

Maintaining a windbreak comprises of periodic pruning to maintain porosity in the range of 50 to 80 per cent. In the first few years after planting, pruning for shaping should be done on an annual basis. Gap filling and replanting is a part of this exercise. Wooded strips left after clearing should be maintained with restorative works or through improvements such as reforestation, removing dead trees and fertilization as per requirement.

Shelter Belts

A shelterbelt as already stated is a wider barrier than a windbreak. It has more than two rows, planted at right angle to the direction of prevailing winds.

A shelterbelt has minimum of three rows of trees and shrubs. For wind speed not exceeding 20 km/hr, shelterbelt may have 3-7 rows. In coastal areas, shelterbelts having more than 20 rows have been used. Field shelterbelts offer extra protection against wind erosion. They are especially useful in dry years when crops offer insufficient residue cover because of poor growth. These are also able to trap snow. The expected yield increase offsets yield loss associated with land going out of cultivation in a shelterbelt.

How Shelterbelts Work?

As wind approaches the belt, some of it goes around the end of the belt; some goes through the belt and most goes over the top of the belt. As the air moves around or over the barrier, the streamlines of air are compressed (Figure 28.2 top). While air pressure builds up on the windward side, it decreases on the leeward side. The difference in pressure so created drives the shelter effect and determines how much reduction in wind speed occurs and how much turbulence is created. The upward alteration of flow begins at some distance windward of the shelterbelt and creates a protected zone on the windward side of the windbreak as a result of reduced wind speed. This protected zone extends windward for a distance of 2 to 5 h, where h is the height of the barrier. A much larger region that typically extends for a distance of 10 to 30 h experiences reduced wind speed in the leeward side. The air pressure

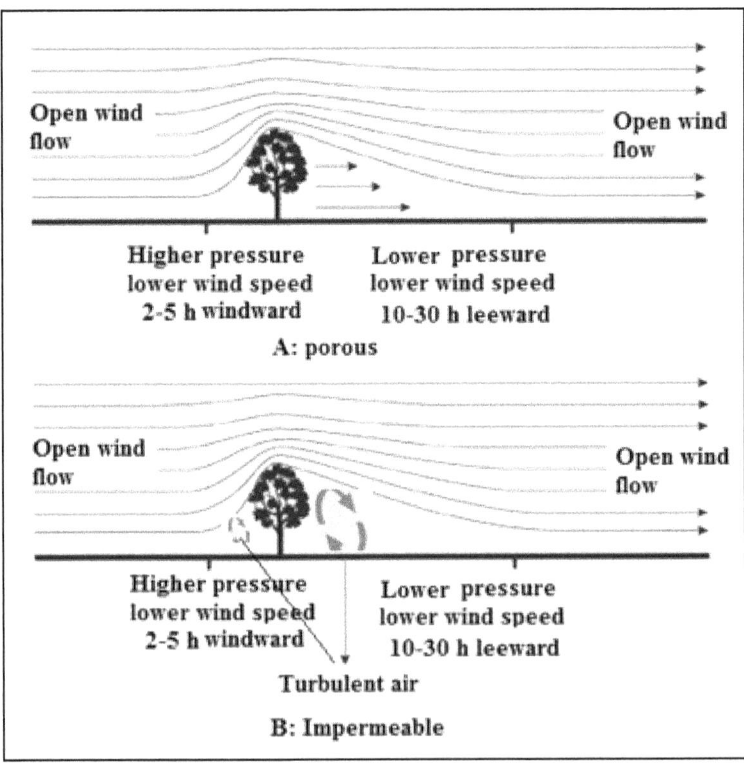

Figure 28.2: Schematics of Shelterbelt Principle for Porous (A top) and Impermeable (B bottom).

difference so created is determined by the structure of the shelterbelt. The denser the shelterbelt, the greater is the difference in air pressure but lesser the effectiveness because of development of eddy currents (Figure 28.2 bottom).

Design of Shelterbelt

Key elements of an effective shelterbelt design are composition of shrubs and the species, height, length, density, location, number of rows, and spacing between trees.

Plant Species

Select plant species native to the area. The use of good quality plant species from a local provenance helps to achieve this. The species used for the taller row should be fast growing so as to achieve maximum height in the least possible time. Some of the promising tall and bushy species are as follows.

Tall Trees

- *Acacia arabica- Babul*
- *Dalbergia sissoo – Shisham*
- *Eucalyptus camaldulensis – Eucalyptus*
- *Azadirachta indica -Neem*
- *Albizzia lebbeck -Siris*

Bushy Shrubs

- *Cassia auriculata*
- *Dodonaea viscosa*
- *Gliricidia maculata*
- *Saccharum munja*
- *Zizyphus species.*

Height

It is important to maximize the height of a windbreak, as its height will determine the area over which the windbreak has a positive impact. Using the tallest shelter species in at least one row of the belt will increase the eventual area over which a windbreak is effective.

Woodruff and Zingg (1952) proposed the following relationship between the distance of full protection (d) and the height (h) of windbreak or shelterbelt.

$$d = 17h\left(\frac{Vm}{V}\right)\cos\theta \qquad (28.3)$$

Here, d is the distance of full protection, m, h is the height of the wind barrier (wind break or shelter belt), m, Vm is the minimum wind velocity at 15 m height required to move the most erodible soil fraction, m/s, V is the actual velocity at 15 m height, and θ is the angle of deviation of prevailing wind direction from the perpendicular to the wind barrier. The value of *Vm* for a bare smooth surface is about 9.6 m/s. This relationship is valid only for wind velocities below 18 m/s.

Shape

Normally, shelterbelts have triangular shape such that control row consist of tall trees and the outside rows on both sides consist of smaller dense shrubs (Figure 28.3 top). However, the shelterbelt profile can be modified to a right angle triangle wherein tallest tree is put on the windward side boundary followed by trees of decreasing height in successive rows (Figure 28.3 bottom).

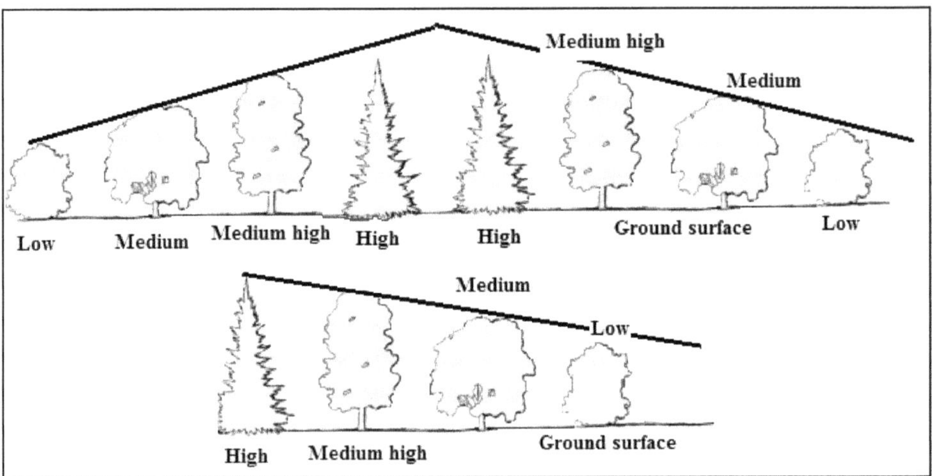

Figure 28.3: Arrangement of Trees of Various Heights in Various Shapes of the Shelterbelt.

Density and Width

Normally the width of the shelterbelt is 10 times the height of the shelterbelt. Shelterbelts up to 50 m width are considered ideal under Indian conditions. The density requirement is similar to the windbreak discussed earlier in this chapter. The shelterbelts can be fully penetrable, partially penetrable (porosity 40-50 per cent) and impermeable. Partially penetrable shelterbelt is desirable since a certain degree of penetration by wind is always preferred.

Length

The length of the shelterbelt depends upon the width of the area to be protected as well as the climatic conditions. The minimum length of a shelterbelt should be about 25 times the trees height. The area of the land protected increases with the length of the shelterbelt (Figure 28.4). To protect the whole field, the length should exceed the field boundaries. It means stakeholder participation may be essentially required to develop a good shelterbelt program.

Maintenance

Shelterbelts offer several advantages but their effectiveness is maintained

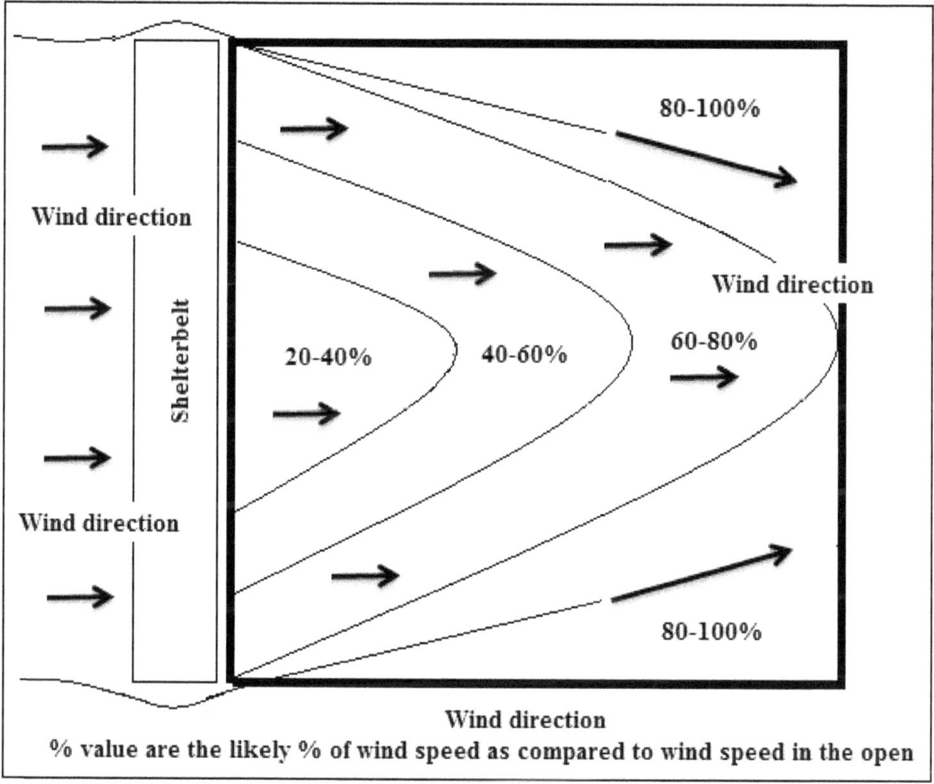

Figure 28.4: Effect of Length on Degree of Protection at Various Points.

only if proper maintenance activities are in place. In fact, all shelterbelts require some level of maintenance to ensure that they continue to perform the objectives for which these are established. Following is the summary of short-term and long-term maintenance activities that help to maintain a shelterbelt in good condition.

- ☆ The dead trees and shrubs need to be replaced as quickly as possible in order to avoid holes that may change the permeability of the belt and allow direct entry to the air.
- ☆ Weeding is a key activity that ensures proper plant growth due to minimal competition for water, light and nutrients.
- ☆ Irrigation and fertilization are necessary in the first year after planting and periods of prolonged drought.
- ☆ In arid regions, rodents are a nuisance. They may nibble the trunk and roots of certain trees, leading to their death. Protective steps are required to save the trees.
- ☆ Livestock can also harm trees. Protect trees from animals especially trees susceptible to grazing.
- ☆ Protection against insects and disease.

- ☆ The overall structure of a shelterbelt determines its effectiveness. Therefore structural management should take into consideration the ideal cross sectional profile, height and density of a belt. Regular pruning is required from second year onwards. It helps to retain the structure of the shelterbelt and from becoming too wide.
- ☆ Long-term maintenance requirement comprises of replanting and renovation of old shelterbelts.

Sand Dunes

The word dune is derived from English word Dun meaning hilly topographical feature. Apparently the sand dune means a hilly topographical feature made of sand. As per an assessment, sand dunes cover an area of about 141740 km^2 in the states of Rajasthan and Haryana. They are formed over many years when windblown sand is trapped by grass or other stationary objects. Sand dunes of many different shapes and sizes have been recognized in the Thar desert, the major ones being the longitudinal (linear), transverse, parabolic, network, star, barchan (convex crescent towards windward side) and barchanoid (dune intermediate between barchan and transverse dunes). The height ranges from nearly 2 to 50 m or even more. Low sand streaks and sandy hummocks are also numerous. On the basis of their age of formation, sand dunes have been classified as the dunes of 'old system' and 'new system, the high sand dunes usually being of the old system. These were formed in an earlier dry climate and have a greater stability than the dunes of new system of recent origin. The old sand dunes are mostly cultivated.

Sand dunes are not only found in deserts, but are also a part of the coastal ecology. A sand dune in that respect is a small ridge of hill of sand found on top of a beach. When they form on a beach, they are typically above the normal maximum reach of the waves. These sand dunes besides being a problem also protect inland areas from storm surges, hurricanes, flood-water, and wind and wave actions that can damage property. Sand dunes support an array of organisms by providing nesting habitat for coastal bird species including migratory birds. Sand dunes are also habitat for coastal plants.

In any setting, three essential prerequisites for sand dune formation are:

- ☆ An abundant supply of loose sand in regions generally devoid of vegetation
- ☆ A wind energy source sufficient to move the sand grains
- ☆ A topography whereby the sand particles lose their momentum and settle down.

Stabilization of Sand Dunes (Desert)

Various methods employed to stabilize sand dunes are based on the principle of dissipating the erosive power of wind, so that the detachment and transportation of soil particles does not take place. Some of the methods employed are:

- ☆ Vegetative measures

- Mechanical measures
- Straw (checkerboard and bales)/mats and netting
- Chemical sprays

Sand dune stabilization programme of the old dunes is mostly directed towards raising the production potential. Central Arid Zone Research Institute, Jodhpur has perfected a technology that can be described as per the following steps.

- Protection of the area from human and livestock encroachment
- Creation of micro-windbreaks on the dune slopes, using locally available shrubs either in a checker board pattern or in parallel strips
- Direct seeding or transplantation of indigenous and exotic species
- Plantation of grass slips or direct sowing of grass seeds on leeward side of micro-windbreaks
- Management of vegetated sites
- Bio-fencing, using locally non-palatable species, a cheaper and more effective form of barrier.

The technology for the stabilization of new shifting sand dunes comprises of the following steps.

- Protection by raising a barbed wire fence or trench
- Raising of micro-wind barriers of mulching in a chess- board design or as rows 3-5 m wide across the direction of wind and
- Planting of grasses and trees at the time of onset of monsoon season.

Using this technology, a sand dunes, get stabilized within 2-3 years. Thereafter, it can be opened to limited grazing and cyclic felling of trees for fuel. The technology is also economically viable with a benefit-cost ratio of 1.8 to 3.6: 1.

Chemical Spray

Sand dunes have been fixed through spraying of byproducts of petroleum on dune surface, use of chemical mulches and sealants (Kaul, 1996). In regions where oil is available, byproducts of oil industry have been used to stabilize the dunes. Products include: i) low gravity asphaltic oil (heavy oil), ii) high gravity (light) waxy oil, and iii) crude oil. The asphaltic oil forms a thin, non-sticky sheet over the dune surface, but is easily damaged by animals. Repeated sprays are often required for any worthwhile benefits. It can be used along the roads where it is not possible to go for tree plantation. High gravity waxy oils are relatively more effective and long lasting. Such oils penetrate 10 to 20 cm below the soil surface. Crude oil is often sprayed to stabilize the dune surface. The oil is heated to 50 °C and sprayed on the dune at the rate of 4 m^3/ha. It may be effective for 2 to 4 years by which time vegetation may take up the primary role of dune stabilization. This method is costly and suitable only for small areas. These chemical products are available in the market and may be employed to stabilize the surface for short periods or till some effective vegetal cover develops. In Mastung valley in Pakistan, sand dunes have been stabilized by spreading soil binding materials like fine clay, mud and water mixed with carbonate-rich fine clay.

Estimation of Soil Loss Due to Wind Erosion

An equation similar to the universal soil loss equation has been developed and can be used to estimate soil loss by wind erosion. However, the evaluation of the constants in the equation for wind erosion is comparatively difficult than the universal soil loss equation. The equation is written as follows.

$$E = IRKFCWDB \tag{28.3}$$

Here, E is soil loss by wind erosion, I is soil cloddiness factor, R is surface cover factor, K is surface roughness factor, F is soil textural class factor, C is factor representing local wind condition, W is field width factor, D is wind direction factor, and B is wind barrier factor.

Another model to estimate wind erosion often used in USA is that of Woodruff and Siddoway (1965).

$$E = f(I, K, C, L, V) \tag{28.4}$$

Where, E is estimated average annual soil loss. t/ha-yr, I is soil erodibility index, t/ha-yr, K is ridge roughness factor, C is climate factor, L is unsheltered length of eroding field, m, and V is vegetative cover factor. The index I is a based on fraction of non-erodible soil aggregates (particles > 0.84 mm) in the erodible size range. Eq. (28.4) has been widely used to determine both the potential erosion from a particular field or the field conditions (soil cloddiness, roughness, vegetative cover, sheltering by barrier, or width and orientation of field) necessary to reduce potential erosion to a tolerable amount.

The use of eq. (28.4) is not as simple as the Universal soil loss equation. Mathematical relationships have been established between individual variables. However, because of the complexity of these relations, a single equation expressing E as a function of the 5 dependent variables has not yet been derived. Although, it is possible to solve the equation using mathematical relations and/or graphical solutions, it is not included, being too complex and being out of course.

Chapter 29

Runoff Estimation and Grassed Waterways

Surface runoff starts contributing to an outlet of the catchment following a rainfall event after all storage requirements are met. In the early stage, the rate of flow is less. But as the time passes more and more area of the catchment contributes runoff at the outlet such that at one time, it reaches a maximum value. This maximum rate of discharge is called peak flow. Once the rain stops, the contribution decreases and runoff rate decreases and attains a constant value equal to the base flow. Estimates of peak rate of runoff, runoff volume, and the time distribution of flow provide the basis for all planning, design, and construction of drainage facilities. Erroneous results may lead to either undersized, oversized, or out of hydraulic balance systems being planned and built. For the design of grassed waterways, the peak flow must be estimated with reasonable accuracy before initiating the hydraulic design of the drainage system. Commonly peak runoff rate is calculated using rational formula, cook´s formula or curve number technique.

Storm water drains are waterways or drainage watercourse in which runoff collects and drains during the rainfall or a storm event. Waterways are of two types namely the vegetative (grass) and non-vegetative waterways. Grassed waterways are natural or manmade water courses, lined with erosion résistance grasses such as low height rhizome type perennial grasses. These are designed to carry intermittent surface runoff from terrace systems, contour furrows and diversion channels at non-erosive velocities to stable outlets. These also carry the flows from emergency spillways of farm ponds or other structures. These are not safe in areas where one expects continuous flow of water through the channels, as prolonged wetness may result in poor vegetal growth.

This chapter discusses one of the commonly used method *i.e.* rational formula to assess the peak flow. In the following section, the design and construction of grassed waterways is included.

Rational Formula

The rational formula considers the entire drainage area as a single unit and it estimates flow at the most downstream point only. Following assumptions are made in arriving at the formula:

- ☆ Rainfall is uniformly distributed over the drainage area. Whether this assumption is true depends upon the size of the watershed and the rainfall event.
- ☆ Rainfall intensity is constant at all time. Based on rainfall records, this assumption is true for short periods of time (a few minutes), but becomes less true as time increases.
- ☆ Storm duration (D) is equal to or longer than the time of concentration, t_c. This assumption has led to a common misconception that the duration of the storm is equal to t_c. It is theoretically possible but it is much more common for the total storm duration to be considerably longer than t_c (Figure 29.1).
- ☆ The runoff coefficient (C) is constant during the rain storm. Normally, C is a variable and it can take on several different values for the various segments even though the land use remains the same

The Rational formula in metric units is given as:

$$Q_p = C \times I \times A/360 \tag{29.1}$$

Here Q_p is the peak runoff rate, m^3/s; C is the runoff coefficient, dimensionless may vary from 0 to 1; I is the rainfall intensity for a duration equal to time of concentration and for a given recurrence interval, mm/h and A is the drainage area, ha. If area is in km^2, eq. (29.1) takes the following form.

$$Q_p = 0.278 \times C \times I \times A \tag{29.2}$$

It may be noted that rational formula has little rationality as the units on both sides of the equation do not match and therefore, a constant is used for different units. Any change in the units would have to be used with care. For example; the value of the constant 1/360 in eq. (29.1) changes to 0.278 in eq. (29.2) if the area is expressed in km^2 rather than ha (100/360 = 0.278). The use of this method to compute Q_p requires three parameters: C, t_c, I for a duration of t_c and for a probability p or return period T (1/p).

Calculation of C

In rational formula, C represents the ratio of the volume of direct runoff to volume of the total rainfall. On a corrugated-iron roof almost all of the rain would become runoff, so C will be almost 1.0, while a well-drained sandy soil, where nine-tenths of the rain soaks in, would have a C value of 0.1. The paved surfaces,

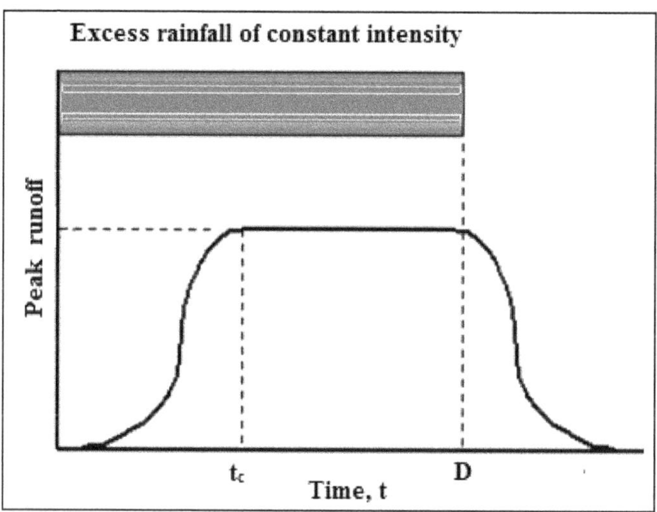

Figure 29.1: Schematic Explanation of Time of Concentration and Duration of Rainstorm.

surface with steep slopes and not so well vegetated surface will have higher values of C (Table 29.1). If the catchment has several different kinds of topography or land use, a weighted average C_w is used by combining the different values in proportion to the area of each land use or kind of surface.

$$C_w = \sum Ci \frac{Ai}{\sum Ai} \tag{29.3}$$

Where Ci is the value of C for an area Ai and C_w is the coefficient applicable to the total area A given by ΣAi.

Time of Concentration

The time of concentration, t_c, is defined as the time required for water to travel from the most hydraulically remote point in a watershed to the point of interest, the outlet. This happens when the time of concentration, the time after which the runoff rate equals the excess rainfall rate, is reached. The Kirpich formula is commonly used to calculate the time of concentration (Kirpich, 1940):

$$t_c = 0.0195 \times L^{0.77} \times S^{-0.385} \tag{29.4}$$

Here t_c is the time of concentration, min; L is the length of main river, m and S is the distance weighted channel slope, m/m. The formula is used for agricultural watersheds up to 80 ha. Although the constant 0.0195 is used for all practical purposes, the value is not sacrosanct. In fact, it is multiplication of two constants, G and k. While G has a value of 0.0078, the value of k varies with ground cover. To derive the constant of 0.0195, the value of k is taken as 2.5 but it can be as low as 0.2 for concrete channels to 2.0 for general overland flow and natural grass channel.

Table 29.1: Values of Runoff Coefficient C for different Topographical and Vegetation Conditions

Topography and Vegetation	*Soil Texture*		
	Sandy Loam	*Clay and Silt Loam*	*Tight Clay*
Woodland			
Flat 0-5 per cent slope	0.10	0.30	0.40
Rolling 5-10 per cent slope	0.25	0.35	0.50
Hilly 10-30 per cent slope	0.30	0.50	0.60
Pasture			
Flat	0.10	0.30	0.40
Rolling	0.16	0.36	0.55
Hilly	0.22	0.42	0.60
Cultivated			
Flat	0.30	0.50	0.60
Rolling	0.40	0.60	0.70
Hilly	0.52	0.72	0.82
Urban areas	30 per cent area impervious	50 per cent area impervious	70 per cent area impervious
Flat	0.40	0.55	0.65
Rolling	0.50	0.65	0.80

After the time of concentration is calculated, the 1 hr rainfall intensity known for Indian conditions is converted to the rainfall intensity of the duration of time of concentration (Figure 29.2). Known one hour rainfall intensity is taken on the y-axis while the rainfall intensity for the time of concentration is read on the x-axis. Predicted peak discharge from the rational formula has the same probability of occurrence (return period) as the rainfall intensity (I) used in the calculations.

Procedure to Determine Peak Runoff

The step wise procedure to determine the peak rate of runoff using rational formula is described as follows.

- Delineate the catchment boundary and determine its area
- Use the Kirpich equation (eq. 29.1) to determine t_c
- Find the rainfall intensity, I, for the design storm using the calculated t_c and the rainfall intensity-duration-frequency curves (Figure 29.2) or formulae
- Determine the average runoff coefficient, C using eq. (29.3) and Table 29.1.
- Calculate the peak flow rate, Q_p using eq. (29.1) or (29.2).

Limitations of the Formula

Some of the limitations of the rational formula are as follows:

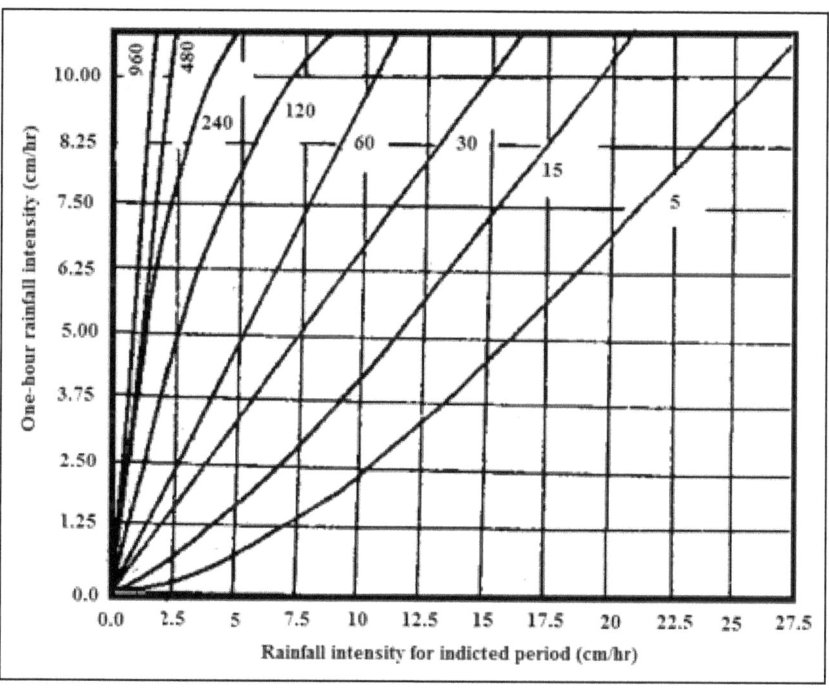

Figure 29.2: Relationship between Intensities of 1-hr and Indicated Durations for Various Return Period (Marked on curves).

A major limitation of the rational formula is that it produces only one point on the runoff hydrograph—the peak discharge rate. When basins become complex and where sub-basins combine, the rational formula will tend to overestimate the actual flow. The over estimation will result in the over sizing of the grassed waterway.

The average rainfall intensities used in the method bear no time sequence relation to the actual rainfall pattern during a storm.

The method assumes that the rainfall intensity is uniform over the entire watershed. This assumption is true only for small watersheds and time periods, thus limiting the use of the formula to small watersheds. Whether "small" means 10 ha or 80 ha is still being debated.

The relationship between rainfall and runoff is linear. If rainfall is doubled then runoff is also doubled. It is not true because all the variables interact and determine the quantum of runoff. In fact, one of the major misconceptions in the use of the formula is that each of the variables (C, i, A) is considered independent and estimated separately. In reality, the variables are interdependent and this interdependence is not recognized in this formula.

Finally, one of the most important limitations is that the results are usually not replicable from user to user. There are considerable variations in interpretation and methodology in the use of the formula. The simplistic approach of the formula permits, a wide latitude of subjective judgment in its application. Each person has its own t_c formula, own table for determining C and his own method for determining which recurrence interval is used in certain situations. Therefore, it is suggested that as far as possible, local graphs or tables should be used, wherever these are available.

Example 29.1

Estimate the peak rate of runoff for a 10 year frequency from a watershed of 35 ha, having 20 ha under cultivation (C = 0.5), 5 ha under forests (C = 0.4) and 10 ha under grass cover (C = 0.45). The distance from the remotest point in the watershed to the outlet is 600 m with a fall of 4.26 m. One hour intensity of rainfall for this site is 10 cm/hr.

Weighted value of C for the watershed = $20 \times 0.5 + 5 \times 0.4 + 10 \times 0.45 = 16.5/25 = 0.47$

For L = 600 m, S = 4.26/600 = 0.0071

Time of concentration, $t_c = 0.0195 \times L^{0.77} \times S^{-0.385}$

$t_c = 0.0195 \times 600^{0.77} \times 0.0071^{-0.385} = 0.0195 \times 136.8 \times 6.71 = 17.9$ min

1 hour rainfall intensity for 10 years frequency = 10 cm/h

Intensity for 18 minutes rainfall = 17 cm/h (Figure 29.2) = 170 mm

Peak runoff rate, $Q_p = CIA/360 = (0.47 \times 170 \times 35)/360 = 7.8$ cumecs

Grassed Waterways

Runoff from sloping land must flow to lower lands in a controlled manner so that there is no risk of gully formation. Vegetated waterways are used to handle runoff and to dispose it of in an orderly manner. Grassed water ways are best suited to areas having land slope less than 20 per cent and when water velocity does not exceed 1.8 m/s. Grassed waterways are preferred over bare channels because of the following reasons:

- Vegetation protects the soil, and retards the channel flow velocity
- Vegetation improve water quality by filtering sediment and by absorbing fertilizer and other chemicals from sheet erosion, and pesticides that enter the grassed channels
- Grass cover protects the waterway from gully erosion
- Provides wildlife grazing and habitat.

Design

The design of grassed waterways is basically a problem of open channel hydraulic engineering. The design comprises of two major steps.

- Hydrologic design *i.e.* to determine the channel capacity
- Hydraulic design *i.e.* to design cross-section and other dimensions

Hydrologic Design, the Channel Capacity

The capacity of the grassed waterways is decided on the basis of expected peak rate of flow. Minimum capacity shall be sufficient to carry peak runoff expected from the 10-year, 24-hr rainfall events. A suitable increase in capacity shall be made to account for the potential loss of channel volume as a result of sedimentation between maintenance activities. Here in the design capacity calculations are made using the rational formula, as explained previously in this chapter.

Hydraulic Capacity of the Channel

The channel must be able to carry the design runoff ($Q \geq Q_p$) at average velocities less than or equal to the permissible or non-erosive velocity. The flow velocity and hence the channel capacity is calculated with the help of the Manning´s formula, a trial and error process as explained in Chapter 13 on irrigation water conveyance through open channels. The readers may refer to that chapter to seek details of the application of this formula for channel design. However, the application has also been illustrated through a solved example at the end of this chapter. Some additional points that need to be considered while designing a grassed waterway are discussed herein.

Location of Waterway

The waterway should normally be located in natural drainage ways where the water can drain from all sides. The soils should be usually deep and fertile and moisture favorable for good grass growth. Waterway should preferably be located along field boundary.

Velocity

Design velocity shall on an average condition of grass cover and channel section should be in the range of 1.5 to 2.0 m/s. The value of Manning´s n is higher in grassed waterways, the minimum being 0.04.

Type of Vegetative Cover	*Velocity (m/s)*
Sparse green cover	0.9 -1.5
Good quality cover	1.5 – 1.8
Excellent quality cover	1.8 – 2.5

Shape

Normally, a parabolic or a trapezoidal (Figure 29.3) or a V- shaped cross section is used. The parabolic cross sections nearly mimic the natural channels. Thus, parabolic cross sections are usually selected for channels in natural waterways. Even under the normal action of channel flow, deposition, and bank erosion, the trapezoidal and triangular sections also tend to become parabolic. 'V'-shaped channels tend to be more erosive, and thus prone to gully formation. As a general guideline, the channel top or bottom width should not be less than 5 and 3 m, respectively. Similarly, channel depth should not be less than 0.30 m (Jedrych, 2017). Too small or too narrow waterways will generate high flow velocity and are prone to gully formation.

Selection of Suitable Vegetation

Selection of proper vegetation forms a major design consideration in the

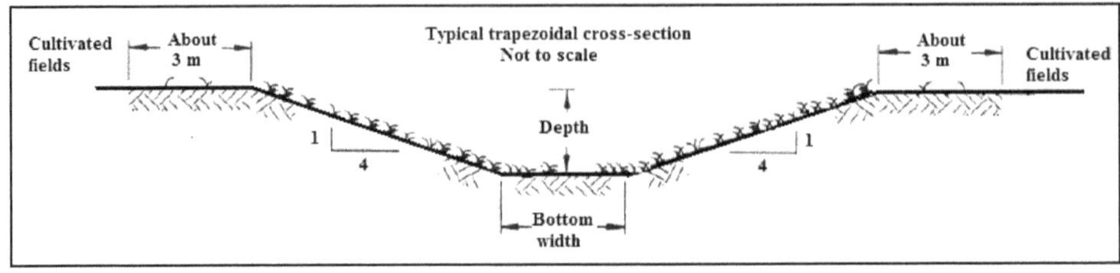

Figure 29.3: Typical Cross-Section of a Trapezoidal Grassed Waterway.

construction of grassed waterways. Soil and agro-climatic conditions are the primary factors in the selection of vegetation. The other factors are volume and velocity of runoff, ease of establishment and time required to develop a good vegetative cover. Furthermore, if the vegetation can be used as fodder, it will add to its economic value. Besides, cost and availability of seeds and redundancy to shallow flows in relation to the sedimentation are few other important factors that assume significance in the selection of vegetation. A grass that has no economic value must not be spreading type lest it spreads to the adjoining fields and cause unnecessary nuisance.

Rhizomatous grasses spread very quickly and provide greater protection to the channel than the bunch type grasses. *Vetiveria lawsonii* (fibrous root system), *Vetiveria zizanioides* commonly known as *khus* grass in India (a perennial grass with thick fibrous adventitious roots), some legumes such as *Canna veritas* (large tropical and sub-tropical perennial herbs with a rhizomatous rootstock) have proved their usefulness in grassed waterways. Bermuda grass (*Cynodon dactylon*) also seems to be a good candidate. Some other grasses such as *cynodon* species, *Axanopus compressus*, *Pogonatherum crinitum* and *Chrysopogon aciculatus* are able to control soil erosion but having weak steam may not suit to areas having heavy rain fall. Some other hilly type grasses *cymbopogon* species, *Sporobolus virginicus*, *Sporobolus spicatus* and *Tripogon* species have good resistance to soil erosion but perform poorly in steep sloped areas.

Several combination of grasses have also been tried and proved successful (Figure 29.4). For example, a combination of *Vetiveria lawsonii*, *Vetiveria zizanoides* and *Canna veritas* provided good resistance to soil erosion as well as helped to reduce the flow velocity (Omkar *et al.*, 2017). A combination of Bermuda grass and *Zoysia* grasses has also been used.

Deep rooted legumes should not be used, because they have the tendency to loosen the soil, making the soil more erodible under the effect of fast flowing water. A light seeding of small grain can be used to develop a quick cover before the grasses are fully established in the waterway. Grassed waterways are generally seeded during non-monsoon season and irrigation is provided, if required for proper germination.

Outlets

All grassed waterways shall have stabilized outlets with adequate capacity to prevent ponding or flooding damage or scour.

Figure 29.4: A Combination of *Vetivar* spp. with *Canna veritas*.

Construction

To construct a grassed waterway in a poorly drained area, a dry period must be identified during which construction can be initiated and undertaken without hindrance (NRCS, 2011). Before beginning to construct a grassed waterway, protect the upland areas from erosion by applying appropriate erosion and sedimentation control practices. It is useful to divert runoff away from the waterways until the vegetation is established. All steep gully banks shall be graded to not steeper than 1H:1V. Remove all brush and rocks larger than 15 cm in diameter and put them away from the waterway. Drive centerline stakes to mark the intended waterway. Offset stakes can help maintain planned grades as well as aid in checking construction. Shape the waterway to the design grade and pre-decided cross section. Fill the gullies gradually. Adequately compact the fill to prevent uneven settling along its length. Plant seed or sod according to the design. If subsurface drains are needed install these along with appropriate outlet control structures before shaping the grassed waterway. If infertile sub-soil is likely to be exposed during construction, strip and stockpile topsoil and spread it over the exposed infertile soil. Protect the channel with mulch or a temporary liner sufficient to withstand anticipated velocities during the initial establishment period.

Care and Maintenance

During the initial years, check the channel after every rainfall event. Assess the condition and operation of temporary check dams, diversions, or other structures while vegetation is established. Repair eroded areas immediately. Also check channel outlets, and any Best Management Practices (BMPs) implemented up-slope of the channel. Remove any accumulated sediment which ends up in the channel, and re-evaluate upland BMPs accordingly. Once the vegetation is established, ensure that waterways have dense, mature vegetation at all times. Repair and reseed any bare or eroded spots as soon as possible. Don't use the waterway as a road. The tracks can lead to the formation of gullies. Regular mowing schedule should be followed to keep the vegetation within the range specified to ensure desired retardance as well as maintain dense vegetative vigor. Maintain outlets to prevent new gullies from forming at the outlet.

Example 29.2

Design a parabolic grassed waterway to carry a peak discharge of 6 m^3/s. Slope of the land is 1 per cent. Take the value of Manning´s n as 0.04. The channel is fairly covered with grass.

In the simple approach, let us assume the limiting velocity as 1.5 m/s (As per guidelines under velocity)

According to Manning´s formula reproduced from Chapter 13, we have

$V = (1/n) \times R^{2/3} \times S^{1/2}$

Since S=0.01 and n =0.04. we have

$V = (1/0.04) \times R^{2/3} \times 0.01^{1/2} = 2.5\ R^{2/3}$

$R = (1.5/2.5)^{3/2} = 0.465$ m

$A = 6/1.5 = 4.0\ m^2$

For a parabolic section, we have d=1.5 R (For best hydraulic section)

$d = 1.5 \times 0.465 = 0.7$ m

Top width, $t = A/0.67\ d = 4/(0.67d) = 8.53$ m

Recheck the capacity

$Q = A.V = 0.67 \times 8.53 \times 0.7 \times 1.5 = 6.0\ m^3/s$

Add 20 per cent freeboard to depth $= 0.7 + 0.2 \times 0.7 = 0.84$ m.

The designed cross-section is shown in Figure 29.5.

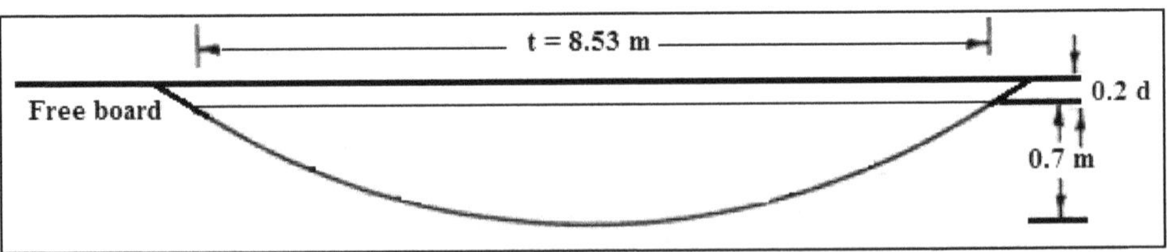

Figure 29.5: Designed Parabolic Cross-Section of a Grassed Waterway.

Chapter 30

Universal Soil Loss Equation

Introduction

Evaluation of soil loss from watersheds is essentially required to assess the severity of soil erosion and its adverse effects on agricultural production or other degradation processes such as deterioration of water quality and environment. For some purposes, meaningful and useful long-term average annual soil loss can be obtained using the Universal Soil Loss Equation (USLE) proposed by. Wischmeir and Smith (1958) and Wischmeir (1959). Derived on the basis of 10,000 plot years of data, it can be adopted under a wide range of conditions. The equation treats the soil erosion as a function of:

- ☆ Rainfall energy and intensity
- ☆ Soil properties: erodibility
- ☆ Topography: slope length and steepness
- ☆ Soil cover
- ☆ Conservation practices

Numerical estimate of each factor for a specific condition is calculated to assess the soil erosion. Since the erosion values reflected by these factors can vary considerably as a result of varying weather conditions, the soil erosion values are treated to represent long-term average of soil loss.

Universal Soil Loss Equation (USLE)

The equation is given by:

$$A = R\,K\,LS\,C\,P \qquad (30.1)$$

where, A is the average soil loss for the given period, R is the rainfall erosivity index, K is the soil erodibility factor, LS is the topographic factor, L being the slope length factor and S being the slope steepness factor (combined in one factor denoted as LS), C is the cropping management factor, and P is the conservation practice factor. Since soil loss is dependent upon the erosivity and erodibility, both these factors are being taken into account in the equation. To explain this, let us rewrite the equation as follows:

A = R (Rainfall erosibility) K (Soil erodibility) LS (Factors affect erodibility) C P (Factors reduce soil erosion)

Average Soil Loss, A

It is an estimate of the soil loss from sheet and rill erosion from rain fall events, normally expressed in tons/ha per unit time usually on an annual basis. Within known limitations, it has been used to compute soil loss for any period and also on probability basis (*e.g.* once in two years, once in five years *etc.*). To arrive at these values, rainfall erosivity index 'R' for the corresponding period is used. The soil loss calculated from USLE does not include losses from stream bank erosion, landslides, gully erosion, and snowmelt erosion *etc.* To use the equation as a management tool, the calculated value is compared with the "tolerable soil loss" limits. If the calculated value is more, it calls for management through changes in the LS factor or C and P factors so that calculated value is equal to or less than the tolerable value.

Rainfall Erosivity Index, R

R factor is defined as the average annual sum of EI parameter for all the storms of a year, such that it is the product of the kinetic energy (KE) of the individual storms (KEi) and the maximum 30 minutes intensity of the rain storm (I). Values may be expressed for any length of period (like daily, monthly or annual) or for any desired probability level. Summation of all storms divided by the number of storms during the pre-decided period (mostly annual) is the value of the R factor for that period. E is calculated according to:

$$KE = 210.1 + 89.0 (\log I) \qquad (30.2)$$

Kinetic energy is expressed in joules/m^2 per cm of rain or metric tons/ha per cm of rainfall and I is in cm/hr. The energy for the amount of rainfall is calculated by multiplying it with the amount of rainfall during that period. Summation gives the total energy of the storm in metric tons/ha. Now R, which is given as EI_{30} is calculated by multiplying the average storm energy with I_{30}. To convert the storm kinetic energy to the dimension in which EI are commonly expressed (Wischmeier and Smith 1959), it is divided by 100, before multiplying with I_{30} such that

$$R = EI_{30}/100 \qquad (30.3)$$

R is the average annual rainfall MJ.cm/ha-hr since E is the kinetic energy, MJ/ha and I_{30} is maximum 30-min rainfall intensity (cm/hr),

Since, the approach to determine R in this fashion is quite cumbersome, it is either related with average annual rainfall or expressed as per geographic location of the place. For example, R factor for Delhi is 100. For some other Indian locations, R factors are given in Table 30.1 (Ram Babu *et al.*, 1978). Depending upon the area of interest, one can select the value of R for use in the USLE.

Table 30.1: Estimated Rainfall Erosion Index of some Geographical Locations of India

Regions on India	*R Values*
Entire Jammu and Kashmir region	100
Northern Himachal Pradesh	400
Eastern and southern Himachal Pradesh, southern and central Uttarakhand, northern UP and entire Bihar	500
Entire Punjab, Haryana, Delhi, western Rajasthan, central UP, central MP, eastern Maharashtra, western Andhra Pradesh, western Tamil Nadu and central Karnataka	400
The region of entire west coast passing through states of Kerala, Karnataka, Maharashtra and Gujarat. Western most boundary region	1500
Rest of the area	500
Western Rajasthan and western Gujarat region	200-300
Central Maharashtra, central Karnataka and the entire states of West Bengal, Jharkhand, Odisha, Chhattisgarh and Andhra Pradesh	600-700
Eastern coastal areas of Tamil Nadu	500
Eastern coastal areas of Odisha, West Bengal	700
Entire states of Sikkim and Arunachal Pradesh	1250
North Assam and Meghalaya region	750
Central Assam and Nagaland region	250-500
Entire Tripura region	1250
Entire Mizoram region	1000

Soil Erodibility Factor, K

Under the conditions of equal slope, rainfall, vegetative cover and soil management practices, some soils may erode more easily than others due to inherent soil characteristics. K is a measure of the susceptibility of soil particles to detachment and transport by rainfall and runoff. Although the principal factor that affects K is soil texture, but K is also affected by soil structure, organic matter content and permeability of the soil. This factor is expressed as tons of soil loss per hectare per unit of erosion index, R for a slope of specified dimensions (9 per cent and 22.13 m long) under continuous cultivated, fallow without the influence of crop cover. As such, K is expressed in t/ha.(ha-h/MJ-mm) or (t.ha-h)/(ha.MJ-mm). The value of K can be obtained in a number of ways as described in the following paragraphs.

1. The estimated K from experimental plots is given by,

   $$K = \text{Total adjusted soil loss (A)/Total EI} \quad (30.4)$$

   Such that A= A0/S. The adjusted soil loss, A is obtained from measured soil loss, A0 divided by the soil slope factor, S.

2. If soil texture and organic matter content is known then K values can also be obtained from Table 30.2. It is seen that K values vary from a low of 0.02 to 0.96 (Table 30.2) the lower value being for sand while the higher value for very fine sandy loam. The values of K for few stations in India are also given in Table 30.3.

Table 30.2: K Factor for Use in USLE

Textural Class	*K Factor, t/ha per unit R*		
	Average OMC[x]	*Less than 2 per cent OMC*	*More than 2 per cent OMC*
Clay	0.49	0.54	0.47
Clay loam	0.67	0.74	0.63
Coarse sandy loam	0.16	–	0.16
Fine sand	0.18	0.20	0.13
Fine sandy loam	0.40	0.49	0.38
Heavy clay	0.38	0.43	0.34
Loam	0.67	0.76	0.58
Loamy fine sand	0.25	0.34	0.20
Loamy sand	0.09	0.11	0.09
Loamy very fine sand	0.87	0.99	0.56
Sand	0.04	0.07	0.02
Sandy clay loam	0.45	–	0.45
Sandy loam	0.29	0.31	0.27
Silt loam	0.85	0.92	0.83
Silty clay	0.58	0.61	0.58
Silty clay loam	0.72	0.79	0.67
Very fine sand	0.96	1.03	0.83
Very fine sandy loam	0.79	0.92	0.74

[x] Organic matter content

Table 30.3: K Values at Selected Stations in India

Station	K Value
Agra	0.07
Dehradun	0.15
Hyderabad	0.08
Kharagpur	0.04
Kota	0.11
Udagamandalam (Ooty)	0.04
Rehmankhera	0.17
Vasad	0.06

3. Nomographs and equations have also been derived to assess the values of K. With the advent of calculators and computers, use of nomographs is getting redundant. For soils containing less than 70 per cent silt and very fine sand, the commonly used nomograph gives the same values as the following equation. Therefore, eq. (30.5) can be used to calculate the value of K.

$$K = 2.8 \times 10^{-7} M^{1.4} (12 - a) + 4.3 \times 10^{-3} (b-2) + 3.3 \times 10^{-3} (c-3) \quad (30.5)$$

Here M is the particle size diameter (per cent silt + per cent very fine sand) (100 - per cent clay), a is the organic matter, per cent, b is the soil structure code (Table 30.4) and c is the profile permeability class (Table 30.4). It also means that the percentage of very fine sand (0.1-0.05 mm) must first be transferred from the sand fraction to the silt fraction to calculate the value of M.

Table 30.4: Soil Structure and Profile Permeability Classes

Soil Structure	*Code*	*Profile Permeability*	*Class*
Very fine granular	1	Rapid	1
Fine granular	2	Moderate rapid	2
Medium or coarse granular	3	Moderate	3
Blocky, platy or massive	4	Slow to moderate	4
		Slow	5
		Very slow	6

Slope Length, L and Slope Factor, S

The longer and steeper the slope, the higher is the risk of erosion. Slope length factor refers to overland flow or run-off from top of ridge to the outlet channel or top of ridge to where deposition begins. The slope length factor, for a given soil erodibility value, is the ratio of soil loss from any length of slope to that from the specified slope length (22.13 m). The slope length factor is calculated from

$$L = (Lp/22.13)^{m} \quad (30.6)$$

Here Lp is the actual unbroken length of slope (m) and m is an exponent. The value of m is 0.5 for slopes equal to or more than 5 per cent, 0.4 for 4 per cent, 0.3 for 3 per cent and 0.2 for 1 per cent.

Steeper the slope, greater is the erosion. The slope gradient factor expresses the ratio of soil loss from a known slope to soil loss from a unit plot under identical conditions. The factor S is calculated by the relation

$$S = (0.43 + 0.30\,s + 0.043\,s^2)/6.613 \quad (30.7)$$

Here s is the slope of the field, per cent.

Wischmeier and Smith (1965) combined the two factors into a single factor denoting it as the topographic factor, LS. It is calculated by the following relation:

$$LS = (\sqrt{Lp})/100)\,(0.76 + 0.53s + 0.076s^2) \quad (30.8)$$

Or

$$LS = ((Lp)^m/100)\,(1.36 + 0.97s + 0.1385s^2) \quad (30.9)$$

The more common equation to calculate LS factor is written as:

$$LS = (Lp/22.1)^m\,(0.065 + 0.0456s + 0.006541s^2) \quad (30.10)$$

Here m is a variable that depends upon the steepness of land slope and varies from 0.5 for slopes greater than or equal to 5 per cent, 0.4 for slopes between 3-5 per cent, 0.3 for slopes between 1-3 per cent to 0.2 for uniform slope of less than 1 per cent. Typical values of LS for a slope length of 120 m for various slopes as calculated with the help of eq. (30.9) are reported in Table 30.5. For other lengths and slope similar calculations can be made.

Table 30.5: LS Factor Calculation

Slope Length (m)	*Slope (per cent)*	*Value of m*	*LS Factor*
120	10	0.5	2.73
	8	0.5	1.98
	6	0.5	1.34
	4	0.4	0.69
	2	0.3	0.30

Crop Management Factor (C)

It is a ratio of soil loss from the field condition of interest to that from tilled continuous fallow. It is an integration of several factors that affects soil erosion.

- ✰ Vegetation cover
- ✰ Crop rotations
- ✰ Length of growing season
- ✰ Land management (tillage practices): Conventional tillage leaves the surface bare therefore susceptible to erosion while conservation tillage leaves residue on surface that protects the soil from rainfall impact and reduces sheet and rill erosion
- ✰ Residue management
- ✰ Soil surface
- ✰ Binding of plant roots

Table 30.6: Crop Type Factor

Crop Type	*Factor*
Grain corn	0.40
Silage corn, beans and canola	0.50
Cereals (spring and winter)	0.35
Seasonal horticultural crops	0.50
Fruit trees	0.10
Hay and pasture	0.02

Crop/vegetation management factor (C) depends upon the crop type and the factor can be selected with the help of Tables 30.6. It also depends upon the tillage method. Its values are given in Table 30.7. Depending upon crop and the tillage method, two factors are multiplied to get the value of C (Stone and Hilborn, 2015). For example cereal crop grown with no till will have factor C as $0.35 \times 0.25 = 0.09$. For limited situations, values of C for Indian conditions can also be directly selected from Table 30.8.

Table 30.7: Tillage Method Factor

Tillage Method	*Factor*
Fall plow	1.0
Spring plow	0.90
Mulch tillage	0.60
Ridge tillage	0.35
Zone tillage	0.25
No-till	0.25

Table 30.8: Crop Management Factor (C) for Various Vegetative Covers at different Locations in India

Crop	*Kota*	*Vasad (Vadodara)*	*Kharag-pur*	*Hydera-bad*	*Agra*	*Lucknow*
Green gram/(*moong*)	0.39	0.47				0.45
Black gram	0.54					
Groundnut	0.41	0.38				0.42
Groundnut + *arhar*	-					
Soybean	0.42					
Cowpea	0.39	0.32	0.17			
Guar	0.59				0.42	
Guar + arhar	-					0.35
Maize	0.50		0.35			
Jowar (Sorghum)	0.62				0.64	
Jowar+ arhar	0.33					0.28
Red gram (Pigeon pea)	-		0.38			
Jowar+ red gram	-				0.32	
Paddy	-	0.28				
Bajra (Pearl millet)	-			0.40	0.61	
Castor	-			0.79		
Til	-				0.51	0.39
Natural cover	0.14					
Grass (*Cynodon dactylon*)	0.22		0.04			
Grass (*Dichanthium annulatum*)	0.01				0.13	

Conservation Practice Factor, P

It reflects upon the effects of conservation practices that reduce the amount and rate of the water runoff and thus reduce the amount of erosion. It is a ratio of the soil loss with the control measures (contouring, strip cropping, terracing, *etc.*) to that with farming up and down the slope. The values of P are generally based on judgment and experience obtained from non-crop situation. Not much data on values of P are available for the Indian conditions. Dhruvanarayana (1993) recommended that values of P may be determined as per information given in Table 30.9.

Table 30.9: Values of Conservation Practices Factor

Slope (per cent)	*Contouring*	*Contouring and Strip Cropping*
1.0 – 2.0	0.60	0.30
3.0 – 8.0	0.50	0.25
9.0 – 12.0	0.60	0.30
13.0– 16.0	0.70	0.35
17.0 – 20.0	0.80	0.40
21.0- 25.0	0.90	0.45

Procedure to Use USLE

- ☆ Determine the R factor (Table 30.1).
- ☆ Based on the soil texture, determine the K value. If there is more than one soil type in a field and the soil textures are not very different, use the soil type that represents the majority of the field. Otherwise find the weighted average or repeat the calculations for various soil types as necessary. Use Table 30.2 or Table 30.3 or eq. (30.4) and Table 30.4.
- ☆ Divide the field into sections of uniform slope gradient and length. Assign an LS value to each section. Use eq. (30.10)
- ☆ Choose the crop type factor and tillage method factor for the crop to be grown. Multiply these two factors together to obtain the C factor (Table 30.6 and 30.7 or Table 30.8).
- ☆ Select the P factor (Table 30.9) based on the support practice used.

Table 30.10: Classification Criteria on the Basis of Soil Loss per Year

Type of Erosion	*Soil Loss per Year (t/ha)*
Slight	0-5
Moderate	5-10
High	10-20
Very high	20-40
Severe	40-80
Very severe	>80

Multiply the 5 factors together to obtain the soil loss per hectare. If the soil loss is higher than the soil loss than desirable (Table 30.10) for maintaining productivity, suitable changes in the LS factor, crop management and conservation practices should be made to reduce the expected soil loss (Table 30.11). In general "tolerable loss level" is 6.7 t/ha-year.

Table 30.11: Management Strategies to Reduce Soil Losses

Factor	*Management Strategies*	*Example*
R	The R factor for a field cannot be altered	–
K	The K factor for a field cannot be altered	–
LS	Terraces may be constructed to reduce the slope length resulting in lower soil losses	Terracing will cause some inconvenience in farming. Adopt some other soil conservation practices first
C	The selection of crop types and tillage methods that result in the lowest possible C factor (will result in less soil erosion)	Consider cropping systems that will provide maximum protection for the soil. Use minimum tillage systems where possible
P	The selection of a support practice that has the lowest possible factor associated with it will result in lesser soil loss	Use support practices such as cross-slope farming. It will allow deposition of sediment close to the origin

Applications of USLE

The USLE is an empirical equation, derived from a large mass of field data. Main applications of USLE are

- ☆ Predict soil loss to identify highly erodible land and to develop farm plans to mitigate the erosion risks. For example
- ☆ Selection of agricultural practices
- ☆ Recommendations on crop management practices
- ☆ Calculate length and gradient through erosion control structures in order to reduce land loss to a tolerable level.

Let us illustrate the application of USLE with the help of following example.

Example 30.1

If values of R, K LS, C and P are180, 0.25, 1.62, 0.20 and 1.0 calculate the soil loss. Show how this loss can be reduced with various interventions. Assume appropriate interventions and the relevant values of the parameters.

With the given values, the soil loss A is given as

$A = 180 \times 0.25 \times 1.60 \times 0.20 \times 1.0 = 14.4$ t/ha per year

Let us now assume that conservation tillage is introduced resulting in the factor C as 0.11. Therefore,

$A = 180 \times 0.25 \times 1.60 \times 0.11 \times 1.00 = 7.92$ t/ha per year

If the conservation tillage is also accompanied by contour cultivation with a factor P as 0.60, then

$A = 180 \times 0.25 \times 1.60 \times 0.11 \times 0.60 = 4.75$ t/ha per year

Change in only LS using terracing (LS = 0.60) will reduce the soil loss as follows.

$A = 180 \times 0.25 \times 0.60 \times 0.20 \times 1.0 = 5.4$ t/ha per year

Example 30.2

In an area subjected to soil erosion, the following information is available.

Rainfall erosivity index = 800; Soil erodibility index = 0.30; Crop factor = 0.60; Conservation practice factor = 1.0 and Slope length factor = 0.1

What will be estimated annual loss? What should be the conservation practice factor to reduce the soil loss below 10 t/ha-year?

Using the Universal soil loss equation, the soil loss is obtained as,

$A = 800 \times 0.30 \times 0.60 \times 1.0 \times 0.1 = 14.4$ tons/ha-year

To reduce the soil loss to 10 t/ha-year, we can calculate the value of P as:

$10 = 800 \times 0.30 \times 0.60 \times 0.10 \times P$

$P = 0.69$

P factor below 0.7 will result in soil loss within the desired limit. Similarly other factors as per details in Table 30.11 can be selected to reduce the soil loss.

Limitations

The USLE has the following limitations:

- An empirical equation that predicts the long-term average annual rate of erosion. It does not apply to individual storms
- It does not include fundamental hydrologic and erosion processes explicitly. For example, the effect of runoff is not expressed directly in the equation
- It has not been tested for steep slopes, slopes steeper than 40 per cent, where runoff is a greater source of energy than rain and where there are significant mass movements of earth
- USLE only predicts the amount of soil loss that results from sheet or rill erosion on a single slope and does not account for additional soil losses that might occur from gully, wind or tillage erosion
- A major limitation of the model is that it accounts for individual factors but neglects certain interactions commonly realized in hydrology. For example, it does not take into account the effect on erosion of slope combined with plant cover, or the effect of soil type on the effect of slope.
- It does not estimate soil deposition or sediment rates; one need to find sediment delivery ratio to assess sediment rates
- Does not adequately predict soil erosion on non-agricultural lands
- In general USLE overestimates at sites with relatively low erosion rates and underestimates at sites with higher erosion rates.

Because of these limitations, the modified USLE (MUSLE) and revised universal soil loss equation (RUSLE) have been designed. RUSLE is currently more widely used than the USLE.

Modified USLE (MUSLE)

To overcome some of the limitations of USLE, Williams (1975) modified the USLE to calculate monthly, seasonal or even storm wise sediment yield. He modified USLE by incorporating another factor that is called runoff factor. Accordingly MUSLE is written as:

$$Y = 11.8\,(Q \times qp)^{0.56}\,K\,LS\,C\,P \qquad (30.10)$$

Here Y is sediment yield in tonnes, Q is volume of runoff in m^3, qp is peak flow rate in m^3/s and K, LS, C and P are, respectively, the soil erodibility, slope length and slope steepness, crop management and soil erosion control practice factors similar to the USLE model. The dimensions are not similar on both sides of the equation, since the model has been empirically and logically developed.

Revised USLE (RUSLE)

Continuous updation of data from field studies and experiments led to the development of the Revised Universal Soil Loss Equation (RUSLE), which is a computerized version of USLE. RUSLE has the same formula as USLE, with improvements in the estimation of many factors. RUSLE can handle more complex combinations of tillage and cropping practices and a greater variety of slope shapes. An enhanced version of the software, known as RUSLE2, can do event-based erosion prediction, although it requires a comprehensive set of input information, which may not always be available at all places.

Chapter 31

Rainwater Harvesting

Rainwater Harvesting: Historical

Archaeological evidences show that the practice of water conservation has been deep rooted in the lives of people and its science has been well developed by the ancient Indians. Excavations show that the cities of the Indus Valley Civilization had excellent systems of water harvesting and drainage. The settlement of Dholavira is a great example of water engineering as it has been laid out on a slope between two storm water channels. Dams built of stone rubble in 3rd millennium BC have been found in Baluchistan and Kutch. Chanakya's *Arthashastra* mentions irrigation using water harvesting systems. Chola King Karikala built the Grand Anicut or Kallanai across the river Cauvery to divert water for irrigation. The system is still functional. King Bhoja of Bhopal built the largest artificial lake covering 65,000 acres, was fed by streams and springs. Rajasthan, a large part of which is covered by the Thar Desert, has had a long tradition of water conservation. Builders of the Bundi and Chittorgarh forts implemented several rainwater harvesting structures to store the runoff coming from natural catchments of undulating hilltops. To name a few traditional rainwater harvesting structures one can mention: *Talab/Bandhis, Johads, Baoris/Bers* (step wells), *Jhalaras* and *Kunds*. The design and structure of each system was decided by the terrain and rainfall pattern of the region. Hence many unique harvesting techniques are encountered in various parts of India. The most potential areas where rainwater harvesting should be useful are:

- ☆ Where groundwater levels are declining on regular basis
- ☆ Where substantial amount of aquifer has been de-saturated
- ☆ Where availability of groundwater is inadequate in lean months
- ☆ Where due to rapid urbanization, infiltration of rainwater into subsoil has decreased drastically and recharging of groundwater has diminished.

Definition

The term water harvesting was first used in Australia by H. J. Geddes (Mondal, 2016). Rainwater harvesting is recently gaining popularity because of widening gap between the water supplies and demand. Water harvesting in its broadest sense is defined as the 'collection of runoff for its productive use'. The rainwater harvesting has also been defined as the collection, storage (surface and subsurface) and use of the precipitation for irrigation or other purposes. The collecting areas known as catchments may be rooftops, land surfaces or rock catchments. Storage may or may not form the part of rainwater harvesting under many conditions although its future use will cover a short span of time. In that event, it is also called *in-situ* rainwater harvesting or short-term rainwater harvesting.

Importance and Benefits

Rainwater is an important water source, may be the only resource in many areas lacking any other conventional supply system. It is best quality water supplied free by the nature. It is also a good source for domestic/industrial use in areas where good quality fresh surface water or groundwater is lacking. The benefits of rainwater harvesting are:

- Reduces the gap between the supply and demand of water resulting from increasing population, urbanization, changes in eating habits and improved living standards of the mankind
- It extends the supply of existing water resources of the region
- Improves moisture content in the soil thus aiding the growth of plants and trees
- Reduces pressure on drainage networks and sewage systems; reduces the risk of sewage overflows; reduces costs and energy needed for water treatment and transport
- Helps to build climate resilience
- Reduces the soil erosion because of reduced runoff
- Mitigates the adverse effects of drought and helps to attain drought proofing
- Decreases in the choking of storm water drains resulting in reduced incidences of floods especially in urban areas
- Recharges the groundwater resulting in groundwater level stabilization or even reversing the trend of declining water levels
- Reduces crack formation in structures
- Improves the groundwater quality as it dilutes the salt content of water
- Saving energy because of reduced need to lift groundwater
- Improves ecology of the area by increasing the vegetation cover *etc.*
- Flexible systems and easy to construct
- Increases green cover in the surrounding areas
- Arrests seawater intrusion into the land

Limitations of Water Harvesting

- ☆ Need for initial investment
- ☆ Quite labour intensive
- ☆ Possibility of high seepage and evaporation losses
- ☆ Floating vegetation may infest reservoir
- ☆ Regular maintenance is required for the operation of a successful rainwater harvesting system
- ☆ Water supply is climate dependent - droughts or long periods with little or no rain can cause serious problems with supply of water
- ☆ The supply of water from a rainwater collection system is not only limited by the amount of rainfall but also by the size of the collection area and storage facilities. Since land is becoming costly, to spare large areas is difficult
- ☆ Additional power required to pump water from the reservoir.

Kinds of Rainwater Harvesting

Water harvesting includes a wide range of methods depending upon the source of water, required storage, duration of storage and intended use of water. On this basis, the term rainwater harvesting is the collective name given to the following kinds of harvesting.

- ☆ Surface runoff harvesting
- ☆ Subsurface runoff harvesting
- ☆ Roof top rainwater harvesting
- ☆ Groundwater harvesting: It may not strictly fall in the realm of the rainwater harvesting, but is included to emphasize that this technical option is a possibility.

Although roof top rainwater harvesting is also a surface runoff water harvesting technique yet it is generally treated as a separate class. Accordingly, surface runoff harvesting is further grouped under the following three categories.

- ☆ *In-situ* rainwater harvesting also known as short-term rainwater harvesting
- ☆ off-site rainwater harvesting also known as long-term rainwater harvesting
- ☆ Flood Water Harvesting

The kinds of commonly used structures for rainwater harvesting are listed in Table 31.1.

Requirements of Rainwater Harvesting

Basic principle of rainwater harvesting comprises of the following three steps.

Catchment

Any land surface or the paved area can be treated as catchment. Even the

Table 31.1: Structures for Rainwater Harvesting

Short-term or *in-situ* rainwater harvesting	Long-term rainwater harvesting	Flood water harvesting	Groundwater harvesting	Rooftop rainwater harvesting
Deep tillage	Dug out ponds	Graded bunds	Subsurface dams	Rooftop water harvesting
Micro-catchment water harvesting	Embankment ponds	Check dams		
Contour farming and ridging	Irrigation Dam	Flood control reservoirs		
Furrows, ridge and furrows and broad bed and furrows	Silt Detention Dam	Storage of flood water in aquifers		
Contour trenches	Sand dam			
Semi-circular hoop	High level pond			
Trapezoidal bunds	Percolation dam or Percolation pond			
Contour bunds				
Graded bunds				
Rock catchments				
Ground catchment				

footpaths and roads can act as the catchment, as these areas too receive the direct rainfall. The coefficient of runoff that determines the potential of a catchment to generate runoff varies from one catchment to another. From this point of view paved surfaces or rooftops are the best because of high coefficient of runoff. Treating the catchment helps to increase the runoff generation.

Conveyance

It is a network of overland flow, open or piped drains which collects the rainwater from the catchment and delivers it to the storage structure. In some cases such as *in-situ* rainwater harvesting, the role of this component is minimal. The conveyance systems are designed appropriately so as to avoid the loss of water during the conveyance process.

Storage

The most important part of the rainwater harvesting is the storage system. It is designed to take into account the amount of water to be stored. It could be a simple storage tank or a pond or an underground storage structure. The site (location) for construction of the storage structure should be properly selected. The design should ensure the hydraulic and structural safety of the structure.

In-situ Rainwater Harvesting

As the name suggests, water is stored at a place where rain falls. It may also include micro-catchment water harvesting, where surface runoff from a part of the area is diverted and stored in the root zone of an adjacent infiltration basin planted with crops, trees or bushes. Clearly, conveyance and storage components have the minimal role. It is achieved by preventing water runoff, allowing the rainwater to stand on the soil surface and infiltrate into the soil. In arid and semi-arid regions, where precipitation is low or infrequent, it is necessary to store the maximum amount of rainwater during the wet season for use over a prolonged period or at a later time, especially for agricultural. Some of the methods adopted are described in the following sections.

Deep Tillage

It is probably the least expensive and easily adoptable method for *in-situ* rainwater conservation. Tillage helps to increase the soil moisture holding capacity through increased porosity and increased infiltration rates. Besides, tillage changes surface micro-relief or roughness which increases the temporary storage of rainwater. These together provide more time for rainwater to infiltrate into the soil profile and reduce surface runoff. Studies have shown that depth of tillage is the most important factor that decides soil moisture pattern in the soil profile following a rain event.

Micro-catchment Water Harvesting

In area where rainfall alone is unable to meet the crop water requirement, some area is left fallow so that runoff from the area contributes water to the cropped area. It helps the crop to complete its life cycle with minimal water stress. Several possible layouts can be designed. Two of such layouts are shown in Figure 31.1. In the first case, strips of land are cultivated along rows leaving appropriate sections of the inter-row space uncultivated. The runoff from uncultivated strips goes to the cultivated strips to meet the crop water requirement (Figure 31.1 left). This system can be used for all types of crops and is easy to mechanize. In the other system, instead of having uncultivated and cultivated strips alternating, the field is divided into two distinct parts, the uncultivated catchment area (CA) and the cultivated area (C). The cultivated area is immediately below the CA (Figure 31.1 right). The cropped area is enclosed by a U-shaped bund to retain the harvested water. To ensure that sufficient water is harvested, the CA should neither be too small such that it fails to meet the water requirement of the crop nor too large such that it becomes uneconomical to cultivate the fields. CA/C ratio commonly known as the Catchment: Basin Area Ratios (CBAR) may vary from 3:1 to as high as 1:40. If CA turns out to be more, then one can either change the crop or treat the area to induce runoff. The treatment may comprise of one or a combination of the following options.

- ✰ Clearing or altering vegetation cover
- ✰ Increasing the land slope with artificial ground cover
- ✰ Reducing soil permeability by soil compaction
- ✰ Application of chemicals

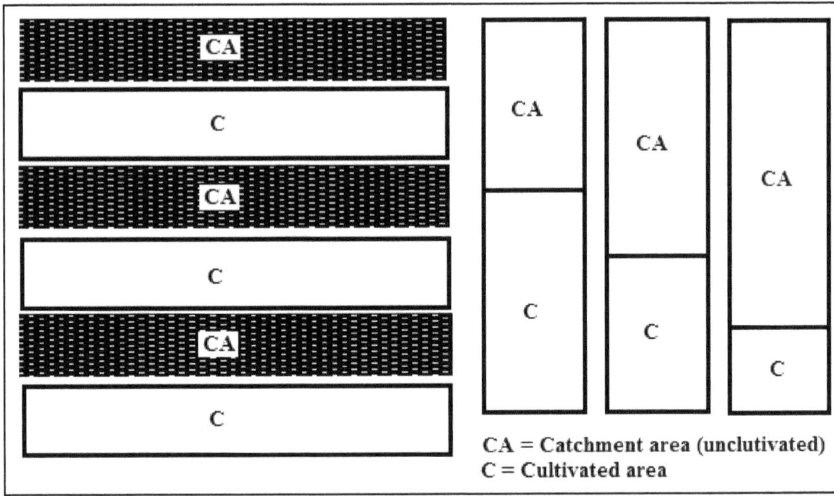

Figure 31.1: Cropping with Runoff Contribution from Uncultivated Strip (Left) and Uncultivated Area (Right).

Depending on effective rainfall and runoff rates, the ratio between the catchment and cropped area (CA/C) can be determined by the equation:

$$\text{Cultivated area} \times (\text{Crop water requirement} - \text{Design rainfall}) = \text{Catchment area} \times \text{Design rainfall} \times \text{Runoff coefficient} \times \text{Efficiency} \quad (31.1)$$

Thus, the minimum ratio (CA/C) can be calculated by:

$$CA/C = (WR - P)/(1 \times P \times E) \quad (31.2)$$

Here WR is the annual crop water requirement, mm, P is the average annual precipitation, mm, and n is the average annual runoff coefficient, fraction. E is the efficiency, which may vary from 0.25 to 0.85.

Example 31.1

Calculate the catchment area required to cultivate a crop of maize needing 600 mm of water to mature. The design rainfall for the area is 400 mm. The runoff coefficient is 0.15. Assume the efficiency factor as 0.7.

$$CA/C = (600\text{-}400)/(400 \times 0.15 \times 0.7) = 4.76 \text{ or approximately } 5$$

Thus, catchment area: cultivated area = 5:1 or for each 1 ha of cultivated area, 5 ha area should be left for water harvesting.

Contour Farming and Ridging

This practice is used when sloped lands having slopes of 3 per cent or more are cultivated. All farm operations such as tilling and weeding are done along the contours so as to form cross-slope barrier to the flow of water. Where this is not enough, it is complemented with ridges, which are sometimes tied to create a high degree of surface roughness to enhance the infiltration of water into the soil. The latter is sometimes referred as tied ridging for *in-situ* rainwater harvesting.

Contour Bunds

In this system, small trash, earth or stone embankments are constructed along the contour lines. These embankments trap the water flow behind the bunds allowing water to infiltrate deeper into the soil profile. The height of the bund controls the net storage behind the structure.

Furrows, Ridge and Furrows and Broad Bed and Furrows

Furrows are commonly used as an *in-situ* means of storing harvested rainwater. They are built prior to or after cropping to store water for future use by the plants. It is a variation on the use of topographic depressions to store rainfall. This method makes use of flattened trenches between the rows of crops to store water. Furrows may have dyked barriers every 2 m to 3 m along the row in order to retain water for longer periods of time. It will help to reduce excessive surface runoff and erosion. Raised beds may also be used to trap the water in the furrows, or uncultivated areas may be left between rows, spaced 1 m apart, to assist in capturing rainwater falling on the land surface between furrows. Broad beds and furrows are also used wherein broad beds are used for cultivation while furrows for irrigation as well as rainwater storage (Figure 31.2).

Semi-circular Hoops/Bunds

These are sometimes referred as half-moon structures. These structures are constructed in series in staggered formation (Figure 31.3 top-left). Runoff water from the area above is collected within the hoop. The contributing area and the depth

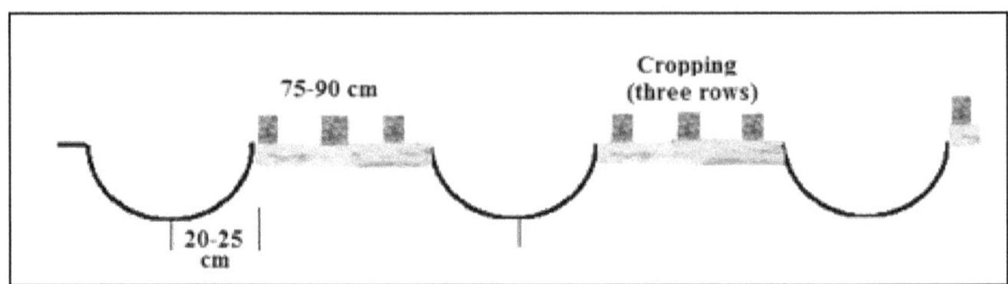

Figure 31.2: Broad Bed and Furrow System.

of water impounded is decided by the height of the bund and the position of the tips. Excess water is discharged around the tips and is intercepted by the second row and so on. Normally the semi-circles are of 4-12 m radius with a height of 30 cm, base width of 80 cm, side slopes 1:1.5 and crest width 20 cm. The percentage of enclosed area which is cultivated depends on the rainfall regime of the area. CA/C ratio may be at least 3:1. These structures are suitable if ground slope is less than 3 per cent and soil depth, at least 1 m.

There is no hard and fast rule, whether it should be a circular hoop. It may be trapezoidal (Figure 31.3 top-right) or even V-shaped (Figure 31.3 bottom) depending upon the ease of construction. Although in few text book, each shape is treated as a separate technique, we refrain from doing so because the basic principle for all these stucures is same.

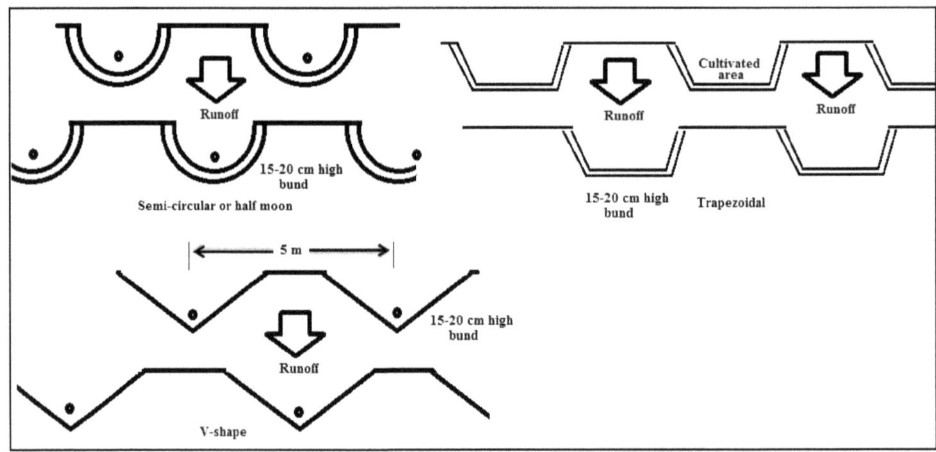

Figure 31.3: Semi-Circular, Trapezoidal and V-Shaped Hoops/*Bunds.*

Graded Bunds

Graded bunds are also referred as off contour bunds or modified trapezoidal bund (the term is explained later in this section). They are either earthen or stone embankments constructed on lands having slope in the range of 0.5 to 2 per cent. The design and construction of graded bunds is quite different than that of a contour bund. They are used where rainfall intensity and soils are such that the runoff water discharged from the field can be easily intercepted. The excess intercepted or harvested water is diverted to the next field though a channel or to a water harvesting structure through the grassed waterway. The height of the graded bund ranges from 0.3 to 0.6 m. The downstream *bunds* consist of wings to intercept the overflowing water from the upstream bunds. Due to this, the configuration of the graded bund looks like an open ended trapezoidal bund.

Rock Catchment

In this case, the exposed rock surfaces make the catchment area. The runoff is collected in a nearby depressed area. Since the runoff potential of a rock catchment is quite high, even small sized catchments say 100 m^2 to few 1000 m^2 may yield enough water. The dimensions of the storage tank are accordingly designed on the basis of anticipated runoff. The water collected in the tank can be used for domestic use or irrigation purposes.

Roaded Catchment

In this method, a large area of ground known as roaded catchment is used to harvest rainwater. A roaded catchment is a water harvesting structure designed to increase the amount of runoff from the catchment. The 'roads' of a roaded catchment are parallel ridges of earth with side slopes that directs runoff to troughs or channels. The surface is lined with clay and compacted to make it smooth to reduce infiltration and increase runoff. It is why, this process is also called as runoff inducement process since the ground is cleared from vegetation and compacted well (to increase or induce runoff). Sometimes, channels are also compacted or lined with gravel to reduce the seepage loss.

Off-Site Rainwater Harvesting

Many times, catchments produce high volumes of runoff that cannot be stored in the soil profile. In such circumstances, the harvested water is stored in dams or ponds located on or around the site or off-site. It is why, these methods are also known as long-term rainwater harvesting. In this case, water is harvested from a large (macro) catchment, transported/diverted to a storage structure for its use later in the season. The catchment areas in these system range from 0.1 ha to thousands of hectares either located near the cropped basin or long distances away. In this system of water harvesting, catchment, conveyance and storage all play an important role. In general, it relies on constructing water storages of appropriate capacity for irrigation, fish farming, electricity generation *etc.* Besides, actions are taken to ensure that water losses due to evaporation and seepage are minimal. Suitable methods for efficient utilization of the harvested water for maximizing crop yield per unit volume of available water are also put in place. Mainly 2 kinds of structures are used:

- Dugout ponds
- Embankment type reservoirs

Dugout Ponds

As the name suggests, dugout or farm ponds are constructed by excavating (digging) the earth from the pre-decided area, normally the lowest portion of the farm. The structures are generally used where embankment type ponds are not technically or economically feasible. The farm ponds are constructed as multipurpose structures as the stored water is used for irrigation, live-stock, water supply needs at the farm and aquaculture *etc.* Therefore, the pond capacity is decided on the basis of water requirement for all these purposes. These ponds are most suited to areas having land slope less than 4 per cent and where water table lies at a shallow depth, as it helps to minimize seepage losses. If water table is deep, the ponds are lined to control seepage losses. The ponds may be fed by surface runoff or groundwater or both. The economic value of the pond improves, if the harvested water helps to mitigate drought situation or in increasing crop production through life saving irrigation. In order to minimize earth work to reduce the construction cost, a depression on the farm is converted into the farm pond. To minimize evaporation losses, for the same capacity of the structure, depth is maximized while surface area is minimized. The major parts of a dugout pond are shown in Figure 31.4. For details, readers may refer chapter 32 of this book.

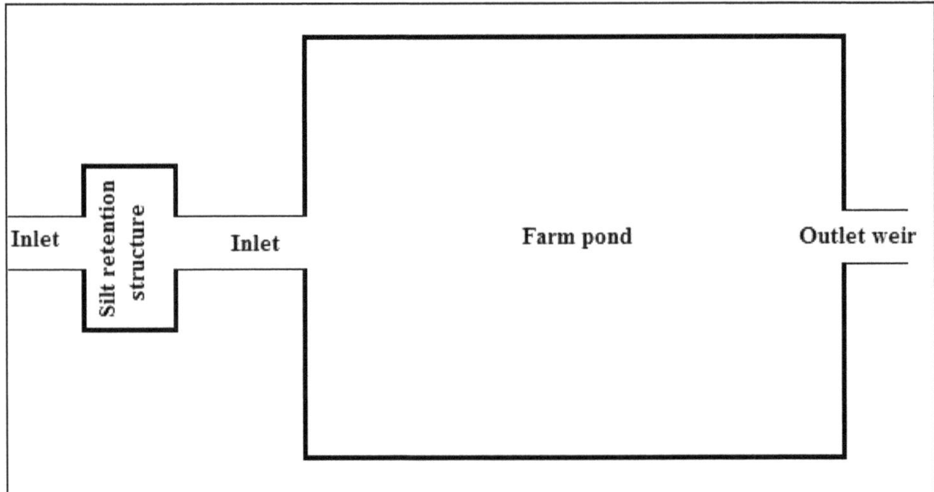

Figure 31.4: Schematic Diagram of a Farm Pond.

Embankment Type Reservoir

These types of reservoirs are constructed by forming a dam or embankment on the valley or depression of the catchment area. The runoff water is collected into this reservoir and is used as per requirement. The storage capacity of the reservoir is determined on the basis of water requirement for various purposes and on the basis of expected surface runoff from the catchment. If there is heavy demand of water, then the storage capacity of the reservoir must be kept sufficient to meet the projected demand for more than a year. Embankment type reservoirs are classified according to the purpose for which they are designed.

Irrigation Dam

The irrigation dams are mainly meant to store the water for irrigation. The capacity is decided on the basis of amount of water available from the catchment and water required for irrigation. A gated pipe spillway is used to take out water for irrigation. The spillway is located near to the bottom of the dam so that some minimum dead storage remains in the pond.

Silt Detention Dam

The basic purpose of a silt detention dam is to detain the silt load of the catchment area coming along with the runoff water and simultaneously harvest the water. The silt laden water stored in the depressed part of the catchment deposits the silt at the bottom and comparatively silt free water is diverted for use. Such dams are located at the lower reaches of the catchment where water enters the valley and silt free water is finally released into the streams. Provision of a suitable outlet is made to take out the water for irrigation. Construction of a series of such dams along the slope of the catchment has given good results.

Sand Dam

The dam is constructed across the valley or depression to reduce the velocity of surface water mixed with sand. As a result of reduced velocity, sand tends to deposit over the bed. After some time the elevation of the dam is raised and the space behind the dam is filled with sand known as sand reservoir. Water flowing in the valley fills the pore spaces of the sand reservoir and thus, creating a water body. The evaporation losses in such dams are much less than from the open surface water bodies.

High Level Pond

Such dams are located at the head of the valley in the shape of a water tank or pond. The stored water in the pond is used to irrigate the area lying downstream. A series of ponds can be constructed in such a way that the command area of the tank located upstream forms the catchment area for the downstream tank.

Percolation Dam or Percolation Pond

A percolation pond, like an irrigation tank, is a structure to impound runoff generated in the watershed. A waste weir to dispose of the surplus flow in excess of the storage capacity of the pond is provided at a suitable location. These dams are generally constructed at the valley head, submerging highly permeable soils in its reservoir so that surface runoff percolates to recharge the groundwater storage. Thus, a large portion of the runoff either goes to the groundwater or is stored in the soil. The recharged area downstream should have sufficient number of wells and cultivable land so that crops can be grown as a result of moist conditions created by the seepage water or by irrigation provided through pumping the recharged water using wells/tube wells.

Flood Water Harvesting

Flood water harvesting is defined as the collection and storage of flood water. The differences between the runoff and flood water harvesting are illustrated in Table 31.2. Flood water harvesting is achieved either by 'flood water harvesting within stream bed', or through 'flood water diversion', In the later case, the water is made to leave its natural course and conveyed to nearby cropped fields. Flood water harvesting is known by different names such as 'large catchment water harvesting' or 'Spate Irrigation'. In the first case, the water flow is dammed and, as a result, water inundates the valley bottom of the flood plain. The water is forced to infiltrate and the wetted area is used for agriculture or pasture improvement. In the second case, the water is forced to leave its natural course and conveyed to nearby cropping areas.

Table 31.2: Differences between Rainwater and Flood Water Harvesting

Rainwater Harvesting	*Flood Water Harvesting*
Usually involves harvesting from micro-catchments or macro-catchments	It refers to harvesting of water from large catchment being many km^2 in size
Water harvesting techniques which harvest runoff from roofs or ground surfaces fall under the term rainwater harvesting	Systems which collect discharges from watercourses are grouped under the term flood water harvesting
Runoff harvesting increases water availability for on-site multiple uses	Flood water harvesting provide a valuable source of water to local as well as downstream water users. It play an important role in replenishing floodplains, rivers, wetlands and groundwater
Runoff harvesting reduces water flow velocity, as well as erosion rate and controls siltation problem	Floodwater enters into the fields through the inundation canals, carrying not only rich silt but also fish which can swim through the canals into the lakes and tanks to provide additional source of livelihood
These are simple structures and can be handled at the local level even with limited knowledge	Relatively high technical input is required since it includes more complex structure of dams and distribution networks

In the second approach some of other ways and means used are described in the following sections.

Graded Bunds

Graded bunds are adopted in areas having more than 800 mm rainfall and soils having low infiltration rates (< 8 mm/hr). Graded bunds are laid along pre-determined longitudinal grade instead of along the contours for safe disposal of excess runoff, which can be taken to a storage site. Gradient given to bunds may vary from 0.4 to 0.8 per cent for light and heavy soils respectively.

Check Dams

A check dam is generally constructed on small streams. Check dams of rock, brick, concrete or other suitable materials are constructed across the stream to check the flow. The height of the check dam is such that even during the highest flood, water does not spill over the banks. It helps to increase the water infiltration from the channel bed to recharge the groundwater. Water stored in the aquifers can be exploited by the farmers of the surrounding areas or even used through surface irrigation for which appropriate arrangements are made. The system does not intercept the peak flows, and the excess water is discharged over the check dam. Where discharge is more, a number of check dams in succession can be constructed at appropriate distances downstream. The spacing between two check dams in such cases should be beyond their water spread.

Besides these, other structures commonly used to harvest flood water are the Gabion structures, cement plug and *nala bunds.*

Flood Control Reservoirs

The reservoirs constructed to control the flood are known as flood control reservoirs. These are multi-purpose reservoirs. Besides flood control, water stored is used for irrigation, aquaculture or for generating hydro-power.

Storage of Flood Water in Aquifers

Storing water in aquifer during times of flood and recovering it later during times of drought provides a cost-effective solution not only for flood management but even for flood water harvesting. Large volumes of water can be stored reducing or altogether eliminating the need to construct expensive surface reservoirs. The aquifers that have experienced long-term decline in water levels due to heavy over-pumping can be used as storage zones. Recharge wells, percolation reservoirs and similar other techniques can be used for artificial groundwater recharge of the flood water.

Groundwater Harvesting

Subsurface Dams

Groundwater harvesting is a rather new term and employed to cover traditional as well as unconventional ways of groundwater extraction. *Qanat* systems, underground dams and some kind of special wells are a few examples of the groundwater harvesting techniques. Groundwater dams or 'Subsurface Dams' is a fine example of groundwater harvesting. Subsurace or groundwater dam is a structures that intercept or obstruct the ephemeral streams in a river bed and provide storage for water underground. They have been used in several parts of the world, notably India, Africa and Brazil.

Subsurface dams are typically used in arid and semi-arid areas where riverbeds are often dry for a portion of the year. But even at this time, subsurface flow may occur, which may often continue throughout the year. To block the flow, a low-permeability barrier (*e.g.* concrete) is inserted into the ground across the riverbed. It can be extended above the surface to bring the water to surface, if needed (Figure 31.5). But more often, the water is stored underground and a water extraction structure (a well) is put on the upstream side of the dam to enable access to water. Such an arrangement minimizes the evaporation losses. Besides, the water in

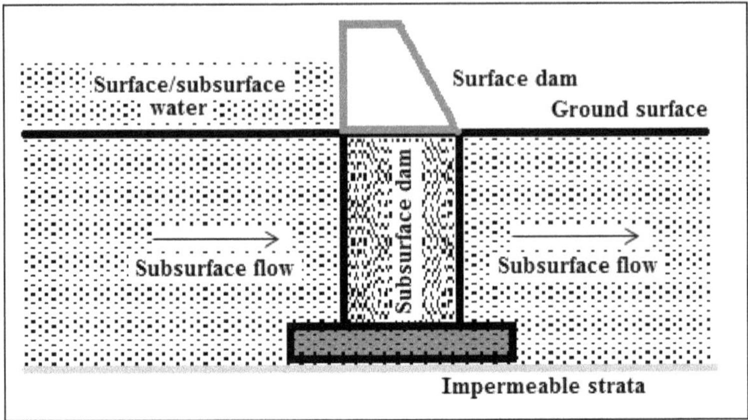

Figure 31.5: Surface-cum-Subsurface Dam in a Stream Course.

such a set-up is of superior quality with no chances of vector/parasite breeding. Groundwater dams cannot be universally applied as these require specific conditions for their efficient functioning. The best sites for construction of groundwater dams are where the soil consists of sands and gravel, with rock or a permeable layer at a depth of a few meters. Ideally the dam should be built where rainwater from a large catchment area flows through a narrow passage. The Central Ground Water Board has sited and constructed a number of subsurface dams in Kerala and now many of these are being constructed in Kutch district of Gujarat.

Rooftop Rainwater Harvesting

The groundwater available in urban localities and metropolis is unable to cope up with the ever increasing demand. It has resulted in the over-exploitation of groundwater with inadequate replenishment as a result of increasing urbanization. Percolation of water to recharge groundwater is meager/insignificant because of pavement of roads, streets and roofs. Therefore, rooftop rainwater harvesting method is use to conserve rainwater and recharge the groundwater aquifer. Roof top rainwater harvesting is a system of catching the rainwater at the place where it falls. Herein, the roof of the building is the catchment from where rainwater is collected. Thereafter, it can either be stored in a tank for domestic use or used for irrigation or diverted to any artificial recharge system. This method is less expensive and very effective. Besides, it helps to augment the groundwater level of the area and improves its quality in the long run.

Components of the Rooftop Rainwater Harvesting

The basic components of a roof top rainwater harvesting system are illustrated through the typical schematic diagram (Figure 31.6). The main components are:

- ☆ Catchment (roof)
- ☆ Gutters and downspouts for transportation
- ☆ First flush

☆ Filter
☆ Storage
☆ Delivery and treatment (Not shown in Figure 31.6)

Catchment

The catchment of rainwater harvesting system is usually a building having flat RCC/stone or sloping roof. Depending upon the kind of material of the sloping roof, the runoff coefficients are used (Table 31.3). Sometimes, these values are assumed as 0.8 and 0.95 for flat and sloping roofs. The water harvested per storm is than given as:

$$\text{Water for harvesting (m}^3\text{)} = \text{Runoff coefficient} \times \text{Roof area (m}^2\text{)} \times \text{rainfall (mm)}/1000 \quad (31.3)$$

For seasonal and annual water harvested, appropriate efficiency factors are used depending upon the total annual rainfall, number of rainy days, intensity and duration of rainfall, temperature and other weather characteristics.

Table 31.3: Runoff Coefficients for Sloping Roofs as Affected by Roof Material

Type of the Roof	*Runoff Coefficient*
GI sheet	0.90
Asbestos	0.80
Tile	0.75
Concrete	0.70
Brick pavements	0.50 – 0.60

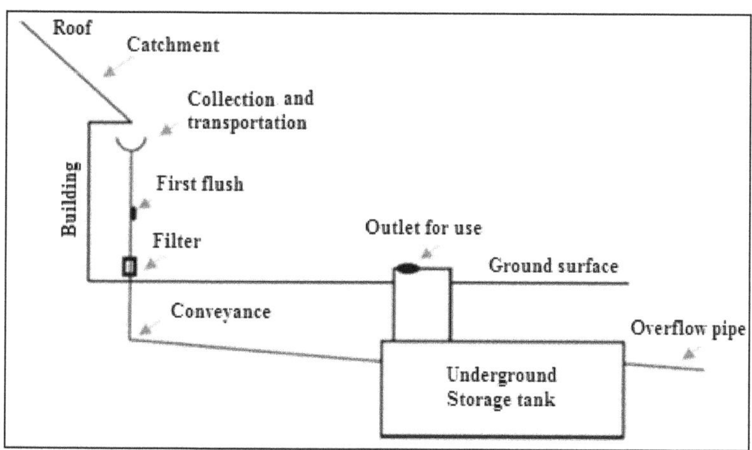

Figure 31.6: Components of a Roof Top Rainwater Harvesting System.

Transportation

The harvested rainwater is collected through a gutter, which is made of different materials and shapes. Usually a plastic pipe cut into two (lengthwise) is used as the gutter. To allow for proper drainage in gutters, a slope approximately 0.2 cm drop per meter toward the downspout is provided to the gutter. At mouth of the each drain, a wire mesh is provided to restrict the entry of floating material that may cause choking of the system. From thereon, rainwater is carried down through water pipes and/or drains to storage/harvesting system. A filter is provided before the water is directed to the storage tank.

First Flush

First flush is a device used to flush off the water received during in first shower. The first shower of rains is flushed to avoid contamination of the storable/ rechargeable water by the probable contaminants of the atmosphere and the roof catchment. Besides, the first flush may also contain silt and other materials deposited on roof during dry season. Provision of first flush valve is made at outlet of each drainpipe. As a guideline 0.2 mm to 2 mm of the runoff should be let off as first flush depending on the quality of water.

Filter

In the systems that are intended for non-potable uses such as irrigation and toilet flushing, screens and first flush diverters are enough to treat the water thereby reducing the cost of the system. On the other hand, for potable use, the collected rainwater requires treatment and disinfection to meet the drinking water standards. Some skepticism is being voiced on the possibility of groundwater contamination if harvested water is used for recharging the aquifers. There is remote possibility of this fear coming true, if proper filtering mechanism is in place.

Storage/Delivery

If the water is used for domestic purposes, it needs to be stored in some storage tank and appropriate arrangements have to be made to draw the water from storage. Leak proof concrete or PVC storage tanks are commonly used to store the harvested water. A pictorial view of storage of roof top harvested rainwater in a *tanka* in Rajasthan is shown in Figure 31.7.

Groundwater Recharge

If the harvested water is not required for domestic use or for irrigation of lawns *etc.*, then it can be used for artificial groundwater recharge. The following structures are used for this purpose.

- ✰ Abandoned dug wells
- ✰ Abandoned hand pumps
- ✰ Recharge pits
- ✰ Recharge trenches
- ✰ Recharge tube wells
- ✰ Recharge shafts

Figure 31.7: The Roof Top Rainwater Harvesting and Storage in a *Tanka*.

It is not an exhaustive list and many local solutions for recharge have been designed and used. Although the scope of this chapter does not include discussions on these structures but for an individual house of < 100 m^2 roof area, a recharge pit of 1 × 1 × 1 m may suffice.

Chapter 32

Farm Ponds

Efficient rainwater management is necessary to improve land and water productivity and protect the natural resource base in regions practicing rainfed agriculture, constituting about 55 per cent of net sown area of India. Water harvesting, from short duration high intense rainfall events, has emerged as one of the key components of successful rainfed farming in arid and semi-arid regions. One of the most successful technologies is to harvesting surplus runoff in dug out ponds and recycle the water for supplemental irrigation to *kharif* crops or pre-sowing irrigation to *rabi* crops besides using the water for the cattle and/or aquaculture activities *etc.* Besides this, it has now emerged that these ponds can play a major role in resolving the drainage problem. If constructed in large numbers, these ponds may prove to be a boon to tackle seasonal flooding commonly encountered in rural and urban centers. The design and construction of farm ponds require a thorough knowledge of the site conditions, watershed characteristics and water requirement at the farm. This chapter describes various issues related to this technology such as types of ponds, their design and the need for lining *etc.*

Types of Ponds

A farm Pond is a dug out structure with definite shape and size having proper inlet and outlet structures for collecting the surface runoff flowing from the farm area. Depending on the source of water and location of the pond with respect to the land surface, ponds are grouped into following three categories and several sub-categories.

- ✰ Dugout ponds
    - ❖ Surface ponds
    - ❖ Spring or creek fed ponds
    - ❖ Groundwater pond

- ✰ Embankment type ponds
  - ❖ On-stream ponds
  - ❖ Off-stream storage ponds
- ✰ Seepage ponds (interception of seepage).

Most common types of farm ponds are the surface water dugout ponds (Figure 32.1 top). These excavated ponds are either fed by surface runoff or springs or creeks or groundwater and are sub-grouped accordingly. While a farm pond can be constructed on any location, is preferred for the construction. Spring or creeks fed ponds are normally constructed in hilly areas. Their construction is subject to the availability of natural springs or creeks. Wherever aquifers are available and is the

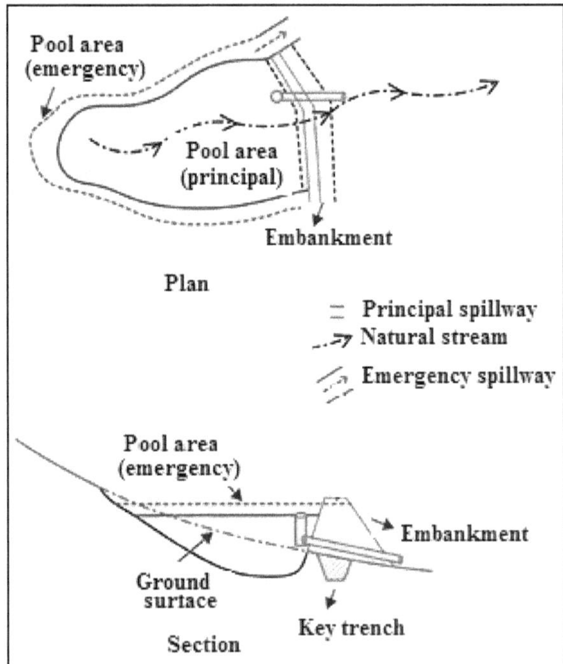

Figure 32.1: A Small Dugout Farm Pond (Top) and an Embankment Pond (Plan Middle) and Section (Bottom).

source of water for the pond, the pond is characterized as groundwater fed pond. Although not a groundwater pond in the real sense, a pond constructed to intercept flowing water through the soil is also designated as a groundwater pond. These ponds are constructed where groundwater flowing from a hilly area emerges at or near the soil surface. The pond constructed along the route intercepts and stores the flowing groundwater. Similarly, ponds constructed along the canal network intercept and store the seepage water to minimize adverse environmental impacts of surface irrigation. The dugout ponds may be fully or partly excavated. In the later case, an embankment may be constructed to increase the storage capacity. For the purpose of classification, even such dugout ponds are classified as embankment-type ponds, if the depth of water impounded against the embankment is about 1m.

Embankment type ponds are constructed across a stream or water course (Figure 32.1 middle and bottom). These are usually built in areas where land slopes range from gentle to moderately steep. Depending upon the storage site, these are categorized as on-stream or off-stream ponds. Water in the former is stored right in the stream bed while in the later; pond is constructed on the side of the ephemeral (seasonal) stream. In the former a site where steam valley is sufficiently depressed is chosen so that maximum storage volume is obtained with least earthwork. In the later case, the seasonal flow in the streams is stored. Suitable arrangements have to be made to convey the water from the stream to the storage site.

Seepage pond is a recent development where groundwater flowing from upstream or even from canal network is intercepted to use it for irrigation or aquaculture. Besides a source of water, such ponds are known to resolve the problems of water logging and soil salinity in close proximity of the canal network.

Advantages of Farm Ponds

- Farm ponds are sources of water to meet multifarious water needs at the farm
- The ponds provide water to sow crops without waiting for rain to occur as well as provide life saving irrigation during rain failure or dry spells between rainfalls. These interventions results in higher sustainable yield
- Pond *bunds* can be used to grow vegetables and fruit trees ensuring household nutritional security and be an additional source of income
- Enables the farmers to apply adequate farm inputs, perform farming operations at the appropriate time and drain excess water resulting in increased productivity
- Helps to check soil erosion and minimizes siltation of waterways and reservoirs
- Promotes aquaculture resulting in higher income
- Source of groundwater recharge
- Over the years silt brought into the pond can be used to enrich soil in the fields, raising farm levels and for the construction of farm roads.

Limitations of Farm Ponds

- ☆ Initial investment is high besides being a labour intensive activity. Government subsidy can significantly reduce these investments
- ☆ The ponds may reduce the water flow to other people's water resources situated on the down stream
- ☆ They occupy a large portion of farmers' lands. However, it may not be a serious limitation as the ponds can be used for aquaculture and *bunds* for growing vegetables and/or tree plantations, *etc.*

Components of a Farm Pond

A typical farm pond consists of the catchment, the storage area, earthen dam, mechanical spillway and an emergency spillway. The mechanical spillway also known as principal spillway maintains the pond water at a desired level (Figure 32.1). Whenever water enters the pond and the level rises, water first exits the pond through this spillway. They are mostly made of pipe or any type of concrete structure (Figure 32.2). These also serve as an outlet for taking out the water for irrigation. The emergency spillway is provided to safeguard the earthen dam. Whenever inflows exceed the designed value and principal spillway is unable to take care of the flow, then the emergency spillway provides a stable pathway for the excess water to flow around and not over the top of the dam (Figure 32.1 middle).

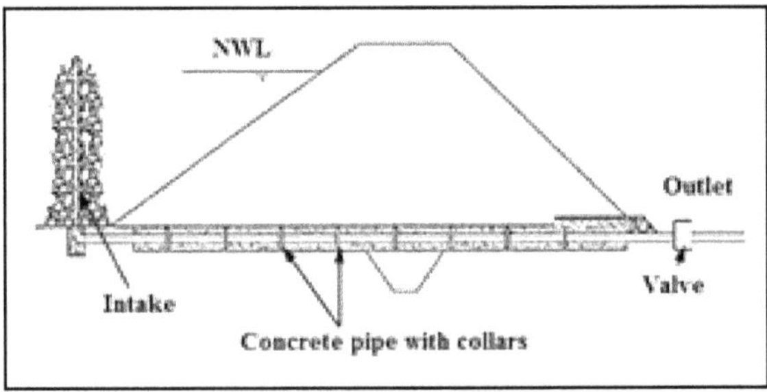

Figure 32.2: A Principal Spillway Serving as an Outlet for Irrigation.

Design of Farm (Dugout) Pond

A well designed pond is a valuable asset to improve crop productivity and to introduce integrated farming system to increase farmer's income. The design of a farm pond consists of:

- ☆ Selection of site
- ☆ Assessment of the storage capacity of the pond
- ☆ Design of the embankment
- ☆ Design of the mechanical spillway
- ☆ Design of the emergency spillway
- ☆ Pond lining: need and materials for seepage control

Traits of a Good Pond Site

The selection of a suitable pond site should begin with preliminary studies of possible sites and topographical survey. The slope of the land and the slope's direction must be carefully evaluated. Where more than one site is available, each should be studied with a view to select the one that is most practical and economical. The selection of a site for a farm pond is critical in maximizing its storage capacity so as to minimize construction cost. The site for the pond is selected keeping in view the following guiding points.

- ☆ The site should be easily accessible. It should be near to farmland so that it can be used for various activities including protective irrigation.
- ☆ The pond must be located at a site where all the rainwater from the catchment converges. The topographical survey and a contour map of the farm helps to identify the most appropriate site. For small catchments of 1-5 ha land, a reconnaissance may suffice to identify the location.
- ☆ A farm pond should be so located that it does not hinder other farm operations. It must be located at least 3 m away from other farmers' fields or common lands.
- ☆ Sufficient catchment area on the upstream side of the pond should be available as per the capacity requirement.
- ☆ Limited geological investigations should be made to see that the soil strata up to 3 m depth are not in the form of disintegrated rock or hard rock. It may pose several problems during digging. See that strata are suitable for the foundation of the dam.
- ☆ From economic point of view, the site should be such that largest storage volume is available with the least amount of earth excavation/fill. A narrow section of the valley with steep side slopes is preferable.
- ☆ Large areas with shallow water table should be avoided as these will favour water weeds to grow and result in excessive evaporation losses.
- ☆ The site conditions should not cause excessive seepage losses so that water can be stored for a longer period. It will also obviate the investments on lining.
- ☆ Where water must be conveyed for use elsewhere, such as for irrigation, pond should be located as close to the point of use as is practical. Moreover, water should flow to all areas proposed to be irrigated through gravity.
- ☆ The site selected should be away from pollution sources. The site should be far away from farmsteads, feeding lots, sewage lines, mine dumps and similar areas so that drainage water from these sites does not reach the pond.
- ☆ A natural waterway should be present so that it can be used as an emergency spillway.
- ☆ To take the benefit of groundwater recharging from the pond, the farm pond should be on the upstream side of a well on the farm.

Capacity of the Pond

Capacity of the pond is designed on the basis of the following three criteria.

Capacity of the pond is decided on the basis of water demand for various activities on the farm. It is assessed by summing up the water requirements of different activities such as household requirement, animal rearing, aquaculture and irrigation. As an illustration, if the structure being designed to provide protective irrigation to the *kharif* crops, 1000 m^3 of storage will normally suffice for 1 ha area. For aquaculture, 1.5 to 1.8 m water depth must be available during the period when fishes are in the pond. It ensures proper temperature and environment to the fishes. Besides, storage losses must be added to the capacity so calculated. In many cases, losses may constitute huge volume of the water stored.

The capacity of the pond is further governed by the catchment characteristics such as its area, type of vegetation, land use, soil, topography, *etc.* The rainfall amount, duration and intensity along with rainfall-runoff relation will determine the size of catchment required to harvest the required amount of water. Although subject to these conditions, at least 3-4 ha catchment may be required to harvest about 1000 m^3 of water in a semiarid region. Clearly, either the catchment area required to generate the water required should be available or the capacity of the pond should be reduced as per the catchment area. If need be, catchment may be treated to induce runoff. In a embankment type of pond, the design storage capacity is decided on the basis of available surface runoff for a pre-decided return period. Frequency analysis procedures are used to assess the value.

Size of a pond is usually dictated by the availability of adequate land in the vicinity of the site. In many cases, we may not have the option to design and build a pond of a desired size to meet the water requirements of the community. For irrigation purposes, normally 5-20 per cent (average 10 per cent) of the land can be allocated for the construction of the farm pond.

The design capacity is the least capacity calculated by these procedures. For example, if the farm pond is served by a smaller catchment than the desired area then it may not have sufficient runoff and will be prone to the risk of being dried during dry spell. Similarly, pond with large catchment may dispose of water through the emergency spillway but will require large water control structure and will be difficult to manage. Generally the area under dug out farm pond on an average should be about 10 per cent (Range from 5-20 per cent) of the farm catchment.

Volume of Storage

The volume of water that can be stored in the pond is determined from a contour survey of the site at which the pond is constructed. From the contour plan of the site the capacity for different depth stages is calculated using the trapezoidal or Simpson's rule discussed in Chapter 4. For a trapezoidal dugout pond, the capacity can be calculated knowing the dimensions and side slopes of the pond. The volume of excavation required can also be estimated with sufficient accuracy with the help of prismoidal formula, which is being reproduced from Chapter 4.

$$V = (A+4B+C) \times D/6 \qquad (32.1)$$

Here V is the volume of excavation, m^3, A is the area of the excavation at the ground surface, m^2, B is the area of the excavation at the mid-depth point, m^2, and C is the area of the excavation at the bottom of the pond, m^2 and D is the average depth of the pond, m.

Example 32.1

Calculate the storage capacity of the pond 4 m deep having the bottom width of 12 m, and a bottom length of 35 m. The side slope may be assumed as 2:1 (z:1).

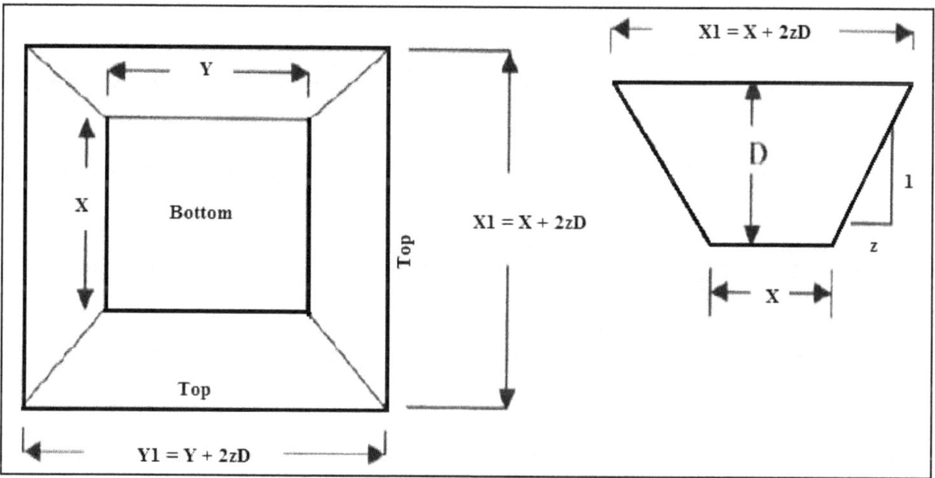

Figure 32.3: The Schematic Diagram of a Trapezoidal Farm Pond: Pan (Left) and Section (Right).

From the schematic diagram of the trapezoidal pond (Figure 32.3), the top length and width are calculated as follows.

Top length = 4 (2) + 4(2) + 35 = 51m

Top width =4(2)+ 4 (2)+ 12 = 28 m

$A = 51 \times 28 = 1428\ m^2$

Mid-length = 2(2) + 2 (2) + 35 = 43 m

Mid-width = 2(2) + 2(2) + 12 = 20m

$4B = 4(43 \times 20) = 3440\ m^2$

$C = 12 \times 35= 420\ m^2$

Therefore, $(A + 4B + C) = 1428+ 3440+ 420 = 5288\ m^2$

Then $V = 5288 \times 4/6 = 3{,}535\ m^3$

Assuming that the normal water level in the pond is at the ground surface, the volume of water that can be stored in the pond is ~3500 m^3.

Example 32.2

A dugout pond is to be constructed in a semi-arid area having red soils. It is proposed to provide two supplemental irrigations of 8 cm depth (including losses) to an area of 1.5 ha. Design the pond dimensions.

Water required for two irrigations is given as

$1.5 \times 8 = 12.0$ ha-cm

Assuming 25 per cent of storage losses (evaporation and seepage),

Losses = $12 \times 0.25 = 3$ ha-cm

Designed capacity of the pond = 12.0 + 3.0 = 15.0 ha-cm = 1500 cu-m

It is presumed that the pond will have sufficient watershed that is capable to fill the pond.

Assuming

Depth of pond = 4.5 m

Side slopes = 1:1

Bottom width = 10 m

Shape = Trapezoidal

Length = ?

Top length = $L + (4.5 \times 1)\, 2 = L+9$ m

Top width = $10 + (4.5 \times 1)\, 2 = 19$ m

Area of pond at top (A2) = $(L+9) \times 19\ m^2$

Mid-length = $L + (2.25 \times 1)2 = L + 4.5$ m

Mid-width = $10 + (2.25 \times 1)2 = 14.5$ m

Area of the pond at d/2 depth below the top of pond (A1) = $(L+4.5) \times 14.5\ m^2$

Area of pond at bottom (A0) = $L \times 10\ m^2$

Using eq. (32.1), Volume (V) is given as

$V = (D/6)(A + 4C + B)$

$V = (4.5/6) \times (10\,L + 4 \times 14.5\,(L+4.5) + (L+9) \times 19) = 1500\ m^3$

$10\,L + 58L \times 19L + 263.25 + 171 = 1500 \times 6/4.5 = 2000$

$87\,L = 1565.75$

$L = 18$ m

Shape

The excavated ponds can be of any shape but most common ones are (a) square, (b) rectangular, circular (Figure 32.4) and c) inverted cone. Since curved shapes are difficult to construct, ponds are either square or rectangular. For example, ponds shaped as inverted cone with circular cross section may be theoretically cheaper, yet are difficult to construct and manage. Besides, lining of a farm pond having inverted

Figure 32.4: A Circular Pond in Meghalaya for Rice-Fish Culture.

cone shape requires more lining material compared to a square or a rectangular farm pond of the same capacity. It will increase the cost of curved shaped farm ponds.

Depth and Side Slope

The depth of a farm pond is normally decided considering soil depth, soil type and equipment used for excavation. Though, the evaporation loss component can be minimized by increasing the depth but, from practical point of view the ideal depth is limited to 3 to 3.5 m. Any depth beyond 4.0 m will be uneconomical, if excavated manually. This depth also suits the requirement of aquaculture and irrigation. While water in the upper 1.5 m depth may be used for irrigation, remaining continues to be in store for aquaculture. An appropriate freeboard to prevent water from flowing into nearby fields during flash floods must be provided. It refers to the difference between the base of the waste weir and the top of the dugout portion of the structure.

Side slopes of the pond should not be steeper than the natural angle of repose of the excavated material. Since the angle of repose is less for wet than dry soils, flatter gradient is kept on the waterside of the earth fill. Flatter side slope helps to avoid slippage due to saturation particularly in unlined pond. For small ponds, uniform slopes on both upstream and downstream sides are provided as it is easier to manage during construct. The recommended side slope of farm pond for different soils vary from 1:1 to 2:1 for clay to 3:1 for sandy soils.

Once the depth, side slopes and capacity of the pond are known, surface area can be calculated depending upon the shape and dimensions of the plot available for the construction of farm pond.

Inlet and Outlet Channels

The inlet channel to the farm pond ensures that the surface runoff from the catchment enters into the pond. The inlet is designed as a chute spillway so as to divert the runoff into the pond in a controlled manner (Figure 32.1). The inlet channel is designed to have safe water velocity to avoid channel erosion. It could be designed as a grassed waterway. The inlet channel can also be lined with boulders or stone pitching can be adopted to make it safe against erosion. The entry section is designed as a rectangular broad crested weir (0.3 to 0.5 m depth of flow and 1 to 1.5 m wide). A silt trap is constructed before the channel enters the pond It is simply a trench or a boulder check or a brick and mortar structure.

Outlet/Waste Weir

The outlet or waste weir is generally located at the lower end of the pond. It is designed to remove the surplus runoff over and above the designed capacity. The flow through outlet should be maintained at non-erosive velocity to minimize the erosion. The outlet is kept at a slightly lower (up to 30 cm) than the inlet elevation to avoid back water flow. The discharge capacity of the outlet is assumed to be half of the inlet capacity. The emergency spillway is designed using the weir formula

$$Q = CLH^m \tag{32.2}$$

Here Q is the discharge in m^3/s, H is the head on crest, m, L is the length of weir´s crest, m, and m is an exponent whose value is 3/2 for trapezoidal and rectangular weirs. The value of coefficient of discharge C is 0.0186 for trapezoidal and 0.0184 for rectangular channel.

Construction of Farm Pond

Site Clearing, Leveling and Layout

After the site selection, the pond site is cleared of all stones and woody vegetation. Remove all vegetation, roots, and organic matter from the fill area. Scrape the upper 30 cm of the soil to get rid of the excessive roots and all tree stumps all along the area that is to come under the fill. The design layout is then transferred to the ground. Stakes are used to mark the limits of the excavation and areas where spoil is to be placed. The depth of cut from the ground surface to the pond bottom is indicated on the stakes.

Key Trench for Bonding

A 1.5 m wide bottom key trench with 2:1 side slopes is provided to ensure a good bondage of the fill material with the original earth (Figure 32.1 bottom). Lay the earthwork in horizontal layers of not more than 8 cm at a time. Water the layers and use sheep-foot roller for maximum compaction at about 14 per cent moisture content or as determined from soil compaction tests. Place the conduit pipe of the mechanical spillway before starting the earth fill.

Stepping Method of Construction

The step method is used to ease the operation of excavation. In this method step like cross section representing rectangular shape of the bottom is adopted for excavation (Figure 32.5).

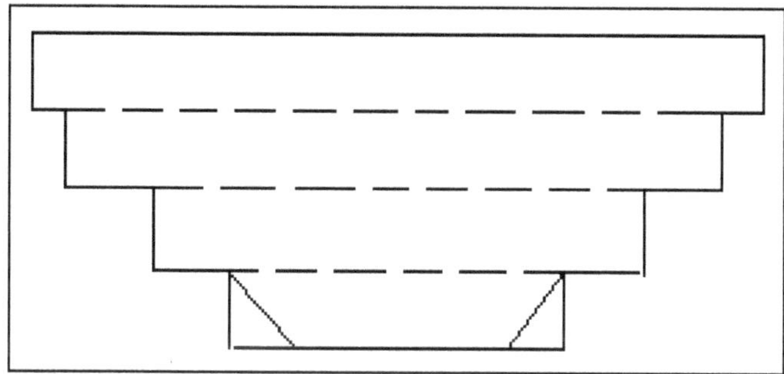

Figure 32.5: Stepping Method of Construction of Farm Pond.

Spoil Bank

Excavated material should be disposed of in a useful manner rather than dumping the material here and there. The best way is to use it for the construction of *bund* around the pond. In some cases farmers prefer to raise the level of their fields using the excess earth provided it has no soil related problems. The excess soil can also be placed as a spoil bank after leaving the space for berm of about 0.5 m along the side length of farm pond.

Silt Trap

It is provided at the beginning before the water enters the pond. It helps to retain the silt as otherwise it may cause sedimentation in the pond requiring frequent desiltation.

Maintenance of Farm Ponds

To ensure efficient service throughout the expected life of a water harvesting structure, a well thought out maintenance plan is needed. One of the key elements is fencing. Fencing must be erected around the farm pond to prevent the entry of wildlife, stray dog *etc.* Barbed wire fencing with stones can be used as it may reduce the cost of fencing. Cost effective vegetative hedges using henna, shallow rooted fruit trees, glyricidia *etc.*, may be planned as a protective live fence. A sign board must be put up near the farm pond indicating that swimming is prohibited and taking animals into the pond is not allowed.

The pond and the nearby area should be inspected periodically especially following heavy downpours. Various components should be inspected and maintenance activity should be initiated as soon as some problem is noticed or is likely to occur. Following major issues are addressed in the short-term and long-term maintenance.

- Maintenance of inlet and outlet
- Maintenance of shoulder bunds
- Cleaning of silt trap
- Maintenance of depth gauge

- Control of water pollution
- Control of aquatic weed growth
- Desiltation as and when required

Construction of Embankment Pond

Embankment is the most important component of an embankment pond. This section deals with various components of an earthen embankment but mainly emphasizes on the construction of embankment (Figure 32.1 middle and bottom).

Foundation including Key Trench or Cut-off

Foundation of an embankment is the key in its structural stability and long life. Although technologies are there through which we can construct a stable earthen *bund* on any foundation, one should avoid sites requiring relatively expansive measures to construct foundation. The most satisfactory foundation can be constructed when the soil profile is underlain by a thick layer of relatively impervious consolidated material at a shallow depth. Such foundations cause very little or no stability issues. It may suffice to remove the top soil (with vegetation and roots) and plough the area to ensure a good bond with the new fill material. Where the impervious layer is overlain by pervious material (sand), a compacted clay cut-off extending from the surface of the ground into the impervious layer is required to prevent excessive seepage and to prevent possible failure by piping. The most common type of cutoff is one constructed of compacted or puddled clay material. A trench, also called key-trench, is cut parallel to the central line of the bund to a depth that extends well into the impervious layer. The trench should have a bottom width of not less than 1.5 m but adequate to allow the use of mechanical equipment for proper compaction. The sides of the trench should be filled with puddled clay or with successive thin layers of relatively impervious material each layer being properly compacted.

Height of the Embankment

The height of the embankment depends upon the volume of runoff to be stored and topography of the reservoir area. The height of the embankment should be so selected that its cost per unit of storage (cum volume) is minimum. While calculating the cost corresponding to any height, allowance for settlement and free board should be given due consideration. The minimum freeboard depends upon the length of the pond. For lengths up to 400 m, a freeboard of 50 cm is sufficient. It is increased to 75 cm for length of pond up to 800 m and to 100 cm for lengths exceeding 800 m.

Settlement Allowance

This includes the consolidation of the fill and the foundation materials due to the weight of the embankment and high moisture content resulting from the storage of water. While settlement allowance for hand compacted (manually constructed) fill is taken as 10 per cent of the design height, it is 5 per cent of the design height, if the embankment is machine compacted

Top Width of Embankment

Top width of the embankment should be adequate so that it can be used as road way and communication routes for the adjoining villages or watersheds. In general, for heights up to 5 m, a minimum width of 2.5 m is recommended. A simple formulae to design top width (TW) as a function of height (H) may be used.

$$TW = 1.105\ H^{1/2} + 0.91 \quad (32.3)$$

Side Slopes

The side slopes of the earthen dam are decided considering several factors such as height of the dam, kind of foundation material and nature of the fill material. In general, the side slopes on upstream and downstream side are decided on the basis of information given in Table 32.1. To protect the upstream side from the wave action, the slope should be protected either by riprap or by rock pavement. A good vegetative cover should be established on the downward side.

Table 32.1: Guidelines on Side Slopes of Embankments (Dam height 15 m)

Nature of Fill Material	*Upstream Slope*	*Downstream Slope*
Average fill material	3:1	2:1
Coarse fill material	3:1	4:1
Well graded soil	2:1	2:1

Maintenance

A properly designed and constructed embankment well protected by sod requires least maintenance. To establish grass quickly, it may be useful to use the topsoil at the top of the embankment. For this purpose, top soil during construction may be heaped in one corner of the construction site. It is also necessary to apply recommended doses of fertilizers so as to quickly cover the earth fill with grass. Special attention is required to avoid surface erosion, development of seepage areas on the downstream face near the toe of the dam, piping, wave action and damage by cattle and human beings. Do not let trees or bushes come up on the embankment. Corrective steps should be taken in time as per requirement.

Seepage Control

Water losses through seepage, percolation and evaporation may be as high as 60 per cent of the gross water storage. The relative proportion of seepage and percolation to that of evaporation may vary from one agro-climatic condition to another. If the soils of the water impounding area are too permeable to hold water, seepage losses are high. As per the website Agritech.tnau.ac.in,the seepage losses under different soil classes can range from 1.21 to 10.54 m^3/million m^2 of wetted surface area (Table 32.2). A reasonable guess of the anticipated losses can also be made from the average infiltration rates of various soil groups (Table 32.3; NABARD, 2012). In light textured soils, it is necessary to apply some kind of sealing or lining to minimize excessive seepage. It can be achieved through several techniques as described in the following sections.

Table 32.2: Seepage Losses in different Soil Classes

Sl.No.	*Type of Soil*	*Water Loss through Seepage (cumec/million m^2 of wetted surface)*
1	Heavy clay loam	1.2
2	Medium clay loam	2.0
3	Sandy clay loam	2.9
4	Sandy loam	5.1
5	Loose sandy soil	6.0
6	Porous gravelly soil	10.5

Table 32.3: Infiltration Rates as Affected by Soil Texture

Soil Type	*Infiltration Rate (cm/hr)*
Coarse sand	2.0-2.5 (2.2)*
Fine sand	1.2-2.0 (1.5)
Fine sandy loam	1.2
Silty loam	1.0
Clay loam	0.8
Clay	0.5

*Values in parenthesis are average values

Clay Blanket

Seepage and percolation losses from a newly dugout pond are quite high. Under this condition, it may be useful to apply clay blanket either naturally or artificially. In the former, no attempt is made to stop the sediments entry to the pond. The sediments deposits on the pond bed will act as a clay blanket to reduce percolation losses from the pond bed. To achieve this, either the construction of silt trap may be postponed or it may be made non-functional for about a year or so.

The artificial clay blanket is a well-graded material containing at least 20 per cent clay. It can be obtained from a nearby borrow area or from an old village pond. The entire area over which water is to be impounded as well as the upstream slope of the embankment is covered with clay blanket. The thickness of the blanket depends on the depth of water to be impounded, the minimum being 30 cm for all depths of water under 3 m. The thickness is increase by 5 cm for each 30 cm of water over 3 m. The clay blankets should not be allowed to dry to safeguard it against cracking. Blanket should not be subjected to freezing and thawing to save it from rupture. It is useful to spread a 30-40 thick cover of gravel over the blanket below the anticipated water level.

Local alternate solutions are also used to minimize seepage although basic principles remain the same. For example, in Kerala, gunny bags are slit opened, soaked in cement and fixed along the sides and bottom of an excavated pond. It is reported that it completely checked seepage and percolation at a fraction of the cost compared to stone lining.

Compaction

Pond soils that contain a wide range of particle sizes (small gravel or coarse sand to fine sand) and enough clay (10 per cent or more) and silt can be made relatively impervious by compaction alone. The seepage through such a compacted soil will be much less than without compaction. To apply this technology, clear the pond area of all the vegetation. Fill all stump holes, crevices, and similar areas with impervious material. Scarify the soil to a depth of 40-50 cm with a disk, roto-tiller, pulverizer, or any other similar equipment. Roll the loosened soil under optimum moisture conditions in a dense, tight layer with four to six passes of a sheep foot roller in the same manner as is used to compact earth embankments. Make the compacted seal no less than 30 cm thick where less than 3 m of water is to be impounded. Since seepage losses are directly related to the depth of water impounded, increase the thickness of the compacted seal proportionately if the depth of impounding is more.

Application of Bentonite

Adding bentonite, fine-textured colloidal clay, is another way to reduce excessive seepage in soils containing high percentages of coarse-grained particles with little clay. When wet, bentonite absorbs several times its own weight of water and, at complete saturation, swells as much as 8 to 20 times its original volume. Mixed in the correct proportions with well-graded coarse-grained material, thoroughly compacted and then saturated, the particles of bentonite swell until they fill the pores to the point that the mixture is nearly impervious to water. To apply, thoroughly mix the bentonite with the surface soil to a depth that will result in a 15 cm compacted layer. However, bentonite returns to its original volume leaving large cracks upon drying. For this reason, sealing with bentonite is not recommended for ponds in which water level is likely to fluctuate widely.

Chemical Additives

Because of the structure or arrangement of the clay particles, seepage is often excessive in fine-grained clay soils. It happens because these soils are well aggregated. Application of small amounts of certain chemicals, collectively called dispersing agents, to these porous aggregates may result in collapse of the open structure and rearrangement of the clay particles. This dispersed structure reduces soil permeability. To be effective, the soils in the pond area should contain more than 50 per cent fine-grained material (silt and clay) and at least 15 per cent clay. Although many soluble salts are dispersing agents, sodium polyphosphates and sodium carbonate are commonly used. Of the sodium polyphosphates, tetrasodium pyrophosphate and sodium tripolyphosphate are most effective. Mix the dispersing agent with the surface soil and then compact it to form a blanket.

Brick Lining/Concrete Lining

Conventional lining method, like brick, stone and concrete lining, are commonly used to line farm ponds but are too expensive or not sufficiently effective because of poor workmanship of the village masons. Since these are thick materials, allowance for the thickness of these materials needs to be kept in view while digging the ponds so as to meet the design storage requirements. Over time, concrete may crack and has to be fixed with silicone cement. If cement concrete blocks are used, then suitable sealants must be used to minimize seepage losses.

Waterproof Linings

Since the water in the reservoir/pond is stagnant, low density polyethylene (LDPE) film lining with 30-60 cm soil cover is adequate for pond lining. Polyethylene film lining alone or in combination with conventional lining has proved to be a seepage proof barrier between the soil and the water. Depending on the circumstances and soil conditions, single tile lining with LDPE film or double tile lining or LDPE film lining with soil cover can be used on the sides. Polyethylene, vinyl, butyl-rubber membranes, and asphalt-sealed fabric liners are gaining wide acceptance as linings materials for ponds. If properly installed they virtually eliminate seepage. Thin films of these materials are structurally weak, but if not broken or punctured they are almost completely watertight. Black polyethylene films are less expensive and have better aging properties than vinyl. Vinyl, on the other hand, is more resistant to impact damage and is readily seamed and patched with solvent cement. Polyethylene can be joined or patched with special cement. The life of such lining may vary from 15 to 20 years.

The HDPE (High density polyethylene) film and geo-membrane (reinforced HDPE) are also used to line pond. The HDPE and geo-membrane film are used for lining bigger ponds where capacity is more than 1000 m^3. The quality and strength of film for lining depends on the capacity of pond and depth of water impounded. For higher capacity say 200-1000 m^3 with 3 to 3.5 m depth, minimum 500 micron thick film weighing 300-350 g/m^2 is required (Reddy *et al.*, 2012). However, for smaller pond (capacity less than 200 m^3 and depth limited to 2.0 m), 200-250 micron film weighing 200 g/m^2 will serve the purpose. Film roll are available in specific lengths and widths as per requirement of the designed section of the ponds. It helps to avoid wastage and minimize the joints.

Method for Film Lining of Farm Pond

- ☆ Calculate the film requirement for dugout pond. Choose the film as per BIS/ISI mark. Use minimum 500 micron black PE film.
- ☆ Clear the pond area of all undesired vegetation. Fill all holes and remove roots, sharp stones, or other objects that might puncture the film. If the material is stony or very coarse texture, then cover it with a layer of fine-textured material before lining. Sand is commonly used as a cushion layer below the film. The sand shall be of fine quality, passing through 53 mm IS Sieve.
- ☆ Some plants may penetrate both the vinyl and polyethylene films. If such plants are present, then all sub-grades, especially the side slopes, should be sterilized. Several good chemical herbicide or weedicide are commercially available for this purpose.

- ☆ Plastic films are manufactured in panels of standard widths. Therefore, convert the film into a single sheet. Join either mechanically by heat-sealing machine like hot air fusion welding machine or manually (by overlapping 15 cm of the edge of two sheet and scrubbed lightly using emery paper or sand paper (120 grade) and joining using bitumen/Synthetic Rubber adhesive No-998 made by fevicol) so that it exactly fits in to the pond. As far as possible, avoid longitudinal joints.
- ☆ Monitor the film in sun light for puncture holes. If any holes are observed, seal them with bitumen/adhesive or by heat – sealing process. Take care that no wrinkles are formed.
- ☆ To avoid any damage to the film, 7.5 cm layer of earth free from gravel or granular material should be laid and compacted over the film. The cover over the film may also consist of any of the following: (a) *In-situ* cement concrete M10 of IS: 456 (b) Pre-cast cement concrete tiles confirming to IS: 3860 (c) Stone slabs (d) Brick and (e) Burnt clay tiles conforming to IS: 3367. The excavated earth from adjacent section can be used to cover the film in the previously laid section. Thus, by the time, earth cover is provided in one section, the adjacent section is ready to receive film lining. Rest of the earth cover should be spread over it in 15 cm layers, watered, and compacted by using light roller or manual templates.
- ☆ Extra length of the film shall be placed in trench at top of the embankment and covered with earth. Anchor the lining by burying it in a trench dug around the pond at or above the normal water level. The anchor trench may be 20-30 cm deep and about 30 cm wide.
- ☆ The embankment may be then raised to designed level

Evaporation Control

Another major loss of water from farm ponds in arid and semi-arid regions is through the process of evaporation. It has been reported that under many situations more than half of the water stored in small ponds is evaporated. Several techniques have been devised to reduce evaporation from the water surfaces. These techniques have been categorized as physical and chemical techniques.

Physical

The main focus of physical methods of evaporation control is on reduction in the open area where vaporization occurs and preventing direct rays of sun from heating the water surface. For the same capacity of reservoir, lesser the open area exposed to the atmosphere lesser will be the evaporation. Thus, increasing the depth in relation to open surface area will reduce the evaporation rate. Shading the structure helps to reduce the energy available for evaporation. It reduces wind speed over the water surface and trap humid air under the cover. Construction of wind breaks around the ponds also reduces water evaporation. The effectiveness however, is limited because 25 per cent reduction in wind velocity will reduce the pond evaporation loss only by 5 per cent. Covering the water surface with barriers that prevents vaporization such as a cover of polyethylene, foam and polystyrene also reduces the water evaporation loss. Employing blocks, rafts, and beads capable of floating over the water surface also cuts the energy of the sun to reach the water surface. Besides, problems related to growth of algae and submerged aquatic weeds are also reduced. Out of the materials mentioned, polystyrene raft, plastic sheet and foamed butyl rubber have proved most effective since they are capable to reduce the evaporation loss by more than 90 per cent.Needless to mention, provision of evaporation control should be made in the design itself so that it can be put in place simultaneously with the construction of pond. More complex methods include storage of water in underground and sand-and-rock-filled dams.

Chemical

A non-toxic chemical capable of forming a thin mono-molecular film on the water surface can be used to prevent water evaporation. Fatty alcohols, known as long chain alcohols, of different grades such as Cetyl alcohol or Hesadecanol (C16-H33-OH), Stearyl alcohol or Octadecanol (C18-H37-OH) and Biphenyl alcohol or Docosanol (C22-H45-OH) or a mixture of these compounds when applied over the water surface form a film. National Chemical Laboratory (NCL), Pune has developed one more compound by synthesizing Alkoxy Ethanols. These films help to reduce evaporation from the water surface. To get the desired properties of monolayer, these alcohols should be 99 per cent pure. Moreover, a major problem with these films is that these accumulate on one side in the direction of the wind. It reduces its efficiency. Another point that must be kept in view is that the film so created must allow passage of air through it so that aquatic life in the pond remains safe. Wax has also been used to form a thin layer but so far has been tested at experimental scale.

Cost of Farm Ponds

Cost components of a dugout farm pond include digging, lining and making relevant structures as discussed in this chapter. The digging can be accomplished through manual or mechanical means. Similarly, there are a number of lining materials that can be used. Even rates of various items including labour vary from one location to another. Therefore, It is difficult to assign a good value to the cost of a farm pond. Collation of data reveal that the cost may be around Rs. 150-200 per m^3 of water storage capacity with a silpaulin film of 500 micron for lining (Table 32.4). It may increase to 400-500 per m^3 with concrete lining and Rs. 300-400 per m^3 for brick lining. A pond capable to store 3000 m^3 of rainwater was constructed at a cost of Rs 1 lakh in Ambalavayal in Wayanad (Kerala), costing about Rs. 330 per m^3

Table 32.4: Construction Cost of Lined Farm Ponds using Machines for Digging

Shape and Capacity (m^3)	*Total Cost of Lined Pond (Rs.)*	*Cost per m^3 of Storage (Rs.)*
Square, 741	98,386	133
Square, 1765	1,83,770	104
Square, 489	75,364	154
Inverted cone, 229	54,834	239
Inverted cone, 582	94,252	162

of water storage capacity. The technology employed to line the sides and bottom of the pond comprised of used gunny bags of jute soaked in a cement mixture, which were spread over the surface. The cost per unit of water storage decreases with increasing size of the storage structure having similar construction.

Objective Type Questions

Mark the correct or most appropriate answer out of the given choice (A, B, C and D)

Part I: Engineering Survey

1 **True shape or curvature of earth is taken into account**

A. Plane survey
B. Geodetic survey
C. Both of these
D. None of these

2 **SI unit for angle is**

A. Radians
B. Degree
C. Minutes
D. None of these

3 **Chain survey is not recommended, if**

A. The ground surface is level
B. A small area is involved
C. The area is crowded with many details
D. The formation of well conditioned triangles is easy

4 **An Engineer's chain is**

A. 100 ft long
B. 100 m long
C. 66 ft long
D. None of these

5 **The standard length of a Gunter chain is**

A. 100 ft long
B. 100 m long
C. 66 ft long
D. None of these

6 **A clinometer is used to measure**

A. Distances
B. Areas
C. Angles of ground slope
D. None of these

7 **Erroneous length of chain is classified in**

A. Systematic (cumulative) error
B. Random (compensating) error
C. Personal error
D. None of these

8 **Correction is always subtracted for error due to**

A. Slope or verticality of the line
B. Absolute length of the chain
C. Temperature
D. Tension or pull

9 Line ranger is used for

A. Direct ranging
B. Indirect ranging
C. Both A and B
D. None of these

10 Indirect ranging is used

A. When two points are inter-visible to each other
B. When two points are not inter-visible to each other such as across a valley
C. When two points are not inter-visible to each other such as rising ground between them
D. Both B and C

11 Area of a right angle triangle is given as (b is the base length and h is the length of the side making the right angle with the base)

A. b.h
B. 0.5. b.h
C. b.h^2
D. None of these

12 The perpendicular offsets were taken at 10 m interval from a survey line to an irregular boundary line. The ordinates measured at midpoint of the division are 9, 12, and 18 m. The area enclosed by the midpoint ordinate rule is.

A. 130 m^2
B. 9.75 m^2
C. 390 m^2
D. None of these

13 The perpendicular offsets were taken at 10 m interval from a survey line to an irregular boundary line. The ordinates measured at each of the division are 12, 12, and 18 m. The area enclosed by the average ordinate rule is.

A. 140 m^2
B. 105 m^2
C. 420 m^2
D. None of these

14 In which method of area calculation, boundary between the ordinates is considered as straight

A. Simpson's rule
B. Trapezoidal rule
C. Both A and B
D. None of these

15 In which method of area calculation, number of ordinates should be odd

A. Simpson's rule
B. Trapezoidal rule
C. Both A and B
D. None of these

16 In which method of area calculation, irregular boundary is taken as series of parabolic arcs

A. Simpson's rule
B. Trapezoidal rule
C. Both A and B
D. None of these

17 Which of the following instrument is used to measure the areas

A. Planimeter
B. Hygrometer
C. Anemometer
D. None of these

18 The common type of figures used in a triangulation system are

A. Double chain of triangles
B. Braced quadrilaterals
C. Centered triangles and polygons
D. All or any of these

19 In triangulation, the sum of the interior angles should be (2N-4) × 90°, such that

A. N is the number of sides of the figure
B. N is equal to 3
C. N is equal to number of sides of the figure minus 1
D. None of these

20 When 1 cm on a map represents a relatively small distance, the map is a

A. Large scale map
B. Small scale map
C. Both A and B
D. None of these

21 A map is usually termed as large scale map if the representative fraction (RF) is

A. More than 1/500
B. Less than 1/500
C. More than 1/10000
D. None of these

22 A perpendicular offset can be drawn by using

A. Setting a right angle in the ratio of 3:4:5
B. Swinging tape for the object to the chain line
C. Cross staff or optical square
D. Any of A, B and C

23 Plane table survey is used where

A. High accuracy is required
B. High accuracy is not required
C. There is no such limitation of accuracy
D. None of these

24 Orientation of plane table is achieved through

A. Keeping the plane table always parallel to the position occupied at the first station
B. Keeping the plane table always perpendicular to the position occupied at the first station
C. In any position A or B
D. None of these

25 Approaches to conduct plane table survey include

A. Radiation method B. Intersection method

C. Traverse method D. All of these

26 An angle of 1 minute in degree units is equal to

A. 1/60 degrees B. 1/3600 degrees

C. 1/100 degrees D. None of these

27 The angle 39°42′ is equal to

A. 39° B. 39.42°

C. 30.7° D. None of these

28 Azimuth is given by

A. The angle between the magnetic meridian and a line

B. The angle between the true meridian and the magnetic meridian

C. The angle between the true meridian and a line

D. None of these

29 Whole circle bearing ranges from

A. 0 to 360 degres

B. 0 to 90 degrees with quadrant specified

C. 0 to 90 degrees

D. None of these

30 Reduced bearing is

A. Whole circle bearing

B. Whole circle bearing converted to quadrantal bearing

C. Both A and B for respective cases

D. None of these

31 Back bearing (BB) is given as

A. BB = FB (Fore bearing) B. BB = FB +90

C. BB = FB ± 180 D. None of these

32 Fore bearing of line AB is 40°, what is its BB?

A. 40° B. 140°

C. 220° D. None of these

33 Fore bearing of line AB is 280°, what is its BB?

A. 100° B. 270°

C. 380° D. None of these

34 Tick mark the most appropriate statement for BB = FB ± 180

A. + sign when FB< 180°
B. - sign when FB > 180°
C. + sign when FB > 180°
D. Both A and B

35 Agonic lines are

A. Lines passing through points of equal declination
B. Lines passing through points of zero declination
C. Both A and B
D. None of these

36 Local attraction can be due to

A. Steel structures
B. Electric cables conveying current
C. A bunch of keys placed near the compass
D. Any or All of these

37 The sum of measured interior angles of a geometric figure is given by (N is the number of sides of figure)

A. (2N-4) × 90°
B. (2N+4) × 90°
C. (N+4) × 90°
D. None of these

38 The map scale can be expressed as

A. Verbal scale
B. Representative fraction (RF)
C. Bar or graphic scale
D. Any of these

39 Least count of prismatic compass is

A. Less than surveyor's compass
B. More than surveyor's compass
C. Equal to the surveyor's compass
D. None of these

40 If the bearings of two lines AB and BC meeting at B are taken from B then the included angle is given as

A. Subtract FB of line BC from BB of previous line AB
B. Add FB of line BC from BB of previous line AB
C. Subtract BB of previous line AB from FB of line BC
D. None of these

41 The permanent adjustments in a dumpy level comprises of

A. Vertical axis
B. Axis of the level tube
C. Line of sight
D. All of these

42 Back sight is the

A. Last reading before the change point

B. First reading taken at a point of known elevation

C. Any reading between the first and the last reading from a setting station

D. None of these

43 Fore sight is the

A. Last reading before the change point

B. First reading taken at a point of known elevation

C. Any reading between the first and the last reading from a setting station

D. None of these

44 Intermediate sight raeding is the

A. Last reading before the change point

B. First reading taken at a point of known elevation

C. All readings between the first and the last reading from a setting station

D. None of these

45 What is the height of the instrument if the reading at the point is 1.85 and assumed RL of the point is 100?

A. 100 B. 101.85

C. 98.15 D. None of these

46 Reduced level of a given point is given as

A. Height of instrument minus foresight or intermediate sight reading

B. Height of instrument plus foresight or intermediate sight reading

C. Height of instrument minus back sight

D. None of these

47 In height of instrument method, check is made using the formula

A. $\Sigma BS + \Sigma FS$ = Last RL - First RL

B. $\Sigma BS + \Sigma FS$ = Last RL + First RL

C. $\Sigma BS - \Sigma FS$ = Last RL + First RL

D. $\Sigma BS - \Sigma FS$ = Last RL - First RL

48 In rise and fall method, check is made using the formula

A. $\Sigma BS - \Sigma FS = \Sigma rise - \Sigma fall$ = last RL – first RL

B. $\Sigma BS - \Sigma FS = \Sigma rise - \Sigma fall$ = last RL + first RL

C. $\Sigma BS - \Sigma FS = \Sigma rise + \Sigma fall$ = last RL – first RL

D. $\Sigma BS + \Sigma FS = \Sigma rise - \Sigma fall$ = last RL – first RL

49 While surveying, the upper and lower readings were noted as 2.5 and 1.5. What is the distance of this point from the station at which instrument is set-up? Assume the value of constant as 100 for the instrument.

A. 375 B. 400

C. 100 D. None of these

50 If the benchmark is too far from the surveying area, which kind of levelling would be most suitable

A. Simple B. Differential

C. Fly D. Profile

51 If a river intersects the area and the possibility of setting the levels in between the two points is not feasible, which kind of levelling would be most suitable

A. Reciprocal B. Differential

C. Fly D. Barometric

52 Which kind of levelling would be most suitable to survey ground surface at small intervals along a line

A. Simple B. Differential

C. Profile D. Barometric

53 The process of levelling in which elevations are measured from the observed horizontal distances and vertical angles is known as

A. Trigonometric B. Differential

C. Profile D. Barometric

54 The horizontal distance between two consecutive contours is known as

A. Horizontal equivalent B. Contour interval

C. Any of A and B D. None of these

55 The constant vertical interval between two consecutive contours is known as

A. Horizontal equivalent B. Contour interval

C. Any of A and B D. None of these

56 Contour lines cross the ridge or valley lines at right angles. If the higher values are inside then it is

A. A hill B. A valley

C. A flat area D. None of these

57 Contour lines cross the ridge or valley lines at right angles. If the higher values are outside then it is

A. A hill B. A valley

C. A flat area D. None of these

58 Contour lines can merge or cross one another only in case of

A. Hills
B. Overhanging cliff
C. Plane area
D. Vertical cliff

59 Contour lines can run into one another only in case of

A. Hills
B. Overhanging cliff
C. Plane area
D. Vertical cliff

60 Tacheometer is a kind of

A. Dumpy level
B. Plane table
C. Theodolite
D. None of these

61 Relatively speaking which one of the following methods is most accurate to interpolate contours

A. Visual estimation
B. Arithmetical calculations
C. Graphical technique
D. All are equally accurate

62 A theodolite is used to measure

A. Horizontal angles
B. Vertical angles
C. Both A and B
D. None of these

63 The operation of bringing the vertical circle from one side of the observer to the other side is known as

A. Permanent adjustment
B. Changing face
C. Transiting
D. None of these

64 The telescope of a theodolite is said to be inverted

A. If its vertical circle is to the left of the observer and the bubble is upward
B. If its vertical circle is to the right of the observer and the bubble is downward
C. Both A and B
D. None of these

65 Natural errors may occur in levelling because of

A. Refaction during high temperature
B. Vibration of the instrument due to high winds
C. Sun shining on the instruments affecting observations
D. All of these

66 Personal errors may occur in levelling because of

A. Inaccurate centering
B. Inaccurate levelling
C. Parallax not properly eliminated
D. All of these

67 Axis of rotation of the telescope in the vertical plane is known as

A. Horizontal axis
B. Vertical axis
C. Both A and B
D. None of these

Answers

1	B	2	A	3	C	4	A	5	C
6	C	7	A	8	B	9	B	10	D
11	B	12	A	13	B	14	B	15	A
16	A	17	A	18	D	19	A	20	A
21	B	22	D	23	B	24	A	25	D
26	A	27	C	28	C	29	A	30	B
31	C	32	C	33	A	34	D	35	B
36	D	37	A	38	D	39	B	40	A
41	D	42	B	43	A	44	C	45	B
46	A	47	D	48	A	49	C	50	C
51	A	52	C	53	A	54	A	55	B
56	A	57	B	58	A	59	D	60	C
61	B	62	C	63	B	64	B	65	D
66	D	67	A						

PART II: On-Farm Development

1 Land levelling should be undertaken carefully or some other alternate option should be considered if

A. The soil is shallow
B. Unproductive sub-soils are likely to be exopsed
C. Excessive earth moving is involved
D. All of these

2 On the basis of soil types, the longitudinal slope is

A. More in a sandy than clay soil
B. More in a clay than sandy soil
C. It does not depend upon soil type
D. None of these

3 **Which one of the method of land levelling is known as analytical method**

A. The profile method

B. The plan inspection method

C. The plane method

D. The contour adjustment method

4 **The centroid of a rectangular field is located at the point of**

A. Intersection of its diagonals

B. Halfway on the longer side

C. Halfway on the shorter side

D. None of these

5 **The cut-fill ratio is more than 1.0 because of**

A. Compaction from equipment in the cut area

B. Compaction in the fill area

C. Crowning at fills because of optical illusion of a dip

D. All of these

6 **Cut-fill ratio should in general be**

A. More than 1.0

B. Less than 1.0

C. Equal to 1.0

D. None of these

7 **Which of the soil will require a higher cut-fill ratio**

A. Organic soils

B. Fine textured (clayey)

C. Medium textured (clay-loam)

D. Coarse textured (sandy)

8 **Which of the soil will require the lowest cut-fill ratio**

A. Organic soils

B. Fine textured (clayey)

C. Medium textured (clay-loam)

D. Coarse textured (sandy)

9 **While using the plane method and assigning the average elevation to the centroid of the area, one ends up with a cut-fill ratio of**

A. More than 1.0

B. Equal to 1.0

C. Less than 1.0

D. None of these

10 **While using the plane method and assigning the average elevation to the centroid of the area, one ends up with a cut-fill ratio of 1.0. To get a higher cut-fill ratio, centroid should be**

A. Lowered

B. Raised

C. Neither lowering nor raising but bringing soil from nearby borrow pits

D. None of these

11 **A grid with an area 800 m^2 has the total cuts (C)and fills (F) at its four corners as : C 0.153 m and F 0.177 m. The volume of cut is**

A. 122.4 m^3 B. 107.27 m^3

C. 92.72 m^3 D. None of these

12 **A grid with an area 800 m^2 has the total cuts (C)and fills (F) at its four corners as : C 0.153 m and F 0.177 m. The volume of fill is**

A. 122.4 m^3 B. 107.27 m^3

C. 92.72 m^3 D. None of these

13 **As per definition, the coefficient of land uniformity indicates perfect levelling when its value is**

A. Zero B. One

C. More than one D. None of these

14 **As per definition, the levelling index indicates perfect levelling when its value is**

A. Zero B. One

C. More than one D. None of these

15 **The aquifer is porous and may contain groundwater but cannot transmit it**

A. An aquifuge B. An aquiclude

C. An aquitard D. None of these

16 **The aquifer is porous and may contain some groundwater as well as transmit water at a relatively low rate compared to aquifers**

A. An aquifuge B. An aquiclude

C. An aquitard D. None of these

17 **Geologic formation that can neither store nor transmit water**

A. An aquifuge B. An aquiclude

C. An aquitard D. None of these

18 **Which is the aquifer type(s)**

A. Groundwater or unconfined aquifer or non-artesian aquifer B. Confined aquifer

C. Perched aquifer D. All of these

19 **A flowing artesian well is the characteristic of**

A. Groundwater or unconfined aquifer or non-artesian aquifer

B. Confined aquifer

C. Perched aquifer

D. All of these

20 A cavity well is

A. Fully penetrating well
B. Partially penetrating well
C. Non-penetrating well
D. None of these

21 A partially penetrating well designed to skim fresh water layer floating on saline water is generally

A. Cavity well
B. Fully penetrating well
C. Partially penetrating well
D. None of these

22 Essential requirement for an hydraulic ram to function is

A. A source of water situated above the pump
B. A source of water situated below the pump
C. Electrical energy
D. None of these

23 Which is known as the displacement pump

A. Reciprocating pump
B. Airlift pumps.
C. Centrifugal pump
D. All of these

24 The discharge of a double acting reciprocating pump is

A. Nearly twice that of a single acting
B. Nearly half that of a single acting
C. Nearly three times that of a single acting
D. None of these

25 To minimize or avoid priming in a centrifugal pump, we use

A. Reflux valve
B. Foot valve
C. Any of A and B
D. None of these

26 The pump characteristic curves provides the relationships

A. Head capacity curve
B. Brake horse power and capacity
C. Net positive suction head required over the capacity range
D. All of these

27 1 meters of water column is equal to

A. 10 kg/cm^2
B. 1 kg/cm^2
C. 0.1 kg/cm^2
D. None of these

28 Water hammers are usually causes by

A. Trapped air in long runs of pipe
B. Sudden valve closure
C. Reverse flow when pumps stop
D. All or any of these

29 Friction head loss can be assessed using

A. Darcy-Weisbach equation
B. Hazen-William equation
C. Blacius equation
D. Any of these

30 For direct driven pump Brake Horse Power is equal to

A. Shaft Horse Power
B. Shaft Horse Power/Drive efficiency
C. Shaft Horse Power/Pump efficiency
D. None of these

31 Energy consumed by a motor in Kilowatts is

A. IHP × 746
B. IHP × 0.746
C. IHP/0.746
D. None of these

32 If the impeller diameter of a centrifugal pump is held constant, the doubling the speed of the impeller will result in

A. Double the original discharge
B. Eight times the original discharge
C. Four times the original discharge
D. None of these

33 If a pump creates perfect vacuum, the water can be theoretically sucked up to _____m at sea level

A. 5.34 m
B. 8.34 m
C. 10.34 m
D. None of these

34 In a submersible pump

A. Motor is under water
B. Pump is under water
C. Both motor and pump are under water
D. Both motor and pump are above the water

35 The pump in which a part of the water is sent back to draw water is

A. Air lift pump
B. Jet pump
C. Reciprocating pump
D. None of these

36 If pump is working but no liquid is being delivered, then it could be due to

A. Pump not primed
B. Strainer or foot valve on the suction line clogged
C. End of suction line is not submerged in water
D. Any or all of these

37 A pipe flowing half full is a case of

A. Pipe flow
B. Open channel flow
C. Mixed flow
D. None of these

38 Natural channels in general are

A. Prismatic channels
B. Non-prismatic channels
C. Rigid channels
D. None of these

39 Hydraulic radius of a rectangular channel is given as

A. Area/Wetted perimeter
B. Wetted perimeter/Area
C. Area/Bottom width
D. None of these

40 For a trapezoidal channel the hydraulically most efficient cross section is given by

A. $B = 2\ d\ \tan\theta/2$ (b is bottom width, d is depth of flow and θ is the angle formed by side slope)
B. $B = 2\ d\ \tan\theta$
C. $B = 2\ d$
D. None of these

41 Which is the Manning's equation (V is the velocity of flow, R is hydraulic radius, S is the slope and n the roughness coefficient)

A. $V = (1/n)\ (R^{1/2})\ (S^{1/2})$
B. $V = (1/n)\ (R^{1/2})\ (S^{2/3})$
C. $V = (1/n)\ (R^{2/3})\ (S^{1/2})$
D. None of these

42 Which is not a structure of the underground pipeline system

A. Pump stand
B. Air vent
C. End plug
D. Man hole

43 Air vent is used at places such as

A. Near the pump stand
B. Summit points if system has significant changes in relief
C. Every 150 m for long pipelines
D. All of these

44 The joint for pipe with plane ends is

A. Collar joint
B. Tongue and groove joint
C. Spigot joint
D. Any of these

45 Horizontal drainage is also categorized as

A. Tile drainage
B. Pipe drainage
C. Pipe less drainage
D. All of these

46 Interceptor drainage is used

A. To lower the shallow water table
B. On sloping lands
C. On canal networks to arrest seepage
D. Both B and C

47 Drained lands are

A. Warmer than waterlogged lands
B. Cooler than waterlogged lands
C. No difference in temperature of drained and waterlogged lands
D. None of these

48 Surface drainage coefficient has the units of

A. mm/day
B. Cumecs/hectare
C. mm
D. Both A and B

49 Bed width in bedding system of drainage

A. Decreases with increase in hydraulic conductivity
B. Increases with increase in hydraulic conductivity
C. Does not depend on hydraulic conductivity
D. None of these

50 Which one of the surface drainage system is used to lower the water table as well

A. Random system
B. Parallel lateral open ditch system
C. Parallel field drain system
D. None of these

51 Horizontal hydraulic conductivity is found out by

A. Auger hole method
B. Inverse auger hole method
C. Both by A ad B depending upon the depth to water table
D. Using the infiltrometers

52 Which one of drainage design criterion is most realistic to represent field conditions

A. Steady state theory
B. Non-steady state theory
C. Fluctuating water table theory
D. None of these

53 In which system, drains enter the main drain at an angle

A. Herringbone system
B. Double main system
C. Parallel system
D. None of these

54 Which is not a rigid conduit

A. Corrugated plastic pipe
B. Concrete tile
C. Asbestos pipe
D. Stoneware pipe

55 The most favourable features where vertical drainage is promising

A. Good quality aquifers are available
B. Groundwater is of good quality
C. Potential of using drainage water exist in near viscinity
D. All of these

Answers

1	D	2	A	3	C	4	A	5	D
6	A	7	A	8	D	9	B	10	A
11	C	12	B	13	B	14	A	15	B
16	C	17	A	18	D	19	B	20	C
21	C	22	A	23	A	24	A	25	C
26	D	27	C	28	D	29	D	30	A
31	B	32	A	33	C	34	C	35	B
36	D	37	B	38	B	39	A	40	A
41	C	42	D	43	D	44	A	45	D
46	D	47	A	48	D	49	B	50	B
51	C	52	C	53	A	54	A	55	D

Part III: Irrigation Water Management

1 Water potential of various river basins in India is assessed as

A. 1850-1890 km^3
B. 850-890 km^3
C. 3900-4000 km^3
D. None of these

2 The ultimate irrigation potential of India is assessed at about

A. 100 M ha
B. 120 M ha
C. 140 M ha
D. None of these

3 A major irrigation project has a culturable command area of

A. More than 2000 ha but less than 5000 ha
B. More than 5000 ha but less than 10000 ha
C. More than 10000 ha
D. None of these

4 A minor irrigation project has a culturable command area of

A. Less than or equal to 2000 ha
B. Less than 5000 ha but less than 10000 ha
C. More than 10000 ha
D. None of these

5 1 cumecs is equal to

A. 1 m^3/s
B. 1000 l/s
C. Both A and B
D. None of these

6 Which is not the velocity-area method of water measurement

A. Float method
B. Current meter
C. Weir
D. All of these

7 A Pigmy current meter is ______ version of Price AA meter

A. 1/5
B. 2/5
C. 3/5
D. 4/5

8 To know the average flow velocity in a shallow channel, the velocity should be measured at a depth of ______ m from the water surface (d is the depth of channel, m)

A. 0.2 d
B. 0.6 d
C. 0.8 d
D. At the water surface

9 To know the average flow velocity in a channel, the velocity measured at a depth of ______ m and ______ m from the water surface (d is the depth of channel, m) is averaged

A. 0.2 d and 0.6 d
B. 0.6 d and 0.8 d
C. 0.2 d and 0.8 d
D. 0.6 d and at the water surface

10 The velocity measured at a depth of 0.2 d m and 0.8 d m from the water surface (d is the depth of channel, m) are 0.7 and 0.5 m/s. The average velocity of water is

A. 0.7 m/s
B. 0.6 m/s
C. 0.5 m/s
D. 0.35 m/s

11 A fully suppressed weir is

A. Contracted at both the ends
B. Contracted at one end
C. Contracted at bottom
D. Not contracted at all

12 For the same head of water flow, the flow through a contracted weir at one end is more than

A. A fully suppressed weir
B. A contracted weir at both ends
C. Both A and B
D. None of these

13 **In the discharge formula, $Q = CLH^m$, the value of m is 3/2 for a weir**

A. Rectangular
B. Trianagular
C. Trapezoidal (Cipolletti)
D. Both A and C

14 **In the discharge formula, $Q = CLH^m$, the value of C is 0.0186 for a weir**

A. Rectangular
B. Trianagular
C. Trapezoidal (Cipolletti)
D. None of these

15 **In the discharge formula, $Q = CH^m$, the value of m is 2.48 for a weir**

A. Rectangular
B. Trianagular
C. Trapezoidal
D. None of these

16 **The correction for contraction is not required in**

A. Rectangular
B. Trapezoidal (Cipolletti)
C. Both A and B
D. None of these

17 **The discharge formula, $Q = CLH^m$, for the triangular weir with m is 2.48 is true if**

A. Triangular weir is fully contracted
B. Triangular weir is fully suppressed
C. Triangular weir is contracted from bottom
D. None of these

18 **A Parshall flume has**

A. A converging section
B. A constricted section, the throat
C. A diverging section
D. All of these

19 **The size of a Parshall flume is given by**

A. The water entry width at the converging section
B. Water leaving width at the diverging section
C. Width of the throat
D. Any of these

20 **To measure the discharge using a Parshall flume having the submerged flow, measurement of head is taken at**

A. At a point at the converging section
B. Two points one at the converging section and other at a point in the throat
C. At a point in the diverging section
D. None of these

21 To assess whether the flow in a Parshall flume is submerged or not, we use guidelines on

A. Head at a point at the converging section exceeds a pre-decided depth

B. Ratio of heads at a point in the throat and head at a point on the converging section

C. Head at a point at the diverging section exceeds a pre-decided depth

D. None of these

22 A flume with a no or zero throat is

A. Parshall flume
B. Long throat flume
C. Cutthroat flume
D. None of these

23 The flume having the flat bottom is

A. Parshall flume
B. Cutthroat flume
C. Both A and B
D. None of these

24 The slope of which section of the Parshall flume is inclined downward

A. Converging section
B. Throat
C. Diverging section
D. None of these

25 The slope of which section of the Parshall flume is level

A. Converging section
B. Throat
C. Diverging section
D. Nonc of these

26 The slope of which section of the Parshall flume is inclined upward

A. Converging section
B. Throat
C. Diverging section
D. None of these

27 Area of channel flow is not required o determine the discharge in the case of

A. Current meter measurement
B. Float method
C. Tracer dilution technique
D. All of these

28 The length of the border is usually more in the case of

A. Sandy soils
B. Loam
C. Clay soils
D. Equal in all of these

29 The slope of the border is usually more in the case of

A. Sandy soils
B. Loam
C. Clay soils
D. Equal in all of these

30 Unit stream size is given as

A. The stream size diverted in one border strip

B. The stream size diverted to a group at one time (Size of the stream in the chanel)

C. The stream size diverted in one border strip per unit width of the border

D. None of these

31 Water saving in furrow irrigation can be made using

A. Surge irrigation

B. Alternate furrow irrigation

C. Both A and B

D. None of these

32 The discharge of a tube well is 12 l/s. A stream of 10 l/s reaches the field boundary. The conveyance efficiency in per cent is nearly

A. 83

B. 120

C. 17

D. None of these

33 The total volume of water diverted from a tube well in an hour is 43,200 liters. If the conveyance efficiency is 80 per cent, how much water reaches the field.

A. 34,560 liters

B. 8,640 liters

C. 43,200 liters

D. None of these

34 The total volume of water reaching the field head is 43,200 liters. If the water stored in the root zone is 36,000 liters, then nearly what per cent of water goes as deep percolation?

A. 83.3 per cent

B. 16.7 per cent

C. None of A and B

D. None of these

35 The water use efficiency is expressed in units

A. kg/ha

B. kg/ha-mm

C. kg/mm

D. None of these

36 Distribution efficiency in a sprinkler irrigation system is defined as the ratio of

A. Average low quarter depth of water infiltrated to average depth of water applied

B. Average high quarter depth of water infiltrated to average depth of water applied

C. Average low quarter depth of water infiltrated to average high quarter depth of water infiltrated

D. None of these

37 Rain gun is a

A. Low pressure sprinkler system
B. High pressure sprinkler system
C. Both A and B
D. None of these

38 Discharge from a sprinkler nozzle is directly proportional to

A. Water pressure at the nozzle head
B. Square root of the water pressure at the nozzle head
C. Cube root of the water pressure at the nozzle head
D. Square of the water pressure at the nozzle head

39 Application rate of sprinkler system should be

A. Less than the infiltration rate of the soil
B. More than the infiltration rate of the soil
C. Equal or more than the infiltration rate of the soil
D. None of these

40 Fertigation in sprinkler irrigation is equivalent of

A. Foliar application
B. Soil application
C. A combination of A and B
D. None of these

41 If sprinklers do not turn, what could be the possibility

A. Low system pressure
B. Nozzle may be blocked
C. Swing arm is not functioning properly
D. All or any of these

42 Sprinkler heads are

A. Oil lubricated
B. Water lubricated
C. Air lubricated
D. None of these

43 If pump does not develop pressure, what could be the possibility

A. Suction lift is more than the desired
B. Air leaks in suction pipe
C. Strainer on the foot valve is blocked
D. All or any of these

44 Fixed costs comprise of

A. Depreciation
B. Interest
C. Taxes, insurance and shelter
D. All of these

45 Net present worth is

A. Net present worth of benefits
B. Net present worth of costs
C. Net present worth of benefits minus the present worth of costs
D. Net present worth of benefits plus the present worth of costs

46 The internal rate of return is the discount rate at which

A. B:C ratio is one
B. Net Present Worth of the project is zero
C. Both of A and B
D. None of these

47 Salts in a drip irrigation system accumulate at

A. Just below the dripper
B. Wetting front
C. Both A and B
D. None of these

48 Settling basin is essentially required if water supply is

A. From the canal
B. From the tube well
C. For both A and B
D. None of these

49 Hydrocyclone filter works on the principle of

A. Centrifugal force
B. Gravity force
C. Centripetal force
D. None of these

50 Pressure compensating emitter are normally used

A. In flat terrain
B. In undulating terrain when water is supplied upslope
C. In undulating terrain when water is supplied down slope
D. All of these

51 Wetted diameter at the soil surface in a drip irrigation is largest in the case of

A. Sand
B. Loam
C. Clay
D. No effect of soil texture, will be same in all cases

52 Diammonium phosphate is designated as 18-46-0 fertilizer. What it means?

A. 18 per cent N, 46 per cent P_2O_5 and 0 K_2O
B. 18 per cent P_2O_5, 46 per cent N and 0 K_2O
C. 18 per cent N, 46 per cent P and 0 K
D. 18 per cent N, 46 per cent P and 0 K_2O

53 One wishes to mix calcium nitrate and potassium sulphate. Mark the appropriate statement out of the following statements.

A. Can be mixed
B. Incompatible and so should not be mixed
C. Likely precipitation of calcium sulphate
D. B because of C

54 **Which of the following equipment for fertigation will result in most non-uniform application of nutrients?**

A. By-pass Pressure tank

B. Venturi

C. Direct injection pump

D. All will give equally uniform application

55 **Which treatment is used to inhibit growth of organisms like bacteria or algae**

A. Acid treatment

B. Chlorine treatment

C. Application of pesticides

D. Any of these

Answers

1	A	2	C	3	C	4	A	5	C
6	C	7	B	8	B	9	C	10	B
11	D	12	B	13	D	14	C	15	B
16	B	17	A	18	D	19	C	20	B
21	B	22	C	23	B	24	B	25	A
26	C	27	C	28	C	29	A	30	C
31	C	32	A	33	A	34	B	35	B
36	A	37	B	38	B	39	A	40	C
41	D	42	B	43	D	44	D	45	C
46	C	47	B	48	A	49	A	50	B
51	C	52	A	53	D	54	A	55	B

Part IV: Soil and Water Conservation Engineering

1 **Which is not a part of the gravity erosion**

A. Splash erosion

B. Landslide erosion

C. Landfill erosion

D. Creep erosion

2 **Soil erosion comprises of**

A. Detachment

B. Transportation

C. Deposition

D. All of these

3 **The soil loss nearly balances the soil formation process**

A. Accelerated erosion
B. Geological erosion
C. Gully erosion
D. None of these

4 **Which is also known as raindrop erosion**

A. Gully erosion
B. Splash erosion
C. Sheet erosion
D. Rill erosion

5 **Splash erosion is affected by**

A. Drop size
B. Terminal velocity of drops
C. Amount and intensity of rain
D. All of these

6 **The raindrop velocity becomes independent of the height of fall after about**

A. 5.6 m
B. 2.5 m
C. 10.5 m
D. None of these

7 **Terminal velocity is proportional to the raindrop diameter as**

A. Directly proportional to drop diameter
B. Inversely proportional to drop diameter
C. Proportional to square root of the drop diameter
D. Proportional to cube root of the drop diameter

8 **Kinetic energy of the falling raindrop is proportional to**

A. Diameter of the equivalent spherical drop
B. Square of the diameter of the equivalent spherical drop
C. Cube of the diameter of the equivalent spherical drop
D. Square root of the diameter of the equivalent spherical drop

9 **Units of kinetic energy in KE = 210.3 + 89 log I are (I is intensity of rain in cm/hr)**

A. Metric tons per hectare per cm of rain
B. Metric tons per hectare
C. Metric tons per hectare per storm of rain
D. None of these

10 **Rills are shallow channels less than**

A. 30 cm deep
B. 50 cm deep
C. 100 cm deep
D. 5 cm deep

11 **Detachment of soil particles is proportional to the velocity of flowing water as**

A. Cube of the velocity (v^3)
B. Square of the velocity (v^2)
C. Velocity raised to five (v^5)
D. None of these

12 Transportation of soil particles is proportional to the velocity of flowing water as

A. Cube of the velocity (v^3)
B. Square of the velocity (v^2)
C. Velocity raised to five (v^5)
D. None of these

13 Tunnel erosion is a kind of

A. Water erosion
B. Wind erosion
C. Both water and wind erosion
D. None of these

14 Detachment of soil particles in water erosion occurs due to

A. Hydraulic action
B. Abrasion
C. Attrition
D. All of these

15 Which is not the transportation process of soil particles in water erosion

A. Solution
B. Suspension
C. Surface creep
D. Attrition

16 Temperature affects water erosion because of

A. Its influence on vegetal cover
B. Its impact on soil organic matter
C. Both A and B
D. None of these

17 Erosivity is a characteristic of

A. Rainfall
B. Soil
C. Both A and B
D. None of these

18 Erodibility is a characteristic of

A. Rainfall
B. Soil
C. Both A and B
D. None of these

19 In rainfall erosivity index, R = KE × I30 such that KE is the kinetic energy, then I30 is

A. Maximum rainfall intensity during the 30 minute of the storm
B. Rainfall intensity during the first 30 minute of the storm
C. Rainfall intensity during the middle 30 minute of the storm
D. None of these

20 If the surface and sub-soils are equally erodible, the gully to be formed are likely to be

A. U shape
B. V shape
C. Trapezoidal
D. None of these

21 Which is not a temporary gully control structure

A. Drop spillway
B. Woven wire check dam
C. Loose rock dam
D. Log check dam

22 Buffer strip cropping is

A. Field strip cropping
B. Contour strip cropping
C. Contour strip cropping with uniform crop strips and grasses and legumes crops in the remaining area between the contour and uniform crop strips
D. None of these

23 Alley cropping is

A. Mixed cropping
B. Cover cropping
C. A crop rotation
D. Food crops grown between hedge rows of trees or shrubs

24 Which kind of bund is used in relatively high rainfall areas

A. Contour bund
B. Graded bund
C. Any of the two
D. None of these

25 In contour bunding, vertical (VI) and horizontal interval (HI) are related as

A. VI = HI/S here S is the land slope
B. VI = HI × S
C. HI = VI/S
D. None of these

26 Which kind of bund is used on sloping land having a land slope of 8 per cent

A. Graded bund
B. Contour bund
C. Any of the two
D. None of these

27 Discontinuous contour trenches can be

A. In line trenches
B. Staggered trenches
C. Any of A and B
D. None of these

28 Which is not a bench terrace

A. Levelled table top
B. Sloping outward
C. Puertorican
D. Broad base graded terrace

29 Area lost is negligible if batter slope is

A. Vertical
B. 1:1
C. 1/2:1
D. Equal area is lost in all

30 Area vulnerable to wind erosion are

A. Desert areas (Arid regions)
B. Semi-arid regions
C. Coastal sandy areas
D. All of these

31 If the wing velocity is doubled, erosive power increases

A. 2 folds
B. 8 folds
C. 4 folds
D. None of these

32 Erosive power of the wind increases with increase in velocity being proportional to

A. V
B. V^2
C. V^3
D. None of these

33 Wind erosion easily erodes

A. Sand
B. Silt
C. Clay
D. All of these equally

34 Once eroded, the soil particles that are transported the farthest are

A. Sand
B. Fine sand
C. Clay
D. All of these equally

35 Jumping and falling movement is noticed when the soil particles are transported by wind through the process of

A. Suspension
B. Saltation
C. Surface creep
D. None of these

36 The diameter of soil particles transported through surface creep range from

A. 0.5 mm to 2.0 mm
B. 0.05 mm to 0.5 mm
C. 0.02 mm to 0.10 mm
D. None of these

37 Wind breaks comprises of

A. One row of trees
B. Two rows of trees
C. Three rows of trees
D. A or maximum B

38 A wind breaks offers protection to the area in the horizontal direction up to a distance of

A. Equal to the height of wind break
B. 5 times the height of wind break
C. 10-20 times the height of wind break
D. None of these

39 The width of the shelterbelt is normally

A. Less than its height
B. More than its height
C. Equal to its height
D. May be C but normally B

40 In India sand dunes are found in

A. Rajasthan
B. Haryana
C. Some coastal areas
D. All of these

41 Shifting sand dunes can be stabilized through

A. Protection through fencing
B. Raising micro-wind barriers
C. Planting of grasses
D. All of these as a step wise technology

42 Peak runoff rate is calculated using

A. Rational formula
B. Cook's formula
C. Curve number technique
D. Any of these

43 In the rational formula peak discharge, Q = CIA/360, A is area in ha and C is the runoff coefficient, I is

A. Average rainfall intensity for the duration of storm
B. Highest rainfall intensity for a duration of 15 minutes
C. Rainfall intensity for a duration equal to time of concentration
D. None of these

44 The value of runoff coefficient in the rational formula will be equal to about 1.0

A. For a highly permeable soil
B. For almost impermeable soil or concrete floor
C. Both A and B
D. None of these

45 Time of concentration is obtained with the use of

A. Darcy's equation
B. Kirpich equation
C. Cook's formula
D. None of these

46 If the value of runoff coefficient is 0.7, I is 150 mm/hr and area is 0.1 km^2, the peak runoff is nearly

A. 0.29 m^3/s
B. 2.92 m^3/s
C. 0.03 m^3/s
D. None of these

47 In a grassed waterway, permissible velocity is higher if vegetation in the waterway is

A. Sparse
B. Good
C. Excellent
D. Not affected by vegetation and is similar in all cases

48 For a parabolic section, the condition to design best hydraulic section is

A. Depth of flow is twice the hydraulic radius

B. Depth of flow is 1.5 times the hydraulic radius

C. Depth of flow is 0.5 times the hydraulic radius

D. None of these

49 The unit of soil erodibility factor are

A. t/ha per unit rainfall erosivity factor B. t/ha

C. t/ha per unit area D. None of these

50 Soil erodibility factor is mainly affected by

A. Soil texture B. Organic matter content

C. Both A and B D. None of these

51 If the tillage factor is 0.3 and crop type factor as 0.5, then crop management factor is

A. 0.3 B. 0.15

C. 0.5 D. 0.6

52 For a conservation practice factor of 0.9, the soil loss was 15 t/ha per annum. If the soil loss is to be limited to 5 t/ha, what should be the value of conservation management factor?

A. 0.3 B. 0.9

C. 0.5 D. None of these

53 Modified universal soil loss equation incorporates

A. Volume of runoff B. Peak flow rate

C. Both A and B D. None of these

54 Which of the equations can predict event based soil loss

A. Universal soil loss equation

B. Modified universal soil loss equation

C. Revised universal soil loss equation

D. Both B and C

55 Which is not the short-term rainwater harvesting technique

A. Deep tillage B. Semi-circular hoops

C. Rock catchments D. Irrigation dams

56 Dugout pond is a

A. Short-term rainwater harvesting B. Long-term rainwater harvesting

C. Flood water harvesting D. None of these

57 Percolation ponds are designed for

A. Irrigation of cropped lands
B. Recharge the groundwater
C. Silt detention
D. None of these

58 Subsurface dams are used for

A. Flood water harvesting
B. Rooftop rainwater harvesting
C. Groundwater harvesting
D. None of these

59 First flush valve is used in rooftop rainwater harvesting to

A. Store water of the first flush in tanks
B. To flush the water harvested from the first shower of rain
C. To take the water to recharge the groundwater
D. None of these

60 Amongst the various components of a large farm pond, we provide

A. Mechanical spillway
B. Emergency spillway
C. Both A and B
D. None of these

61 Areas at the bottom, mid point and top of a trapezoidal pond are 10, 12 and 14 m^2. If the depth of the pond is 4 m, what is the storage capacity of the pond as per trapezoidal rule?

A. 48 m^3
B. 24 m^3
C. 144 m^3
D. None of these

62 As per trapezoidal rule, volume of the trapezoidal pond is give as (A,B, C are the areas at the bottom, middle and top of the structure and D the depth of the pond.

A. (A+4B+C). D/6
B. (A+2B+C). D/4
C. (A+2B+2C). D/6
D. None of these

63 Seepage is controlled by providing

A. Clay blanket
B. Application of chemicals
C. Application of bentonite
D. Any of these

64 LDPE stands for Low _____ Polyethylene

A. Degree
B. Density
C. Dense
D. None of these

65 The quality and life expectancy of the polyethylene depends upon

A. Thickness of the sheet (microns)
B. Weight of the sheet (g/m^2)
C. Both A and B
D. None of these

Answers

1	A	2	D	3	B	4	B	5	D
6	C	7	C	8	C	9	A	10	A
11	B	12	C	13	A	14	D	15	D
16	C	17	A	18	B	19	A	20	A
21	A	22	C	23	D	24	B	25	A
26	A	27	C	28	D	29	A	30	D
31	B	32	C	33	A	34	C	35	B
36	A	37	D	38	C	39	D	40	D
41	D	42	D	43	C	44	B	45	B
46	B	47	C	48	B	49	A	50	C
51	B	52	A	53	C	54	D	55	D
56	B	57	B	58	C	59	B	60	C
61	A	62	A	63	D	64	B	65	C

Glossary

Accelerated Erosion: Erosion intensified by human (anthropogenic) activities.

Acre: A unit of area equal to 4840 yd^2 or 0.404686 ha.

Active Root Zone: The depth of soil in which a plant extracts most of the water it needs for evapotranspiration.

Advance Time: Time required for a given surface irrigation stream to move from one point in the field to another Or time required for a given stream of irrigation water to move from the upper end of a field to the lower end.

Aerial Surveys: A survey using aerial photographs.

Afforestation: The establishment of a forest on land that has not previously, or not recently, been timbered.

Agro-forestry: The integration of commercial tree growing with farming enterprise. The aim is to ensure a long-term viable enterprise based on fuel wood, timber and other products, as part of the overall farm operation.

Air/Vacuum Relief Valve: Small device used on some subsurface drip irrigation that allows air into the dripper line when the system is not running. It eliminates any vacuum that may result in soil particles, debris or contaminants being sucked into the piping system.

Alfalfa Valve: Outlet valve having an adjustable lid or cover to control water flow attached to the top of a pipeline riser.

Allowable Depletion: The amount of water, in irrigation management, that is allowed to be depleted from the plant available water before an irrigation event is scheduled. As a rule of thumb, 50 per cent depletion is normally allowed. Also referred to as Management Allowed Depletion (MAD).

Altitude: The vertical angle between the plane of the horizon and the observed line to the object. In photogrammetry, altitude applies to elevation above a datum of points in space.

Anthropogenic: Resulting from human activity

Application Efficiency: Ratio of the average depth of the irrigation water stored in the root zone to the average depth of irrigation water applied.

Application Efficiency of Lower Quarter: See distribution uniformity.

Aquifer: Underground geological formation, or group of formations, containing usable amounts of groundwater that can supply wells or springs for domestic, industrial, and irrigation uses.

Artificial Drainage: Drainage constructed artificially by men in contrast to natural drainage which develops naturally in a region.

Aspect: The direction that a slope faces, measured at right angles to the contour.

Available Water: Portion of water in a soil that can be readily absorbed by plant roots. It is the amount of water released between the field capacity and the permanent wilting point. Expressed in cm of water per m depth of the root zone.

Avalanching: The process of self-perpetuation that increases the intensity and severity of wind erosion at the down-wind side. It may be initiated by saltation.

Azimuth: The number of degrees from north (or other reference direction) that a line runs, measured clockwise.

Backflow: An unwanted reversal of the flow of water in a piping system, caused by back pressure or back siphonage, which can pollute or contaminate the potable water supply.

Back Pressure: Any condition that creates pressure in the discharge (outlet or downstream) side of the piping system that is greater than that of the supply side of the system.

Back Sight: A back sight is a reading taken on a position of known elevation or coordinate(s) usually a benchmark of sort. Since a survey progresses from a point of known position to points of unknown position, a back sight is a reading looks "backward" along the line of progress.

Baseline: A long line surveyed and established with utmost precision. The baseline accumulates distances throughout a triangulation network, extending to other baselines, providing further integrated control.

Basin Irrigation: Irrigation by flooding areas of level land surrounded by dikes. Usually refers to smaller areas than borders.

Batter: The excavated or constructed face of a dam wall, embankment or cutting produced as a result of earthmoving operations involving cutting and filling. In describing batter grade, 1:3 means a fall (or rise) of 1 vertical meter in a horizontal distance of 3 meters.

Bearing: An angle measured clockwise from a north line of 0° to given surveyed line.

Bench Mark: The name given to a definite, permanent accessible point of known height above a datum to which the height of other points can be referred. A benchmark denoted as BM can be natural or artificial, and it can be either permanent or temporary.

Benefit-cost Ratio: The discounted values of benefits divided by the discounted values of costs adjusted to a common time basis.

Bentonite: A highly plastic clay consisting of the minerals, montmorillonite, and beidellite that swell extensively when wet. Often used to seal soil surface to reduce seepage losses.

Berm: A narrow shelf or flat area that breaks the contiguity of a slope.

Best Efficiency Point: Highest efficiency point on a pump characteristic curve.

Border Irrigation: Irrigation by flooding strips of land, generally rectangular in shape and cross levelled bordered by dikes. Border strips having no down field slope are referred as level border and those with longitudinal grade as graded border systems.

Breakeven Point: In general, the point at which gains equal losses. For agricultural machines, the cost of owning the machine/h is equal to cost of the hiring the machine per hour.

Bubblers: A type of sprinkler head mounted on a riser and puts out a very small umbrella-shaped adjustable flow of water.

Buffer Strips: A strip of perennial vegetation designed to slow run-off and increase deposition of suspended materials.

Cadastral Map: A map depicting land parcels and associated nomenclature.

Capillary Water: Water held in the capillary or small pores of the soil, usually with soil water pressure (tension) greater than 1/3 bar.

Cartesian Coordinates: Numbers expressing the location of a point in two or three dimensions as the perpendicular distances from two or three orthogonal axis.

Cavitation: A process where pressure in the suction line falls below the vapor pressure of the liquid, and the vapors formed move with the liquid flow. These vapor bubbles or "cavities" collapse when they reach regions of higher pressure on their way through pumps.

Chain: Unit of length usually understood to be Gunter's chain, but possibly its variants. A land surveyor's measure - 66 feet or 100 links

Change Point: Change points are points of measurement which are used to carry the measurements forward in a run. Each one will be read first as a foresight, the instrument position is changed, and then it will be read as a back sight.

Check Basin Irrigation: The basin is a small check. Water is applied rapidly to relatively level plots surrounded by dikes.

Check Valve: A valve which permits water to flow only in one direction.

Chemigation: Application of chemicals to crops through an irrigation system by mixing them with the irrigation water. When fertilizers are applied, it is called fertigation.

Chord: The straight line connecting the end points of an arc.

Chute: A high-velocity, open channel (usually paved) for conveying water down a steep slope to avoid erosion.

Cipoletti Weir: The Cipoletti weir has side slopes in the vertical to horizontal ratio of 4 to 1. It is also called trapezoidal weir because of the trapezoidal shape. It is spelled variously as Cipoleti, Cipolleti, Cipolletti, Cippoleti, Cippolleti, Cippoletti, and Cippolletti.

Class (pipe): Term generally used to describe the pressure rating of PVC or other pipes. For example, a class 200 PVC pipe has a pressure rating of 200 psi.

Clear-cutting: Logging practice in which most or all trees are cut down.

Clinometer: An instrument used to determine the angle of elevation or depression; used to set out slopes and gradients in the construction of paths, tracks and roads.

Closed Traverse: A closed traverse either closes back upon its starting point, or begins and ends on stations of known positions. See Traverse.

Coastal Erosion: Loss of sand, soil and rock material to the sea in response to wave action,

Compass: The magnetic compass has a pivoting magnetized needle that always points to magnetic north unless geological features influence the readings. The compass circumference is divided into degrees from which a bearing of a chosen direction from magnetic north can be determined. A compass magnetic bearing is then converted to a grid bearing.

Conservation Agriculture: Conservation agriculture (CA) is defined as a concept for resource saving agricultural crop production that strives to achieve acceptable profits together with high and sustained production levels while concurrently conserving the environment.

Contour Farming: The performance of farming operations such as tillage, sowing and harvesting on the contour, mainly to reduce erosion hazard.

Contour Interval: It is the difference in elevation between adjacent contours as delineated on a map or is a predetermined difference in elevation (vertical distance) at which contour lines are drawn.

Contours: Lines joining points of equal height as shown on a topographic map.

Contour Line: An imaginary line on the ground, all points of which are above or below a specified datum.

Contour Map: A map that depicts relief by means of contour lines.

Control Points: These are fixed points of known coordinates and are determined by high-accuracy surveys.

Conveyance Efficiency: Ratio of the water delivered to the total water diverted or pumped into an open channel or pipeline at the upstream end.

Corrasion: That part of natural erosion processes in which earth and rock materials are worn away due to their abrasion when carried in flows of water, air or ice.

Corrugation Irrigation: Method of surface irrigation similar to furrow irrigation, in which small channels, called corrugations, are used to guide water across a field. No attempt is made to confine the water entirely to the corrugations.

Cost Recovery: Fee structures to cover the cost of providing a service.

Coupler (sprinkler): Device that connects the ends of two lengths of pipe or pipe to a hose.

Cover: Vegetation or other material providing protection to the ground surface against erosive agents.

Cover Cropping: Vegetation established between cash crops with the intention of erosion control and/or soil improvement.

Crop Coefficient: Coefficient used to modify reference evapotranspiration to water use by a particular plant or group of plants. A ratio of the actual crop evapotranspiration to its potential (or reference) evapotranspiration.

Crop Rotation: The growing of different crops and pasture in recurring succession on the same parcel of land with a view to increasing soil fertility and structural stability, and reducing plant disease and soil erosion hazard.

Cumecs: A term used for rates of flow, representing the unit of cubic meters per second. In written communications it is written either in full or abbreviated as m^3/s.

Cusecs: A term used for rates of flow, representing the unit of cubic feet per second. In written communications it is written either in full or abbreviated as ft^3/s.

Cutback Irrigation: Reduction of the furrow (sometimes even in border) inflow stream after water has advanced partially or completely through the field in order to reduce runoff.

Cut-fill Ratio: Volume of cut to volume of fill. In land grading, it should be more than 1.0.

Dam: A barrier to confine or impound water for storage or diversion, to prevent gully erosion, or to retain soil, sediment, or other debris.

Datum: A known position from which all heights are relatively measured.

Declination: The difference between magnetic north and geographic (true) north.

Deep Percolation: Movement of water downward through the soil profile below the root zone that cannot be used by plants.

Deficit Irrigation: Irrigation water management to save water where the soil in the plant root zone is not refilled to field capacity in all or part of the field. Also application of irrigation water at critical growth stages only for the purpose of saving or managing limited water.

Deflation: The removal of fine particles from soil by wind.

Depreciation: Reduction in the value of an asset over time, due in particular to wear and tear.

Depth of Irrigation: Depth of water applied expressed in mm or cm.

Deposition: The process through which mobile sediments loose kinetic energy and accumulate at a place.

Design Storm: A selected storm event, described in terms of the probability of its occurrence once within a given number of years, for which drainage or flood control improvements are designed and built.

Detachment: The fragmentation of soil structural units into sediment (primary particles or smaller aggregates) resulting in greater potential for transport.

Detention: Managing storm water runoff by temporary holding it for controlled release at a later time.

Diameter of Throw: A term used in sprinkler irrigation characterizing average diameter of the area wetted by the sprinkler operating in still air.

Differential Levelling: Differential levelling is the term applied to any method of measuring directly with a graduated staff the difference in elevation between two or more points.

Digital Elevation Model: The digital cartographic representative of the surface of the earth or a subsurface feature through a series of three-dimensional coordinate values: a continuous variable over a two dimensional surface by a regular array of z values referenced to a common datum. Digital elevation models are typically used to represent terrain relief.

Direction: Angular measure by reference to a defined line.

Discharge: Volume of fluid passing a point per unit time commonly expressed as m^3/s, or a similar unit.

Discounting: The use of interest factors to adjust cash flows to a common time base.

Discounted Benefits: Actual benefits occurring in future adjusted through the use of discount rates to a common time base.

Discounted Costs: Actual costs occurring in future adjusted through the use of discount rates to a common time base.

Distribution Efficiency: A measure of distribution uniformity. It indicates the extent to which water is uniformly distributed along the run.

Distribution Uniformity (DU): The uniformity of the distribution of water over a given irrigated area. DU is the ratio of the average depth of water of the driest 25 per cent (lowest one quarter) of catch cans by the average reading of all the catchments, multiplied by 100. Other measure is Christiansen uniformity coefficient.

Disdrometer: An electronic instrument used to measure the size distribution of raindrops.

Diversion: A channel with a supporting ridge on the lower side constructed at the top, across, or bottom of a slope for the purpose of controlling surface runoff.

Diversion Dike: A barrier built to divert surface runoff.

Divide (Drainage): The boundary between watersheds.

Drain: (a) A channel, such as open ditches or drain tile, to remove excess water by surface or by internal flow. (b) To lose water (from the soil) by percolation.

Drain Depth: For open surface drains, it is the depth from the ground surface to the drain bottom. For subsurface drains, it is the distance between the ground surface and the drain tube centre.

Drain Spacing: The horizontal distance between the centrelines of adjacent parallel drains.

Drain Tile: Concrete or ceramic pipe used to construct subsurface drainage to conduct water from the soil.

Drainage: Process of removing groundwater or surface water by artificial or natural means.

Drainage Coefficient: It is the amount of water to be removed in 24 hours to provide favourable conditions for crop growth. It is usually expressed in mm/d. It is also known as drainage modulus.

Drainage System: A system comprising a main drain, its branches and subsidiary drains, used to collect and dispose of excess surface or subsurface water.

Drip Irrigation: The slow application (low volume) of water to the specific root zone area of the plant material using very low pressure.

Drop Inlet: A structure in which water drops through a vertical riser connected to a discharge conduit.

Drop Spillway: A structure in which the water drops over a vertical wall onto an apron at a lower elevation.

Drop Structure: A structure for dropping water to a lower level and dissipating its surplus energy without erosion.

Dune: A moderately inclined to very steep ridge or hillock built up by wind action, Dune formation may be initiated by the entrapment or transported material by wind reducing vegetation or structures.

Earth Dam. A dam constructed of compacted suitable soil materials.

Earth Embankment: It is made up of deposit of soil, rock, or other material often used to form an impoundment.

Earth Work: It involves total volume of shifting of soil in a field (plot) through cut and fill operations involved during land levelling. Also the total volume of earth handled for which payment is released to the contractor.

Economic Analysis: Assessment of an investment's profitability to the whole society or economy regardless of who in the society contributes to the costs and receives the benefits, conventionally measured in 'opportunity' rather than 'market' prices.

Effective Rainfall: Portion of the rainfall which is available for plant growth.

Electronic Distance Measurement, EDM: EDM, a relatively new technique employs either electro-optical instruments, which use laser or infrared light, or microwave instruments to measure distance.

Elevation: The height above mean sea level.

Emergency Spillway: Usually a vegetated earth channel used to safely convey flood discharges around an impoundment structure.

Emitters: A small mechanical device (used in drip irrigation) to limit water flow to a drip or slow trickle. Discharge rate is usually expressed in mm/hr.

Erodibility: Susceptibility of the soil to erosion.

Engineer's Chain: A 100 foot chain containing 100 links of one foot each.

Erosion: The wearing away of the land surface by water, wind, ice, gravity, or other geological agents.

Erosivity: Potential ability of eroding agent to cause erosion. The term is commonly applied to rainfall.

Error: The difference between a measured value and the true value. Error in measurement is inherent, but is separate and distinct from a blunder (a mistake).

Evapotranspiration (ET): The sum of water that evaporates from the soil and the water that is transpired from the plant material. It is the total amount of water needed by the plants to grow.

Fertigation: Application of nutrients through an irrigation system.

Field Books: Field books are standard forms (notebooks) for recording of survey data as it is collected. There are different types of field books for different types of surveys.

Field Capacity: This is the amount of water retained in the soil after the soil has been saturated and allowed sufficient drainage (usually several days after a rainfall). It is the moisture content of the soil at 1/3 atmosphere.

Field Notes: Field notes are a permanent record of field procedures and the data collected in those procedures. Field notes should be prepared carefully.

Financial Analysis: Assessment of an investment's profitability to those people or entities contributing capital and directly benefitting from the returns, conventionally measured at market prices.

FIRBS: Furrow Irrigated Raised Bed Planting System is the system of planting wherein the crops are grown at the top of the beds and the furrows are used as irrigation channels.

Flood Irrigation: Method of irrigation where water is applied to the soil surface without any flow controls. Also known as wild flooding.

Flush Valves: A necessary component in drip irrigation systems to keep lines free of debris. Fitted at the end of sub-mains.

Foresight: A foresight is a reading taken on a position of unknown coordinate(s). Since a survey progresses from a point of known position to points of unknown position, a foresight is a reading looking "forward" along the line of progress. Foresights may be taken on the "main circuit" of the survey or on additional points of interest. Readings on additional points of interest are sometimes called side shots or intermediate foresights to distinguish them from the readings that form the main circuit of the survey.

Freeboard: A vertical distance between the elevation of the design high-water and the top of a dam, diversion ridge, or other water control device.

Friction Loss: Pressure loss due to friction as water flows through all components such as meter, backflow device, zone valves, piping, *etc.* of a system. Friction loss increases with increasing roughness of inside surfaces.

Furrow Irrigation: Method of surface irrigation where the water is supplied to small ditches or furrows for guiding across the field.

Gabion: A wire mesh cage, usually rectangular, filled with rock and used to protect channel banks and other sloping areas from erosion.

Gated Pipe: Portable pipe with small gates installed along one side for distributing irrigation water to corrugations or furrows.

Gauge: 1) A device for measuring precipitation, water level, discharge, velocity, pressure, temperature, *etc.* 2) A measure of the thickness of metal.

Gauging Station: A selected section of a stream channel equipped with a gauge, stage recorder, or other facilities for determining stream stage and discharge.

Geodetic Survey: A survey that takes into account the size and shape of the earth (as distinguished from a plane survey, in which the surface of the earth is considered a plane).

Geographic Coordinates: Geodetic coordinates, state plane coordinates, and rectangular survey coordinates, local coordinates. Any coordinates that relate to a specific geographical network of coordinates. Usually means geodetic coordinates.

Geologic Erosion: Background or natural rates of lowering of earth surfaces without human influence.

GIS: An acronym for Geographic Information System, a system designed to store, manipulate, analyze, manage, and present geo-referenced data.

Global Positioning System: GPS is a satellite based navigation system originally developed by the United State's Department of Defense. A GPS receiver calculates a position by measuring distances to four or more satellites of a possible 24 that orbit the Earth at all times.

GPS: An acronym for Global Positioning System, a navigation system using radio communication with a network of 4 or more satellites to determine the location, velocity, and time 24 hours a day anywhere on or above the earth's surface.

Grade: 1) The slope of a road, a channel, or natural ground. 2) The finished surface of a canal bed, roadbed, top of embankment, or bottom of excavation; any surface prepared to a design elevation for the support of construction, such as paving or the laying of a conduit. 3) To finish the surface of a canal bed, roadbed, top of embankment, or bottom of excavation, or other land area to a smooth, even condition.

Grassed Waterway: A natural or constructed waterway having a good cover of erosion-resistant grasses, may be broad and shallow. These are used to safely conduct surface water from an area.

Grid: A group of parallel lines that run perpendicular to another group of parallel lines to form a map coverage of squares.

Grid Coordinates: A point on a map given as an easting and northing reading. The values are given in meters.

Grid-iron System: A system in which parallel laterals enter into the branch or main drain only from one side.

Ground Survey: A survey made by ground methods. These surveys may or may not include the use of photographs.

Gully: An open incised erosion channel in the landscape generally greater than 30 cm deep.

Gully Head: The upstream end of a gully where runoff from the catchment above falls to the gully floor. It is the exposed part of the gully upon which erosive forces, including water flow, splash and seepage, act to cause the gully to extend upstream by head ward erosion.

Gunter's chain: A distance measuring device composed of 100 metal links fastened together with rings. The length of the chain is 66 feet. A mile is 80 chains. An area one chain wide by ten chains long will be exactly an acre.

Head, Total (Dynamic): Head required for pumping water from its source to the point of discharge. It is equal to the sum of static, pressure, friction and velocity head that a pump works against while pumping at a specific flow rate.

Head, Total (Suction): Head required to lift water from the water source to the centerline of the pump plus velocity head, entrance losses and friction losses in suction pipeline.

Head, Vapor Pressure: Pressure head at which the liquid (water) will vaporize or boil at a given temperature.

Hectare: Metric unit of area equal to 10,000 m^2, or 2.471 acres.

Hand Level: A hand level is a small scope fitted with a spirit level that is visible while looking through the scope. It is used to make rough estimates of relative elevations.

Height of Collimation: Height of collimation is the elevation of the optical axis of the telescope at the time of the setup. The line of collimation is the imaginary line at the elevation.

Herringbone System: A drainage layout in which drain laterals enter the collector from both sides at an angle to ensure adequate slope.

Horizontal Drainage: A method of drainage to control water table through tile or pipes placed horizontally beneath the soil surface.

Hydrant: Outlet, usually portable, used for connecting surface irrigation pipe to an alfalfa valve outlet.

Hydraulic Conductivity: Soil-water characteristic describing the ability of water to flow through a particular soil.

Hydraulic Radius: It is ratio of cross sectional area of flow to the wetted perimeter.

Hydrographic Surveying: The measurement and description of the physical features offshore and adjoining coastal areas with special reference to their use for the purpose of navigation.

Hygroscopic Water: Water that is held tightly by the soil and inaccessible to the plants.

Infiltration: Downward movement of water in to the soil is known as infiltration.

Infiltration Rate: The rate at which the water can infiltrate the soil. Also referred to as absorption rate or percolation rate.

Infiltrometer: An apparatus for measuring infiltration. Single or double ring infiltrometers are used.

Instrumental Errors: From imperfections or faulty adjustment of the instruments or devices with which measurements are taken.

Intake Rate (Basic): Rate at which water percolates into the soil after infiltration has decreased to a low and nearly constant value.

Intangible Benefits: All the benefits (direct and indirect) that cannot be expressed in monetary terms.

Interception: Part of precipitation or water sprayed from the sprinkler irrigation system caught on the vegetation and prevented from reaching the soil surface.

Interceptor Drain: (i) Surface/subsurface drains or their combination, installed to collect subsurface flow before it resurfaces (ii) Surface/subsurface drains to intercept seepage from canals.

Internal Rate of Return: The rate of discounting at which the present value of benefits becomes equal to the present value of costs.

Irrigation Efficiency: It is the ratio expressed in percentage of water stored in the root zone depth of the soil to the water delivered in the field from the farm supply source.

Irrigation Interval: Time between irrigation events. Usually considered the maximum allowable time between irrigations during the peak ET period.

Irrigation Period: Time that it takes to apply one irrigation to a given design area during the peak consumptive-use period of the crop being irrigated.

Irrigation Scheduling: A way to determine when to irrigate and how much water to apply, based upon measurements or estimates of soil moisture or crop water used by a plant.

Isoline: A line joining points of equal value. Examples of these include height, contours on a map or isobars showing atmospheric pressure on a weather map.

kilopascal (kPa): The unit of pressure. Multiply by 7 to convert pounds per square inch (psi) to kilopascals, which gives an approximate conversion to about 2 per cent accuracy. In irrigation practice, pressure is often expressed as 'meters head' that can be converted to kPa by multiplying by 9.8.

kilowatt (kW): The unit of power for electric motors and internal combustion engines. The pre-metric equivalent is the 'horsepower' which is a smaller unit.

Land Degradation: The decline in quality of natural land resources, commonly caused through improper use of the land by humans.

Land Forming: Changing the micro-topography of the land to meet the pre-decided requirements say of drainage or irrigation.

Land Grading: A process of land levelling to reshape the field surface to a planned grade.

Landmark: A monument or material mark or fixed object used to designate a land boundary on the ground.

Landslide: A general term used to encompass the types of mass movement where the material is displaced down slope and along distinct surfaces of separation. The term encompasses a wide variety of materials but relates specifically to slope failures that involve the 'sliding' of the moving material over the ground surface.

Land Uniformity Coefficient (LUC): It represents the magnitude as well as frequency of occurrence of successively larger undulations in the field. It is mathematically described as:

$$LUC = 1 - [\Sigma \,|(LD - LE)|\, / \Sigma LD\,]$$

Here LUC is the coefficient of land uniformity, LD is the desired elevation at a grid point, LE is the existing elevation at that grid point. Summation is made over the total number of grid points used in the field. The summation term between the two lines means that all values irrespective of the signs are to be added. The LUC may range from 0.0 (very poor graded land) to 1.0 (A precisely graded land without any deviations from the desired level).

Lateral Drain: It is a field drain to remove excess water from root zone. It can be unlined open channel, mole drain, tile drain, perforated pipe drains, *etc.*

Latitude: The angular distance along a meridian measured from the Equator, either north or south.

Leaching: The process of removal of soluble material from soil by the passage of water through the soil. A primary step in the reclamation of saline soils.

Levelling: 1) The process of surveying to determine the difference in elevation between two or more points by measuring the vertical distance between two points; the determination of elevation of points above a datum. 2) The tillage operation in which the soil is moved to establish a desired soil elevation slope.

Levelling Index (LI): Levelling index is an indicator for the levelness of the land surface, which is the difference in desired and achieved levelness of the field.

Levelling Rods: Rods marked in feet and decimals of a foot, or in meters and decimeters, used with levelling instruments to determine distances in elevation.

Level Surface: A level surface is a surface which is everywhere perpendicular to the direction of the force of gravity. An example is the surface of water in a still lake.

Longitude: The angular distance measured from a reference meridian, Greenwich, either east or west.

Magnetic Declination: Magnetic declination is the horizontal angle between true north (*i.e.*, the geographic meridian) and magnetic north (*i.e.*, the magnetic meridian).

Map Scale: The relationship between a distance on a map and the corresponding distance on the earth's surface.

Matric Potential: Formerly called capillary potential, it is a dynamic soil property and is near zero for a saturated soil. Matric potential results from capillary and adsorption forces.

Mean Sea Level: Mean sea level is the mean height of the surface of the ocean taking into account all stages of the tide. Used as a reference for elevations.

Meridian: A straight line connecting the North and South Poles and traversing points of equal longitude.

Motor Efficiency: Ratio of the power delivered to the pump by the power unit to the input power to the motor.

Mulch: A natural or artificial layer of plant residue or other materials covering the land surface which conserves moisture, holds soil in place, aids in establishing plant cover, and minimizes temperature fluctuations.

Mulching: The process of application or retention of a mulch.

Net Present Value: The present value of benefits minus the present value of costs.

Net Positive Suction Head (NPSH): Minimum pressure head required at the eye of an impeller in a pump to prevent cavitation.

Nutrient-uptake Efficiency: It refers to kilogram of nutrients uptake by the crop per kilogram of plant nutrient added through fertilizers.

Nutrient Use Efficiency: It refers to the part of the applied nutrients recovered by the crop.

Offset: A short distance usually measured at a right angle to a line, to preserve the position of the line when it is anticipated that points marking the line itself may be disturbed.

Offset Line: An offset line is a supplementary line close to, and usually parallel to a main survey line to which it is referenced by measured offsets. When the line for which data is desired is in such position that it is difficult to measure over it, the required data is obtained by running an offset line in a convenient location and measuring offset from it to salient points on the other line.

Open Traverse: An open traverse does not close on either itself or a station of known position. Although open traverses should generally be avoided, but if used, the procedure should be repeated to provide a check of accuracy. See Traverse.

Operating Pressure: Pressure at which a system of sprinklers operates, usually measured at the base or nozzle of a sprinkler.

Opportunity Time: Time that water remains on the soil surface with an opportunity to infiltrate.

Orientation: Orientation of a point or a text feature measured in degrees anticlockwise from grid east.

Orifice: Opening with a closed perimeter through which water flows. Certain shapes of orifices are calibrated to measure flow rates.

Origin: The zero point in a system of rectangular Cartesian Coordinates.

Orthogonal: At right angles to each other.

Osmotic Potential: Potential attributable to the presence of solutes in the soil-increases with increasing salt concentration of the soil solution. It is given as: Osmotic potential (Mpa) = EC $\times$ 0.36.

Overlap: Area which is watered by two or more sprinklers.

Pan Coefficient: Factor to relate reference crop evapotranspiration to the rate of water evaporation from a free water surface in a shallow pan.

Parallax: A parallax is a change in positions of the image of an object with respect to the telescope cross hairs when the observer's eye is moved. This can be practically eliminated by careful focusing.

Parallel Field Ditch System: A system similar to bedding, except that the drains are spaced farther apart and have a greater capacity than the dead furrows.

Parallel Lateral Open Ditch System: A system similar to field ditch system, except that the ditches are deeper and cannot be crossed with machinery.

Payback Period: The amount of time taken to break even on an investment. Since this method ignores the time value of money and cash flows after the payback period, it provides only a partial picture of whether the investment is worthwhile. Lesser the payback period, better the investment.

Permanent Wilting Point (PWP): Moisture content, on a dry weight basis, at which plants can no longer obtain sufficient moisture from the soil to satisfy water requirements. Plants will not fully recover when water is added to the crop root zone once permanent wilting point has been experienced.

Permeability: Quantitatively, the specific soil property designating the rate at which gases and liquids can flow through the soil or porous media.

Photogrammetric Survey: A method of surveying that uses either ground photographs or aerial photographs;

Pitot Tube: A small tube which can be connected to a pressure gauge to measure the water pressure at the nozzle of a rotary irrigation head.

Plane Surveying: It is a subset of the general field of surveying in which it is assumed that a Cartesian Coordinate system is applicable or appropriate. The methods of plane surveying are appropriate for most construction and planning tasks that are relatively small in scale.

Polyethylene (PE): Flexible (usually black) plastic material used to make irrigation pipe and other items. It is further categorized as high density (HDPE) and low density (LDPE) material.

Polyvinyl-chloride (PVC): Semi-rigid plastic material used to make irrigation and drainage pipes and other items.

Precision: It denotes relative or apparent nearness to the truth.

Precision Farming Techniques: Farming techniques that conserve natural resources such as soil, water and energy in agriculture such as zero tillage, residue management, direct seeded rice, minimum tillage *etc.* See conservation agriculture.

Pressure: The force exerted over a surface divided by its area

Pressure Compensating: A name given to an emitter that maintains constant discharge regardless of inlet pressure. Pressure compensating emitters greatly improve the uniformity of the water applied.

Pressure Relief Valve: A valve which will open when the inlet pressure exceeds a pre-set pressure.

Primary Benefits or Primary Effects: Benefits from project including the value of any reduction in costs due to the project and of any increase of farm products marketed or consumed by the farm family.

Principal Spillway: A dam spillway generally constructed of permanent material and designed to regulate the normal water level, provide flood protection, and/or reduce the frequency of operation of the emergency spillway.

Project Efficiency: Efficiency (overall) of irrigation water use in a project to the water diverted.

psi: Acronym for pressure in pounds per square inch.

Pump: Mechanical device that converts mechanical forms of energy into hydraulic energy *e.g.* reciprocating and centrifugal pumps.

Pump Efficiency: Ratio of the water power produced by the pump, to the power delivered to the pump by the power unit.

Radius of Throw: For heads with circular wetting patterns, distance from the sprinkler head to the farthest point of water application.

Raindrop Splash: The result of the break up and dispersion of raindrops when they hit the ground surface such that soil particles are dislodged and scattered over a considerable distance, due to the energy of the raindrops impact. Also known as raindrop or splash erosion.

Recharge: Replenishment of groundwater by infiltration and transmission from the outcrop of an aquifer or from permeable soils.

Recharge Basin: A basin to increase infiltration for the purpose of replenishing groundwater.

Reconnaissance Survey: An exploratory study of the area to identify problems, degree and extent of the problem and even possible solutions although only qualitatively.

Reduced Level: A reduced level is the vertical distance between a survey point and the adopted level datum.

Revetment: Facing of stone or other material, either permanent or temporary, placed along the edge of a stream to stabilize the bank and protect it from the erosive action of the stream.

Rill: A small intermittent watercourse with steep sides, usually less than 30 cm deep that can be managed with normal tillage operations.

Rill Erosion: The movement of soil by flowing water forming relatively closely spaced small channels that can be filled by normal tillage operations

Riprap: Broken rock, cobble, or boulders placed on earth surfaces, such as the face of a dam or the bank of a stream, for protection against the action of water (waves).

Riser: Length of pipe usually affixed to a lateral or sub-main to support a sprinkler or anti-siphon valve.

Rod (or Pole): It is a surveyor's lineal measure of 16.5 feet or 1/4 of a chain.

Rough Levelling: Removal of abrupt irregularities such as mounds, dunes or filling of pits, depressions and gullies.

Runoff: The portion of precipitation that flows from a drainage area on the land surface, in open channels, or in storm water drains.

Saltation: Particle movement in water or wind where particles skip or bounce along a stream bed or land surface. The preferred usage relates to the bouncing movement of particles of soil or sand grains across the land surface when being moved by wind.

Salvage value: It is the price at which a machine/equipment can be sold at the time of its disposal.

Scheduling: The process of calculating and managing the time and amount of water to apply to a crop or landscape.

Scouring: Localized erosion of a bank or channel which typically occurs due to excessive slope, turbulence or flow velocity.

Sediment Delivery Ratio: The fraction of the soil eroded from upland sources that actually reaches a stream channel or storage reservoir.

Seepage: The passage of water or other fluid through a porous medium, such as water through an earth embankment.

Self-mulching: The condition of a well aggregated soil in which the surface layer forms a shallow mulch of soil aggregates when dry.

Sheet Erosion: Detachment of soil particles primarily by raindrop impact and their movement by water flowing overland as a sheet instead of concentrated flow.

Shelterbelt: Living trees and/or shrubs established in more than 2 rows and maintained for the protection of crops and grazing animals from adverse climatic conditions. Shelterbelts may also serve as windbreaks.

Shutoff Head: Pressure head on the outlet side of a pump at which the discharge drops to zero or a maximum pressure a pump will develop at a given speed.

Sill: A horizontal section at the outlet of a soil conservation or hydrologic structure which spreads water flowing from the structure, hence preventing it from concentrating and forming gullies.

Silt Trap: Also called sediment trap, denotes a structure, usually relatively small, designed specifically to collect sedimentary material from the water.

Slash and Burn: Also known as *Jhoom* cultivation in India, is a traditional form of agriculture that involves cutting and burning forest vegetation to prepare for several seasons of crop production before the forest is allowed to regenerate

Slope: The inclination of the soil surface from the horizontal.

Slope Length: The distance from the point of origin of overland flow to the point where either the slope gradient decreases enough that deposition begins or the runoff water enters a well-defined channel that may be part of a drainage network or a constructed channel.

Smoothening: An operation required nearly every cropping season, particularly where cultivation and harvesting processes significantly disrupts the field surface. It is also a final process in the land levelling to smooth the topographic variations that remains following the operation.

Spile: Conduit, made of lath, pipe or hose, placed through channel banks to transfer water from an irrigation channel to a field.

Sod (Turf): A piece of earth containing plants with matted roots used to revegetate the critical areas where a stable vegetative cover is required for erosion control.

Sodding (Turfing): A method of vegetation such that pieces of sod are placed together over an area to quickly establish a stable vegetative cover to control erosion.

Soil Erosion: The movement of surface soil by water or wind that may be accelerated by human activities.

Soil Abrasion: The process by which moving soil particles erode the soil surface or rocks, which may subsequently move along, thus aiding soil erosion.

Soil Conservation: The prevention, mitigation or control of soil erosion and degradation through the application of cultural, vegetative, structural and land management measures, either singly or in combination to any land being eroded or likely to be eroded.

Soil Erodibility: The susceptibility of a soil to the detachment and transportation of soil particles by erosive agents.

Soil Survey: The systematic examination, description classification, and mapping of soils in an area. Soil surveys are classified according to the kind and intensity of field examination.

Soil Texture: The texture is the term indicating the coarseness or fineness of the soil. It is decided on the basis of the amount and quantity of each of the grain size group of particles sand, silt and clay that constitute the soil.

Specific Speed: It is the speed (RPM) of a geometrically similar impeller needed to operate in order to deliver 1 L/s of fluid at a head of 1 m

Spillway: An open or enclosed channel, or a combination of both, used to convey excess water from a dam or similar storage, gully control structure or detention structure. See emergency spillway or principal spillway.

Staking: The placement of markers on a site to identify certain locations (such as the corners of a building, the right-of-way of a road, the extent of the slope faces of a dam, *etc.*) with corresponding information (such as cut or fill for earthmoving) is the process of staking out a project. It is the transfer of information from the plan to the actual site.

Static Lift: Vertical distance between water source and discharge water levels in a pump installation. Also known as total static head.

Station: The term station refers to a point on a baseline that is at a known distance from a starting/reference point. The starting point is usually referenced as 0 + 00, but there are occasions where another value might be assigned.

Storage Efficiency: Ratio of the amount of water stored in the root zone during irrigation to the amount of water needed to fill the root zone to field capacity.

Strip Cropping: The conservation farming technique of growing crops in rotation in a systematic arrangement of strips at right angles to the direction of water flow on lands of low slope (< 2 per cent). It is also used on generally level areas to control wind erosion, the strips being orientated at right angles to the prevailing winds.

Stubble Mulching: A conservation farming practice whereby stubble is retained on the surface of the soil, thereby protecting the soil from erosion. The tillage operation is done in a manner with such implements which cause minimum surface disturbance and stubble burial.

Subsurface Drainage: The removal of surplus or excess groundwater accomplished by means of natural or by artificial means using tile and pipe drains beneath the soil surface.

Subsurface Drains: Subsurface channels used primarily to remove subsurface water from soil. Also known as tile drains or pipe drains.

Subsurface Drip Irrigation: Application of water below the soil surface through emitters, with discharge rates generally in the same range as drip irrigation. The method of water application is different from sub-irrigation where the root zone is irrigated by water table control.

Suction (Static) Lift: Vertical distance between the elevation of the surface of the water source and the center of the pump impeller.

Surface Creep: The rolling or sliding of larger particles, too heavy to be lifted, along the ground under the influence of wind.

Surface Drainage: The removal of surplus or excess surface water collected on land accomplished by natural or artificial means, such as open ditches and terracing.

Surface Irrigation: Type of irrigation in which water is distributed to the plants by a surface distribution network. Also broad class of irrigation methods in which water is distributed over the soil surface by gravity flow.

Surge Irrigation: Surface irrigation technique wherein flow is applied to furrows (or less commonly, borders) intermittently during a single irrigation set.

Survey: The act or operation of making measurements for determining the relative positions of points on, above, or beneath the earth's surface; also, the results of such operations.

Suspension: Particle movement in water or air where particles are kept dispersed by fluid motion in currents, by turbulence and/or by molecular motion of the surrounding medium.

Tangible Benefits or Tangible Effects: All the benefits that can be expressed in monetary terms.

Tapes/Taping: A tape is a flexible device used to measure linear distances. There are tapes made of many materials, such as cloth, kevlar, steel, and invar.

Terminal Velocity: The final steady-state velocity of falling raindrops before they hit the soil surface,

Terrace: An embankment or ridge of earth constructed across a slope to control runoff and minimize soil erosion.

Theodolite: A precise optical instrument used to measure horizontal and vertical angles.

Thematic: Depicting one or more specific topics or subjects *e.g.* Land use, rainfall, population density.

Tie Line: A survey line that connects a point to other surveyed lines.

Tile Drainage: The use of subsurface drain lines, originally made of tiles but now normally made of plastic, to accelerate lateral movement of soil water.

Tilth: Physical condition of the soil in relation to plant growth.

Time of Concentration: It is equivalent to the time required for runoff to flow from the most remote part of the catchment to the specified point, normally the catchment outlet.

Topography: The study of the physical features of the earth depicted by the presence of hills, mountains or plains. For example flat or steep topography.

Topographic Map: A detailed representation of cultural, hydrographic relief and vegetation features. These are depicted on a map on a designated projection and at a specified scale.

Topology: Properties of geometric forms that remains invariant when the forms are deformed or transformed by bending, stretching or shrinking.

Total Station: A survey instrument that combines a theodolite and distance meter.

Transpiration: The process of plant material giving off water vapor from the leaves.

Traverse: 1) any line surveyed across a parcel and 2) A series of such lines connecting a number of points, often used as a base for triangulation.

Triangulation Station: A permanently marked and fully documented control station whose position on the earth's surface has been established to a high accuracy both absolutely and in relative terms to other adjacent stations by means of angular or electronic distance measurement.

Trigonometrical Survey: A concise method of surveying in which the stations are points on the ground located at vertices of a chain or network of triangles. The angles of the triangles are measured instrumentally and the sides are derived by computation from selected sides termed as baselines.

True North: The direction to the Earth's geographic North Pole.

Uniformity Coefficient (Christiansen's): Measure of the uniformity of irrigation water application. The average depth of irrigation water infiltrated minus the average absolute deviation from this depth, all divided by the average depth infiltrated. Mathematically, it is written as:

$$CU = 100 \times \left(1.0 - \frac{\sum |yi - m|}{\sum Yi}\right)$$

Here CU is the coefficient of uniformity, per cent, yi is depth of water at a given point or water measured in each container in sprinkler irrigation, n is the number of points or cans used for measurements (as such i varies from 1 to n), m is mean depth of water applied or collected in all the cans,. Vertical lines states that absolute values of deviations are used.

Uniformity of Application: A general term designating how uniform the application of water is over an irrigated area.

Vertical Error of Closure: The error of closure of a levelling survey refers to the cumulative error of the entire circuit. If a circuit is run from a point of known (or assumed) elevation back to that point, then the starting elevation and the ending elevation should be the same. The discrepancy, expressed as a raw vertical distance (in units such as feet or meters), is the error.

Vertical Drainage: (1) Disposal of drainage water through wells into a porous layer of earth or an open-rock formation. (2) The term vertical drainage is also used for a drainage system in which tube wells are used to lower the water table.

Vertical Interval: The vertical distance from one soil conservation structure to another. The term also describes the vertical distance between contour lines on a map.

V-notch: A weir having the shape of V

Water Application Efficiency: Water application efficiency is the measure of efficiency with which water delivered to the field is stored in the root zone.

Water Distribution Efficiency: The extent to which water is uniformly distributed in an irrigation system. Or, it is defined as the percent of difference from unity of the ratio between the average numerical deviations from the average depth stored during the irrigation.

Water Hammer: A shock wave in the piping system usually created by excessive flow velocity and or as a result of sudden closing of valve.

Water Horsepower (WHP): Energy added to the water by a pump.

Water Logging: It refers to the condition of land in which it cannot be put to its normal use as a result of either or both of shallow water table and surface ponding. In arable lands depending upon the extent and duration of water logging, the crop yield may reduce substantially or even become zero.

Water Productivity: Ratio of the gross income earned per hectare to the amount of water used to earn the gross income. Units are Rs./m^3.

Watershed: The region drained by or contributing water to a specific point that could be along a stream, lake or other storm water facilities. Watersheds are often broken down into sub-areas for the purpose of hydrologic modeling.

Water Storage Efficiency: See storage efficiency.

Water Table: Upper surface of a saturated zone below the soil surface where the water is at atmospheric pressure.

Water Use Efficiency: The amount of dry matter that can be produced from a given quantity of water. Ratio of the yield per unit area to the applied irrigation water per unit area

Waterway: Lake, pond, river, stream, creek, canal, *etc.*

Weir: A structure to measure or regulating the flow of water.

Well: An artificial excavation (pit, hole, tunnel) often walled (drilled, dug, driven, bored or jetted) sunk into the ground to such a depth so as to penetrate water yielding rock or soil and allow the water to flow or to be pumped to the surface.

Wetted Perimeter: Length of the wetted contact of a liquid, measured along a plane at right angles to the direction of flow.

Wind Break - one or at the most two rows of trees or shrubs or other materials which reduces the velocity of the wind near the soil surface, thus protecting the soil from wind erosion.

References

Agarwal, M.C. and Goel, A.C. 1981. Effect of field leveling quality on irrigation efficiency and crop yield. Agricultural Water Management, 4: 457-464.

Agarwal, M.C. and Khanna, S.S. 1983. Efficient Soil and Water Management in Haryana. Haryana Agricultural University, Hisar. 118 p.

Ambast, S.K. 2006. Land leveling: an on farm water management strategy for improving crop productivity in saline environment. In: Sustainable Irrigated Agriculture through Command Area Development (Ambast *et al.*, Eds.). Central Soil Salinity Research Institute, Karnal, 16-25.

Ambast, S.K., Gupta, S.K. and Sharma, D.K. 2014. Laser Land Leveling. Technical Bulletin, Central Soil Salinity Research Institute, Karnal, 40 p.

Anderson, J. M. and Mikhail, E. M. 2012. Surveying: Theory and Practice, 7th Edition. Tata McGraw-Hill Education Pvt. Ltd., New Delhi. 1204 p.

Armstrong, D., O'Donell, D. and Thompson, C. 2001. Irrigation Equipment and Techniques, Module Notes, Irrigation Systems, Department of Primary Industries, Parks, Water and Environment, Australia. Accessed on 15th September 2015 from www.dpiw.tas.gov.au/ inter.nsf/Attachments/JMUY-5FLVYP/$FILE/3 per cent 20Irrigation per cent 20Systems per cent 20V3.pdf

Basak, N. N. 2014. Surveying and Levelling. McGraw Hill Education (India) Pvt. Ltd., New Delhi. 580 p.

Bhaskar, K.S., Rao, M.R.K., Mendhe, P.N. and Suryavanshi, M.R. 2005. Micro Irrigation Management in Cotton. CICR Technical Bulletin 31. Central Institute for Cotton Research Nagpur. 43 p.

Bhavikatti, S. S. 2010. Surveying Theory and Practice. I.K. International Publishing House Pvt Ltd. New Delhi. 896 p.

Booher, L. J. 1974. Surface Irrigation. FAO Agricultural Development Paper no. 95. Food and Agriculture Organization of the United Nations, Rome. 160 p.

Bos, M. G. (Ed). 1976. Discharge Measurement Structures. Publication 20, International Institute for Land Reclamation and Improvement, Wageningen.

Bos, M. G. (Ed). 1989. Discharge Measurement Structures. Publication 20, Third Revised Edition, International Institute for Land Reclamation and Improvement, Wageningen. 401 p.

Bouwer, H. 1978. Groundwater Hydrology, McGraw-Hill Book, New York. 480 p.

Bouyoucos, G. J. 1935. The clay ratio as criterion of susceptibility of soils to erosion. Journal of the American Society of Agronomy. 27: 738-741.

Calder, T. 2005. Efficiency of Sprinkler Irrigation Systems. Department of Agriculture, Australia.www.agric.wa.gov.au/1we/water/irr/fn048_1992.pdf,1-3.(Accessed on 18.01.2014).

CDoT. 2004. Drainage Design Manual. Chapter 9: Culverts. Colorado Department of Transportation, Colorado, USA. 9.1-9.70

CGWB. 1995. Ground Water Resources of India. Central Ground Water Board, Govt. of India, New Delhi.

Chow, V. T. 1959. Open-Channel Hydraulics. McGraw-Hill Book Company, New York. 680 p.

Christiansen, J. E. 1942. Irrigation by Sprinkling. California Agricultural Experiment Station, Bulletin No. 670. University of California, Berkeley. 124 p.

CWC. 2010. Water Statistics. Central Water Commission (CWC), Ministry of Water Resources, GoI, New Delhi.

CWC. 2013. Water Statistics. Central Water Commission (CWC), Ministry of Water Resources, GoI, New Delhi. 201 p.

CWC, Ministry of Water Resources. 2017. Guidelines for Planning and Design of Piped Irrigation Network part I and 2. Central Water Commission, Ministry of Water Resources, New Delhi. 124 p.

Davoren, A. 1995. Spray Irrigation: Is your System up to Scratch. School of Isolated and Distance Education (SIDE), New Zealand. http/www.side.org.nz/IM_custom/Contentstore/Assets/7/39/136a843f72640. (Accessed on 18.01.2014).

Dhruvanarayan, V. V. 1993. Soil and Water Conservation Research in India. Indian council of Agricultural Research, New Delhi.

Director of Agriculture. 2016. UGPL 2015-16. Letter Dated 11.02.2016

Finkel, H. J.1983. Handbook of Irrigation Technology. Vol.II, CRC Press. Florida, USA. 223 p.

Fortier, S. and Scobey, F. C.1926. Permissible Canal Velocities. Trans. ASAE. 89: 940-948.

Garg, S. K. 1996. Irrigation Engineering and Hydraulic Structures, Khanna Publishers, Delhi, 12th Edition.1296 p.

Geyik, M. P. 1986. FAO Watershed Management Field Manual: Gully Control. FAO Conservation Guide 13/2. Food and Agricultural Organization, Rome

GoI. 2004. Report of the National Task Force on Micro-irrigation. Ministry of Agriculture, Government of India. 330 p.

Govt. of Rajasthan. 2016. State Specific Technology Manual For Watershed Development. Maharana Pratap University of Agriculture and Technology, Udaipur, Rajasthan. 217 p.

GSWMA (Gujarat State Watershed Management Agency). 2011. Technical Manual for IWMP Gujarat State Watershed Management Agency. Commissionerate of Rural Development, Gandhinagar.198 p.

Gupta, S.K. 2011. Case studies on subsurface drainage systems and guidelines for design and evaluation. In: Salinity Management for Sustainable Agriculture in Canal Commands (Dey, P. and Gupta, S.K. Eds.). Central Soil Salinity Research Institute, Karnal. 11-22.

Gupta, S.K. 2014. Response of crops to surface stagnation and ameliorative strategies. In: Agricultural Land Drainage in India (Gupta, S.K. Ed.) Agrotech Publishing Academy Udaipur, 25-48.

Gupta, S.K. 2019. Drainage Engineering: Principles and Practices. Scientific Publishers (India), Jodhpur. 531 p.

Gupta, S. K., Minhas, P. S., Sondhi, S. K., Tyagi, N. K. and Yadav, J. S. P. 2000. Water resources management. In: Natural Resources Management for Agricultural Production in India, (Eds.) Yadav and Singh. Int. Conf. on Managing Natural Resources for Sustainable Agriculture Production in the 21st Century. Indian Society of Soil Science, New Delhi.137-244.

Gupta, S. K., Sharma, D. P., Tyagi, N. K. and Dubey, S. K. 2002. *Lavaniya-Kshariya mirdaon ka Sudhar aur Parbandh*. CSSRI, Karnal. 153 p.

Gupta, S.K., Tejwani, K.G. and Babu, R. 1971. Drainage coefficients for surface drainage of agricultural lands for different parts of the country. J. Irrig. Pwr. 28: 53-60.

Hagin, J., Sneh, M. and Lowengart-Aycicegi, A. 2003. Fertilization through Irrigation. IPI Research Topic 23, International Potash Institute, Basel, Switzerland. 81 p.

Hansen, V. E., Israelsen, O. W. and Stringham, G. E. 1981. Irrigation Principles and Practices. John Wiley and Sons, Hoboken, New Jersey, USA.

Hooghoudt, S. B. 1940. Bijdragen tot dekennis van enige naturrk undige grootheden van de ground. Versi Land bouwk Onderz. 7: 449 p.

https:/beeindia.gov.in/sites/default/files/3Ch6.pdf, Pumps and Pumping System. 113-134.

https:/civilseek.com/chain-surveying/. Chain Surveying: Its Procedure, Instruments, and Principles. Opened on 23.10.2018.

Hudson N. 1971. Soil Conservation. B.T. Batsford Ltd., London. 320 p.

Hurd, C.J. 1969. Sprinkler Irrigation Guide Book. US Agency for International Development. Washington- DC.

ICAR-NAAS. 2010. Degraded and Wastelands of India: Status and Spatial Distribution. Indian Council of Agricultural Research and National Academy of Agricultural Sciences, New Delhi. 158 p.

ICID. 1979. Amendments to the Constitution, Agenda of the International Council Meeting at Rabat. International Commission on Irrigation and Drainage (ICID), Morocco, ICID, New Delhi, India, A-156-163.

ICID. 1982. ICID Standard 109, Construction of Surface Drains. ICID Committee on Irrigation and Drainage Construction Techniques, ICID Bulletin. 31: 47-57.

INCID. 1998. Sprinkler Irrigation in India. Indian National Committee on Irrigation and Drainage, New Delhi.

Irrigation Commission. 1972. Report of the Irrigation Commission. Vol. I. Ministry of Irrigation and Power, New Delhi. 430 p.

IRRI. 2013. Laser land Leveling Training Manual. International Rice Research Institute, Manila, Philippines. 28 p.

ISI. 1974. Criteria for Design of Cross-section for Unlined Canals in Alluvial Soil- IS: 7112 – 1973. Indian Standards Institution, New Delhi.

Israelsen, O. W. and Hansen, V. E. 1962. Irrigation Principles and Practices. John Wiley and Sons, Inc. Third Edition. 341p.

James, L. G. 1988. Principles of Farm Irrigation System Design, John Wiley and Sons, Inc., New York. 543 p.

Jat, M.L., Chandna, P., Gupta, R., Sharma, S.K. and Gill, M.A. 2006. Laser Land Leveling: A Precursor Technology for Resource Conservation. Rice-Wheat Consortium Technical Bulletin Series 7: Rice-Wheat Consortium for the Indo-Gangetic Plains, New Delhi, India. 48 p.

Jedrych, T. A. 2007. Grassed Waterway Improvement and Gully Restoration. Alberta Agriculture, Food and Rural Development. https:/www1.agric.gov.ab.ca/$department/deptdocs.nsf/all/agdex 1344/$file/573-5.pdf?OpenElement. Opened on 05.03.2018.

Kaul, R. N. 1996. Sand dune stabilization in the Thar desert of India: A synthesis. Annals of Arid Zone. 35: 225-240.

Khepar, S. D., Chaturvedi, M. C. and Sinha, B. K. 1982. Effect of precise leveling on the increase of crop yield and related economic decision. Journal of Agricultural Engineering. 19: 23-30.

Kindsvater, C. E. and Carter, R. W. C. 1957. Discharge characteristics of rectangular thin plate weirs. J. Hydraulic Div. ASCE. 6: 83.

King, J., Tiffin, D., Drakes, D. and Smith, K. 2006. Water use in agriculture: establishing a baseline Project WU0102. http:/randd.defra.gov.uk/Document.aspx?Document=WU0102_4274_FRP.doc)

Kranz, W. L., Yonts, D. and Martin, D. 2005. Operating Characteristic of Centre Pivot Sprinklers. University of Nebraska – Lincoln. NebGuide G1532. 4 p.

Kumar, A. and Singh, A. K. 2002. Improving nutrient and water use efficiency through fertigation. J. Water Management. ISWAM. 10: 42-48.

Kumar, P. 2014. Ground Water and Well Drilling: A Reference Book on Ground Water and Wells. CBS Publishers and Distributors Private Ltd., New Delhi.

Kumar, R., Singh, R. D. and Sharma, K. D. 2005.Water resources of India. Current Science. 89: 794-811.

Kundu, D. K., Neue, H. U. and Singh, R. 1998. Comparative effects of flooding and sprinkler Irrigation on growth and mineral composition of rice in an Alfisol. Proceedings of the National Seminar on Micro-irrigation Research in India: Status and Perspective for the 21st Century, Bhubaneswar, July 27-28, 1998.

Michael, A.M. 1978. Irrigation: Theory and Practice. Vikas Publishing House Pvt. Ltd., New Delhi. 801p.

Michael, A. M. 2008. Irrigation Theory and Practice, Second Edition Vikas Publishing House Pvt. Ltd, Noida, U.P. India.772 p.

Michael, A.M., Khepar, S.D. and Sondhi, S.K. 2008. Water Wells and Pumps. Tata McGraw-Hill Company Limited, New Delhi, 690 p.

Miller, R. W. 1984. Flow Measurement Engineering Handbook. 2nd Ed. McGraw-Hill Book Co., New York, N.Y. 960 p.

Ministry of Agriculture and Farmers Welfare. 2017. Operational Guidelines of Per Drop More Crop (Micro Irrigation) Component of PMKSY. Division of Rain-fed Farming System (RFS), Ministry of Agriculture and Farmers Welfare, New Delhi. 54 p.

Mishra, C. B. 2018. Conventional symbols, chain survey and offset measurements. Gujarat Tcchnological University, Vallabh Vidyanagar- Gujarat. https:/www. gtuelibrary.edu.in/ALCE/Civil/Conventional per cent 20symbols per cent 20_ per cent 20Chain per cent 20survey.pptx. Opened on 23.10.2018.

Mishra, P. K. and Sharma, S. N. 1994. Theoretical considerations in design of farm ponds for minimizing evaporation and seepage losses. Indian J. Dryland Agric. Res. and Dev. 9: 114-120.

Mondal, J. 2016. Water harvesting and recycling. International Journal of Environment, Agriculture and Biotechnology (IJEAB). 1: 623-626.

MOWR. 1999. Report of the Working Group on Water Availability for Use. National Commission for Integrated Water Resources Development Plan, Ministry of Water Resources, Government of India, New Delhi.

MPUA&T (Maharana Pratap University of Agriculture and Technology). 2016. State Specific Technology Manual for Watershed Development. National Rainfed Area Authority, Ministry of Agriculture and Farmers Welfare, New Delhi and MPUAT, Udaipur. 210 p +Annexures.

Mridha Md. A. K. 1993. Performance of Concrete Buried Pipe Distribution Systems for Surface Irrigation under Farmers' Management in Tangail, Bangladesh. M.Phil. Thesis, Loughborough University, UK. https:/dspace.lboro.ac.uk/2134/10840 Publisher: 244 p.

Murthy, V.V.N. and Jha, M. K. 2018. Land and Water Management Engineering. Third Edition Kalyani Publishers, New Delhi

NABARD. 2012. Model bankable scheme for sprinkler irrigation systems. http:/www.nabard.org.

NARC (National Agricultural Research Center). 1992. Handbook of Sprinkler Irrigation Systems. Part I Sprinkler Irrigation Technology: Hydraulics and Design of Rain-gun Systems. Water Resources Research Institute, Islamabad (Pakistan). 66 p.

National Rainfed Area Authority (NRAA), Government of India. 2011. Common Guidelines for Watershed Development Projects- Rev. Ed. – 2011. NRAA, Planning Commission, GOI, New Delhi. 60 p.

NCPA. 1990. Status, Potential and Approach for Adoption of Drip and Sprinkler Irrigation Systems. National Committee on the Use of Plastics in Agriculture, Pune, India.

NRCS, Natural Resource Conservation Service. 2011. Grassed Waterway. Practice Standard # 412. Technical Guide Section IV. Natural Resource Conservation Service, Washington DC.

Publication Division. 2011. India-2011. Ministry of Information and Broadcasting, Government of India, New Delhi. 1233 p.

Oldeman, L.R. 1992. Global extent of soil degradation. Bi-Annual Report 1991-1992. International Soil Reference and Information Centre, Wageningen, The Netherlands.19-36.

Oldeman, L. R., Hakkeling, L.A. and Sombroek, W.G. 1991. World Map of the Status of Human Induced Soil Degradation: An Explanatory Note. International Centre and United Nations Environment Programme, Wageningen. The Netherlands and Nairobi, Kenya.

Omkar, D., Vallapu, R. K. and Joyce, M. 2017. Vegetative water ways. Proceedings of SARC International Conference, 3rd December, 2017, Chennai, India. 41-44.

Praveen Rao, V. 2002. Drip irrigation and its application in farmers' fields of India. In: Recent Advances in Irrigation Management for Field Crops. Chinnamuthu, C.R., Velayutham, A., Ramasamy, S., Sankaran, N., Sunder Singh (Eds.). S.D. Centre of Advance Studies in Agronomy, Tamil Nadu Agric. Univ., Coimbatore. 66 – 70.

Phocaides, A. 2000. Technical Hand Book on Pressurized Irrigation Techniques. Food and Agriculture Organization of the United Nations, Rome.

Pillusbury, A. F. 1968. Sprinkler Irrigation. Food and Agricultural Organization of the United Nations, Rome. 179 p.

Planning Commission, Government of India. 2009. Report of the Task Force on Irrigation. Planning Commission, New Delhi. 59 p.

Raghunath, H.M. 1987. Groundwater. New Age International Publishers, New Delhi. 563 p.

Raghuwanshi, R. 2016. Design of Pipe Distribution Network in the Command of a Canal Outlet for Optimal Crop Planning. M. Tech. Thesis Indira Gandhi *Krishi Vishwavidyalaya*, Raipur, CG. 89 p.

Raman, S. 2010. State-wise micro-irrigation potential in India-an assessment. Unpublished paper- Seminar Natural Resources Management Institute, Mumbai.

Ram, Babu, Tejwani, K. G., Agarwal, M. C. and Bhushan, L. S. 1978. Distribution of erosion index and iso-erodent map of India. Indian J. Soil Cons. 6: 1-14.

Reddy, K.S., Manoranjan Kumar, Rao, K.V., Maruthi, V., Reddy, B.M.K., Umesh, B., Ganesh Babu, R., Srinivasa Reddy, K., Vijayalakshmi and Venkateswarlu, B. 2012. Farm Ponds : A Climate Resilient Technology for Rainfed Agriculture. Planning, Design and Construction. Technical Bulletin 3/2012. NICRA, Central Research Institute for Dryland Agriculture, Hyderabad. 67 p.

Rickman, J.F. 2002. Manual for Laser Land Leveling. Rice-Wheat Consortium Technical Bulletin Series 5. Rice-Wheat Consortium for the Indo-Gangetic Plains. New Delhi, India, 24p.

Robinson, A. R. and Chamberlein, G. E. 1960. Trapezoidal flumes for open channel flow measurement. Trans. Am. Soc. Agricultural Engineering. 3: 120-124

Saxena, C.K. and Gupta, S.K. 2004. Drip irrigation for water conservation and saline/sodic environments in India: a review. In: Proceedings of International Conference on Emerging Technologies in Agricultural and Food Engineering. IIT, Kharagpur. December 14-17, 2004. Natural Resources Engineering and Management and Agro-Environmental Engineering. Amaya Publishers. New Delhi. 234-241.

Schwab, G. O., Fangmeier, D. D., Elliot, W. J., and Frevert, R. K. 1993. Soil and Water Conservation Engineering. John Willey and Sons, Inc., New York, USA. 269 p.

Sharma, B. R. and Paul, D. K. 1998.Water Resources of India. In: 50 Years of Natural Resource Management Research. Singh and Sharma (Eds.). Indian Council of Agricultural Research, New Delhi. 31-49.

Singh, A. K. 2002.Water resource conservation and management. *Kheti*. 8:17-23.

Sivanappan, R. K. 1998. Irrigation water management for sugarcane. In VSI: Vol. II: 100-125.

Sivanappan, R. K. 1999. Status and perspectives of micro-irrigation research in India. Proceedings of the National Seminar on Micro-Irrigation Research in India: Status and Perspectives for the 21st Century, Bhubaneswar. July 27-28, 1998. Institution of Engineers (India) : 17-29.

Slastikhin, V. V. 1964. Problems of the amelioration of slopes in Moldavia (Russia). Izd Kartaya Kishinev. Quoted in Holy, 1990. 41 p.

Smedema, L. K. and Rycroft, D. W. 1983. Land Drainage: Planning and Design of Agricultural Drainage Systems. Cornell University Press, Ithaca, New York. 375p.

Smith, R. M. and Stamey, W. L. 1964. How to establish erosion tolerance? J. Soil and Water Conservation. 19: 110-111.

Srivastava, R.C., Mohanty, S., Singandhuppe, R.B., Mohanty, R.K., Behera, M.S., Ray, Lala, I.P. and Sahoo, D. 2010. Feasibility evaluation of pressurized irrigation in canal commands. Water Resources Management. 24: 3017-3032.

Thokal, T. R., Mahale, D. M. and Powar, A. G. 2012. Drip Irrigation System: Clogging and its Prevention. Pointer Publishers, Jaipur.

Stone, R. P. and Hilborn, D. 2015. OMAFRA Factsheet, Universal Soil Loss Equation (USLE), http:/www.omafra.gov.on.ca/english/engineer/facts/12-051.htm opened on 15.02.2018.

Tyagi, N. K. 1984. Effect of land surface uniformity on irrigation quality and economic parameters in sodic soils under reclamation. Irrigation Science. 5: 151-166

Tyagi, N. K. and Singh, O. P. 1979. Investigation on the effect of soil moisture on performance of land leveling implements. J. Agric. Eng. 16: 67-72.

UNEP (United Nations Environment Programme). 1997. World Atlas of Desertification, 2nd edition. N. Middleton and D. Thomas (Eds). London. 182 p.

USDA. 2007. Hydraulic Ram Pumps. Technical Note No. 26, Natural Resources Conservation Service, U.S. Department of Agriculture, Portland, Oregon. 9 p.

USDA, NRCS. 2011. National Engineering Handbook, Part 650 (Engineering Field Handbook) Chapter 8.

USDA, SCS. 1954. A Manual on Conservation of Soil and Water, Handbook for Agricultural Workers No. 61.USDA, Washington, D.C. 208 p.

USDA-SCS. 1962. Measurement of Irrigation Water. SCS National Engineering Handbook, Section 15, Chapter 9, USDA. Washington, DC.

USDA, SCS. 1983. National Engineering Handbook - Irrigation -Section 15 Land Leveling (Chapter 12). US Government Printing Press, Washington. 59 p.

USDI Bureau of Reclamation. 2001 (rev.). Water Measurement Manual. USDA. U.S. Government Printing Office, Washington, DC

US SCS. 1953. Farm Planner´s Engineering Handbook for Upper Mississippi Region. US Soil Conservation Service.

Venkatramaiah, C. 1996. Textbook of Surveying. University Press, Hyderabad. 676 p.

Venugopala Rao, P. and Akella, V. 2015. Textbook of Surveying. PHI Learning Pvt. Ltd., Delhi. 248 p.

Williams, J. R. 1975. Sediment yield prediction with universal soil loss equation using runoff energy factor. In: Present and Prospective Technology for Predicting Sediment Yields and Sources. ARS- S-40. Proceedings of the Sediment-Yield Workshop, November 28-30, 1972. USDA Sedimentation Laboratory, Oxford, Mississippi.

Wishmeier, W. H. 1959. A rainfall erosion index for universal soil loss equation. Proc. Soil Sci. Soc. Am. 23: 246-249.

Wischmeier, W. H. and Smith, D. D. 1958. Rainfall energy and its relationship to soil loss. Trans. AGU. 39: 285-291.

Wishmeier, W. H. and Smith, D. D. 1965. Predicting Rainfall Erosion Losses- A Guide to Conservation Planning. USDA Handbook No. 282. Washington, DC.

Woodruff, N. P. and Siddaway, F. H. 1965. A Wind Erosion Equation. Soil Science Society of America Proc. 29: 602-608.

Woodruff, N.P. and Zingg, A.W. 1952. Wind-tunnel Studies of Fundamental Problems related to Windbreaks. USDA, SCS-TP-112, USDA, Washington DC.

WWW.midh.gov.in/AtGlance/MI-AT-A-Glance.docx. Note on Centrally Sponsored Scheme on Micro- irrigation under Pradhan Mantri Krishi Sinchayee Yojana (PMKSY) midh.gov.in/AtGlance/MI-AT-A-Glance.docx.

www.nptel.iitm.ac.in/courses/105107059/module1/./lecture1.pdf.

Index

D

H

I

J

K

L

R

S

T

www.ingramcontent.com/pod-product-compliance
Ingram Content Group UK Ltd.
Pitfield, Milton Keynes, MK11 3LW, UK
UKHW021435280726
14060UKWH00001BA/94